Foundation
Engineering

The Intext *Series in*

Civil Engineering

Consulting Editor

Russell C. Brinker
New Mexico State University

Bouchard and Moffitt – SURVEYING, 5th ed.
Brinker – ELEMENTARY SURVEYING, 5th ed.
Clark, Viessman, and Hammer – WATER SUPPLY AND POLLUTION CONTROL, 2d ed.
Ghali and Neville – STRUCTURAL ANALYSIS: A UNIFIED CLASSICAL AND MATRIX APPROACH
Jumikis – FOUNDATION ENGINEERING
McCormac – STRUCTURAL ANALYSIS, 2d ed.
McCormac – STRUCTURAL STEEL DESIGN, 2d ed.
Meyer – ROUTE SURVEYING AND DESIGN, 4th ed.
Moffitt – PHOTOGRAMMETRY, 2d ed.
Salmon and Johnson – STEEL STRUCTURES: DESIGN AND BEHAVIOR
Ural – MATRIX OPERATIONS AND USE OF COMPUTERS IN STRUCTURAL ENGINEERING
Wang – MATRIX METHODS OF STRUCTURAL ANALYSIS, 2d ed.
Wang and Salmon – REINFORCED CONCRETE DESIGN
Winfrey – ECONOMIC ANALYSIS FOR HIGHWAYS

Foundation Engineering

ALFREDS R. JUMIKIS

Professor of Civil Engineering

Rutgers University, The State University of New Jersey

 INTEXT EDUCATIONAL PUBLISHERS

Fourth Printing
COPYRIGHT © 1971, BY INTERNATIONAL TEXTBOOK COMPANY

All rights reserved. No part of the material protected by this copyright notice may be reproduced or utilized in any form or by any means, electronic or mechanical, including photocopying, recording, or by any informational storage and retrieval system, without written permission from the copyright owner. Printed in the United States of America by The Haddon Craftsmen, Inc., Scranton, Pennsylvania. Library of Congress Catalog Card Number: 74-122539.

ISBN 0-7002-2311-8

INTEXT EDUCATIONAL PUBLISHERS
666 FIFTH AVENUE
NEW YORK, N.Y. 10019

Preface

This volume is prepared as a text for a one semester course in foundation engineering at the junior and/or senior levels. Foundation engineering today requires the knowledge of a large number of fundamental ideas and principles for the safe and economical design of structural foundations and their elements. Hence the object of this book is to try to set down some of the important principles and methods of design as first steps for students of civil engineering. The material is consistent with practical soil mechanics.

It was impossible to comprehend the entire field of foundation engineering and many of its design details because of space limitations. However, it is believed that the content of this book reflects the minimum necessary amount of knowledge to introduce the reader to the fundamentals of foundations engineering. This content includes:

1. A brief survey of some historical highlights in foundation engineering.
2. A description of the preparation of excavations and cofferdams for laying foundations in the dry.
3. A presentation of the principles of shallow foundations.
4. Analyses of deep foundations—open, floating, and pneumatic caissons and piles.
5. A brief discussion of some special topics in foundation engineering such as scour of river beds around bridge piers and effects of earthquakes on foundations.

With an undergraduate core of mathematics and mechanics the civil engineering student should have little difficulty in grasping and comprehending the principles of foundation design. The basic analytical concepts are applied in this text in order to provide a sound basis of design of structural foundation elements.

Although primarily intended for civil engineering students, the book may also be used successfully by civil engineering technicians working in the construction field. Likewise it may be used for self-study of foundation engineering, as well as for ready reference in practice. Each chapter includes worked-out examples of design. There is also a list of all significant, bona fide references of interest.

It is hoped that the book will be useful to readers in acquiring their first knowledge of the important civil engineering discipline of foundation engineering, and in encouraging them to continue their studies along these lines. Engineering cannot remain static.

ALFREDS R. JUMIKIS

University Heights
New Brunswick, New Jersey
May, 1971

Acknowledgments

The author expresses his sincere thanks to Dr. R. L. Handy, Professor of Civil Engineering and Director of the Soil Mechanics Research Laboratory of the Iowa State University of Science and Technology at Ames, Iowa, for giving generously of his time in carefully reading and constructively criticizing the author's manuscript and for suggesting improvements in clarity of presentation.

Also, the author appreciates the efforts of Professor R. C. Brinker of the Department of Civil Engineering of the New Mexico State University at University Park, New Mexico and Consultant to Intext Educational Publishers for reviewing the manuscript and making comments and corrections.

It is a pleasure for the author to add a word of appreciation to Dr. E. C. Easton, Dean, College of Engineering and Director of Bureau of Engineering Research, and to Dr. M. L. Granstrom, Chairman, Department of Civil and Environmental Engineering, both of Rutgers University, The State University of New Jersey, for their interest and encouragement.

The author is especially, grateful to the following authors, firms, publishers and copyright holders for granting permissions for the use of their material in this book:

Dr. A. Casagrande, Professor at Harvard University, Cambridge, Mass.

Dr. M. T. Davisson, Professor at The University of Illinois, Urbana, Ill.

Mr. S. B. Palmeter, Manager of Construction of the Jersey Central Power and Light Company, Morristown, N.J.

Mr. R. Crimmins of the Thomas Crimmins Constructing Company, New York, N.Y.

Monsen and Thorsen A/S. Civil Engineers and Contractors, Copenhagen, Denmark.

Bethlehem Steel Corporation, New York, N.Y.
Boston Public Library, Boston, Mass.
Dravo Corporation, Pittsburgh, Pa.
Griffin Wellpoint Corporation, New York, N.Y.
Hercules Concrete Pile Company, Ridgefield, N.J.
Moretrench Corporation, Rockaway, N.J.
Raymond Concrete Pile Division of Raymond International Inc., New York, N.Y.
Steinman, Boynton, Gronquist and London, Consulting Engineers, New York, N.Y.
United States Steel Corporation, Pittsburgh, Pa.
Wilhelm Ernst und Sohn, Publishers, Berlin, Germany.
Springer Verlag, Berlin, Germany.
Bureau of Engineering Research, College of Engineering, Rutgers University, The State University of New Jersey.
Civil Engineering (ASCE), New York, N.Y.
Highway Research Board, Washington, D.C.
McGraw Hill Book Company, New York, N.Y.
Rutgers University Press, New Brunswick, N.J.
Van Nostrand-Reinhold Company, New York, N.Y.
John Wiley & Sons, Inc., New York, N.Y.

Sincere thanks are also expressed to Mrs. Ruth Ahrens for her skillful editing of the author's handwritten manuscript and for reading the proofs.

The author's thanks go to the publisher for publishing this technical book for engineering education, and to all concerned in its production.

A. R. J.

Contents

PART I
INDOCTRINATION INTO FOUNDATION ENGINEERING

Chapter 1 Introduction... 3

1-1.	Legal Responsibilities of Civil Engineers	3
1-2.	Why Study Foundation Engineering?	3
1-3.	Definition of Foundation	5
1-4.	Purpose of Foundation	5
1-5.	Foundation Terminology	5
1-6.	Foundation Engineering	7
1-7.	General Classification of Foundations	7

Chapter 2 Some Historical Notes About Foundation Engineering... 10

2-1.	General Notes	10
2-2.	Foundation Engineering in Ancient Times	10
2-3.	Foundations in Roman Times	12
2-4.	Foundations in The Medieval Ages	16
2-5.	Foundations in the Period from the 15th to the 17th Centuries	16
2-6.	Foundations from the 18th Century to Date	17
2-7.	Closing Remarks on Historical Review of Foundation Engineering	17
	References	18

Chapter 3 Choosing the Kind of Foundation... 19

3-1.	Requirements of Foundations	19
3-2.	Choosing the Kind of Foundation	20
3-3.	General Procedure for Designing of Foundations	28
	References	32
	Problems	32

Chapter 4 Forces... 36

4-1.	Some Possible Forces Acting on a Foundation	36
4-2.	Lateral Earth Pressures	37
4-3.	Unit Weight of Soil	39
4-4.	Angle of Internal Friction. Cohesion. Wall Friction	41
4-5.	Uplift Pressure. Artesian Water Pressure. Seepage Pressure.	44
4-6.	Ice Pressure	51
4-7.	Seismic Forces.	51
4-8.	Soil and Rock Safe Bearing Capacity	53
4-9.	Bearing Capacity	55
4-10.	Specialization of Ultimate Bearing Capacity Equation	57
4-11.	Soil Bearing Capacity of Deep Foundations.	59
4-12.	Allowable Soil Bearing Capacity	60
4-13.	Standard Penetration-Resistance Test	61
4-14.	Foundations on Sand. Shallow Foundations. Individual Footings.	67
4-15.	Foundations on Clay. Shallow Foundations. Individual Footings on Clay	76
4-16.	Foundations on Silt. Shallow Foundations. Individual Footings on Silt	84
4-17.	Foundations on Rock	87
4-18.	The Factor of Safety in Foundation Engineering.	89
	Conclusions.	97
	References.	98

PART II
EXCAVATIONS AND COFFERDAMS

Chapter 5 The Excavation... 103

5-1.	Excavation—Important Element in Laying Foundations	103
5-2.	Maximum Unsupported Depth of an Open, Stepped Excavation	105
5-3.	Protection of Excavation Banks	108
5-4.	Sheet Piling	111
5-5.	Lateral Earth Pressure on Sheeting	119
5-6.	Spacing of Braces for Equal Axial Compression	122
5-7.	Statics of Strutting	122
	References	125
	Questions.	125
	Problems.	125

Chapter 6 Sheet Piling... 129

6-1.	Introduction	129
6-2.	Concerning Analyses of Sheet Piling	131
6-3.	Anchored Sheet Piling.	134
6-4.	Calculations of Free-Cantilever Sheet Piling (Conventional Method).	134

Contents   xi

6-5.	Calculations of Free-Cantilever Sheet Piling (H. Blum Method)	136
6-6.	Cantilever Sheet Piling Loaded with Active Earth Pressure	140
6-7.	Calculation of Cantilever Sheet Piling Loaded with Active Earth Pressure According to Blum	141
6-8.	Anchored Sheet Piling	146
6-9.	Method of Free Earth Support	146
6-10.	Pressures	148
6-11.	Moment Reduction	155
6-12.	Method of Fixed Earth Support	158
6-13.	Concept of Equivalent Beam	159
6-14.	Determination of Reaction R_J	159
6-15.	Determination of d_1	160
6-16.	Bending Moment	162
6-17.	Tschebotarioff's Anchored Sheet-Piling System	163
6-18.	Anchor Tension	164
6-19.	Placing of Anchor Blocks	165
6-20.	Forces Acting on Anchor Wall	168
6-21.	Frictional Force R_L on Vertical Side of Passive Wedge	170
6-22.	Stability of Sheet Piling-Anchor Wall-Soil System	172
6-23.	Length of Anchor	174
6-24.	Wale	174
	References	174
	Problems	175

Chapter 7 Cofferdams... 177

7-1.	Definition. Classification	177
7-2.	Earth Cofferdams. Rock-Fill Cofferdams	178
7-3.	Stability Requirements of Earth Cofferdams	180
7-4.	Single-Wall Sheet-Piling Cofferdams	180
7-5.	Land Cofferdams	183
7-6.	Double-Wall Sheet-Piling Cofferdams	185
7-7.	Underwater Concreting	191
7-8.	Cofferdam on Rock	195
7-9.	Cofferdam on Soil	198
7-10.	Driving Depth of Sheet Piling	204
7-11.	Deflection of Sheet Piling	206
	References	209
	Problems	210

Chapter 8 Cellular-Type Cofferdams... 211

8-1.	Types and Use of Cellular Cofferdams	211
8-2.	Segment-Type Cellular Cofferdams	211
8-3.	Circular-Type Cellular Cofferdam	214

xii   Contents

8-4.	Static Systems and Needed Information for Design of Cellular Cofferdams	216
8-5.	Cellular Cofferdam on Rock	219
8-6.	Stability of a Cellular Cofferdam Constructed on Thick Deposit of Sand	231
8-7.	Stability of a Cellular Cofferdam Constructed on a Thick and Homogeneous Clay Deposit	233
8-8.	Cloverleaf Cellular Cofferdam	240
8-9.	Example of Analysis of a Circular Cellular Cofferdam	240
	References	246
	Questions	246
	Problems	246

Chapter 9   Dewatering of Excavation. . . 248

9-1.	Methods of Dewatering	248
9-2.	Amount of Influx of Water	249
9-3.	Lowering the Groundwater Table	256
9-4.	Pump Capacity	266
9-5.	Electro-osmosis	273
9-6.	Soil Densification by Hydrovibration	277
9-7.	Soil Stabilization	279
9-8.	Grouting	282
9-9.	Chemical Solidification of Soil	285
9-10.	Artificial Freezing of Soil	290
9-11.	The Slurry-Trench Method	295
9-12.	Sand Piles	300
	References	302
	Problems	304

PART III
SHALLOW FOUNDATIONS

Chapter 10   Shallow Foundations and Footings. . . 309

10-1.	Classification of Foundations and Footings	309
10-2.	Classification of Shallow Foundations	309
10-3.	Depth of Shallow Foundations	317
10-4.	Calculating the Minimum (Critical) Depth of Foundation	321
10-5.	Laying of Shallow Foundations	322
10-6.	Calculation of Centrically Loaded Plain Concrete Square Footings	329
10-7.	Calculation of Eccentrically Loaded Footings	331
10-8.	Inclined Resultant Load	337
10-9.	Stability Against Sliding	338
10-10.	Two-Way Eccentricity	345

Contents   xiii

    References 348
    Problems 348

Chapter 11   Calculation of Rectangular Plain Concrete Footings... 350

11-1.   Procedure of Design. 350
11-2.   Critical Sections 353
11-3.   Calculation of Moment 356
11-4.   Critical Section for Shear. 360
11-5.   Calculation of Shear Stress 360
11-6.   Rectangular Plain Concrete Footings 367
11-7.   Plain Concrete Wall Footing. 367
11-8.   Calculation of Stresses in a Massive, Plain Concrete Stepped or Pyramidal Strip Footing 374
    References 380
    Problems 381

Chapter 12   Calculations of Rectangular Reinforced-Concrete Footings and Mats... 382

12-1.   Centrically Loaded Square Reinforced-Concrete Footings. 382
12-2.   Centrically Loaded Rectangular Reinforced-Concrete Footings 388
12-3.   Centrically Loaded Reinforced-Concrete Strip Footings 392
12-4.   Eccentrically Loaded Reinforced-Concrete Footings 395
12-5.   Eccentrically Loaded Reinforced-Concrete Wall Footings. 399
12-6.   Footings on Piles 400
12-7.   Foundations Restricted by Property Line 412
12-8.   Trapezoidal Footing at Property Line 417
12-9.   Combined Footings 419
12-10.   Combined Trapezoidal Footings at Property Line 422
12-11.   Connected Footings 422
12-12.   Continuous Foundations; Mats 426
12-13.   Uses of Mat Foundations 426
12-14.   Design of Mat Foundations 427
12-15.   Soil Pressure Distribution Under Mats 430
12-16.   Steps for Conventional Design of Rigid Mat Foundations. 431
12-17.   Design of Rigid, Uniform Mat Based on Statics 432
12-18.   Mats Resisting Hydrostatic Uplift Pressure 442
12-19.   Waterproofing of Foundations. 444
12-20.   Grillages. 451
12-21.   Timber Grillage. 451
12-22.   Steel Grillages 452
12-23.   Grade Beams. 452
    References 455

Questions		455
Problems		456

Chapter 13 Calculations of Circular Footings... 459

13-1.	Circular Footings	459
13-2.	Static System	459
13-3.	Equations for Calculating a Circular Footing of Uniform Thickness, Centrically and Symmetrically Loaded	461
13-4.	Eccentrically Loaded Circular Footings of Uniform Thickness	467
	References	477

PART IV
DEEP FOUNDATIONS

Chapter 14 Open Caissons... 481

14-1.	Definition and Description	481
14-2.	Types of Caissons	482
14-3.	Open Caissons	483
14-4.	Advantages and Disadvantages of Open Caissons	484
14-5.	Elements of Open Caissons	485
14-6.	Form of Open Caissons	486
14-7.	Size of Open Caisson	487
14-8.	Excavation; Sinking of Open Caisson	488
14-9.	Forms of Cutting Edges	490
14-10.	Thickness of Concrete Seal	491
14-11.	Open Caissons with Dredging Wells	493
14-12.	Dome-Capped Caissons	497
	References	500

Chapter 15 Open Caisson Statics... 501

15-1.	Forces on Open Caisson	501
15-2.	Thickness of Shell of Round Open Caisson	506
15-3.	Thickness of Wall of Rectangular Open Caisson	507
15-4.	Soil Lateral Earth Pressure on Vertical, Circular, Cylindrical Caisson	509
15-5.	Pressure Distribution Around Cutting Edge (Soil Reaction)	509
15-6.	Static Conditions of Open Caissons	511
15-7.	Artificial Sand Island	512
15-8.	Special Caissons	514
15-9.	Problem	515
	References	517
	Problems	518

Contents

Chapter 16 Floating Caissons... 521

16-1.	Description	521
16-2.	Buoyancy of Caisson	522
16-3.	Stability of a Floating Body	523
16-4.	Launching of a Floating Caisson	527
16-5.	Completion of Caissoned Foundation	529
16-6.	Little Belt Bridge Caissons	529
16-7.	Static Calculations for Floating Caissons	533
	References	544

Chapter 17 Pneumatic Caissons... 545

17-1.	Description	545
17-2.	Construction of Pneumatic Caisson	547
17-3.	Constituent Parts of a Pneumatic Caisson	549
17-4.	Work in a Pneumatic Caisson	552
17-5.	Calculations of Pneumatic Caissons	553
17-6.	Advantages and Disadvantages of Pneumatic Caissons	563
	References	564
	Problems	564

Chapter 18 Piles... 568

18-1.	Function of Piles	568
18-2.	Classification of Piles	568
18-3.	Negative Mantle Friction	572
18-4.	Pile Material	573
18-5.	Timber Piles	573
18-6.	Concrete Piles	578
18-7.	Reinforced-Concrete Pipe Piles	582
18-8.	Cast-In-Place Concrete Piles	585
18-9.	Metal Piles	587
18-10.	Composite Piles	593
	References	595
	Questions	595
	Problems	596

Chapter 19 Bearing Capacity of Piles... 598

19-1.	Methods of Determining Bearing Capacity of Piles	598
19-2.	Some Notes About Dynamic Pile-Driving Formulas	599
19-3.	Pile Driving	599
19-4.	Summary of Some Dynamic Pile-Driving Formulas	601
19-5.	Michigan Studies on Piles	606
19-6.	Theoretical Calculations of Pile Static Bearing Capacity	608
19-7.	Forces on Pile	608

19-8.	Bearing Capacity of a Pile Tip in Soil	614
19-9.	Bearing Capacity of a Pile Tip Considering Soil Lateral Resistance	617
19-10.	Minimum Tip Resistance	620
19-11.	Bearing Capacity of a Conical Pile Tip Considering Critical Pressure on Soil	621
19-12.	Static Pile Bearing Capacity Formulas	623
19-13.	Empirical Rules	624
19-14.	Safe Bearing Capacity of Piles	625
19-15.	Tension Piles	625
19-16.	Embedment of Piles by Vibration	630
19-17.	Electro-osmotic Pile Driving	633
	References	635
	Questions and Problems	636

Chapter 20 Pile Foundations . . . 639

20-1.	Use of Pile Foundations	639
20-2.	Arrangement of Piles	640
20-3.	Efficiency of Piles in a Pile Group	644
20-4.	Some Considerations in Pile Group Design	645
20-5.	Tension Piles and Batter Piles	646
20-6.	Static Pile-Group Systems	650
20-7.	Statically Determined Pile Systems	651
20-8.	Statically Indeterminate Pile Systems	657
20-9.	Culmann's Method	657
20-10.	Trapezoidal Method for Distributing Loads on Piles	659
20-11.	Eccentric Loadings on Vertical Piles from Resultant Vertical Load	665
20-12.	Bearing Capacity and Stability of Pile Foundations	670
20-13.	Theoretical Minimum Spacing of Friction Piles	671
20-14.	Settlement of a Pile Foundation	676
20-15.	Summary on Settlement of a Pile Group	678
20-16.	Elastic Deformations of Piles	680
20-17.	Horizontal Displacement of Pile Bent Tops	682
20-18.	Precise Calculations of Pile Groups	684
20-19.	Offshore Towers	684
20-20.	Foundations of Tall Structures	685
	References	692
	Questions	693
	Problems	694

PART V
SPECIAL TOPICS

Chapter 21 Scour . . . 701

21-1.	Bridge Foundations	701

Contents　　　　　　　　　　　　　　　　　　　　　　　　　　　　　　xvii

21-2.	Loads on Bridge Piers and Abutments	701
21-3.	Scouring of Riverbeds	702
21-4.	Protection of Foundations Against Scour	705
21-5.	Flow Velocities for Scour	706
21-6.	Depth of Scour	707
	References	716
	Problems	717

Chapter 22　Lateral Pressure in Deep Foundations... 718

22-1.	Introductory Notes	718
22-2.	Assumptions	718
22-3.	Parabolic Soil Lateral Stress Distribution	721
22-4.	Equilibrium Equations	722
22-5.	Maximum and Minimum Edge Pressures	723
22-6.	Tangential Friction Force T and Horizontal Force H	724
22-7.	Position of Pivot Point O	725
22-8.	Soil Lateral-Pressure Ordinates	725
22-9.	Critical Load H_c	726
22-10.	Embedment Depth	726
22-11.	Soil Bearing Capacity for Deep Foundations	727
	Reference	729
	Problem	729

Chapter 23　Crib-Wall Cofferdam... 730

23-1.	Rock-Filled Timber Crib	730
23-2.	Stability of a Crib Wall	732
23-3.	Forces Acting on Crib Elements	733
	References	735
	Problems	735

Chapter 24　Seismic Effects on Foundations... 738

24-1.	Earthquake Acceleration	738
24-2.	Effect of Earthquakes on Structures, Foundations, and Earthworks	739
24-3.	Earthquake Effect on Earth Retaining Walls	741
24-4.	Effect of Earthquake Action on Bridge Piers	741
24-5.	Effect of Vibration of Sand Behind Bulkhead	741
24-6.	Mechanical Vibrations of Machine Foundations	742
24-7.	Effective Measures in Aseismic Design	743
24-8.	Earthquake Considerations in the Design of the San Francisco-Oakland Bay Bridge Piers	746
24-9.	Effect on Shock of Water on Dams	748
	References	749
	Problems	750

Appendix I	Greek Alphabet... 753	
Appendix II	Key to Signs and Notations... 754	
Appendix III	Conversion Factors of Units of Measurements... 770	
Appendix IV	Commonly Used Soil Tests and Their Applications... 781	
Appendix V	Bibliography for Further Reading... 786	
	Index... 789	

Foundation Engineering

part I

INDOCTRINATION INTO FOUNDATION ENGINEERING

chapter **1**

Introduction

1-1. LEGAL RESPONSIBILITIES OF CIVIL ENGINEERS

Prudence requires that structural foundations be of safe and economical design. In the legal sense, this implies that the civil engineer has a twofold responsibility:
1. The designing and constructing of stable, safe, and functional structures.
2. According to most engineering licensing requirements—the safeguarding of the lives of the people using or passing by these structures.

Hence civil engineers should be familiar not only with structural theory, methods of design, properties of construction materials and their strength, but also with his natural "opponents" such as soil and water. The opponent water in many instances turns out to be particularly troublesome in connection with building of foundations and earthworks when it is necessary to lay their footings below the groundwater table. The foundation engineer must therefore have a solid knowledge not only of foundation engineering but also of soil mechanics as well.

1-2. WHY STUDY FOUNDATION ENGINEERING?

The discipline foundation engineering comprehends a number of problems associated with the establishment of all theoretical, natural and technological relationships in connection with the design and construction of foundations.

Contemporary civil and hydraulics engineering structures are now becoming larger in size and/or taller in height, and thus heavier than the relatively light superstructures of the past. The enormous weight of such structures requires the strongest and the most permanent foundations. Therefore foundations must be well designed and well laid, because the life and performance of the structure depends upon the strength of the soil or rock upon which the structure is founded. In this respect it is well to remember the saying frequently heard that the foundation is no better than its supporting soil, the structure no better than its foundation. Because most structures rest on soil, the performance of soil as a foundation-supporting material is of paramount importance.

Another important point in designing foundations is to assess their importance relative to the structure as a whole, to estimate what may happen if the foundation fails either totally or partially, and what can be done to remedy or avoid such failure.

Mistakes made in the design of foundations are seldom due to mistakes in advanced theories or wrong assessment of data, but are mostly simple matters of oversight or nonobservance of basic principles involved in analyses, in design, or in construction.

Failures of engineering structures occur usually when designs are extended beyond previous experience. In designing and constructing a foundation, it is wise to face the disclosure of dangerous conditions and to overcome them by remedial measures and/or redesign rather than to hold obstinately to preconceived plans in order to avoid the expense and embarrassment of revising the plans. It is difficult later to relay a poor foundation and repair the damaged superstructure without excessive cost. To quote Sir Francis Bacon, "Defence of his own ignorance is the worst cause of man's error." The remembrance of quality and a safe, economical structure remains long after the expenses are forgotten.

Thus from this discussion the following maxim in foundation engineering comes to the fore: *safety first*, economy second. This, however, does not mean that one cannot build economical and safe structures. With reference to failures, it may be said that a structure which is plausibly safe is also economical.

Although not an inexact science, foundation engineering is often an engineering science of the inexact, and the design of foundations often involves more inexact information than is encountered in the design of the superstructure. While all construction materials must meet some specifications, in foundation engineering the soil underneath the foundations is such as Mother Nature has made it. It must be used as it occurs at a given site and cannot be made to order. Also, it should be added that unanticipated superior or irresistible forces, the so-called *force majeure* (or acts of nature), have caused distress and disasters which conceivably could have occurred regardless of any kind of theoretical analyses. In these respects, one should be aware that nature makes no contracts with us. She does not follow textbooks, design standards, nor does she comply with any building codes.

Introduction

1-3. DEFINITION OF FOUNDATION

A structure usually consists of two major parts, the upper, known as the superstructure (a truss of a bridge, for example), and the lower, known as the substructure (piers, abutments, and their supporting bases), known in a broad sense as the foundation to hold the superstructure in place and to transmit all loads of the superstructure to the underlying supporting soil or rock.

A foundation is the connecting link between the structure proper and the soil. It is not visible to the public eye. The true value of a foundation lies out of sight in the substructure below the ground surface.

The term "foundation" originates from the Latin *fundatio*, or from *fundare* = to found, to set, to place, i.e., the act of building. Thus "foundation" means the artificially laid base on which a structure (superstructure) stands, or on which any erection is built up. Generally, foundations refer to all those parts of the weighty structure laid below the ground surface, and upon which a structure of any kind is to be built and supported. Hence engineers speak of a foundation as an artificial structural element and of the laying of foundation. The foundation is usually below the ground surface, viz., below the water table, as in the case of a bridge pier or other hydraulic structure.

This definition of a foundation is in agreement with the spirit of most of the languages of Western culture.

1-4. PURPOSE OF FOUNDATION

The foundation is a special, important, artificially built constituent part of a structure having for its purpose
1. To receive and transmit structural loads and loads (wind, vibration) externally applied to the superstructure to the soil (rock) at a given depth below the ground (water) surface, whichever is the case.
2. To distribute stresses at the base of the footing of the foundation to an intensity allowable on the soil (rock).

A footing is a unit of foundation designed to transmit structural loads to the soil or rock.

1-5. FOUNDATION TERMINOLOGY

For effective studies and discussion of foundation engineering it is first necessary to establish a mental picture of the meaning of some of the most commonly used terms in this discipline. The appropriate terms may be most readily learned and classified from an illustration like Fig. 1-1, for example, which shows a direct, shallow strip foundation. A footing or a foundation is termed *direct* when it rests in direct contact with the soil (no piles). In this soil-foundation system the dead and live loads acting on the superstructure of a construction are transmitted down into its substructure,

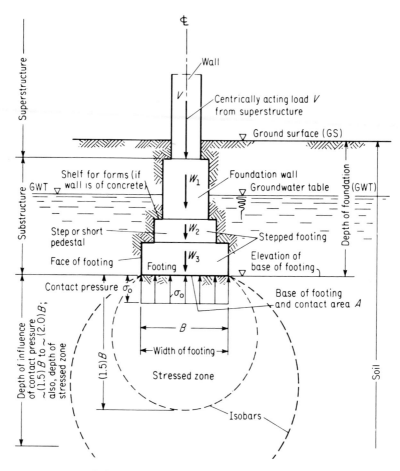

FIG. 1-1. Foundation terminology illustrated.

viz., foundation, by walls and interior columns (if any). Terms pertaining to pile foundations, bridge foundations and foundations of other structures will be explained in the text where they first occur.

The base of the footing of the foundation in contact with the soil transfers and distributes the load to the soil.

Introduction

1-6. FOUNDATION ENGINEERING

Foundation engineering concerns the dual problem of soil bearing capacity and the design and construction of the foundation, viz., the mutual interaction of the soil-foundation system as an integral unit. Earth's gravity —this materially natural property of all terrestial bodies—is the bond that ties or couples the soil-foundation system as an integral unit.

The stability of the soil-foundation system depends not only on the strength of the foundation proper as an artificial construction element, but also on the strength of the soil or rock beneath the foundation, whichever the case. Soils may vary considerably in their bearing capacity. Hence, success in foundation engineering requires among other things a solid knowledge and good understanding of the physical properties of soils and rocks utilized for supporting structures. A full knowledge of the geology of the site, of soil performance, cofferdamming and soil dynamics (vibration and earthquake) and hydrodynamics (groundwater flow, seepage forces, pore water pressure) is essential in foundation engineering. Thus foundation engineering deals with academic-theoretical-technological-practical matters, but has no political repercussions. The term *foundation engineering* has now become a universally accepted collective engineering concept which comprehends the science and technology of the various methods for the safe and economic laying of foundations of structures.

In order to solve foundation engineering problems one must not only know the discipline of construction of structures and the technology and management of construction methods, but must also have some knowledge of engineering geology, hydraulics, soil mechanics, and other pertinent engineering disciplines. This is to say that foundation engineering, dealing with the mutual interaction of soil-foundation systems, requires the synthesis of engineering knowledge more than any other construction specialty, and involves some of the greatest engineering skills and experience. Also, the laying of a foundation may sometimes demand some of the most difficult work of all operations carried on in connection with the erection of an engineering structure.

From the foregoing discussion it becomes apparent that foundation engineering represents a complex field of endeavor which discloses itself to the engineer only after many years of experience in which theory and practice are indivisible.

1-7. GENERAL CLASSIFICATION OF FOUNDATIONS

For the purpose of systematic classification and study, foundations may be broadly classed into three major groups:

(A) ordinary or shallow foundations
(B) deep foundations
(C) special foundations

Ordinary or Shallow Foundations

These foundations are sometimes called *spread* foundations. They are foundations which are laid in relatively shallow depths on a soil of capable bearing capacity below the ground surface, above as well as below the groundwater table.

Shallow foundations in turn may be classed into (a) footings and (b) mats or rafts.

A *footing* is a unit of foundation designed to transmit safely structural loads to soil and/or rock. Specifically, a footing is the enlargement of the base of a column or wall to distribute the load on soil to the limit of or below its allowable bearing capacity.

A mat foundation is a single slab, laid directly on the soil, usually reinforced, cast of uniform thickness, or with ribs. A mat supports many columnar and/or wall loads.

Combined footings are intermediate between single spread footings and mats. In essence, a combined footing is one where several individual footings are merged to form a little mat. When several footings are joined together in a straight line by means of a slab or beam, they are called *strapped* footings.

Deep Foundations

Deep foundations are laid at relatively considerable depth below the ground surface and/or below the lowest part of the superstructure. One should be aware that this classification of the foundations is a relative one: the shallow foundations group blends into the group of deep foundations, because there are no sharp demarcations between the shallow and deep groups of foundations.

Deep foundations may be laid in open foundation pits (excavations), in dry, or below the water table. To the group of deep foundations there belong open caissons, pneumatic caissons, floating caissons, pile foundations (end-bearing pile foundations and friction pile foundations).

Special Foundations

To this group the following foundations may belong:
- foundations for tall structures (smoke stacks, radio and television towers, lighthouses)
- foundations for subsurface and overland pipe lines
- foundations for port and maritime structures
- machine foundations

Introduction

- vehicular and aqueous tunnels
- foundations on elastic support
- other foundations of specific nature.

The type of a foundation to use depends upon the bearing capacity of the soil beneath the foundation, not only that at the base of the footing but also of those forming the underlying strata.

chapter 2

Some Historical Notes About Foundation Engineering

2-1. GENERAL NOTES

In order to elucidate the foundation engineering discipline it is necessary to review briefly some highlights in its historical development.

History, to paraphrase Leibnitz, is a useful thing, for its study not only gives men of the past their just due but also provides those of the present with a guide for orienting their own endeavors. This applies to foundation engineering as well, and helps one to appreciate its gradual evolution. The past is prologue, therefore study the past.

It has been said that most mathematical theories of importance to structural engineering were anticipated and used before applied mechanics supplied the logic. But it is also true that until this theoretical basis for design came to be used progress was slow, and really outstanding achievements were accomplished only by the few.

The applied mechanics of soils now being practiced by civil engineers all over the world is complex because of the variable physical properties of the soil material itself. However, a knowledge of the relationship between cause and effect does enable today's engineer to deal successfully with the mutual interaction of soil-foundation systems.

In contemporary times, when every day brings us new technical knowledge, it is of interest to survey how foundation engineering problems were solved in the past.

2-2. FOUNDATION ENGINEERING IN ANCIENT TIMES

Work in foundation engineering dates far back in time; building foundations is one of the oldest arts of man.

From references to foundations in the ancient Egyptian hieroglyphs, and information about the importance of foundations in the scriptures (Luke vi: 47-49), one may deduce that the problem of foundations has been paramount for the stability of a structure. The saying that "a house must be built on a rock foundation" (meaning a solid support for the structure), where rock is available, of course, comes down to us from Biblical times, and in the Old Testament one reads that King Solomon built his temple on cedar piles. As a matter of interest the Bible mentions foundations 165 times. It is also known that prehistoric lake dwellers built their huts on piles driven into the lake bottom.

Likewise one may perceive that in the large cities of early civilization there were many large edifices and fortifications that presented to their builders foundation problems as complex as those met today.

With the increase of man's civilization and cultural activities, and settlement of people on seashore and along waterways, it was necessary to build not only on rock but also on "soft" material such as sandy, silty, or clayey soil—good soil and poor, firm soil and soft—in short, whatever type of soil was encountered. As a result of erosion and degradation processes of the mountains inland, the fine-particled soil materials, such as fine sand, silt, and clay, are transported by water to the sea. Thus these fine-particled soil materials are always found at the river deltas. Here the velocity of flow of water is usually small, thus permitting the fine-particled soil material to settle and to form sediment. Historically, people have usually settled at the river deltas, where civilization developed. During the course of development, foundations for structures were built and soil-foundation problems encountered. Settlement of soil and structure and soil rupture led builders to determine safe loads.

The long span of life of the oldest structures of the ancient eastern and western worlds, which have survived till present time and still are found in a fair state of preservation, may be attributed to the inherent soundness of their foundations (and also probably partly to favorable climatic conditions).

It is interesting to note that in Egypt about 2000 B.C. cutting-edges of caissons were known and used for sinking of shafts through sand and conglomerate.[1]* The cutting edge was made of a circular limestone slab by making a hole in it. The sinking of the open caisson was accomplished by the weight of its brick wall. To reduce frictional resistance during penetration, the outside mantle surface of the caisson was made smooth.

The Cheops pyramid, 750 ft × 750 ft at the base and 480 ft high, was built about 3000 B.C.[2] on a mat of limestone blocks, the latter being supported by limestone bedrock. Thus it can be seen that builders of ancient Egypt and other nations of antiquity were aware that sound rock is more stable for supporting structures than the loose, windblown, traveling desert sand.

---

* Numbers indicate References at end of chapter.

The first under-river tunnel was built in Babylon from approximately 2180 to 2160 B.C.[3] Tunneling for the Persian underground galleries, thousands of miles in length and built to tap and/or transport groundwater, or to drain land, was accomplished as early as 800 B.C.[3]

As evidenced by the Babylon's Hammurabi (2130-2088 B.C.) code of laws, the code required substantial construction work and imposed great liabilty with severe penalties on builders. For example, a passage from Hammurabi's laws reads:[4] "If a builder builds a house for a man and the house which he has built collapses and causes the death of the owner of the house, that builder shall be put to death." This implies that the builders were required to possess certain professional skill for safe building.

According to Steinman and Watson,[5] bridge piers in China (from 200 B.C. to approximately A.D. 220) were founded as follows: "Part of the river was closed by a double-row cofferdam made of bamboo piles fastened together with ropes. Then bamboo mats were put on each row of piles, the intervening space being filled with clay. The whole cofferdam was curved against water pressure. The water was pumped out by means of crude wooden tread pumps worked by pairs of men."

Technical problems such as placing of the bridge piers in the shifting river bed of mud and under changing conditions of sand, and sudden torrents during the rainy season were also solved in ancient China.

The early Greeks were also noted for their engineering skill. They built harbors, tunnels, aqueducts (on the island of Samos) and other hydraulic structures.

As interesting as the achievements in engineering technology of the ancients may be, the ancients did not have scientific and engineering knowledge beyond our own. But sometimes it happens that a useful invention is made, lost, and rediscovered centuries later.

From the professional point of view it is interesting to note that the engineer held a position of power and influence in all ancient communities.

2-3. FOUNDATIONS IN ROMAN TIMES

It is known that the Romans built notable engineering structures, such as harbors, breakwaters, aqueducts, bridges, large edifices and public buildings, sewage lines, and a vast network of durable and excellent roads.[6] About foundations, the Roman engineer Vitruvius (first century B.C.) in his *Ten Books on Architecture* wrote:[7]

"Let the foundations of those works be dug from a solid site and to a solid base if it can be found. But if a solid foundation is not found, and the site is loose earth right down, or marshy, then it is to be excavated and

Some Historical Notes About Foundation Engineering

cleared and remade with piles of alder or of olive or charred oak, and the piles are to be driven close together by machinery, and the intervals between are to be filled with charcoal."

To this J. Gwilt[8] translated one more sentence: "The heaviest foundations may be laid on such a base."

Familiar, isn't it?

In Roman times the building of maritime structures was considered to be a very great accomplishment. For constructing foundations in water, single-wall and clay-puddled cofferdams were used. These were pumped out by means of screw-type pumps or by bucket elevators, thus permitting the masonry foundations to be laid in the dry. Sometimes foundations were concreted under water. Figure 2-1a shows a single-walled Roman cofferdam as illustrated by Perrault in 1684.[9] At that time Perrault translated Vitruvius' *Ten Books of Architecture*.

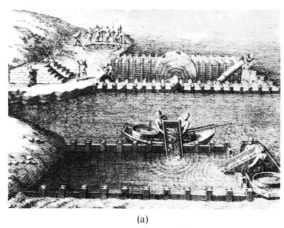

(a)

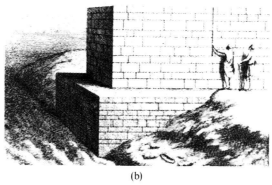

(b)

FIG. 2-1. Roman methods of laying foundations according to Perrault. (a) A cofferdam and concreting under water; (b) Construction of a mole.

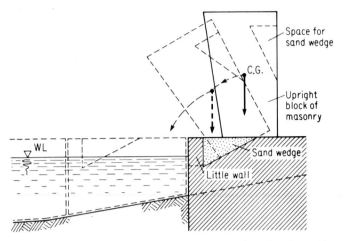

FIG. 2-2. Roman principle of constructing moles after Perrault's description.

According to Perrault, laying the foundations of a mole was done in the Roman days as follows (Fig. 2-1b and 2-2). A mole is a massive work of masonry or of large stones, laid in the sea, and is frequently used as a breakwater for the protection of a harbor from the action of high waves. The Romans built on the shoreline a masonry block (foundation) on the upper face of which there was built an upright block of masonry in such a position that the line of the resultant weight passed through the sand wedge. After the upright block had hardened, the sand-surrounded wall was removed, whereupon the waves of the sea could wash the sand away until finally the block tipped and fell into the water. This procedure was repeated until the entire necessary length of the mole was completed. This method of construction of moles in those days represented real ingenuity on the part of the builders.

The method of building a mole as described by Vitruvius found a renewed and very spectacular application in 1930 in Canada,[10,11] where no reference to Vitruvius' name was made. Repetition of this method of construction, forgotten during the many past centuries, utilized new construction materials, namely, portland cement concrete, and steel. To erect a closure (viz., river-diversion dam) across the swift Saguenay River (6 m/sec of flow) on the bare rock bottom of the river, a concrete monolith 28 m high, steel-reinforced, was cast standing on its end at the side of the main river channel (Figs. 2-3a and 2-3b). The height of the monolith block corresponded to the closing width of the river. One of the faces (= sides) of the block toward the channel had the contoured form of the irregular rocky river bed profile to fit the bottom of the river channel. Upon blasting away the little retaining wall

Some Historical Notes About Foundation Engineering 15

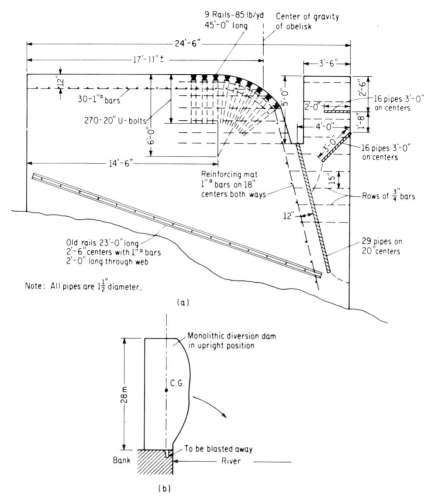

FIG. 2-3. (a) Monolith support for diverting the Saguenay River, Canada. (b) Concrete monolith cast standing on end at the side of the main river channel.[10,11]

on the block-supporting foundation, the whole 28 m high monolith tipped down into the main channel of the swift water. 8.55 m deep, on July 23, 1930. The dump and thus the successful closure took only 5.75 sec. The final position of the monolith was about 2.5 cm off the intended position. In a way, this closure of the rapids may be regarded as a massive cofferdam, permitting the pursuit of work during times of floods. This river diversion operation was made for the Aluminum Company of America in connection with the building of a hydroelectric plant of 200,000 kw capacity. This example indicates that the study of old, long-forgotten methods of construction is worthwhile, for such methods can occasionally be adopted even today.

2-4. FOUNDATIONS IN THE MEDIEVAL AGES

In the medieval period (about A.D. 400 to 1400), the only structures of significance associated with laying of foundations were (1) castles, (2) city fortification walls, (3) cathedrals, and (4) campaniles (bell towers).[12]

As evidenced by some medieval structures, many of the foundations of medieval cathedrals have been founded on poor soil. The heavy Königsberg Dom in East Prussia (about A.D. 1330), for example, was placed on peat,[12,13] bringing about large settlements—more than 5 ft.

As magnificent as the cathedrals are, their foundations were so poorly designed that they usually are at insufficient depth to transmit the heavy loads of the superstructures of the cathedrals to a firm soil at proper depth.

The Leaning Tower of Pisa, 179 ft in height, was started in A.D. 1174 and completed in 1350. In 1910 the tower had a visible slant and its top was 16.5 ft out of plumb. According to Terzaghi,[14] the cause of leaning of this campanile was due to the consolidation of the clayey soil material under a layer of sand on which the tower was founded.

2-5. FOUNDATIONS IN THE PERIOD FROM THE 15th TO THE 17th CENTURIES

Some of the noteworthy examples of foundations built in these centuries are those of the Rialto single-arch bridge in Venice, Italy (completed in 1591), and some of the bridges in Paris built in the sixteenth and seventeenth centuries. Because of the marshy site and adjacent large buildings, pile driving for the Rialto bridge foundations presented a problem of great acuteness and responsibility.

The Pont Neuf in Paris was built at the end of the 16th century, and stood for more than 300 years. Concerning construction of the foundation of the Pont Neuf,[15] a report describes the cofferdam used for laying these foundations.

"It consisted of two enclosures: the inner one was made up of two rows of closely jointed sheet piles; the outer one consisted of but one row of sheet piling, with an earth embankment outside that. The space between the two walls was filled with well-puddled clay. In the building of the foundations for the long arm of the bridge, the force of the current necessitated deeper excavation, and the engineers in charge were directed to make models in wood, the earliest reference to this interesting and very modern

Some Historical Notes About Foundation Engineering 17

and useful device for bridge design. But, because the builder did not employ piles for pier foundations, the river current immediately began to cause trouble. Foundations had to be repaired before the bridge was scarcely finished."

The foundations of the Pont Royal (a seventeenth century bridge in Paris) were built for the very first time by means of the open caisson, with watertight timber sides. After the excavation, the caisson was sunk to the bed, but the top was kept above water level. The masonry work of the pier was then built up inside the chamber.

2-6. FOUNDATIONS FROM THE 18th CENTURY TO DATE

The eighteenth century may be considered as the real beginning of civil engineering, when science became a basic factor in structural design. Similar to other structural engineering sciences, the science of foundations was organized as an independent discipline during the nineteenth century. The French engineer Triger is credited for using in 1841 pneumatic caissons for laying of deep foundations in water-logged soils.

In 1870 the Brooklyn Bridge in New York City represented the first large-scale use of pneumatic caissons for bridge foundations in the United States. The method of laying foundations by means of the pneumatic caisson is limited to a depth of about 120 ft, which is the endurance limit, about 4 atmospheres, for man to work under pressure. For greater depths, open caissons are used.

In the late 1920's the artificially built island method was used for sinking caissons 170 ft deep for the Huey P. Long bridge across the Mississippi at New Orleans.

The piers of the Transbay bridge (completed in 1936) between San Francisco and Oakland were sunk by means of dome-capped caissons 240 ft below the water surface—the greatest foundation depth achieved until 1941.

Today, in the early 1970's, foundation engineering has attained a high degree of perfection.

2-7. CLOSING REMARKS ON HISTORICAL REVIEW OF FOUNDATION ENGINEERING

Comparing the methods practiced for laying of foundations in the old days with those practiced today, one cannot escape the striking observation that the principles and types of foundations in the old days were the same as those used today. These embraced knife-edges for open caissons, piles, earth-retaining walls, cylindrical shafts, densification of soil by piles, cofferdams (Chinese and Roman), modeling of structures. The difference between the early foundations and the contemporary ones may be seen only in the difference of size, scope and perhaps depth of foundations, as well as in present-day materials such as concrete and steel. Timber has been used before as well as now. For excavation of foundation pits and pile driving today we have no doubt far better and more efficient construction equipment and machinery than our predecessors. Thus it is apropos to cite the French saying: *Plus ça change, plus c'est la même chose* (the more things change, the more they remain the same).

However, today the civil engineer has a much better knowledge of soils to enable him to encounter the two opponents—soil and water—in foundation engineering than had his colleagues only some forty years ago.

REFERENCES

1. Anon., *Engineering News-Record*, Dec. 7, 1933, p. 675.
2. *The History of Herodotus*, trans. by George Rawlinson. London: Dent; New York: Dutton, 1945, Vol. 1, p. 178.
3. S. G. W. Benjamin, *Persia and the Persian People*. Boston: Ticknor, 1887.
4. R. F. Harper, *The Code of Hammurabi, King of Babylon* (about 2550 B.C.), 2nd ed. Chicago: U. of Chicago Press, 1904, Secs. 228-233, p. 81.
5. D. B. Steinman and S. R. Watson, *Bridges and their Builders*. New York: Putnam, 1941.
6. S. B. Platner, *Topography and Monuments of Ancient Rome*. Boston: Allyn and Bacon, 1904.
7. Vitruvius, *Ten Books on Architecture*, trans. by Frank Granger. New York: Putnam, 1934.
8. Vitruvius, *The Ten Books of Vitruvius*, trans. by Joseph Gwilt. London: Lockwood, 1874, Book III, Chap. III, p. 72.
9. C. Perrault, *Les dix livres d'architecture de Vitruve, corrigés et traduits nouvellement en français, avec des notes et figures*. Paris, 1684.
10. C. P. Dunn, "Blasting a Precast Dam into Place," *Civil Engineering*, May 10, 1930, p. 464.
11. "J. E.," "Barrage Provisoire, en Béton Armé, construit par basculement d'un bloc en travers, de la Saguenay River (province de Québec, Canada)." *Le Génie Civil*, Vol. 99, No. 1 (July 4, 1931), pp. 14-15.
12. A. R. Jumikis, *Soil Mechanics*, Princeton, N. J.: Van Nostrand, 1962, p. 15.
13. B. Tiedemann, "Die Bedeutung des Bodens im Bauwesen," in E. Blanck (ed.), *Handbuch der Bodenlehre*, Berlin: Springer, 1932, Vol. 10.
14. K. Terzaghi, "Die Ursachen der Schiefstellung des Turmes von Pisa," *Der Bauingenieur*, No. 1/2 (1934), pp. 1-4.
15. D. B. Steinman and S. R. Watson, *Bridges and their Builders*. New York: Putnam, 1941, pp. 87-88.

chapter 3
Choosing the Kind of Foundation

3-1. REQUIREMENTS OF FOUNDATIONS

Like any element of a structure, the foundation must satisfy certain stability requirements. Among the many requirements for a solid foundation the following are important:
1. The foundation should be laid at a proper depth below the ground surface to avoid soil lateral expulsion from underneath the base of the foundation, to avoid damage to it by freezing and thawing (where it applies), and to protect it from scour and washout by erosion of soil by water.
2. The foundation should be resistant to groundwater and any other aggressive water relative to foundation material.
3. The foundation should be strong in its details as well as a whole. This is to say that the deformations of the foundations should be no larger than those allowable under the condition of its exploitation.
4. The foundation should be stable against any lateral (horizontal) sliding, against any rotary movement, and against intolerable differential settlement to avoid any distress to the structure.
5. The soil-foundation system must be safe against rupture of soil (groundbreak); this requirement pertains to the exhaustion of the shear strength (viz., bearing capacity) of the soil, and may be regarded as a natural consequence of the requirement mentioned under point 1 above.
6. The foundations must be designed and laid with the view of future excavations of the soil around the foundations for eventual repairs of foundations, for installation of service ducts and pipes (for example), and for foundations of new additions to the initial structure. Hence the stability of the soil-foundation system must be analyzed with no foundation backfill.

7. The foundation should be durable and function properly during its assigned service period.
8. The foundation should be economical and subject to mechanization of work in laying the foundation.

Obviously, the requirements pertain equally to the foundation-supporting soil as well as to the structural foundation as an integral soil-foundation system.

To summarize, in designing structures such as foundations, the engineer must satisfy two independent foundation stability requirements, which must be met simultaneously:

1. There should be adequate safety against a shear failure within the soil mass (not to exceed the bearing capacity of the soil being built upon).
2. The probable maximum and differential settlements of the soil, viz., various parts of the foundations, must be limited to a safe, tolerable magnitude.

3-2. CHOOSING THE KIND OF FOUNDATION

The use of a particular kind of foundation depends upon the character of the soil and the presence of water at the site. The following compilation in Tables 3-1, 3-2, 3-3, and 3-4 may serve for the purpose of orientation and as a suggested guide for tentatively choosing, based on preliminary soil exploration, a certain general kind of foundation suitable for various soil and water conditions at the site. The kind of foundation chosen dictates, then, the nature of a more detailed site exploration, the kind of soil-sampling program to follow, and the kinds of soil tests to be performed. A table in the Appendix lists the commonly used soil tests aiding in the study of soil properties.

Depending upon the type of structure under consideration and upon the conditions at the building site, one chooses that type of foundation which is not merely safe but also economical. The influx of "public enemy number 1"—water—into the foundation excavation considerably increases the cost of laying foundations.

The following tables of compilation on suitable kinds of foundations for various sites and general type of soil and water conditions at the site are not to be construed as cookbook recipes for a final choice of the foundation type to use. The tables merely may assist one to narrow down the choice in selecting a type of foundation, eliminating those which will not be suitable in solving a particular foundation problem. From experience, it is a well-established fact that usually there may be several acceptable types of founda-

Choosing the Kind of Foundation

TABLE 3-1
Suitable Kinds of Foundations for Various Site Conditions

Site Conditions		Suitable Kinds of Foundations	Nature of Excavation and Coping with Water
Soil	Water Conditions Relative to Excavation		
1. Firm soil near the ground surface or at moderate depth	(a) Open, dry excavation	Direct, shallow foundations: individual footings, combined footings, strip footings, mats	Open excavation with vertical or sloped walls
	(b) Excavation through open water:		
	1. Shallow depth of water	Direct foundations. High-piled grillage	Open excavation between sheet-piling enclosure or cofferdam
	2. Great depth of water	Caissons	
		Long tubular piles; shafts	
2. Firm soil at attainable depth	(a) Open, dry excavation and excavation with moderate influx of water	Direct foundations	Dewatering of excavation by pumping
	Seams and springs	Reinforced concrete piles; pier foundations	Sealing off springs
	(b) Excavation in groundwater and through open water	High-piled grillage with sheet-pile enclosure; reinforced concrete slabs	
		Caissons	Expulsion of water by compressed air
		Direct foundations between steel sheet piling or cellular cofferdams	Open excavation between sheet piling or cofferdam. Lowering of groundwater table
3. Firm soil, overlying soft material	(a) Open, dry excavation	Direct foundations	
	(b) Open excavation with water present	Mat Friction piles	Dewatering by pumping. Lowering of groundwater table

TABLE 3-1 (*Continued*)

Site Conditions		Suitable Kinds of Foundations	Nature of Excavation and Coping with Water
Soil	Water Conditions Relative to Excavation		
4. Weak soil overlying firm soil	(a) Open, dry excavation	Bearing piles or piers	
	(b) Open excavation with water present	Bearing piles or piers	
5. Thick stratum of weak soil present. Firm soil unattainable	(a) Open, dry excavation	Mat; reinforced concrete slab Slab grillage Soil stabilization Friction piles Concrete and bulb-end reinforced concrete piles	
	(b) Construction in and through open water and groundwater	Mat; piled slab grillage Caissons Foundations in sheet-pile enclosures	Cofferdams Expulsion of water by compressed air Sheet-pile enclosures or cofferdams

TABLE 3-2

Selection of Kind of Foundation Depending Upon Soil Condition and Structural Load

All foundations may be classed into two principal groups, namely: (a) Shallow Foundations, (b) Deep Foundations

Kind of Foundation	Soil Conditions	Type of Structure
	(a) Shallow Foundations	
1. Footings; combined footings; strip foundation; timber grillage; concrete slab; caissons	All kinds of soil, cohesive and noncohesive ones. Caution in silt soils	Factories, machine foundations; masts; bridge piers
2. Reinforced-concrete mats with and without ribs Inverted vaults	Cohesive, weak load-bearing soils; caution in silt	Edifices, low warehouses

Choosing the Kind of Foundation

TABLE 3-2 (*Continued*)

Kind of Foundation	Soil Conditions	Type of Structure
I. Single Foundation	(b) Deep Foundations	
1. Piers	Where there is a possibility to transmit load on firm, load-bearing layers; caution in silt soils	Tall structures; buildings, towers, chimneys, beacons, machine foundations, which should be free of intolerable vibration and settlement
2. Floating foundations	Rock at unattainable depth	Structures which are insensitive to settlement
3. Piling: load transmission to firm, bearing stratum	Near rock on economically feasible depth of firm bearing soil	Bridge foundations; beacons; tall buildings; heavy machine foundations
4. Pile types: timber piles, concrete piles; in situ; prefabricated	The prefabricated concrete piles are suitable for floating foundations in soils with many intervarying layers	All structures
	Bulb-end pile on firm soil can transmit heavy loads Timber piles must be permanently under groundwater table	
5. Screwpiles	Cohesive soils	Used where piles are subjected to tensile forces, and in temporary structures
6. Steel piles	In soils, where steel is not aggresively attacked, i.e., in dense, corrosion-free soils, and in soils having a large coefficient of friction between steel and soil	All structures
7. Soil densification piles	Suitable for sandy and clayey-sand soils	
8. Open caisson	For transmitting loads to firm, load-bearing stratum, or where rock is attainable	Tall houses; heavy machine foundations which are free of vibration; bridge foundations
II. Sheet Piling	In fine-particled soils with high water pressure; in soils with high flow velocity of groundwater Where expulsion of soil from underneath the base of the footing can be anticipated	Water-impounding dams; to decrease the length of seepage path, viz., to decrease the hydraulic gradient

TABLE 3-3

Guide to Tentative Selection of Foundation Types

Hydrological Conditions at Site

Construction Site not Covered with Water		Construction Site Covered with Water Up To		
Groundwater Absent	Groundwater Present	~12ft (4.0 m) deep	~25ft (8.0 m) deep	>25ft (>8.0 m) deep
1. Simple shallow foundation with or without widening of footing	1. Simple shallow foundation with or without widening of footing	1. Simple foundation within a cofferdam	1. Floating caisson or bottom-closed box on or with no piles	1. Floating caisson or bottom-closed box
2. Reinforced concrete slab	2. Reinforced concrete slab	2. Timber piles in a cofferdam	2. Open caisson	2. Open caisson
3. Concrete or reinforced concrete piles. Steel piles	3. Timber piles at high groundwater table; steel piles	3. Floating caisson or bottom-closed box	3. Pneumatic caisson	3. Pneumatic caisson
	4. Concrete or reinforced concrete piles at low-lying groundwater table, steel piles	4. Floating or open caisson		
	5. Open caissons	5. Pneumatic caisson where obstacles in soil are encountered		

TABLE 3-4

Tentative Selection of Foundation Types Based on Site Condition

Soil Conditions at Site	Types of Structures	
	Light, Flexible	Heavy, Rigid
Thick, firm stratum	Individual footings Strip footings Combined footings	Individual footings Strip footings Combined footings Mats

Choosing the Kind of Foundation

TABLE 3-4 (*Continued*)

Soil Conditions at Site	Types of Structures	
	Light, Flexible	Heavy, Rigid
Firm stratum over soft stratum	Individual footings Strip footings Combined footings Light surface mats	Mats Friction piles
Thick, soft stratum	Friction piles Mats	Mats Friction piles
Soft stratum overlying economically and feasibly attainable firm stratum	End-bearing piles Piers Caissons	End-bearing piles Piers Caissons

tions to choose from and to use, and that all may render a correct solution depending upon the designer's approach consistent with safety and economy. It is also well to remember that no foundation engineering problem is exactly like another, and that therefore each foundation engineering problem must be treated and solved according to its own requirements and merits.

There are no set, rigid rules available for choosing the "right" kind of foundation. The choice of a particular kind of foundation depends upon the character of the soil and on the hydrological regimen at the site. The choice is also a matter of one's judgment based upon soil-exploration data and experience gained in executing foundation work.

The choice of a correct kind of foundation is one of the most important and difficult problems in foundation engineering. A foundation problem may have several correct solutions, but each of them will be different. Usually such solutions differ qualitatively in that some solutions may be simpler than others or more expensive than others. Therefore, after all pertinent design data have been obtained, a second step would be to make several alternate foundation designs or variances.

In forming an opinion about the "right" decision, consideration should be given to the following practical factors:

1. Complete clarity about stratification of soil and its physical properties, as well as about the site proper, and about the area adjacent to the contemplated construction site.
2. Foundation and subsurface engineering practice.
3. Definite knowledge of the pertinent building codes, construction

regulations, ordinances, and traditional practices in laying foundations by the various organizations of the construction industry.
4. Careful investigation of available construction machinery and construction materials and their use.
5. Economic factors.

Depending upon the nature of the soil and purpose of the structure, the design calls for a certain depth of the foundation footing to be laid below the ground surface. Thus, at that depth the designer faces the problem of bearing capacity of the foundation-supporting soil. If the soil there has an adequate bearing capacity, prudence and economy usually require the use of direct, shallow footing foundations. Hence one of the basic principles in designing foundations observed by engineers is placement of the base of a foundation footing as near to the level of the ground surface as possible (subject to stability and building code requirements, of course), if by doing so,

1. One can save in dewatering expenses of the excavation.
2. The stresses on the weaker, deeper-lying compressible soil layers can be reduced to an allowable magnitude.
3. An unallowable weakening of the clay layer resisting water pressure can be avoided.
4. Excavation and concrete work can be kept out of water.

A shallowly based footing is also used when a relatively thin compressible layer of soil (peat, muck, very plastic clay) lies below a groundwater table that is difficult to lower. In such cases soil contact pressure must be reduced, or instead of a contemplated strip footing a mat foundation may be used. By these means the compression of the sandwiched compressible soil layer may be reduced. Because of the greater width of the mat, the stressed zone (viz., stress isobars) will reach deeper than under the strip foundation, and will extend over and through the lower weak layers. This is so because the stress distribution in and settlement of soil depends not only upon the intensity of the soil contact pressure but also on the size and form of the foundation itself, as well as upon the depth coordinate of the soil layer in question. Foundations should not be laid directly on peat or organic matter.

However, if a load-bearing soil or rock stratum is encountered at a greater depth below the ground surface, the use of deep foundations becomes unavoidable, especially when

1. The structural loads cannot be reduced,
2. The dewatering of the excavation brings about difficulties in foundation work (persisting danger of erosion),

Choosing the Kind of Foundation

3. In clayey soils, a strong tendency for the structure to settle exists,
4. The artificial stabilization of the interbedded weak layers of soil would turn out to be very uneconomical,
5. It can be anticipated that in the future cuts will be made in the soil or deep foundations may be laid near the contemplated structure.

In such cases the structural loads are transferred through the weak soil layers by means of special constructions, such as pile foundations or open or pneumatic caissons, for example, to a deep-lying firm soil or rock, if technically attainable and economically feasible. This applies to deep foundations.

However, the decision as to whether in adhering to the required safeties a shallow or a deep foundation is more economical cannot always be unerringly made, but requires the working out of comparative variations of several types of foundation designs, depending on the kind, nature, stratification and thickness of the firm soil at the construction site, upon the groundwater condition, and upon the physical properties of the soil, as well as on the available construction materials to be used for the construction of the deep foundations.

As to the position of the groundwater table below ground surface, a desirable stratum well below the water table may sometimes be reached easily in clay, whereas it is difficult to attain in fine, running sand. In the case of clay, piles, piers or open caissons may be suggested. In the case of running sand, piles may be indicated, or at shallow depths a footing confining sheet piling has been utilized.

Some of the factors affecting pile foundations are

1. Unimpregnated timber piles must remain below the lowest level of the fluctuating water level to prevent decay (Figs. 18-5 and 18-9),
2. In some harbors, marine borers attack wood piles,
3. Consolidation of compressible soil layers below the pile bearing stratum will cause the pile foundation to settle,
4. The pile-bearing stratum must be checked to see whether it will support the entire pile system,
5. Consolidation of soil layers along the pile due to surcharge on the ground surface causes negative mantle friction on the pile.

Other uses of piles are in anchorage foundations, various kinds of supports, strutting, bracing, underpinning of structures and stabilization of soils.

The feasibility of different designs of pile foundations is shown in Figs. 18-9, 18-12, 19-9, 19-10, 20-11, 20-27, 20-28, 20-36, Question 20-4,

Prob. 20-3, Fig. 21-1, and Fig. 24-4. After several variations of foundation designs have been worked out, their economic aspects are studied.

The matter of selecting the type of foundation to use will become more meaningful to the reader after having read Chapter 4.

3-3. GENERAL PROCEDURE FOR DESIGNING OF FOUNDATIONS

The general procedure in designing foundations comprehends usually three major parts, namely:

1. Soil exploration and soil testing, as well as determination of soil bearing capacity.
2. The selection of type of foundation, and determination of loads on foundations. The suitability of one or more types of foundations may have to be compared.
3. The design of the selected foundation, checking stresses in the foundation and on soil.

Before any design of a structure is commenced, soil exploration should proceed in order to determine the site conditions and the safe bearing capacity of the soil. Further, in the design of a foundation one first determines its external dimensions (=size) such as depth of laying it, and the size of the base area of the footing based on the load intensity and the properties of the soil. After that one determines the structure of the foundation based on its strength, calculates the pressure on soil exerted by the foundation, and checks the stability of the soil-foundation system for the worst loading conditions.

In designing foundations, a question of paramount importance is the depth of laying of foundations. In this respect it is important to remember that the foundation should not be divorced from its supporting soil. The engineer's task is to design economically feasible types of foundations commensurable with their safety.

For the sake of orientation it suffices to say that the depth of foundations depends upon a series of factors such as

1. Technological significance of the structure
2. Nature of geological conditions of the foundation-supporting soil and/or rock (faults, degree of weathering, cavities in karst areas)
3. Position of groundwater table
4. Water regimen at the site (scour, erosion of soil by runoff)

Choosing the Kind of Foundation

5. Depth of frost penetration in soil (where applicable), or depth of thawing in permafrost regions
6. Soil bearing capacity
7. Adjacent structures
8. Method of laying the foundation, and other factors.

The above means that the footing of the foundation should be taken down to firm soil (where available or attainable). Also, if possible, the footing should be kept above the groundwater table, or else drainage of soil and/or the lowering of the groundwater table is pursued. Where firm soil is unattainable, deep foundations are used. Some of the factors listed above are discussed in great detail in the author's books on soil mechanics.[1,2,3,4]

Example
 Given: A soil-strip foundation system as shown in Fig. 3-1. The contact pressure is given as $\sigma_o = 1$ ton/ft² under the 4-ft wide strips as well as under the mat 20.8 ft wide. The allowable bearing capacity of the plastic, compressible layer of clay, 5 ft thick, is given as $\sigma_{all} = 0.5$ ton/ft². Furthermore, it is also given that the $\sigma_z = (0.5) \sigma_o$ isobar just intrudes into the plastic layer of clay. Discuss and compare the choice between one wide strip mat of width $2B = 20.8$ ft and individual, shallow strip footings of width $2b = 4$ feet under each wall load.

For given conditions, is the wide slab (mat) of width $2B$ a satisfactory solution relative to consolidation and settlement of the plastic clay brought about by the given compressive stress on this layer, $\sigma_z = (0.5) \sigma_o = 0.5$ ton/ft², or would you choose three individual strip footings under each wall?

Discussion
 In discussing the choice between the kinds of foundation to use, some of the following factors may have to be considered: width of footings, depth of compressible layer of soil, rigidity of structure, presence of groundwater table, and any unusual factors at the site.

1. Effect of Width of Footings

From soil mechanics[2] it is known that the wider the loaded footing, the deeper its effect, i.e., the deeper is the extent of the $\sigma_z = (0.2) \sigma_o$ isobar, the depth of such an isobar being approximately $(1.5)(2B)$. This isobar in soil encompasses the seat of its settlement where approximately 80 percent of the total settlement of soil takes place at a depth less than approximately $(1.5)(2B)$ viz., approximately $(1.5)(2b)$. If several loaded narrow footings are

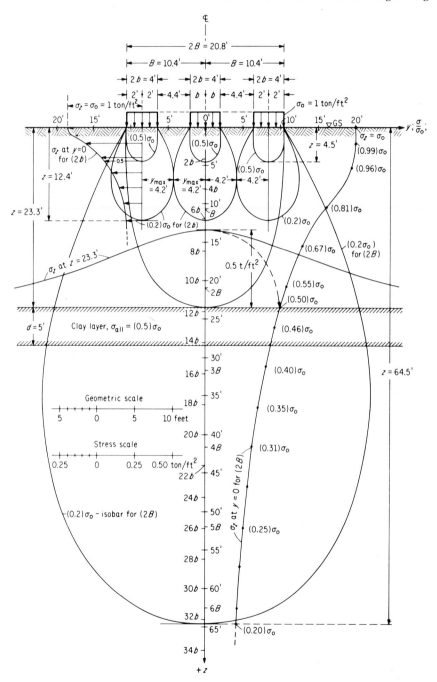

FIG. 3-1. Soil-strip foundation system.

Choosing the Kind of Foundation

spaced closely enough, the individual isobars of each footing in question would combine and merge into one large isobar of the same intensity, reaching approximately $(1.5)(2B)$ deep below the base of the sum of the closely spaced footings into the plastic clay. Because in this problem the vertical pressure $\sigma_z = (0.2)\,\sigma_o$ comprehends the seat of settlement, and because the pressure on the plastic clay is 0.5 ton/ft² = σ_{all}, settlement of the plastic clay layer and of the foundation will take place in addition to settlement of the stressed zone between the base of the footing and the plastic clay layer. The question is: are these settlements tolerable?

Consolidation test results on the clay soils should give the answer as to the tolerability of the expected settlements under the wide strip mat and the narrow strip footings. Obviously, differential settlement can be expected, the greatest one occurring on the center line underneath the wide strip mat and under the center of the narrow footings.

2. Depth, Bearing Capacity, and Settlement

Replacing the given wide strip slab with individual narrow strip footings under each wall, there would be no improvement relative to the extent in depth of the $(0.5)\,\sigma_o =$ isobar, nor to the problem of settlement for the given spacing of the walls. The individual pressure bulbs under each of the narrow footings would merge to result in one combined pressure bulb extending approximately $(1.5)(2B)$ deep.

3. Pressure Distribution by Means of a Rigid Structure

If the individual footings are too closely spaced, earthwork considerations and the problem of differential settlement may point toward the use of a wide strip footing, or mat. Almost uniform compression, consolidation, and hence settlement of the clay may be attained by constructing a rigid foundation-structure system under the entire structure in order to redistribute the load to the outer parts of the foundation. By means of such a rigid structure, equilibrium in settlement of the soil may be quickly attained, and settlement of the rigid structure may become relatively uniformly distributed.

4. Piles

If economically feasible, a pile foundation may be considered as an alternative. The advantage of a pile foundation is that it avoids extensive excavations required by normal strip foundations closely spaced.

5. Groundwater

If there were groundwater present near the ground surface, this, too, speaks for the use of piles.

From this discussion alone it is clear that the choice of the kind of foundation to use may be a very complex problem.

REFERENCES

1. A. R. Jumikis, *The Frost Penetration Problem in Highway Engineering*. New Brunswick, N. J.: Rutgers U. P., 1955.
2. A. R. Jumikis, *Soil Mechanics*. Princeton, N.J.: Van Nostrand, 1962.
3. A. R. Jumikis, *Mechanics of Soils: Fundamentals for Advanced Study*. Princeton, N.J.: Van Nostrand, 1964.
4. A. R. Jumikis, *Thermal Soil Mechanics*. New Brunswick, N.J.: Rutgers U.P., 1966.

PROBLEMS

3-1. Given a soil-footing-load system as shown in Prob. 3-1. The allowable bearing capacity of the medium sand above the groundwater table is given as $\sigma_{all} = 1.5$ ton/ft^2 and that of the sandy-clay layer as $\sigma_{all} = 0.5$ ton/ft^2. Discuss what kind of foundation you would recommend.

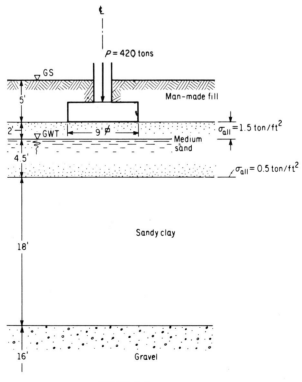

PROB. 3-1

Choosing the Kind of Foundation

3-2. Given a soil profile as shown in Prob. 3-2. If piles were chosen (the load per pile is $P = 300$ kips), how deep should they be driven?
(*Ans.* About 2 feet into the fine sand below the stiff, blue clay, 14 ft thick. Why?)

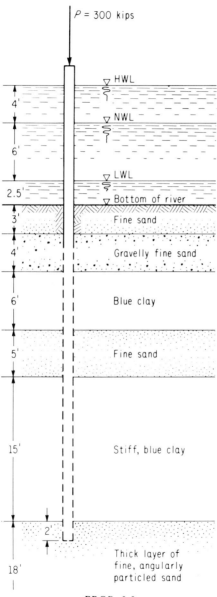

PROB. 3-2

3-3. Decide on the kind of foundation to use (pier, piles, open or pneumatic caisson, or whatever) for a bridge pier support as indicated in Prob. 3-3. Also, determine the depth of the foundation.

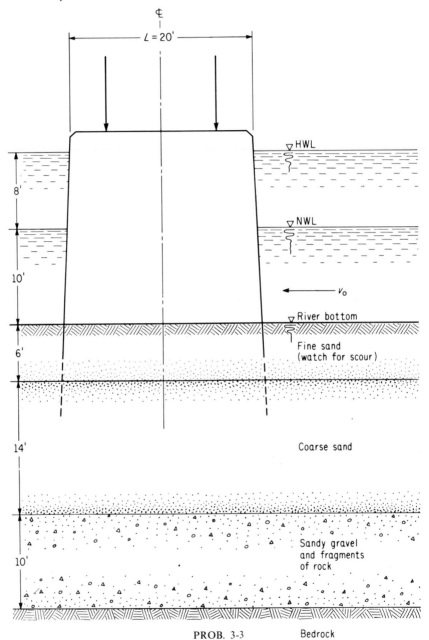

PROB. 3-3

Choosing the Kind of Foundation

3-4. Given a soil profile as shown in Prob. 3-4, and a concentrated load $P = 420$ kips from a column. Recommend tentatively the kind of foundation to use.

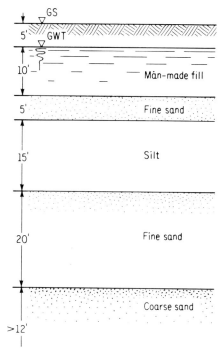

PROB. 3-4

chapter 4
Forces

4-1. SOME POSSIBLE FORCES ACTING ON A FOUNDATION

Some of the forces most commonly encountered in practice and acting on a foundation or earth retaining structure are:
- Weight of the structure
- Lateral earth pressure
- Soil reactions at the base of the footing of the foundation
- Friction forces between soil and foundation
- Wind pressure
- Snow load (where it applies)
- Horizontal loads on the wall (for example, loads attached to a puller, or the pull in tie-rods and cables)
- Hydrostatic pressure of water
- Artesian water pressure
- Hydrostatic uplift force
- Pore water pressure
- Swelling force of soil where it applies
- Hydrodynamic water pressure (wave action, seepage forces)
- Ice pressure in front of dam or wall, or soil moisture behind the wall when frozen
- Dynamic impact (by ships on a quay)
- Seismic forces (earthquake)
- Vibrations induced by machinery, rolling stock, pile-driving operations, explosions and military operations
- Thermal forces (induced by temperature)
- Breaking forces on bridges and possibly other forces.

Loads may act centrically, or eccentrically, constantly or transiently, periodically or at random. The various loads are then combined into a resultant load by way of methods known from graphical statics or analytically. The magnitude and direction of action of the resultant force

Forces

(and moment, where it applies) are of great importance in the design of the foundation and in the stability analysis of the soil-foundation system. The combination of forces usually results in the four types of loading on the foundation (Fig. 4-1).

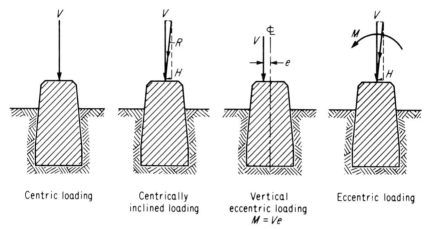

Centric loading   Centrically inclined loading   Vertical eccentric loading $M = Ve$ Eccentric loading

FIG. 4-1. Resultant loads on foundations.

The strength of the foundation depends principally upon the following three quantities:
1. Magnitude and direction of forces acting on the foundation
2. Allowable stresses on the construction materials of which the foundations are built
3. Allowable pressure on the soil.

As a rule, for simple residential flats on good soil, usually there is no need for an elaborate calculation of the foundation. However, with residential flats on poor soil, and with heavily loaded buildings such as commercial buildings, industrial buildings, public buildings and engineering structures (bridges, dams, power plants) the foundations should be analyzed thoroughly, and solidly built.

Sections 4-2 through 4-8 contain discussions on magnitudes of the various possible forces acting on foundations; on the effect of cohesion on stability of foundations, earth retaining structures, sheet piling and dams; on uplift pressures on foundations; on seepage pressure, ice pressure, seismic forces, and safe presumptive bearing capacity values of soil and rock.

4-2. LATERAL EARTH PRESSURES

For calculations of lateral earth pressures, viz., active and passive earth pressures, see Refs. 1-4. To facilitate earth-pressure calculations, earth-pressure coefficient tables such as given by Ref. 4, for example, can be used.

The corresponding active and passive earth-pressure coefficient equations for calculating K_a and K_p are given in Fig. 4-2.

System Nos. and Signs of Angles	Earth Pressure System	COEFFICIENTS OF	
		Active Earth Pressure K_a	Passive Earth Pressure (= resistance) K_p
1.	2.	3.	4.
① $\begin{array}{c\|c} A & P \\ (+)a & (+)a \\ (+)\delta & (+)\delta \\ (+)\phi & (-)\phi \\ (+)\phi_1 & (-)\phi_1 \end{array}$	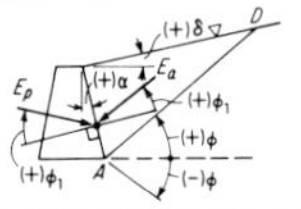	① $K_a = \dfrac{\cos^2(\phi-a)}{\cos^2 a \cdot \cos(a+\phi_1)\left[1+\sqrt{\dfrac{\sin(\phi+\phi_1)\cdot\sin(\phi-\delta)}{\cos(a+\phi_1)\cos(a-\delta)}}\right]^2}$	⑨ $K_p = \dfrac{\cos^2(\phi+a)}{\cos^2 a \cdot \cos(a-\phi_1)\left[1-\sqrt{\dfrac{\sin(\phi+\phi_1)\sin(\phi+\delta)}{\cos(a-\phi_1)\cos(a-\delta)}}\right]^2}$
② $\begin{array}{c\|c} A & P \\ (+)a & (+)a \\ \delta=0 & \delta=0 \\ (+)\phi & (-)\phi \\ (+)\phi_1 & (-)\phi_1 \end{array}$	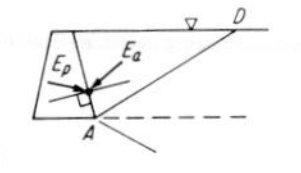	② $K_a = \dfrac{\cos^2(\phi-a)}{\cos^2 a \cdot \cos(a+\phi_1)\left[1+\sqrt{\dfrac{\sin(\phi+\phi_1)\cdot\sin\phi}{\cos(a+\phi_1)\cdot\cos a}}\right]^2}$	⑧ $K_p = \dfrac{\cos^2(\phi+a)}{\cos^2 a \cdot \cos(a-\phi_1)\left[1-\sqrt{\dfrac{\sin(\phi+\phi_1)\cdot\sin\phi}{\cos(a-\phi_1)\cdot\cos a}}\right]^2}$
③ $\begin{array}{c\|c} A & P \\ (+)a & (+)a \\ (-)\delta & (-)\delta \\ (+)\phi & (-)\phi \\ (+)\phi_1 & (-)\phi_1 \end{array}$	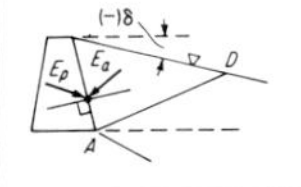	③ $K_a = \dfrac{\cos^2(\phi-a)}{\cos^2 a \cdot \cos(a+\phi_1)\left[1+\sqrt{\dfrac{\sin(\phi+\phi_1)\cdot\sin(\phi+\delta)}{\cos(a+\phi_1)\cos(a+\delta)}}\right]^2}$	⑦ $K_p = \dfrac{\cos^2(\phi+a)}{\cos^2 a \cdot \cos(a-\phi_1)\left[1-\sqrt{\dfrac{\sin(\phi+\phi_1)\sin(\phi-\delta)}{\cos(a-\phi_1)\cos(a+\delta)}}\right]^2}$
④ $\begin{array}{c\|c} A & P \\ a=0 & a=0 \\ (+)\delta & (+)\delta \\ (+)\phi & (-)\phi \\ (+)\phi_1 & (-)\phi_1 \end{array}$		④ $K_a = \dfrac{\cos^2\phi}{\cos\phi_1\left[1+\sqrt{\dfrac{\sin(\phi+\phi_1)\sin(\phi-\delta)}{\cos\phi_1\cdot\cos\delta}}\right]^2}$	⑥ $K_p = \dfrac{\cos^2\phi}{\cos\phi_1\left[1-\sqrt{\dfrac{\sin(\phi+\phi_1)\sin(\phi+\delta)}{\cos\phi_1\cdot\cos\delta}}\right]^2}$
⑤ $\begin{array}{c\|c} A & P \\ a=0 & a=0 \\ \delta=0 & \delta=0 \\ (+)\phi & (-)\phi \\ (+)\phi_1 & (-)\phi_1 \end{array}$		⑤ $K_a = \dfrac{\cos^2\phi}{\cos\phi_1\left[1+\sqrt{\dfrac{\sin(\phi+\phi_1)\cdot\sin\phi}{\cos\phi_1}}\right]^2}$	⑤ $K_p = \dfrac{\cos^2\phi}{\cos\phi_1\left[1-\sqrt{\dfrac{\sin(\phi+\phi_1)\cdot\sin\phi}{\cos\phi_1}}\right]^2}$
⑥ $\begin{array}{c\|c} A & P \\ a=0 & a=0 \\ (-)\delta & (-)\delta \\ (+)\phi & (-)\phi \\ (+)\phi_1 & (-)\phi_1 \end{array}$		⑥ $K_a = \dfrac{\cos^2\phi}{\cos\phi_1\left[1+\sqrt{\dfrac{\sin(\phi+\phi_1)\sin(\phi+\delta)}{\cos\phi_1\cdot\cos\delta}}\right]^2}$	④ $K_p = \dfrac{\cos^2\phi}{\cos\phi_1\left[1-\sqrt{\dfrac{\sin(\phi+\phi_1)\sin(\phi-\delta)}{\cos\phi_1\cdot\cos\delta}}\right]^2}$
⑦ $\begin{array}{c\|c} A & P \\ (-)a & (-)a \\ (+)\delta & (+)\delta \\ (+)\phi & (-)\phi \\ (+)\phi_1 & (-)\phi_1 \end{array}$	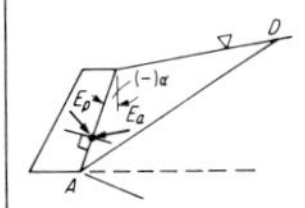	⑦ $K_a = \dfrac{\cos^2(\phi+a)}{\cos^2 a \cdot \cos(a-\phi_1)\left[1+\sqrt{\dfrac{\sin(\phi+\phi_1)\cdot\sin(\phi-\delta)}{\cos(a-\phi_1)\cos(a+\delta)}}\right]^2}$	③ $K_p = \dfrac{\cos^2(\phi-a)}{\cos^2 a \cdot \cos(a+\phi_1)\left[1-\sqrt{\dfrac{\sin(\phi+\phi_1)\sin(\phi+\delta)}{\cos(a+\phi_1)\cos(a+\delta)}}\right]^2}$
⑧ $\begin{array}{c\|c} A & P \\ (-)a & (-)a \\ \delta=0 & \delta=0 \\ (-)\phi & (-)\phi \\ (+)\phi_1 & (-)\phi_1 \end{array}$	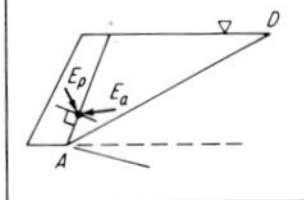	⑧ $K_a = \dfrac{\cos^2(\phi+a)}{\cos^2 a \cdot \cos(a-\phi_1)\left[1+\sqrt{\dfrac{\sin(\phi+\phi_1)\cdot\sin\phi}{\cos(a-\phi_1)\cdot\cos a}}\right]^2}$	② $K_p = \dfrac{\cos^2(\phi-a)}{\cos^2 a \cdot \cos(a+\phi_1)\left[1-\sqrt{\dfrac{\sin(\phi+\phi_1)\cdot\sin\phi}{\cos(a+\phi_1)\cdot\cos a}}\right]^2}$
⑨ $\begin{array}{c\|c} A & P \\ (-)a & (-)a \\ (-)\delta & (-)\delta \\ (+)\phi & (-)\phi \\ (+)\phi_1 & (-)\phi_1 \end{array}$	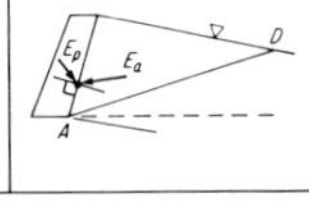	⑨ $K_a = \dfrac{\cos^2(\phi+a)}{\cos^2 a \cdot \cos(a-\phi_1)\left[1+\sqrt{\dfrac{\sin(\phi+\phi_1)\cdot\sin(\phi+\delta)}{\cos(a-\phi_1)\cos(a-\delta)}}\right]^2}$	① $K_p = \dfrac{\cos^2(\phi-a)}{\cos^2 a \cdot \cos(a+\phi_1)\left[1-\sqrt{\dfrac{\sin(\phi+\phi_1)\sin(\phi-\delta)}{\cos(a+\phi_1)\cos(a-\delta)}}\right]^2}$

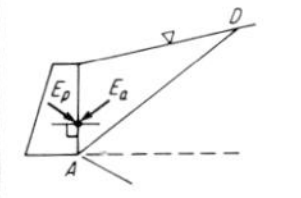

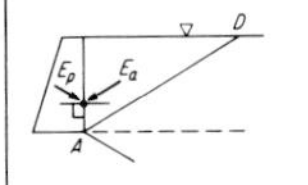

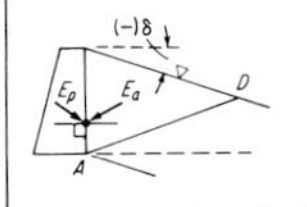

FIG. 4-2. Earth pressure coefficients (Ref. 4).

The total active earth pressure on a certain retaining wall system is calculated as

Forces

$$E_a = (1/2)\gamma_1 H^2 K_a$$

and the total passive earth pressure (resistance) of a certain retaining wall system is calculated as

$$E_p = (1/2)\gamma_1 H^2 K_p$$

where $\gamma_1 = \gamma + \dfrac{2p\cos\delta}{h}$ = surcharged unit weight of soil

H = height of retaining wall
γ = unsurcharged unit weight of soil
p = surcharge intensity
δ = angle of slope of ground surface on backfill side of wall
h = distance of heel point of wall to ground surface or its extension

4-3. UNIT WEIGHT OF SOIL

One of the important parts in all earth pressure and stability calculations of soil-foundation systems is the correct assessment and assumption to use

TABLE 4-1
Approximate Values of Unit Weight of Some Soils and Rocks

Soils and Rocks	Natural, Moist, Unit Weights	
	kg/m^3	lb/ft^3
Basalt	2,400-3,100	150-195
Clay	1,800-2,600	110-160
Clay, very dense	2,000	125
Clay, loose	1,800	115
Clay, silty, dry	1,600	100
Clay, moist	1,800	115
Gneiss	2,400-2,700	150-170
Granite	2,600-2,700	160-170
Gravel, dry, loose	1,400-1,700	85-105
Gravel, dense	1,600-1,900	100-120
Gravel, wet	1,900-2,000	120-125
Limestone	2,000-2,900	125-180
Loess	1,600	100
Peat	1,100	70
Sand, fine, dry	1,600	100
Sand, dry	1,400-1,700	85-105
Sand, wet	1,900-2,000	120-125
Shale	2,600-2,900	160-180
Silt	1,800	115
Stone, crushed	1,600	100
Traprock	2,700-3,400	165-215

of the unit weight and angle of internal friction of soil. For the purpose of one's orientation the approximate unit weights of some soils are given in Table 4-1. This table contains the approximative unit weights of most commonly encountered soil types in their natural deposition. These values reveal that the range of the unit weights of soils may vary between quite wide limits, and that their influence upon the static effects in stability problems in soil and foundation engineering, in some instances, may be relatively large. Therefore it is advisable in each particular foundation engineering problem to determine the unit weight of soil by actual test.

Besides the unit weight of soil in its natural deposition containing moisture, other unit weights are to be distinguished: for example, dry unit weight, γ_d; unit weight of solids by absolute volume, $\gamma_s = G\gamma_w$; saturated unit weight of soil, γ_{sat}, and submerged or buoyant unit weight of soil, γ_{sub}. A summary of unit weight formulas of soil is given in Table 4-2.

TABLE 4-2
Formulas for Calculating Unit Weights of Soil

Soil Condition	Unit-Weight Formulas	
	In Terms of Porosity n	In Terms of Void Ratio e
Dry soil: $S = 0$	$\gamma_d = (1-n)G\gamma_w$	$\gamma_d = \dfrac{G}{1+e}\gamma_w$
Moist soil: $S = S$	$\gamma = (1-n)G\gamma_w + nS\gamma_w$	$\gamma = \dfrac{G+eS}{1+e}\gamma_w$
Saturated soil: $S = 1.0$	$\gamma_{sat} = (1-n)G\gamma_w + n\gamma_w$	$\gamma_{sat} = \dfrac{G+e}{1+e}\gamma_w$
Buoyant (submerged) soil	$\gamma_{sub} = (1-n)(G-1)\gamma_w$	$\gamma_{sub} = \dfrac{G-1}{1+e}\gamma_w$

G = specific gravity of soil e = void ratio of soil
γ_w = unit weight of water S = degree of saturation of soil
n = porosity of soil w = moisture content of soil, by dry weight

For calculation purposes, unit weight of water in the British system of units is taken as 62.4 lb/ft³. In the metric system, $\gamma_w = 1$ g/cm³ = 1 kg/1,000 cm³ = 1,000 kg/m³ = 1 t/m³ = 62.4 lb/ft³.

In approximative calculations, the unit weight of salt water can be taken as 64.0 lb/ft³, or = 1.025 g/cm³ = 1.025 kg/liter = 1.025 kg/m³ = 1.025 t/m³.

For conversion purposes the relationship of pressure 1 ton/ft² ≈ 1 kg/cm² can be used.

For conversion factors of units of measurement between the British and metric systems, see Appendix III.

Forces

The unit weight of a moist soil γ can also be expressed in terms of moisture content of soil w, by dry weight as

$$\gamma = \frac{(1+w)G}{1+e}\gamma_w$$

4-4. ANGLE OF INTERNAL FRICTION. COHESION. WALL FRICTION

In performing earth-pressure calculations it is essential that the shear strength of the soil, viz., angle of internal friction ϕ (and cohesion c), be determined by test on the particular soil material in question. The value os this so-called *angle of internal friction* is a test parameter which depends upon the method of test used for its determination.

The angle of internal friction for coarse-particled (permeable) soils is almost independent of moisture content of the soil. Regardless of whether wet or dry, the coefficient of internal friction of such soils varies considerably between about $\tan\phi = 0.45$ to about $\tan\phi = 0.70$. For clayey soils the $\tan\phi$ values usually range from about $\tan\phi = 0.20$ to about $\tan\phi = 0.58$, depending upon the moisture content and the presence of sand. About 70-75 percent sand is required to make an appreciable difference in ϕ, because lesser amounts tend to "float" in a matrix of clay. Firm clay often has a lower ϕ than soft clay due to preconsolidation—the application of normal pressure only slightly affects the shear strength of a stiff, preconsolidated clay, whereas it increases in soft clay by consolidation. The ϕ for loess is about 25°, and higher for consolidated alluvial silt which is freely draining. The magnitude of angle ϕ for clean sand is practically the same above and

TABLE 4-3
Angle of Friction of Some Soils in Degrees

Soil	Angle of Internal Friction $\phi°$	Coefficient of Friction, $\mu = \tan\phi$
Gravel, dry	35-40	0.700-0.839
Sand,		
Dry	30-35	0.577-0.700
Moist	20	0.364
Saturated	15	0.268
Silt	5	0.087
Clay,		
Firm ($w = 10$ to 15 percent)	20	0.364
Moist	15	0.268
Soft-plastic	7	0.123

below the groundwater table, and is unaffected by water. Some ϕ values of some soils are shown in Table 4-3.

In practice, in earth pressure calculations, and in performing stability analyses on retaining walls by applying Coulomb's earth-pressure theory, the cohesion of a cohesive, frictional soil material is often omitted. The omission of cohesion is often practiced under the pretense of increased degree of safety relative to the stability calculations of the structures. This may be the case when, upon the omission of cohesion in earth-pressure calculations, the true angle of internal friction ϕ is retained and used. Also, upon the increase in soil moisture content, cohesion may decrease considerably. Cohesion also is shear-rate-dependent, and may in effect decrease to zero in fissured or creeping soils. However, if, as is frequently done, a larger angle of friction is used as a substitute for the true one, with the thought behind it of compensating for the reduced cohesion, then such a method will result instead in a decreased degree of safety. This can particularly be noticed with frictional-cohesive soils where, as can be seen from the earth pressure equations for such soils, the frictional forces vary quadratically with H, whereas the cohesive forces vary only linearly with H.

In dry sand, there is practically no cohesion present. It is also of small magnitude in moist and wet granular soils, the particles of which are angularly shaped, such as in the case of some sand and gravel. Cohesion of considerable magnitude is present in cohesive, plastic soils.

The value of cohesion for some soils ranges approximately from $c = 0.01$ ton/ft^2 to about $c = 0.12$ ton/ft^2. Admittedly, exceptions to these values are possible, and c-values are reported as high as $c = 3{,}500$ lb/ft$^2 = 1.75$ ton/ft^2 for a very stiff boulder clay[5] with $\phi = 16°$.

The magnitude of cohesion, viz., shear strength of a soil, must be determined by test on that soil. The test should be performed under conditions which would simulate as closely as possible those most likely to occur in nature.

Changes in moisture conditions in the backfill soil may bring about changes in the values of the apparent cohesion of a soil. Shear strength of a cohesive soil is very much affected by the soil moisture content. This phenomenon gives an idea of the possible danger of rupture of earth masses, such as laterally supported and unsupported embankment slopes.

Cohesion of soil is a very sensitive physical property, especially relative to water. Because soil in the field is very difficult to protect against climatic variations which occur during the various seasons of the year, particularly in respect to changes in moisture content in soil, in practice cohesion is very seldom considered in sheet-piling calculations.

Forces

Wall Friction Angle ϕ_1

The angle of wall friction ϕ_1 depends upon the kind of wall material of the earth retaining structure, the smoothness of the surface of the back face of the wall, the kind of backfill soil material, the density of soil, the hydrologic conditions, vibration of soil and shocks, and possibly other changes and variations.

The smoother the backface surface of the retaining wall, and the smoother the particles of the soil, the less is the angle of wall friction ϕ_1. In hydraulic structure calculations the angle of wall friction is usually set equal to zero when the backfill material is wet, and when the backface surface of the retaining structure is smooth.

Vibrations and shocks in soil have an effect of decreasing the angle of wall friction ϕ_1 (kinetic friction is less than static friction).

The densification of soil by vibration increases the magnitude of the active earth pressure E_a. Generally, for smooth walls with no vibration, the angle of wall friction ϕ_1 is taken from $\phi_1 = (1/3)\phi$ to $\phi_1 = (2/3)\phi$. Upon steel sheet piling the value of ϕ_1 is usually taken as $\phi_1 = (1/2)\phi$, according

TABLE 4-4

Coefficients of Friction of Sand on Solid Surfaces

Nos.	Description of Solid Surface	Pressure kg/cm² from	Pressure kg/cm² to	Coefficient of Friction, f from	Coefficient of Friction, f to	ave
1	Sand and gravel on smooth masonry	—	—	—	—	>0.30
2	Quartz powder on steel, dry	2		0.348	0.400	0.376
3	Quartz powder on steel, water-flooded	2		0.348	0.450	0.416
4	Sand, almost dry, on glass	—	—	—	—	0.47
5	Dry, coarse sand on steel	0.016	0.021	0.48	0.48	0.48
6	Moist, coarse sand on steel	0.010	0.021	0.48	0.51	0.50
7	Dry, fine sand on steel	0.007	0.025	0.57	0.55	0.56
8	Moist, fine sand on steel	0.007	0.022	0.62	0.60	0.61
9	Steel sheet piling against					
	Gravel, sand-gravel mixtures	—	—	—	—	0.40
	Clean sand, silty sand-gravel	—	—	—	—	0.30
	Silty sand, sand with silt or clay	—	—	—	—	0.25
	Nonplastic silt	—	—	—	—	0.20
	Soft clay and clayey silt — Stiff, hard clay and clayey silt — adhesion:			0.0488-0.2900 kg/cm² 0.2900-0.5800 kg/cm²		
10	Steel on steel at sheet piling interlocks	—	—	—	—	0.30
11	Sand and gravel on rough masonry	—	—	—	—	<0.60
12	Moist sand on timber	—	—	—	—	0.35

to Müller-Breslau.[2]   Based on his large-scale experimental research, Müller-Breslau suggested for rough walls and good drainage of the backfill material to set $\phi_1 = (3/4)\phi$.

Table 4-4 gives some coefficients of friction of sand on solid surfaces, and Table 4-5 coefficients of friction on lubricated surfaces.

TABLE 4-5
Coefficients of Friction on Lubricated Surfaces

Nos.	Kind of Lubrication	Coefficient of Friction, f
1	Smoothest and best greased surfaces	0.030 to 0.036
2	Wood on wood, greased with soap	0.040 to 0.200
3	Smooth surfaces, constantly greased	0.05
4	Hardwood on polished metal, greased	0.06
5	Smooth surfaces, occasionally greased	0.07 to 0.08
6	Oak on oak, greased	0.075
7	Iron on oak, greased	0.08
8	Castor oil between glass plates	0.10
9	Leather belts on cast iron pulleys, highly greased	0.12
10	Metal on metal, moist and greasy	0.14

4-5. UPLIFT PRESSURE. ARTESIAN WATER PRESSURE. SEEPAGE PRESSURE

If the footing of a foundation is laid below the water table, then the direct upward hydrostatic pressure of water called the *uplift pressure* acts on its base. Relative to the uplift pressure of water, there are two kinds of structures, namely:

1. Structures the foundations of which are laid below the water table, the latter being at the same elevation on both sides of the foundation. In such a case there prevails a uniform hydrostatic pressure on both sides of the foundation, and the uplift pressure is uniformly distributed over the base of the footing (Fig. 4-3),
2. Structural foundations on both sides of which the water table is at different elevation, viz., when groundwater is in motion (Fig. 4-4). In such a case the uplift pressure of water on the base of the footing is nonuniformly distributed. Besides, if the difference in pressure heads on both sides of the foundation is caused by seepage, i.e., water is in motion, then the pressure is hydrodynamic.

Forces

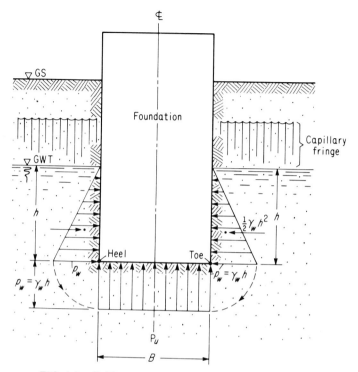

FIG. 4-3. Uplift pressure diagram for groundwater at rest.

Uplift Pressure for Groundwater at Rest

Among the several kinds of water only the free water (in bulk) exerts a hydrostatic pressure. If water gains access under a foundation or a massive dam it causes an upward pressure known as *uplift*. In general, uplift pressure acts against the weight of a foundation or dam, thus increasing the possibility of danger of sliding as well as overturning of the structure. Figure 4-3 shows a vertical uplift pressure diagram for the case of groundwater at rest. The uplift pressure p_w is calculated as

$$p_w = \gamma_w h \qquad (4\text{-}3)$$

where γ_w = unit weight of water
h = immersed depth of foundation below groundwater table

For a section of dam or foundation one unit long, the total uplift pressure P_u is calculated as

$$P_u = p_w B(1) = \gamma_w B h \qquad (4\text{-}4)$$

acting at the center of gravity of the rectangular pressure distribution

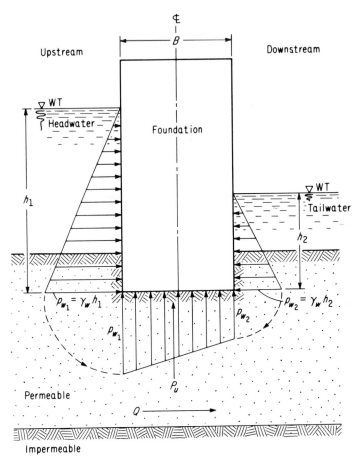

FIG. 4-4. Uplift pressure diagram for groundwater in motion.

diagram.

The capillary water above the groundwater table does not transmit hydrostatic pressure.

Uplift Pressure for Groundwater in Motion

When the water tables on both sides of the immersed foundation are at different levels (groundwater in motion, for example), then the uplift pressure diagram is a trapezoid (Fig. 4-4). For a one-unit-long section of the dam the magnitude of the total uplift pressure P_u is calculated as

$$P_u = \left[\frac{(p_{w1}+p_{w2})}{2}\right] B(1) = \frac{\gamma_w B}{2}(h_1+h_2) \qquad (4\text{-}5)$$

where the symbols are as shown in Fig. 4-4.

Forces

The resultant of this uplift force P_u is applied at the center of gravity of the trapezoidal uplift pressure diagram.

The discharge quantity Q may be determined by means of the flow net theory.[2,3]

Uplift Pressure Underneath the Base of a Massive Dam

Assuming a linear uplift pressure distribution, and referring to Fig. 4-5, the uplift pressure p_w under the edge of the dam on the upstream side has

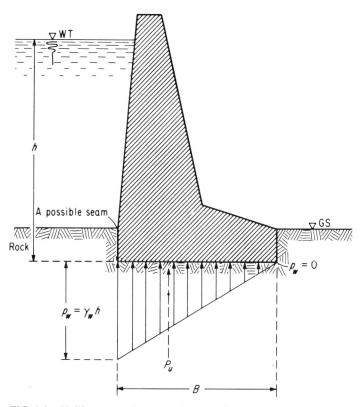

FIG. 4-5. Uplift pressure diagram under a massive water-impounding dam.

its full value of $p_w = \gamma_w h$ (because of seams, joints and fissures in the rock), but on the downstream side it is $p_w = 0$. The uplift pressure diagram is then a triangle, and the total uplift pressure is

$$P_u = (1/2) p_w B(1) = (1/2) \gamma_w Bh \qquad (4\text{-}6)$$

According to Crieger and McCoy,[6] because of grout cutoff at the heel of a homogeneous dam, the pressure ordinate p_{w_1} is calculated as

$$p_{w_1} = \gamma_w [h_2 + \zeta(h_1 - h_2)] \tag{4-7}$$

where ζ = uplift intensity factor depending upon the effectiveness of the grouted cutoff and drains. However, the uplift intensity at the downstream face remains constant at $p_{w_2} = \gamma_w h_2$ and the uplift pressure varies as the line 1-2

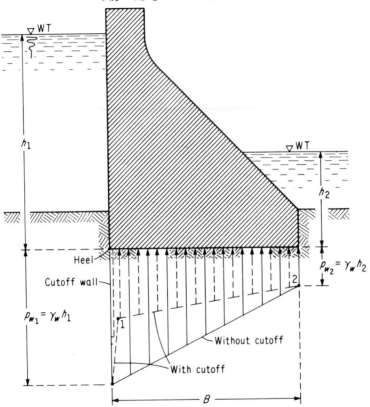

FIG. 4-6. Uplift pressure diagrams under a homogeneous dam with and without cut-off.

(see Fig. 4-6). The total uplift pressure P_u is then equal to

$$P_u = \gamma_w B \left[h_2 + \frac{\zeta}{2}(h_1 - h_2) \right] \tag{4-8}$$

The intensity factor ζ varies for well-grouted and drained foundations from close to 0 to not over 0.5.[7] This uplift intensity is considered to be exerted on 100 percent of the area of the base.

Depending on the nature of the rock and on the degree of its weathering, the magnitude of the uplift pressure of water may vary between 100 percent and 50 percent of the greatest water pressure (depth) at the upstream side.

Forces

The practices of assessing the magnitude of the uplift pressure vary from agency to agency and from country to country. The various assumed uplift pressure lines are shown in Figs. 4-7b and 4-7c.

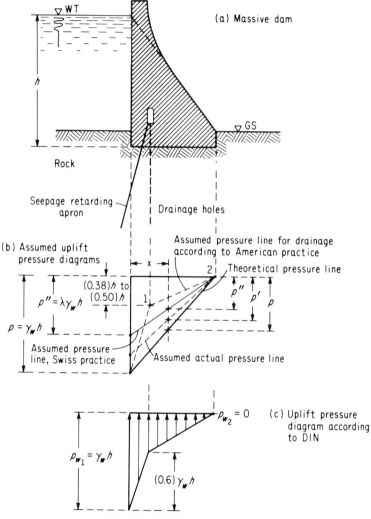

FIG. 4-7. Uplift pressure diagrams.

A linear decrease from full or two-thirds hydrostatic pressure at the heel to tailwater or atmospheric pressure at the toe and extending over the full or upstream half or two-thirds of the base has been postulated in many designs, among other things, upon the extent of internal drainage of the soil, for which provisions have been made.[7]

Uplift pressure can be reduced by means of cutoff walls, seepage-retarding aprons, high-pressure grouting of rocks supporting dams or foundations, by constructing of drainage wells. These methods reduce the length of the seepage path, thus reducing the hydraulic gradient and hence the seepage velocity, and consequently resulting in a lessened danger of overturning.

An instructive description on uplift pressure across a base of a masonry dam on soil may be found in Ref. 8.

Artesian Water Pressure

If an impermeable clay layer below the bottom of the excavation is of inadequate thickness t, the artesian water pressure p, depending upon its intensity and the strength of the clay, may break the bottom of the excavation and inundate it (Fig. 4-8). To avoid such a situation, reliable information about the soil profile and groundwater condition at the site should be procured prior to commencing of the excavation work. This facilitates selection of a proper foundation to use as well as a method of laying it.

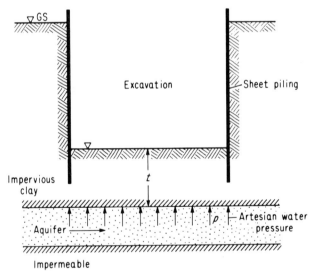

FIG. 4-8. Effect of artesian water pressure on the stability of the bottom of excavation.

Seepage Pressure

The magnitude of seepage pressure,

$$D = \gamma_w i \qquad (4\text{-}9)$$

per unit of volume of soil may be determined (1) by the method of trial and adjustment, (2) by the method of hydraulic analogy, (3) by the method of

Forces

electrical analogy, and (4) analytically. For these methods the reader should consult Refs. 1–3.

4-6. ICE PRESSURE

Sheet ice may exert an undesirable pressure on bridge piers and water-impounding structures, such as dams. Chellis[9] writes that Swedish observations have indicated that the magnitude of the ice pressure on a dam may approach 10,000 lb per linear foot.

Experimental investigations by the Bureau of Reclamation[10] show field measurements of ice thrusts as high as 20,000 lb/(lin ft) on the concrete arch dam 445 ft long, and 125 ft high, impounding Eleven Mile Canyon Reservoir, part of the municipal water supply of Denver, Colorado.

The greatest ice pressure ever recorded in the laboratory or field is 23,000 lb/(lin ft).[11]

The Mackinac Straits bridge piers were designed for an ice pressure of 115,000 lb/(lin ft) of pier width. On certain piers, an ice pressure of 65,000 lb/(lin ft) was reckoned with.[10]

4-7. SEISMIC FORCES

Earthquakes create both horizontal and vertical inertia forces. In regions of seismic activity the stability calculations of soil-foundation systems and earthworks should also include the seismic forces, because they reduce the margin of safety, or may even bring about the collapse of a structure. Seismic forces are applied at the center of gravity of the body, or at the center of gravity of the part of the body above any horizontal plane passed through the structure. Both horizontal and vertical seismic forces must be considered. The magnitude of a horizontal seismic force F is calculated as

$$F = \frac{W}{g} a = ma \tag{4-10}$$

where m = mass of structural body above horizontal plane being studied
W = weight of structural body above horizontal plane being studied
g = acceleration of gravity
a = seismic acceleration, such as acceleration of earthquake wave

In the United States, the value of the seismic factor λ for a is taken as $\lambda = 0.75$, or $a = (0.75)g$ for foundations laid on rock, and $a = (0.10)g$ for foundations laid on sand and other types of soil. If a dam or foundation is constructed near a live, or active fault, higher values of horizontal accelerations are in order.

TABLE 4-6
Mercalli-Cancani-Sieberg Earthquake Intensity Scale

Nature of Intensity of Earthquake	Maximum Acceleration (Cancani), cm/sec^2	Intensity Scale (Mercalli), Degrees	Remarks[21]	Seismic Factor $\lambda = a_{max}/g$, percent
Instrumental	<0.25	I	Microseismic tremors and shocks detectable by instruments only	<0.025
Very light	0.25-0.5	II	Noted only by a few sensitive people at rest	0.05-0.10
Light	0.5-1.0	III	Felt by people at rest; like passing truck	0.05-0.10
Perceptible or moderate	1.0-2.5	IV	Generally perceptible by people in motion; loose objects disturbed; glassware trembles; ceilings creak	0.1-0.25
Strong	2.5-5.0	V	Many awakened; dishes broken; bells rung; pendulum clocks stopped; oscillations of suspended objects	0.25-0.50
Very strong	5.0-10.0	VI	Felt by all; some people frightened; damage slight; some plaster cracked; fall of objects; collapse of chimneys	0.5-1.0
Strongest	10.0-25.0	VII	Noticed by people in autos; damage to poor construction; light damage to brick buildings and light structures; light cracks in walls; falls of chimneys and tiles	1.0-2.5
Destructive	25.0-50.0	VIII	Chimneys fall; cracks in filled panel walls in steel lattice structures having no corner stiffeners; much damage in substantial buildings (1/4 of buildings); heavy furniture overturned	2.5-5.0
Devastating	50.0-100.0	IX $a = (0.1)g$	Great damage to substantial structures and to lattice-type steel structures (not frames); ground cracked; pipes broken	5-10
Disastrous	100.0-250.0	X	Crushing of column heads; many buildings ($\sim 3/4$) destroyed; most of them collapse; rails bent; crevices in soil; damage to dams; landslides; light damage to aseismically designed steel and reinforced concrete structures	10-25

Forces

Vertical inertia force reduces the effective weight of a dam or foundation only momentarily.

After studying all the available data on the effect of earthquakes on structures, American engineers[12,13] have arrived at the following opinion: the maximum acceleration a of a foundation-supporting soil at earthquakes in Japan probably attained a value of half of the acceleration of gravity, i.e., $a = (0.5)g$, where $g = 981$ cm/sec^2 = 32.2 ft/sec^2. For example, the very severe Japanese earthquake of September 1, 1923, had an acceleration of $a = (0.33)g$. According to Briske,[14,15] the Tokyo Bridge Building Code requires a vertical acceleration of $(1/6)g$ coupled with a horizontal acceleration of $(1/3)g$. In California the horizontal acceleration a is calculated with a seismic factor of $\lambda = 0.25$, i.e., $a = (0.25)g$. It is interesting to note that structures which in Japan were designed for $a = (0.10)g$ have survived the most severe earthquakes.[16]

For the design of San Francisco-Oakland bridge, the engineers reckoned on a horizontal acceleration of the soil of $a = (0.10)g$ with a period of oscillation of 1.5 sec, corresponding to an oscillation of 56 mm.

Although usually vertical acceleration on structures due to earthquakes is neglected in stability analyses of structures, and regardless of the casual figures for seismic effects here cited, it is imperative that in regions of seismic activity the design of any structure should meet all of the seismic requirements prescribed by the corresponding seismic building codes for aseismic design.[17-19]

Oscillatory variation in hydrostatic pressure, earthquake-induced, is assumed to equal $(0.555)(a/g)\gamma_w H$ and act at a distance $(4H)/(3\pi)$ above the bottom of a reservoir.[20]

Table 4-6 presents the Mercalli-Cancani-Sieberg earthquake intensity

TABLE 4-6 (*Continued*)

Nature of Intensity of Earthquake	Maximum Acceleration (Cancani), cm/sec^2	Intensity Scale (Mercalli), Degrees	Remarks[21]	Seismic Factor $\lambda = a_{max}/g$, percent
Catastrophic	250.0-500.0	XI	Few structures left standing; damage to bridges and dams; wide cracks in ground; rock and landslides	25-50
Violent catastrophe	500.0	XII	Total destruction of all structures, including foundations; nothing stands which is created by man's hands; ground deformations to a large extent; lakes run out; rivers are diverted	50

scale. Some of the information in this table is compiled from J. M. Trefethen, *Geology for Engineers*.[21]

The XIth degree intensity of this scale has not been exceeded yet in historical times in large cities.

The Sieberg acceleration scale from traffic is ten times larger than that given by Cancani (Column 2, Table 4-6) for earthquakes.

Earthquakes are possible in almost any area, but the range of probability varies enormously.

4-8. SOIL AND ROCK SAFE BEARING CAPACITY

The foundation has for its purpose the transmission of structural loads to the soil safely without bringing about any distress to its superstructure. The superstructure exerts a resultant load R on the foundation. Thus the foundation must be designed so that the resultant does not (1) bring about any lateral sliding of the foundation base on soil, (2) bring about rotation, viz., tilting of the structure, (3) cause any groundbreak, and (4) produce any intolerable settlement. Also, the stresses within the foundation materials and those of the loaded soil should not exceed the allowable limits.

The sources for obtaining soil-and rock-bearing capacity values for designing foundations are: (1) building codes, official regulations, and civil engineers' handbooks; (2) soil-loading tests in place; (3) laboratory testing of soils and rocks, and (4) analytical methods.

When a construction is contemplated within the jurisdiction of a building code, the character of the soil and rock and their safe bearing capacity upon which soil and rock the structure is to be erected must be ascertained by the methods as directed by the code.

Building codes usually contain tables of safe bearing capacity values for various types of soils and rocks. Table 4-7 contains a compilation of the safe bearing capacity values (or "presumptive" bearing capacities) for various types of soils and rocks. The values listed in this table are the maximum allowable soil and rock bearing capacity values. It must, however, be cautioned that any table of allowable soil bearing capacity values should be used with great discretion and in accordance with what experience has shown to be safe, allowable values in certain regions and climatic and hydrological conditions.

The presumptive allowable bearing capacity values may be used for preliminary calculations, or when elaborate soil exploration and testing is not justified. Presumptive bearing capacity values may be unreliable for foundations laid on very soft to medium-stiff, fine-particled soils. In such instances the soil bearing capacity values should be checked by theoretical

TABLE 4-7

Presumptive Bearing Capacity Values of Some Soils and Rocks, $kg/cm^2 \approx ton/ft^2$

Clay, medium soft	1-1.5
Clay, dense	2
Clay, hard	5
Sand, loose	1
Sand, fine	2
Sand, coarse, dense	3
Gravel, dense	4-6
Rock, soft	8
Rock, medium hard	40
Rock, sound	60
Basalt, diabase, gneiss, traprock	20-100
Limestone	10- 20
Marble	10- 20
Sandstone	10- 20
Shale	8- 20

soil bearing capacity calculations substantiated by tests of soils, especially by shear tests.

4-9. BEARING CAPACITY

Soil Ultimate Bearing Capacity Equations for Shallow Foundations

The approximate, theoretical, ultimate soil bearing capacity equations for shallow foundations in terms of compressive stress q_{ult} have been put forward by Terzaghi.[22] His equations and the specializations that follow are given for the following conditions:

1. The soil is uniform.
2. The soil properties are γ, ϕ, c.
3. The depth D of the footing below the ground surface is less than its width B, i.e., $D \leq B$.
4. The groundwater table is below the soil wedge to be expelled from underneath the base of the footing.
5. The resultant vertical load on the horizontal ground surface is centric.
6. The resistance of soil due to soil friction and cohesion (adhesion) in the vertical side faces of the footing is ignored.

The ultimate soil bearing capacity equations now follow.

1. For continuous (strip) footing with rough base and general shear in soil.

$$q_{ult} = cN_c + \gamma DN_q + (0.5)\gamma BN_\gamma \qquad (4\text{-}11)$$

where c = cohesion of soil
  γ = unit weight of soil; if soil is submerged below groundwater table, the submerged soil unit weight should be used
  D = depth of footing below ground surface
  B = width of footing
  q = index for surcharge

N_c, N_q, and N_γ are the so-called soil bearing capacity factors (dimensionless), and are functions of the angle of internal friction ϕ of soil. One notes that the ultimate soil bearing capacity q_{ult} in Eq. 4-11 is equal to the sum contributed by cohesion c, surcharge γD and self-weight γ of the soil. The N_c, N_q, and N_γ values are given graphically (solid curves) after Terzaghi in Fig. 4-9.

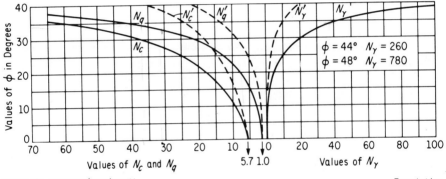

FIG. 4-9. Soil bearing capacity factors (after Terzaghi, Ref. 22).

2. For continuous footing with rough base and local shear of soil.

$$q'_{ult} = (0.67)\, cN'_c + \gamma DN'_q + (0.5)\, \gamma BN'_\gamma \qquad (4\text{-}12)$$

This formula is derived by introducing into Eq. 4-11 reduced soil coefficients such as

Forces 57

$$\tan \phi' = (2/3) \tan \phi \qquad (4\text{-}13)$$

and

$$c' = (2/3)c \qquad (4\text{-}14)$$

Here ϕ' and c' pertain to the total stress. The N'_c, N'_q, and N'_γ factors are given graphically in Fig. 4-9 (dashed curves).

For smooth base of footing ($\phi = 0$) at the ground surface the ultimate soil bearing capacity value according to Prandtl[23] is

$$q_{\text{ult}} = (5.14)c = (2.57)q_u \qquad (4\text{-}15)$$

where q_u = unconfined compressive strength of cohesive soil.

Foundation depth D, eccentricity of resultant load on footing, form of foundation plan, inclined base of footing, sloped ground surface, and other foundation parameters may be taken into account in the general soil bearing capacity formula, Eq. 4-11, by applying to the latter Brinch Hansen's correction or shape factors.[24] Such correction factors (20 pages of graphs) may also be found in Ref. 25 on retaining walls published by the Swiss Association of Highway Specialists. This source gives corrected N values.

3. General and local shear.

The case "general shear" pertains to the assumption that the sliding surface in soil can fully develop upon exhaustion of the soil's shear strength. This condition is more particularly fulfilled in dense soils. If the soil is in a very loose state, then local shear in soil (viz., local sliding zones) may set in because of its compressibility. This phenomenon and shear process are known as "local shear."

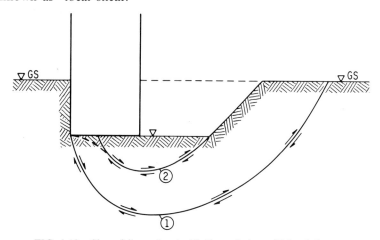

FIG. 4-10. Shear failure of soil: (1) General shear; (2) local shear.

A local shear may also occur when a trench is excavated adjacent to the foundation (Fig. 4-10).

4-10. SPECIALIZATION OF ULTIMATE BEARING CAPACITY EQUATION

The general ultimate soil bearing capacity formula, Eq. 4-11, was specialized by Terzaghi[22] for various kinds of shallow footings and for $D \leq B$ as follows.

(a) Frictional-cohesive ($\phi - c$) soils:

Rectangular footing:

$$q_{ult} = cN_c \left(1 + 0.3 \frac{B}{L}\right) + \gamma D N_q + (0.4) \gamma B N_\gamma \qquad (4\text{-}16)$$

Square footing, $B = L$; $B \times B$:

$$q_{ult} = (1.3) cN_c + \gamma D N_q + (0.4) \gamma B N_\gamma \qquad (4\text{-}17)$$

Circular footing of radius r, on dense or stiff soil:

$$q_{ult} = (1.3) cN_c + \gamma D N_q + (0.6) \gamma r N_\gamma \qquad (4\text{-}18)$$

(b) Noncohesive (ϕ) soil ($\phi > 0$; $c = 0$).

Continuous footing:

$$q_{ult} = \gamma D N_q + (0.5) \gamma B N_\gamma \qquad (4\text{-}19)$$

Rectangular and square footings:

$$q_{ult} = \gamma D N_q + (0.4) \gamma B N_\gamma \qquad (4\text{-}20)$$

Circular footings:

$$q_{ult} = \gamma D N_q + (0.6) \gamma B N_\gamma \qquad (4\text{-}21)$$

(c) Cohesive (c) soil ($\phi = 0$; $c > 0$):

Continuous footing:

$$q_{ult} = cN_c + \gamma D \qquad (4\text{-}22)$$

Rectangular and square footings:

$$q_{ult} = cN_c \left(1 + 0.3 \frac{B}{L}\right) + \gamma D \qquad (4\text{-}23)$$

Forces

Circular footing:

$$q_{ult} = (1.3)\,cN_c + \gamma D \tag{4-24}$$

If $\phi = 0$ and $c > 0$, the soil bearing capacity from surcharge (γD) is exactly compensated by the weight of soil removed by excavation. Therefore, in footing design the net bearing capacity q_{net} is

$$q_{net} = q_{ult} - \gamma D \tag{4-25}$$

Skempton[26] put forward the following net bearing capacity formula for a rectangular footing on clay:

$$q_{net} = 5c\left(1 + 0.2\,\frac{D}{B}\right)\left(1 + 0.2\,\frac{B}{L}\right) \tag{4-26}$$

where D = depth of footing $[D \leqq (2.5B)]$
B = width of footing
L = length of footing

For circular footings, $D = B = L$.

The allowable soil bearing capacity σ_{all} is here obtained by dividing the net soil bearing capacity by a factor of safety:

$$\sigma_{all} = \frac{q_{net}}{\eta} \tag{4-27}$$

where η = factor of safety. Usually, the factor of safety here is not to be less than $\eta = 3.0$.

4-11. SOIL BEARING CAPACITY OF DEEP FOUNDATIONS

The soil ultimate bearing capacity of a deep foundation such as a pier (Fig. 4-11), for example, may be considered to be composed of two parts, namely

$$P_{ult} = P_L + F_s = q_{ult}A + UsD \tag{4-28}$$

where $q_{ult}A = P_L$ is the ultimate soil bearing capacity underneath the entire base area A of the pier
$UsD = F_s$ is the mantle or skin resistance force
U = perimeter of pier
s = average value of skin friction between soil and mantle surface of pier at failure
D = embedment depth of foundation

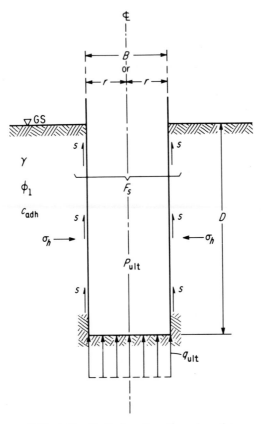

FIG. 4-11. Vertical section through a deep foundation.

The skin resistance s is expressed as

$$s = \sigma_h \tan \phi_1 + c_{adh} \tag{4-29}$$

where σ_h = average horizontal pressure on the vertical mantle surface of the embedded part of the pier at failure

ϕ_1 = angle of friction between soil and pier material

c_{adh} = adhesion of soil to foundation wall material, all in consistent units

4-12. ALLOWABLE SOIL BEARING CAPACITY

The safe, allowable soil bearing capacity values are based upon two considerations, namely:

1. The factor of safety against ultimate shear failure of the soil must be adequate.

Forces 61

✗2. Settlement under allowable soil bearing pressure should not exceed tolerable magnitudes.

To obtain safe, allowable soil bearing capacity value, apply to ultimate soil bearing capacity an appropriate factor of safety η (between 2 and 3 for dead load plus normal live load):

$$\sigma_{all} = \frac{q_{ult}}{\eta} \quad \eta = 3 \Rightarrow usually \quad (4\text{-}30)$$

4-13. STANDARD PENETRATION-RESISTANCE TEST

Description

The standard soil-penetration-resistance test (SPR test) is a widely used, ordinary routine, empirical exploration method to determine the compactness or hardness (viz., density) and consistency of soil in situ. This test is widely practiced in America, and is also used in other countries.

In essence, the standard soil-penetration-resistance test may be described as follows.[27]

A hammer, the standard weight of which is 140 lb, is dropped through a height of 30 in. to exert blows on the drill rod to drive a standard 2-in. O.D. split-barrel soil sampler 12 in. into the soil at the bottom of a soil bore-hole, and the number of blows N to do this is counted and recorded.

The standard penetration-resistance tests are performed at frequent intervals, say one every 5 ft in a uniform soil, and at least one test in each of the different soil strata along the depth of the boring.

After the blow counts are made, the soil sampler is withdrawn from the bore hole, the soil sample thus extruded is secured, and the penetration-resistance testing is resumed all over again.

The number of blows N is considered to be the penetration resistance of the soil, or a measure of the relative density of the soil. For example, any number of blows less than $N=4$ indicates a loose sand. Any number of blows greater than $N=50$ indicates a very dense sand.

The penetration resistance test results N can be empirically and approximately correlated with various physical properties of soil, such as relative density, angle of internal friction, and consistency, for example, as shown in Table 4-8. They can also be correlated with the bearing-capacity factors N_c, N_q, and N_γ, or else with approximate allowable bearing capacity values σ_{all}, or the unconfined compressive strength of soil. By means of the N-values the bearing capacity factors may be ascertained, and the soil bearing capacity for a footing estimated. Thus from this rather vague

TABLE 4-8

Correlation of Number of Blows N with Some Properties of Soil

State of Relative Density of Granular Material	Sand				Clays	
	Relative Density $D = \dfrac{e_{max} - e}{e_{max} - e_{min}}$ for sand	Standard Penetration Resistance. Number of Blows ft. N	Angle of Internal Friction of Soil. ϕ Degrees		Standard Penetration Resistance. Number of Blows N	State of Consistency
Very loose	0.00	4	30		2	Very soft
Loose		4-10	30-35		2-4	Soft
Medium dense	0.33	10-30	35-40		4-8	Medium
Dense	0.67	30-50	40-45		8-15	Stiff
Very dense		50	45		15-30	Very stiff
	1.00				30	Hard

e_{max}, e, and e_{min} are void ratios of soil.

measure of quality some estimates may be made of the ability of the soil to carry load.

Advantages of the SPR Test

1. The test apparatus is simple and relatively inexpensive.
2. The SPR test is simple to perform.
3. The SPR test can be performed relatively quickly.
4. Although the SPR test is an empirical test, rendering only approximate results, it affords a simple soil identification procedure and a rough comparison between some soil properties of different strata.

Disadvantages of the SPR Test

Regardless of some of the relatively meager advantages the SPR test affords, many engineers have reservations as to the plausibility of the SPR test method for use as a standard test in important and responsible work such as foundation engineering because of some of the disadvantages which are inherent in this test method.

Some disadvantages of the SPR test method are:

1. Although the penetration resistance test correlation with sand relative density is considered to be fairly reliable by some engineers, others, however, have had the experience in the SPR test method that the correlation between the relative density of granular soil and the number of blows N does not render very reliable results in gravel

and gravelly sand. Divergent N-values result from performing the SPR test in such soils.
2. In loose gravel the split-barrel soil sampler tends to find its way into the large void spaces of the gravel, resulting in a low number of blows, viz., low penetration resistance.
3. Rotation of round pebbles by the soil sampler upon its penetration into the large voids may result in a low number of blows N.
4. Blocking of the sampler by large sizes of gravel may result in a large number of blows.
5. When a rock fragment or a large piece of gravel becomes wedged inside the sampler a high number of blows may result. Hence in gravel and gravelly soils the SPR test is anything but standard, and must thus be regarded as a rough estimate only.
6. Blow counts in sand are increased by the presence of large particles of gravel.
7. In general, the unconfined compressive strength of fine-particled soils for a given number of blows N is greater for the more plastic soil materials.
8. Correlation of N to the unconfined compressive strength of fine-particled (cohesive) soils has thus far shown a great scatter, which condemns the SPR test method as unreliable for such soils.
9. Penetration resistance in clays does not reflect fissures, hair cracks and slickensides, which may be important in characterizing the strength of such materials.
10. Standard penetration resistance tests performed at shallow depths usually render too low a number of blows N.
11. The test performed at greater depths in the same soil with the same relative density renders, on the other hand, a high number of blows N, viz., a higher resistance to penetration mainly because of the overburden (surcharge) pressure around the bore hole.
12. Irregularities in the drop of the hammer may consume the entire energy of the hammer. Also, the friction between the drill rod and the wall of the bore hole or its lining should not be overlooked.
13. Also, much of the standardized equipment is not comparable in many instances, which prevents performing statistical analysis of the data obtained from the SPR tests. It would probably be proper to set up new norms for this test by specifying a new sampler, driving rod and a different method of guiding the rammer.
14. The several corrections and modifications for the counted number of blows N calling in some instances for reduction of the counted N

values by as much as 50 percent are merely a proof that the widely used SPR test method of soils coupled with correlation of certain properties of soil is a very vague empirical "standard" method indeed. These and possibly other disadvantages (soil samples are not "undisturbed") show that the SPR test method is very vulnerable and sensitive to various incidental effects, random and casual factors, and various possibilities of chance.

Thus one gathers that the initial intent of this standard penetration test has become completely lost. The fact about this test is that by means of the SPR test one cannot obtain information about all of the important soil properties necessary for the design of soil-foundation systems. Therefore the SPR test results must be used with a great deal of mature judgment and discretion. Especially, one should never rely upon the SPR test results for the purpose of designing foundations for important structures.

However, the fact is that the SPR test method is practiced extensively here and is picking up momentum in use also in some countries abroad for the empirical determination of the approximative value of the allowable uniform soil pressure.

Example 1

Given: The soil penetration test result is $N = 16$ blows/ft on coarse sand. The unit weight of this soil is $\gamma = 110$ lb/ft^3 = 0.055 ton/ft^3. A strip footing 5 ft wide is to be laid $D = 4.0$ ft below ground surface.

Required: Calculate the ultimate bearing capacity of this soil.

Solution: Here $c = 0$ (a noncohesive soil). For $N = 16$ blows/ft (medium density of soil) obtain from Table 4-8 a value of $\phi \approx 37°$. Now, by Terzaghi's equation for $c = 0$, the ultimate bearing capacity formula for such a soil is

$$q_{net} = \gamma D N_q + (0.5) \gamma B N_\gamma \qquad (4\text{-}19)$$

where $N_q \approx 55$ and $N_\gamma \approx 70$ (from Fig. 4-9).

$$q_{ult} = (0.055)(4.0)(55) + (0.5)(0.055)(5.0)(70)$$
$$= 12.10 + 9.63 = \underline{21.63 \ (\text{ton/ft}^2 \approx \text{kg/cm}^2)}$$

Example 2

Given an 8 ft × 8 ft footing, laid at the ground surface ($D = 0$) on a cohesive soil; $\gamma = 0.055$ ton/ft^3; $\phi = 15°$; $c = 0.15$ ton/ft^2.

Required: Determine the ultimate bearing capacity of the soil.

Solution: By Eq. 4-20,

Forces

$$q_{ult} = (1.3)cN_c + \gamma DN_q + (0.4)\gamma BN_\gamma \quad (4\text{-}20)$$

With $D=0$ and Terzaghi's bearing capacity factors $N_c = 13.0$ and $N_\gamma = 2.0$ (from graph in Fig. 4-9), obtain

$$q_{ult} = (0.195)(13.0) + (0.176)(2.0) = \underline{2.90 \ (\text{ton/ft}^2)}$$

Example 3

Given a strip foundation $B = 10$ ft wide and laid in a frictional-cohesive soil $D = 6$ ft below ground surface. The unit weight of this soil is $\gamma = 0.06$ ton/ft³; $\phi = 10°$, and $c = 0.30$ ton/ft².

Required: (1) calculate the ultimate bearing capacity of the soil; (2) determine the safe bearing capacity of the soil. Use a factor of safety, $\eta = 3.0$.

Solution: Because $D = 6$ ft is less than $B = 10$ ft, the footing may be considered as a shallow one. The ultimate bearing capacity by Eq. 4-11 is

$$q_{ult} = cN_c + \gamma DN_q + (0.5)\gamma BN_\gamma \quad (4\text{-}11)$$

or, with $N_c = 9.0$; $N_q = 3.0$, and $N_\gamma = 1.0$ (from graph in Fig. 4-9),

$$q_{ult} = 2.70 + 1.08 + 0.18 = \underline{3.96 \ (\text{ton/ft}^2)}$$

$$q_{safe} = \frac{q_{ult}}{\eta} = \frac{3.96}{3} = \underline{1.32 \ (\text{ton/ft}^2)}$$

Pressure relief by an excavation 6 ft deep:

$$D = (0.06)(6.0) = \underline{0.36 \ (\text{ton/ft}^2)}$$

Ultimate bearing capacity:

$$q_{ult} = 3.96 + 0.36 = \underline{4.32 \ (\text{ton/ft}^2)}$$

Safe bearing capacity:

$$q_{safe} = 1.32 + 0.36 = \underline{1.68 \ (\text{ton/ft}^2)}$$

Example 4

Given a shallow strip footing $B = 4$ ft wide. The footing is laid in clay $D = 3.0$ ft deep below the ground surface. The unit weight of this soil is $\gamma = 0.055$ ton/ft³. The unconfined compressive strength of the clay specimen is $q_u = 4.00$ ton/ft².

Required: (1) calculate the ultimate bearing capacity of the clay soil; (2) determine the safe bearing capacity of clay for a factor of safety $\eta = 3.0$.

Solution: (1) The ultimate shear strength $\tau_{ult} = c$ of the clay is one half of the unconfined compressive strength q_u:

$$\tau_{ult} = c = \frac{q_u}{2} = \frac{4.00}{2} = \underline{2.00 \text{ (ton/ft}^2)}$$

For a pure clay, only the first two terms in Terzaghi's ultimate bearing capacity equation are pertinent:

$$q_{ult} = cN_c + \gamma D N_q \tag{4-22}$$

where $c = 2.00 \text{ ton/ft}^2$
$N_c = 5.7$ (from graph in Fig. 4-9, at $\phi = 0°$)
$N_q = 1.0$ (from same graph, at $\phi = 0°$)

Thus,
$$q_{ult} = (2.00)(5.7) + (0.055)(3.0)(1.0)$$
$$= 11.40 + 0.16 = \underline{11.56 \text{ (ton/ft}^2)}$$

(2) Safe bearing capacity for $\eta = 3.0$:

$$\sigma_{all} = q_{ult}/\eta = 11.56/3.0 = \underline{3.85 \text{ (ton/ft}^2)}$$

The relief given by the excavation must now be added to the ultimate bearing capacity:

$$q_{ult\ total} = q_{ult} + \gamma D = 11.56 + (0.055)(3.0) = \underline{11.72 \text{ (ton/ft}^2)}$$

Safe bearing capacity:

$$\sigma_{safe} = \frac{q_{ult}}{\eta} + \gamma D = \frac{11.56}{3.0} + (0.055)(3.0) = \underline{4.00 \text{ (ton/ft}^2)}$$

Example 5

Given a ϕ-c soil; $\gamma = 0.05 \text{ ton/ft}^2$; $\phi = 10°$; $c = 0.125 \text{ ton/ft}^2$. Factor of safety to use: $\eta = 3.0$. Structural load on footing: $P = 50$ tons.

Required: determine the size of the square footing at a depth of $D = 10$ ft below the ground surface.

Solution: Using Terzaghi's ultimate bearing capacity equation (4-11) for shallow foundations, and with $N_c = 9.0$, $N_q = 3.0$, and $N_\gamma = 1.0$, (from graph in Fig. 4-9), obtain:

$$\frac{P}{B^2} = q_{ult} = 1.44 + 1.50 + (0.02)B = \frac{50}{B^2}$$

or

$$B^3 + (147)B^2 - 2500 - 0$$

Forces

rendering
$$B = 4.07 \approx 4.1 \text{ (ft)}$$
$$q_{ult} = \underline{3.0 (\text{ton}/\text{ft}^2)}$$

Safe bearing capacity:
$$\sigma_{safe} = q_{ult}/\eta = 3.00/3 = \underline{1.00 \text{ (ton}/\text{ft}^2)}$$

Area A of footing for safe bearing capacity:
$$A = B^2\eta = (4.1)^2(3) = \underline{50.4 \text{ (ft}^2)}$$
$$B_{safe} = \sqrt{50.4} = \underline{7.1 \text{ (ft)}}$$

$P/A_{safe} = 50/7.1^2 \approx \underline{1.00 \text{ (ton}/\text{ft}^2)}$ (safe bearing capacity)

SUMMARY CONCERNING FOUNDATIONS ON SAND, CLAY, SILT, AND ROCK

4-14. FOUNDATIONS ON SAND. SHALLOW FOUNDATIONS. INDIVIDUAL FOOTINGS

Ultimate Bearing Capacity

The ultimate bearing capacity q_{ult} of sand depends mainly (a) upon the width B of the footing, (b) the depth D of surcharge surrounding the footing (viz., depth of base of footing below ground surface), (c) the relative density of the sand, and, (d) the position of the groundwater table.

The ultimate bearing capacity of sand is calculated theoretically by Eq. 4-19. The necessary base area A is calculated based on the *net soil pressure* q:

$$q_{ult_{net}} = q_{ult} - \gamma D = (0.5)\gamma B N_\gamma + \gamma D N_q - \gamma D$$
$$= (0.5)\gamma B N_\gamma + (N_q - 1)\gamma D \tag{4-31}$$

where the bearing capacity coefficients N_γ and N_q are functions of the angle of internal friction ϕ of sand, and in its turn, ϕ is a function of relative density of sand.

Allowable Bearing Capacity

The safe, allowable bearing capacity σ_{all} on sand is obtained by applying to the ultimate bearing capacity a factor of safety $\eta \geq 3$. This allowable bearing capacity, viz., safe soil contact pressure for footing design in sand, is based on the lesser of the two pressure values obtained either for one inch settlement or for failure of soil by groundbreak (viz., exhaustion of shear strength of sand), i.e., the worst soil condition—the least bearing capacity value—governs the design.

Determination of Bearing Capacity

The bearing capacity of sand may be determined through a direct measurement of relative density of sand in situ by means of the standard

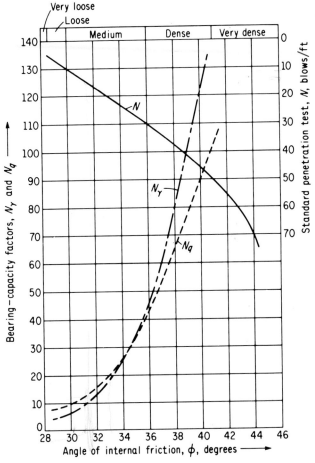

FIG. 4-12. N_γ and N_q values (Ref. 29).

penetration resistance test (N), and in conjunction with N by using the safe soil pressure charts and the chart for soil pressure corresponding to 1 in. settlement of footings on sand (Figs. 4-12, 4-13, and 4-14). If ϕ is determined

Forces

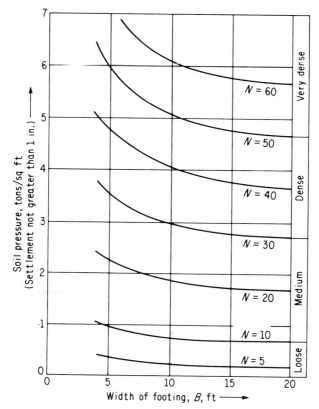

FIG. 4-13. Chart based on water table not closer than B below base of footing (Ref. 29).

by methods other than the standard penetration resistance test, use Fig. 4-9 (N_q, N_γ).

The sand bearing capacity may also be evaluated based on the stress-strain characteristics of sand obtained from field loading tests of sand. The bearing capacity can also be evaluated by means of theoretical formulas (see Refs. 2, 3, and 28).

Position of Groundwater Table

If the position of the groundwater table is below the depth B ($B =$ width of footing) below the base of the footing, the safe allowable soil pressure σ_{all} is that of its normal value.

If the position of the groundwater table is near, at or above the base of the footing, the normal, safe, allowable soil pressure is to be decreased by 50 percent.

If the position of the groundwater table is within the depth B below the

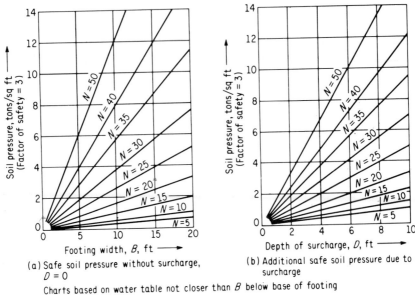

FIG. 4-14. Soil pressure as a function of B and N (Ref. 29).

base of the footing, the normal, safe allowable soil pressure is to be decreased proportionally by interpolation.

For fine sand below the groundwater table the relative density (N) values are referred to as N', i.e., $N \equiv N'$. Therefore, before use of the N-diagrams, the N'-values must be converted into N values (diagram values) by the following correction formula:[29]

$$N = 15 + (0.5)(N' - 15) \tag{4-32}$$

Settlement

Settlement of individual footings depends on the relative density of sand and the position of the groundwater table. Variations in sand density at the contemplated construction site should be ascertained to avoid excessive, intolerable differential settlements.

Settlement of sand is almost instantaneous upon application of a structural load, and almost ceases during the construction time.

The design load consists of dead load + weight of footing less that of the surrounding surcharge + maximum live load (including wind pressure and weight of snow where applicable). Their single application will bring about full settlement. However, relative to design loads, there must be adherence to building codes.

For sands in the natural state whose relative density $N > 5$, the allowable soil pressure σ_{all} should be such that the maximum settlement will not exceed 1 in. This pertains also to sand which has been compacted into a state of relative density whose $N > 5$.

Special Considerations

Vibration of sand may bring about its densification. Overvibration may cause loosening of the sand. If a saturated and very loose ($N < 5$) sand is impacted by a sudden shock, the sand may spontaneously liquefy. Liquefaction may bring about the collapse of a structure founded on such a sand. If, nevertheless, footings must be laid in loose, saturated sand whose $N < 10$, the footings are empirically designed for pile support, or else, such sand must be densified until $N > 10$.

MAT FOUNDATIONS

If the sum of the areas of the individual footings is greater than the total plan area of the structure, then generally a mat foundation is designed.

Ultimate Bearing Capacity

Foundation engineering experience has indicated that there exists a little danger for a large mat to bring about groundbreak in sand accompanied with expulsion of soil laterally from underneath the base of the mat. This means that the factor of safety η against groundbreak in sand is usually very large–usually $\eta \geq 3.0$. Therefore mats are not designed for bearing capacity failure of soil, but for settlement only.

Allowable Bearing Capacity

The allowable soil pressure on sand under mats is usually about twice as large as for individual footings. It is computed by the following empirical formula for most conditions found in practice:[29]

$$\sigma_{all} = \frac{N-3}{5} \quad (\text{ton/ft}^2) \tag{4-33}$$

provided that the width B of the footing is $B \geq 20$ ft. All symbols in Eq.4-33 are the same as before.

If the sand at the site is of varying relative density N, an average N is to be used in Eq. 4-33.

Terzaghi and Peck[30] proposed allowable bearing capacity values σ_{all} for mats on sand as a function of relative density N of the sand. They are compiled in Table 4-9. These allowable bearing capacity values are based on maximum settlement of not more than 2 in. (≈ 5 cm), and are given for a sand layer the thickness of which is greater than the width B of the mat.

TABLE 4-9
Proposed Allowable Bearing Values for Mats on Sand[30]

Relative Density	Number of Blows per 1 ft in Standard Penetration Test, N	Proposed Allowable Bearing Value σ_{all}, ton/ft² ~ kg/cm²
Loose........................	< 10	Requires compaction
Medium......................	10-30	0.7-2.5
Dense........................	30-50	2.5-4.5
Very dense	> 50	> 4.5

Determination of Bearing Capacity

Allowable soil pressure is determined from the minimum average N-value of the mat below the elevation of the base of the mat.

Position of Groundwater Table

The position of the groundwater table is an important factor to be considered in mat foundation design on any soil.

If the position of the groundwater table is between the base of the mat and a depth B below the base, the normal allowable soil pressures should be reduced from 0 to 50 percent by interpolation. If the groundwater table is at a depth $< B/2$ below the base of the mat, or is at an elevation higher than the base of the mat, or is at an elevation higher than the base of the footing, the normal, safe, allowable soil pressure is to be decreased by 50 percent.

The N' values for fine sand in groundwater are to be corrected by means of Eq. 4-32.

Settlement

Intolerable differential settlements of mats on sand are usually much less than those of individual footings. Therefore a larger soil pressure is allowed under mats than under individual footings.

When the depth D of the base of the mat, viz., surcharge, is about 8 ft or more, differential settlements of soil and mat rarely exceed tolerable amounts. If $D < \sim 8$ ft, edges of mat would settle more than the interior of the mat (because of lack of confinement of sand).

If $N < 5$, the sand is too loose for a mat foundation. Or else, the density of such a sand soil could be increased by artificially compacting it until $N > 10$.

Forces

Special Considerations
These are the same as for individual footings.

Design Procedure
The design procedure for mats is similar to that followed in footing design—namely, by the use of a chart, to choose a soil pressure that would bring about a 1-in. settlement of footing on sand. One determines the actual pressure from loads on the mat, and compares the soil pressure from the mat with the allowable one.

DEEP FOUNDATIONS. PIER FOUNDATIONS AND CAISSONS IN SAND

Use of Piers and Caissons
Piers are used, for example, (a) when the upper layers of soil are too soft or compressible; (b) when the overlying material may be eroded away by scour, or (c) if a soft soil material contains obstacles such as sunken logs, trunks of trees, and boulders, that piles cannot pierce through and penetrate.

Piers are founded on the top surface of a stratum of dense sand or gravel encountered at a considerable depth below the ground surface overlaid by a soft, compressible soil material.

Ultimate Bearing Capacity
Because of the surcharge of soil around the piers, the ultimate bearing capacity of sand under piers is greater than that of footings on sand of equal density. The ultimate bearing capacity of a pier is calculated by Eq. 4-28. The effect of surcharge of adjacent soil is neglected in ultimate bearing capacity determination if scouring of the surcharge is anticipated.

Allowable Bearing Capacity
In practice, it is conservative to determine the safe, allowable bearing capacity for piers on sand by the same methods as described for individual footings.

The factor of safety η against groundbreak should be $\eta \geqq 3.0$.

If the N-charts are used, the allowable bearing capacity values are multiplied by a factor of 2. However, if scour may erode away most of the surcharge above the dense sand layer upon which the pier is laid, the soil pressure values should not be increased above that allowed for individual footings.

The allowable sand bearing capacity to use must comprehend an adjustment for the position of the water table.

Determination of Bearing Capacity

The methods of determining the bearing capacity of sand for sinking piers are much the same as those described for individual footings.

Settlement

Because of adjacent surcharge, settlement of sunked caisson-type piers on sand is usually about 50 percent less than that of individual footings of the same size and form on sand of comparable bearing capacity.

Piers laid in open excavations undergo greater settlements than the sunken caisson type piers because of the surcharge removed by excavation.

Position of Water Table

If a pier must be founded in water, corrections for position of water table, i.e., for soil density, viz., safe, allowable soil pressure, must be made as for individual footings.

If in the laying operations of a pier foundation, say open caisson, the water table outside the pier is higher than that within the open caisson, a "quick" condition of sand may set in, boiling up through the bottom of the caisson: the bearing capacity is destroyed. To prevent this from happening, one tries to maintain the water table inside the caisson at a greater height than that of the water table outside the caisson. This is especially possible when a pier is founded in an open excavation. The benefit of the downward flow of water through the sand thus arranged is obvious: the downward flow of water through the bottom of the caisson or a hollow pier strengthens rather than loosens the sand. This principle is also applicable during construction in rainy periods: the level of water in the side drainage ditches in the excavation should always be lower than that in a wide, open, yet unfinished stage of the foundation.

Special Considerations

Pumping in loose, fine sand from an excavation is difficult. To cope with this situation, construction sites are dewatered by means of wellpoint systems prior to excavation.

The design of a dewatering system requires knowledge of the coefficient of permeability.

Loose sand can be densified by methods of vibration and hydrovibration, for example.

Vibrating foundations, in densifying the sand, may bring about intolerable settlement.

Small-diameter piers (caissons) may be founded in sand by the method of vibration.[31] (See Chapter 19 in this book.)

PILE FOUNDATIONS IN SAND

Piles may be driven into loose sand in order to transfer the structural load to a firm soil or rock. Piles may be driven to firm soil below the greatest depth of scour (where applicable, of course). Also, piles may be used for densification of soil.

An empirical method seems to be the only one for designing foundations on friction piles.

Before a pile foundation is finally decided on for use in a poor soil, other methods of laying foundations should be investigated: for example, long tubular piles, caissons, floating foundations, improvement of soil properties.

Piles may also be embedded in sand by the method of vibration.[31] (See Chapter 19 in this book.)

Determination of Bearing Capacity of Piles

The bearing capacity of a pile is derived from its end point (tip) resistance to penetration and from its mantle surface friction (skin friction) in sand.

The bearing capacity of a pile may be determined by dynamic pile driving formulas (see Sec. 19-4), static pile bearing capacity formulas and by pile static loading tests at the site.[2] The most reliable method of determining bearing capacity of piles is by static loading test. In loading tests, piles should be subjected to twice the load (=design load) that is likely to be experienced under the structure.

Tapered piles are most effective for load transfer to soil as compared with cylindrical types of piles.

Uplift resistance (skin resistance) of piles depends upon the relative density of the sand; the length, diameter and taper of the pile; and results from pile pulling tests.

Settlement

Settlement of a pile is determined from the static loading test. Under design load, the penetration of a pile should not exceed 0.75 in. (depending upon the requirements of the local building codes, and upon the nature of the structure, of course). The amount of settlement should not exceed the tolerable one.

Special Conditions

Upon driving, interaction between pile and soil reduces temporary pile bearing capacity, which is regained after some lapse of time, say, approximately in a few days after driving.

Vibration of sand soil and vibration of a pile into sand reduce temporarily pile bearing capacity on account of reduction of skin friction between the pile material and sand due to motion of the vibrating soil particles, because sliding (kinetic) friction is less than static friction.

For a more complete discussion on pile bearing capacity see Chapter 19 in this book and Ref. 2.

4-15. FOUNDATIONS ON CLAY. SHALLOW FOUNDATIONS. INDIVIDUAL FOOTINGS ON CLAY

Ultimate Bearing Capacity of Clay

The bearing capacity of clay is mainly a function of its shear strength. The ultimate shear strength τ_{ult} of a pure clay ($=$ a c-soil) is approximately equal to one half of the unconfined compressive strength q_u of that clay.

The ultimate net bearing capacity $q_{ult_{net}}$ of a pure, homogeneous clay for a rectangular footing may be calculated by means of Terzaghi's equation (Eqs. 4-22, 4-23) for $\phi = 0$, $D = 0$, and $N_\gamma = 0$:

$$q_{ult_{net}} = cN_c \tag{4-34}$$

or, in terms of the unconfined compressive strength q_u for a rectangular footing with a rough base, as

$$q_{ult_{net}} = (2.85)\, q_u \left(1 + 0.3\, \frac{B}{L}\right) \tag{4-35}$$

where $c =$ cohesion of clay

N_c, N_q, and $N_\gamma =$ bearing capacity coefficients

$\phi =$ angle of internal friction of soil (a test parameter)

$B =$ width of footing

$L =$ length of footing

Recall that this equation for the net ultimate bearing capacity of clay is derived from the shear strength of the clay. Hence, the sizing of footings on clay is to be made based on the shear strength of clay.

The ultimate-bearing-capacity equations of clay should be used only when the individual footings are spaced sufficiently far apart so that their stress distribution in soil does not interact or overlap (recall the stress overlapping of two adjacent, proximate pressure bulbs from two adjacent load bearing areas).

The possibility of groundbreak, viz., failure in shear, in clay is greater than in sand.

Equation 4-23 can be specialized for a square footing by setting $B = L$. For a strip footing on pure clay ($\phi - 0$, $N_\gamma = 0$),

$$q_{ult} = cN_c + \gamma D N_q \quad (4\text{-}36)$$

exclusive of the extra support given by the surcharge surrounding a deep footing, where γ = unit weight of the clay, and D = depth of base of footing below ground surface.

Allowable Bearing Capacity of Clay

It is known from soil mechanics that the safe, allowable soil bearing capacity of clay is approximately equal to the unconfined compressive strength q_u of that clay.

Individual footings laid on clay should be sized in such a manner as to result in a factor of safety η against groundbreak (shear failure of clay), or, for a settlement of ≤ 0.75 in., of $\eta \geq 3.0$, whichever soil pressure of the two (for groundbreak or for settlement) governs the design. Thus with $\eta = 3.0$, the net safe, allowable soil pressure against groundbreak for a rectangular footing is calculated from Eq. 4-35 as

$$\sigma_{all_{net}} = (0.95) \, q_u \left(1 + 0.3 \, \frac{B}{L}\right) \quad (4\text{-}37)$$

where q_u is the average unconfined compressive strength of clay at a depth B below the base of the individual footing.

Specializing Eq. 4-37 for a square footing $B = L$, one may deduce that Eq. 4-37 renders the same net allowable soil pressure $\sigma_{ult_{net}}$ on clay regardless of whether the square footing is large or small. Thus safe soil pressures on clay from square footings are practically independent of the width B of the footing.

For long strip footing $L \to \infty$ in Eq. 4-37, and therefore $[(0.3)(B/L)] \to 0$. Hence by Eq. 4-37 the net allowable bearing capacity of clay for a long strip footing is

$$\sigma_{all_{net}} = (0.95) \, q_u \approx q_u \quad (4\text{-}38)$$

That is, the safe net allowable bearing capacity of clay for a strip footing is approximately equal to the unconfined compressive strength q_u of that clay. Also, from these specializations of Eq. 4-35, it can be said formally that the net allowable bearing capacity of clay for a square and for strip footings against groundbreak is independent of their width.

Determination of Bearing Capacity of Clay

The bearing capacity of a homogeneous clay soil is most readily determined by means of the unconfined compression test on undisturbed specimens of that clay. Unconfined compressive strength of clay should be tested on clay samples taken at 6-in. intervals vertically.

Loading tests made on clay in the field do not afford prediction of consolidation settlement of a prototype footing (model law!).

If a stiff clay contains hair cracks, fissures, and slickensides, the ultimate bearing capacity of such a clay may be determined approximately in the field by means of loading tests.

The empirical standard penetration test in clay is a very vague and uncertain method of indicating the compressive strength of clay.

The shear strength of clay can also be tested in situ in the bore hole by means of the shear vane device, as well as by Handy's bore-hole direct-shear testing device.[32]

Settlement

Besides determining the allowable bearing capacity of clay against groundbreak, individual footings on clay must also be analyzed for their settlement. Upon static loading clay consolidates, viz., settles, over a long period of time. The magnitude of settlement depends upon compressibility of the clay.

Settlement of clay is studied based on results obtained from consolidation tests of that clay.

Special Considerations

Shear strength and consolidation characteristics of clay are the principal design factors in designing footings laid on clay.

Frequently the performance of any one footing within a group of footings is significantly influenced in respect to settlement by the presence of adjacent footings. If the sum of the base areas of the individual footings makes up approximately 50 percent of the total plan area of the structure, one then resorts to a mat foundation.

MAT FOUNDATIONS

Ultimate Bearing Capacity of Clay

During the excavation for laying a mat on clay, the overburden γD is removed; hence the extra soil support due to the surcharge surrounding a deep foundation does not exist in the case of an excavation for a mat. Therefore, for $\phi = 0$ and $N_\gamma = 0$, only the first term of Terzaghi's equation 4-11 is used for determining the ultimate bearing capacity q_{ult} of a clay:

$$q_{ult} = cN_c \qquad (4\text{-}34)$$

where all symbols are as before.

Forces

Otherwise, the net ultimate pressure on a thick stratum of clay from a mat at its base may be calculated by means of Eq. 4-35:

$$q_{ult_{net}} = (2.85)q_u \left(1 + 0.3\frac{B}{L}\right) \quad (4\text{-}35)$$

The possibility of groundbreak in clay from a raft is greater than that in sand.

Allowable Bearing Capacity

The factor of safety η against failure of clay beneath the mat should be not less than 3.0 under normal loads, and greater than 2.0 for extreme loading conditions. In the latter respect, the local building code must be adhered to.

The maximum allowable pressure on clay soil at the base of the mat laid on a thick stratum of clay against groundbreak may thus be calculated by Eq. 4-37 as

$$\sigma_{all_{net}} = (0.95)\, q_u \left(1 + 0.3\frac{B}{L}\right) \quad (4\text{-}37)$$

The safe, allowable pressure on clay soil increases with the depth of laying of the mat foundations. Because it is very impractical to extend or project the mat far out beyond the perimeter of the structure, a greater depth of laying the mat foundation rather than a shallow one seems to be the only way to increase the bearing capacity of the clay soil. This is to say that the excavation for the mat foundation is to be made sufficiently deep so that the same weight of the soil is excavated as the weight of the structure to be built.

Settlement

The settlement of a mat depends upon the compressibility of the clay. If the mat is, for example, $A = 20$ ft $\times$ 40 ft $= 800$ ft^2 in size, and the pressure p on clay with regard to settlement is $p = 1.0$ ton/ft^2, then the total pressure on the mat is

$$(p)(A) = (1.0)(800) = \underline{800 \text{ (tons)}}$$

This pressure must be compared with the actual weight and load of the structure. Thus if $c = 0.80$ ton/ft^2, then $N_c = 5.7$ (from Fig. 4-9), and

$$q_{ult} = cN_c = (0.800)(5.7) = \underline{4.56 \text{ (ton/ft}^2)}$$

The factor of safety η for this case calculates as

$$\eta = q_{ult}/p = (4.56)/(1.00) = \underline{4.56 > 3.0 \text{ (O.K.)}}$$

Design Procedure for a Mat

The design procedure for a mat is as follows:
1. Determine loads on mat and the corresponding pressure p on clay soil required by the base area A of the mat.
2. Compare this pressure on soil with the ultimate bearing capacity of clay.
3. The factor of safety $\eta = q_{ult}/p$ should be not less than $\eta = 3.0$ for normal loading, and not less than 2.0 for extreme loading. Consult local building codes.

Determination of Bearing Capacity

The ultimate bearing capacity of clay for shallow mat foundations is most readily determined by means of the unconfined compression test on that clay.

Special Considerations

These are the same as were described for individual shallow footings.

DEEP FOUNDATIONS. PIER FOUNDATIONS AND CAISSONS

Ultimate Bearing Capacity of Clay

For ultimate load of piers founded on clay refer to individual shallow footings on clay. To reduce pressure on soil, piers are sometimes built hollow, consistent with the stability of the foundation system, of course.

For a rectangular foundation ($B \times L$), and for ratios $D/B \leq 2.5$, Skempton's bearing capacity equation (Eq. 4-26) for determining the ultimate load-bearing capacity of the pier may be used:

$$q_{ult_{net}} = 5c \left(1 + 0.2 \frac{D}{B}\right) \left(1 + 0.2 \frac{B}{L}\right) \qquad (4\text{-}26)$$

Otherwise, the net ultimate bearing capacity $P_{ult_{net}}$ of a pier may be calculated as

$$P_{ult_{net}} = cN_cA + UsD \qquad (4\text{-}28)$$

where c = undrained shear strength of the undisturbed clay
  N_c = bearing capacity factor for clay
  A = base area of pier
  U = perimeter of pier
  s = skin friction in the mantle surface of the shaft of the pier
  D = embedment depth of pier foundation

Forces

Allowable Bearing Capacity

For allowable load on a pier founded in clay refer to individual shallow footings on clay. Otherwise,

$$\sigma_{\text{all}_{\text{net}}} = (0.95)\, q_u \left(1 + 0.3\, \frac{B}{L}\right) \tag{4-37}$$

The factor of safety η against groundbreak of clay under a pier should be not less than $\eta = 3.0$ under the worst possible loading conditions that may be normally anticipated, or $\eta \geq 2.0$ under the most severe combinations of loadings. Consult the local building code.

Determination of Bearing Capacity

The safe load on piers, viz., safe bearing capacity of clay under piers, is determined by the same methods and rules that were given for individual footings.

Settlement

Besides safety against groundbreak, it is also necessary to check the pier-clay system against intolerable differential settlement. Predicted settlements of a homogeneous clay from consolidation tests are usually somewhat larger as compared with the actual ones. Differential settlements of piers on clays with erratic structure and distribution, however, cannot be predicted with any degree of accuracy. Considerable judgment is to be exercised on the part of the designer in performing such nonuniform settlement evaluations.

One should not overlook secondary consolidation and settlement of clay where pertinent. Some consolidating clays under structural load may ultimately bring about large settlements after the structure has been in service for many years.

Special Considerations

To prevent the sinking of a pier or caisson through soft clay from becoming stuck at some stage, the necessary sinking loads for sinking operations must be ascertained. For this purpose, it is necessary to know the numerical values of the skin friction of the pier. Average values of skin friction are [33]

Soft clay ...	150- 600 lb/ft²
Stiff clay ...	1,000-4,000 lb/ft²

In stiff clay, straight-shafted or bulbed-end piers may be drilled for load transference to clay.

PILE FOUNDATIONS IN CLAY

General Notes

If a clay soil is too compressible to support individual footings, mats and piers, the structure may be founded on piles, usually friction piles. The greatest benefit from a friction-pile foundation will be derived if the length of the piles is as great as possible consistent with economy, or if the piles are tapered.

Ultimate Bearing Capacity

The ultimate bearing capacity of a group of friction piles or that of a single friction pile varies with time. The strength of the clay soil is usually at a minimum immediately after pile driving.

For a cylindrical pile, the ultimate bearing capacity is approximately equal to the product of shear strength of soil and the mantle area of the imbedded part of the pile in the soil.

The shear strength τ of clay is

$$\tau = (1/2)q_u \tag{4-39}$$

where q_u = unconfined compressive strength of clay.

The shear resistance R_τ of clay soil around the boundary (perimeter of the pile group) must also be checked. It is

$$R_\tau = (1/2)q_u U L_p \tag{4-40}$$

where U = perimeter of pile group
L_p = embedded length of piles

Allowable Bearing Capacity

Friction piles are usually driven to form a group of piles for supporting a structure. The allowable pressure on clay from a pile group may be calculated by Eq. 4-37:

$$\sigma_{all} = (0.95)q_u \left(1 + 0.3\frac{B}{L}\right) \tag{4-37}$$

where B = width of pile group
L = length of pile group

The safe allowable load for a single friction pile in clay may be taken as one third of the failure test load for the maximum probable loading, or as 50 percent of the failure load for the extreme maximum loading, as ruled by the building code in question, whichever is smaller.

Determination of Friction Pile Capacity

Formulas for determining bearing capacity of piles are not reliable (see Sec. 19-5 in this book). The friction pile capacity in soft to medium dense clay is best determined by static load tests on piles or by theoretical static bearing capacity formulas.

If the spacing of piles in a pile group is too close, the capacity of a friction pile group is less than the sum of the capacities of the individual piles. The spacing of the piles in the group determines the efficiency of the pile group. Piles spaced at larger spacing than optimum may have a bearing capacity of the pile group equal to the sum of individual friction pile bearing capacities.

Uplift resistance of a pile is determined by means of a pulling test on the pile.

Settlement

The settlement of pile foundations in clay is discussed in Secs. 20-14 and 20-15.

If a firm clay is overlain by a soft clay, piles may be driven through the soft clay to rest on the firm clay, thus acting as end-bearing piles. Consolidation and settlement analysis here pertains to the firm stratum of clay below the elevation of the tips of these end-bearing piles.

The bearing capacity of end-bearing piles on firm clay is most reliably determined by static load tests. In cohesive soils, the efficiency of an end-bearing pile group increases with pile spacing.

The negative mantle friction of piles due to consolidation process of the upper weak soil should not be overlooked. Uplift resistance of end-bearing piles is determined by pulling tests on those piles.

Piles which bear on a permeable gravel and sand stratum sandwiched between clay transmit the loads to a lower elevation. This is the elevation at which any calculations of bearing capacity and settlement should be made.

Special Considerations

Some consideration should be given to the thixotropic properties of soft clay, especially if the possibility of pile driving (impact, vibrations) for future construction can be anticipated nearby. This applies especially when driving piles in thixotropic clays. Also, attention must be paid to negative mantle friction of piles when clay soil consolidates.

Piles in sensitive clay, unless they bear on a firm stratum, frequently do more harm to the structure than would occur if they had been omitted. Remolding of a sensitive, normally consolidated clay brings about loss in shear strength, resulting in an intolerable settlement. In a very overconsolidated clay piles usually perform satisfactorily.

4-16. FOUNDATIONS ON SILT. SHALLOW FOUNDATIONS. INDIVIDUAL FOOTINGS ON SILT

General Notes

Silts are soils intermediate between sand and clay. The design of foundations on silt is not a routine matter, but very complex and difficult. This explains why no routine, straightforward methods of design of foundations in silt can be found in the technical literature.

Empirically, if in the standard penetration resistance test the number of blows N per one foot required to drive a soil-sampling device is $N < 10$ per foot, the silt is said to be soft and loose. If $N > 10$, the silt is considered to be medium or dense insofar as density is concerned.

As a usual thing, loose silt is incapable of supporting individual footings, combined footings or mat foundations. Instead, pile and/or pier foundations should be considered.

Silts with $N < 10$ are generally regarded as even less suitable than a normally loaded soft clay for direct support of foundations.

Medium and dense silts with $N > 10$ may be grouped into rock flour and plastic kinds of silt. The rock flour kind of silt with a plasticity index P.I. ≈ 0 is assumed to perform as a fine sand; the plastic kind of silt (P.I. > 0) may behave as clay. However, such a subdivision of silt is a great oversimplification, and considerable judgment is needed to deal with the design of foundations in silt.

Ultimate Bearing Capacity

The strength of silt is derived from its internal friction and cohesion. Hence the ultimate bearing capacity of silt for important foundation projects may be calculated by means of the corresponding theoretical soil bearing capacity equations. These equations require knowledge of the appropriate data about the silt and the numerical value of its cohesion. The bearing capacity of a normally loaded silt is generally very low, whereas that of loess may be very high.

Allowable Pressure on Silt

The allowable pressure on the rock flour type of silt may be crudely estimated in a manner similar to that for sand. The allowable pressure on plastic silt may be estimated as being approximately the same as that for clay.

If the ultimate bearing capacity is determined by theoretical formulas, a factor of safety $\eta \geq 3.0$ is usually applied to the ultimate bearing capacity to obtain the allowable bearing capacity of silt.

Forces

For crude preliminary estimates of allowable bearing capacity, to be used in tentative designs, the N-chart may sometimes be used.

Determination of Bearing Capacity

Information on the ϕ and c values for use in bearing capacity equations is to be obtained from triaxial compression tests on undisturbed silt samples. Otherwise the strength tests on silt may be determined by means of unconfined compression tests in the air and by the method of immersing the test specimen in water to exclude the confining effect of the surface tension of water in the voids of the silt specimen.

In general, the test conditions and test parameters sought for the project in question should correspond to those encountered in nature at the construction site.

Groundwater

Loose, saturated silt of any kind is unsuitable for direct support of the static load of a structure on individual footing foundations. Dynamic loads in silt will bring about its liquefaction.

Special Considerations

Foundation engineering experience in laying foundations on soft, loose silt is relatively meager, hence simple, reliable design procedures for foundations on silt have not been dependably established as yet. No general rules for this kind of soil can replace a thorough soil reconnaissance in the field, soil testing and study of the soil's physical and mechanical properties. In order to arrive at a mature judgment an acceptable value of the bearing capacity of the silt encountered based on physical facts is the first requisite.

MAT FOUNDATIONS

Bearing Capacity

The considerations of bearing capacity of silt for mat foundations are generally the same as those described above for individual footings. Reliable data on friction and cohesion for design of mat foundations on silt are to be optained from triaxial tests.

In the design procedure of a mat on silt, the soil pressure exerted by the mat foundation is determined. This pressure is compared to the allowable soil pressure of

$$\sigma_{\text{all}_{\text{net}}} = (0.95) q_u \left(1 + 0.3 \frac{B}{L}\right) \tag{4-37}$$

The factor of safety against groundbreak in silt is checked. It should be $\eta \geq 3.0$ under normal loading conditions and $\eta \geq 2.0$ under extreme loading conditions. In this respect, the local building code should be consulted.

Sometimes mats are constructed on piles.

Special Considerations

To increase the overall safety factor against failure with respect to groundbreak of silt, mats on silt are sometimes laid at such a depth below the ground surface that the weight of the excavated soil material approximately equals that of the designed structure.

Loess Soils

Loess soils exhibit a great many varieties of physical properties. Some of them vary considerably with their water content. Therefore no plausible rules for assessing the allowable bearing capacity of loess have been put forward as yet, either theoretically, or empirically. Load tests on loess in the field in conjunction with laboratory testing of loess soils subjected to the effect of varying moisture conditions are the only means for orientation as to the performance of loess under load, water, and temperature.

DEEP FOUNDATIONS: PIERS AND PILES.
PIER FOUNDATIONS AND CAISSONS IN SILT

General Notes

Piers are often used to penetrate weak silt deposits in order to transfer structural loads to a deeply situated firm stratum of soils or on rock, if attainable. When founded on silt, piers perform much like individual footings on silt, and may be designed in an approximately similar manner.

Groundwater

Excavation in silt of low plasticity and excavation below the groundwater table is difficult and expensive. Rock flour may flow into the excavation like a viscous fluid. Coping with this requires tight sheeting of excavation walls. Such viscous rock flour is termed by local contractors "bull's liver." This rock flour presents great difficulties to contractors in excavation work, because it has a tendency to slump and collapse unsupported walls of excavation pits if the groundwater table is not lowered.

Settlement

Open excavation below the groundwater table in silt may bring about settlement of adjacent ground surface due to drainage of the silt and the accompanying process of consolidation.

Special Considerations

In determining the strength of silt, its water content and plasticity are the main factors requiring careful consideration. Working in a silt of the rock flour type, contractors fear losing machinery operating on such a soil:[34] machinery simply bogs down, sinks, and disappears in the viscous rock flour, and is almost impossible to recover.

PILE FOUNDATIONS IN SILT

General Notes

Piles in silt are usually of the friction type. End-bearing piles are used only when there is a firm stratum below the ground surface to support the piles.

Skin friction is low during the phase of driving the pile into silt, and the decrease in bearing capacity of the silt continues for several days. After this time the silt gradually recovers its strength.

Piles in silt may be dealt with in much the same way as piles in soft clay.

Bearing Capacity

The method of design of pile foundations in silt may be taken as being approximately like that used in soft clay. The bearing capacity of piles in silt may be best determined by the results of static loading tests. Tapered piles have a greater bearing capacity than cylindrical ones.

The bearing capacity of a group of piles depends to a considerable extent upon the perimeter of the group and cannot be increased appreciably by driving additional piles within the boundaries of the pile group.

The ultimate bearing capacity and settlement of a group of friction piles in a thick layer of silt may be explored by the same methods used for friction pile groups in soft clay.

4-17. FOUNDATIONS ON ROCK

Individual Footings

If the most economical design and safe performance of an engineering project founded on rock is to be achieved, adequate information on the properties of the subsurface rock must be available. Rock masses in situ present very bewildering and complex problems. The dividing line between soil and weak, weathered rock is not distinct. Where there is any doubt whether the material is rock or soil, the material may be considered as a soil, and the design is then made on that basis.

Many rocks are at least as strong and as rigid as the concrete that would be used in the construction of a foundation. Yet sound rocks almost always contain zones of weakness such as joints, bedding planes, cracks, fissures, partings, cavities. These discontinuities represent structural defects that may influence the supporting capacity of the rock. Rock slides occur where water seeps into bedding planes, fissures or joints.

To provide for an adequate bond, all rock surfaces receiving a concrete footing or mat must be roughened and thoroughly cleaned; loose rock fragments, spalls, and dirt must be removed. Cleaning is usually done by means of suitably effective tools, such as water and air jets, or wet sandblasting.

In exploration work, boulders should not be mistaken for bedrock. To check which is which, two or three boreholes should be driven at a radius of about 6 to 8 ft from the obstructed boring.

In important engineering structures such as water-impounding dams, hydraulic power plants, locks, and the like, when founded on limestone or in karst areas, the rock may have to be grouted. Also, such areas must be checked for soluble subterranean streams and caves. Porous rocks must be tested for their permeability in situ. Stress conditions for failure of saturated rocks for foundations of hydraulic structures should not be overlooked.

The presumptive allowable bearing capacity values of rocks are governed by building codes.

If the strength of the rock is greater than that of the concrete used in the construction of foundations, the design bearing capacity values of rocks are frequently taken as that of the concrete.

Great attention must be paid in laying foundations in weathered zones of rocks. Lateral loads such as wind loads are transferred by the structure to the weathered rock. Because of this, the lateral loads may dislodge the weathered fragments of the rock, in time bringing about settlement of foundations and even intolerable cracking of structural walls.

Mat Foundations

See notes pertaining to individual footings on rock. The strength and compressibility of a weathered rock requires careful investigation because many weathered rocks, especially in humid or tropical regions, are locally much weaker and more compressible than concrete. Particular attention must be paid to relative settlement of the mat and to rock shifting back and forth from one end of the slab to the other.

Piers

The design of piers on rock follows just about the same general procedure as for individual footings and mat foundations, but the relatively great bearing capacity of the supporting rock material as compared with that of sand, clay and silt allows a greater freedom of design. As noted above in connection with footings and mats laid directly on rock, local discontinuities must be taken into account in determining the strength and bearing capacity of rock.

Piles

Where firm rock, overlain by weak soil, is attainable, bearing piles are driven to rock to provide for their firm support and reliable load transfer to rock. Nothing can really be gained through attempting to drill through rock for setting piles.

4-18. THE FACTOR OF SAFETY IN FOUNDATION ENGINEERING[35]

In designing a soil-foundation-load system, or any other earthwork, engineers organize the loads involved and design foundations so that the supporting soil or rock can withstand several times as much load as will ever be imposed upon it. This is done to allow for any unexpected overloading of the soil or rock.

Overloading of the soil may bring about intolerable settlement of the soil-foundation-load system, or rupture of soil upon exceeding its ultimate bearing capacity, viz., shear strength. To indicate the degree of safety, engineers use the concept of "stability" characterized by the "factor of safety."

The term *factor of safety* η is a deep-rooted concept in the field of civil engineering design, particularly in those branches of structural design which are based on the ultimate strength of the material.

The factor of safety supposedly should safeguard against (a) possible overloading, (b) errors introduced by simplifications and approximations in design methods and procedures, and (c) variation in quality of materials. Materials having unpredictable nonuniformities necessitate a careful evaluation of their strength. This pertains forcefully to soil and rock as a construction material. Here a factor of safety is introduced.

In general, the factor of safety should also take care of (a) the imperfection of human observations and actions (objective uncertainty), and (b) the imperfection of intellectual concepts devised to reproduce physical phenomena (subjective ignorance). Therefore some call the factor of safety the "factor of ignorance." This latter term is, however, used more in the context of uncertainty, for example, when the engineer does not know enough facts, or when exact knowledge about some factors in design is lacking. Thus the factor of safety should provide for contingencies that affect the design and construction of a structure. Besides, the choice between safety and economy is usually a problem with which engineers must cope. The engineer is also confronted with the problem of how much importance should be attached to safety and how much to economy, thereby introducing another factor of uncertainty in the design and stability calculations. In any event, the factor of safety should provide for the contingency of external causes weakening the soil's supporting capacity. In other words, the factor of safety has been used by engineers to cover the unknown gap between the strength of the material and the load applied.

Some of the variables necessitating the introduction in design and stability calculations of a certain margin are:

1. Assumptions made in mode and nature of loading.
2. Nonuniformity in composition of materials.
3. Insufficient knowledge and limited or meager analytical comprehension in testing of soil and rock materials, measurement, knowledge of the (shear) strength properties and deformation of materials and structures, as well as in sampling of undisturbed soil samples. A single component such as the shear strength of soil in turn depends on a large number of other factors such as reconnaissance of the soil, soil sampling, quality of tests made, temperature, pore water present in the soil, and quality of interpretation, to mention but a few.
4. Failure to reproduce field conditions in the laboratory when testing materials and structural prototype models.
5. Uncertainties in simplifying assumptions in static and dynamic calculations of structural and/or soil-foundation-load systems.
6. Errors in and departures from design which inevitably occur in the execution of the construction project.
7. Quality of construction work (very difficult to anticipate).
8. Mode and probability of failure on which the stability analysis and calculation of factor of safety is based. Instead of considering rupture condition, one may also consider plasticity condition, or intolerable deformation in terms of settlement of the structure, thus basing the soil-foundation system's design on the so-called *proportional limit* or *critical edge pressure*.
9. The reduction in quality of the structure and its materials which can occur by aging, weathering and exploitation (service and use).

The mutual relationship of these and other variables to the resultant stability of a structure must be well understood if the structure is to have a reasonable degree of safety.

Loading

In soil engineering frequently the self-weight of the structure is the principal load on the soil. Its magnitude can be determined with relatively satisfactory precision. When variable lateral forces such as wind, waves, ice thrust and earthquakes apply, or where moving repetitional and vibratory loads occur, the assessment of loading, its effects, and the corresponding factors of safety are more difficult to determine than for systems with no such forces. There should be incorporated in the design of a structure as many overall safety precautions as possible against the hazards of earthquakes. Earthquakes are conspicuous examples of vibration effects on structures, and therefore a higher factor of safety is required.

Besides these externally applied loads there also should be considered the variable natural loads, such as pore-water pressure, buoyancy of submerged soils, seepage pressure and the liquefaction phenomenon.

Soil Material

Whereas the properties of a man-made material such as steel are well known, the properties of materials, such as soil, provided by nature are nonuniform and as yet not too well understood. Soil conditions are less reliable than most other materials connected with a structure. The strength of the soil material refers to its resistance to shear and other physical properties associated with it. A single element in the composite factor of safety, such as the shear strength of the soil, for example, which is a fundamental factor in stability analyses, itself depends on a large number of other factors: the quality of field reconnaissance, the quality of the soil sampling, quality and type of shear tests of soil made (unavoidable errors arising from incompleteness of soil shear-testing devices and testing methodology, for example), shape of particles, particle-size distribution, angle of internal friction, moisture content, pore-water pressure, temperature, quality of calculations and interpretation of test results. Many of these elements themselves are functions of other independent factors.

Assumptions made regarding the angle of internal friction of sandy soils, the relative density of a sand deposit, or the compressibility of a layer of clay are more examples of variables affecting the magnitude and quality of the factor of safety. The determination of the coefficient of permeability of the soil is another vulnerable variable, and so is the unreliable information when attempts are made to predict the pore-water pressure in stratified sand layers or in beds of clay containing seams of more permeable material. This is because the pore-water pressures depend on unexplorable structural details of foundation soils and other earthworks.

The properties of soil vary also with season (frozen or unfrozen), position below ground surface, and location and fluctuation of the groundwater table. Clayey soils, fissured clays and silts certainly perform differently from sand under load. The causes for dispersion of stress and shear strength of a soil-foundation-load system may be almost infinite in number.

The foregoing indicates that not only do external variables (loading, for example) affect the soil material, but there are also many internal variables involved. These variables in the properties of soils are unique to them and are not encountered in a material such as concrete or steel. The complexity of the nature of soil is due partly to the fact that it is a combination of solid, liquid, and gaseous phases and partly to the fact that in many cases a significant portion of the solid matter is so finely divided that its particles are of colloidal size, with the characteristic colloid-chemical properties and behavior in water in motion (electrokinetic phenomena) and adhered water films.[2, 36] In any given soil the quantities of solid, liquid and gas are subject to wide variation due to loading, or effects due to the elements such as wetting, drying, freezing, and thawing. Other properties that may vary in a soil include its density, void ratio, moisture content, degree of saturation, soil consistency, and the effects due to groundwater, such as capillary action and permeability.

One sees that the factor of safety is a much more complex problem in soils than in homogeneous materials like steel, and that it will probably never be possible to describe their strength and deformation characteristics as precisely as those of steel. Hence the variable factors, as well as the resultant factor of safety for soils, are, indeed, very difficult to assign. In general, considering these variable factors, the factor of safety here should not be judged too small.

Uncertainties in Simplifying Assumptions

One of the uncertainties in simplifying assumptions underlying theories as applied to soil engineering pertains to the nature and course of stress distribution in soil. Because there is nothing better available, Boussinesq's stress-distribution theory in an ideal, elastic, homogeneous and isotropic medium is also being applied to problems of stress distribution from loads in soil.

Methods of Calculating the Factor of Safety

The method used in calculating the factor of safety also has an effect on its magnitude. The equation $\eta = M_R/M_D$ is a general analytical statement of the factor of safety. However, various authors have put forward their own methods of computing η, which include considering the ratios of shear strength, s, of the soil to the shear stress, τ, or available cohesion, c, of the soil to the necessary cohesion, c_{nec}; available coefficient of internal friction of soil, tan ϕ, to the necessary coefficient of internal friction, tan ϕ_{nec}, and consideration of friction and cohesion being important but acting separately. However, all these methods reduce, in a vague way, to the equation $\eta = M_R/M_D$. What is more important is that all these methods point out that the concept of the factor of safety is a subjective one and depends to a great extent on the individual's ideas on a particular problem. Thus one notes that for the same problem the factor of safety varies from author to author, from method to method of calculation.

Analysis of soil and earthwork stability problems is largely a trial-and-adjustment procedure to determine the safety factor of an assumed design or of an actual soil mass. First, a potential failure surface is assumed, and the shearing resistance acting along the surface is calculated. The forces acting on the segment of soil bounded by the failure surfaces are determined, and then the safety factor of the segment is calculated as follows:

Safety against rotation:

$$\eta_M = \frac{\text{resisting moments}}{\text{moments causing failure}}$$

Factor of safety against any motion:

$$\eta_T = \frac{\text{forces opposing motion}}{\text{forces causing motion}}$$

Theoretically, if a large number of different segments is assumed, the smallest safety factor found for any will be the actual safety factor of the mass.

Quality of Construction; Modification in Value of Structure with Age and Use

The factor of quality of construction (workmanship, construction accuracy and reliable inspection) in earthworks and foundation engineering is very important. It is especially important in soil compaction operations of earth dams, highways and airfields, in artificial improvement of soil properties by means of compacted soil cushions, in installing sand drains, in grouting operations, lowering of the groundwater table, pile and sheet-pile driving, performance of field bearing-capacity tests of soils and other construction pertaining to soil engineering. Variation of the groundwater table due to the introduction of local and regional drainage systems after erection of structures (decrease in buoyancy underneath dams and foundations), intolerable settlements and subsequent cracking of structures, increase in impounded head of water behind the dam (increase in buoyancy, decrease in stability) after it has been finished may modify the service and proper functioning of the structure.

Access of air and/or water to soil may change some soil properties for the worse, particularly decreasing the shear strength of the soil in the slopes of earthworks such as cuts for roads and canals, for example. Improper functioning of drainage structures in soil such as weep holes, French drains, lateral drainage galleries in slopes, and galleries and drainage carpets underneath earth dams may bring about the collapse of earth retaining walls, earth slopes and earth dams.

Numerical Values of the Factor of Safety

At the start of design of an earthwork and/or soil-foundation system there always arises the question of the magnitude of the factor of safety to use in stability calculations of earth masses. Such discussions, as well as a review of the technical literature, reveal that there is no uniformly accepted value of η to use.

For most engineering construction materials and modes of loading the factor of safety is officially regulated in one way or another by building codes, city ordinances, or the like. These codes vary from one edition to another, and since there is no statutory guarantee that their compilers know what they are codifying, the ratio, strength as calculated by one arbitrary code divided by loading stipulated by another code (= factor of safety) is variable and no more reliable than the opinions of those members of the drafting committee who were not too busy to take an active part in drafting the code. In any case it is safe to say that the factor of safety to use in soils engineering is a much more complex problem than in those fields of engineering which deal with homogeneous materials.

In the past the factor of safety was chosen arbitrarily by the designer, utilizing all his skill and years of experience. In general, it was probably a larger factor of safety than actually needed, but it tended to comprehend the "ignorance" and uncertainties of the design. In selecting a factor of safety the designer had to keep an eye on economic considerations as well as the safety of the people who were depending on his skill.

Analyses of existing slopes of earthworks for their stability have revealed relatively small factors of safety as compared with those of other structures. A factor of safety of 1.5 or 2.5 is usually uncommon in structural engineering.

However, larger magnitudes of safety factors, if applied to earthworks, would make their cost so prohibitively high that they probably could not be built. On the other hand, many earth structures having a computed factor of safety as low as $\eta = 1.0$ have been demonstrated to be stable by the test of time (or else, some important "favorable" factors have been forgotten in the stability calculation).

Typical values of the factor of safety for most soil engineering work range between 1.5 and 2.5, as can be readily found in any textbook and in many published papers. It is interesting that almost none of these texts or papers report the origin of the factor of safety range indicated, but merely suggest the range. The values of the factors of safety range between 1.3 and 1.7.

To avoid the various types of soil failure, the following minimum factors of safety have usually been used:

1. The factor of safety against a sliding failure must be at least 1.5.
2. The factor of safety against a deep-seated failure must be at least 1.5.
3. The factor of safety against a shallow shear failure must be at least 2.0.
4. The factor of safety against rotation must be at least 1.5.

It is interesting to note that the factor of safety to use depends also on the size, shape, and importance of the structure (the more important the structure the higher the factor of safety to be used), and the degree of accuracy with which the applied loads and strength of the soil can be ascertained.

Critical Edge Pressure

Besides ultimate failure, the plastic flow condition (proportional limit) may be used as the basis for calculating the safe bearing capacity of soil. For example, Fröhlich [2,3, 31,37] calculates the allowable pressure from the foundation on the soil, which he calls the "critical pressure." By critical pressure is understood that pressure at which soil particles just begin to flow out from underneath the edge of the loaded footing of the foundation. In essence, Fröhlich's critical pressure corresponds approximately to the proportional limit. Hence the soil bearing capacity may be based on the proportional limit of the soil. The calculated critical edge pressure needs no factor of safety.

Probability and Statistics

The philosophical nature and content of the concept of the factor of safety are currently receiving a new look and study by the engineering profession. The most recent ideas put forward for the rational selection of the factor of safety to use are from the discipline of mathematics, particularly that of probability and statistical analysis.[38, 39] Relative to loads and structural behavior, Freudenthal[40] considers a means for estimating failure through the laws of probability (or rather relative frequency). He writes:[41] "The probabilities of failure or of survival for which structures are designed must be referred to the critical operating conditions of the structure; different values of probabilities are necessarily associated with different operating conditions.

"Because the design of a structure embodies certain predictions of the performance of structural materials as well as of expected load patterns and intensities," Freudenthal adds, "the concept of probability must form an integral part of any rational design."[40] Thus, instead of a factor of safety, the mathematics would be concerned with the probability of failure. In this context Svensson[39] writes that "the load and the strength of the material exhibit variations defined by certain probability functions and thus the occurrence of failure will also be defined by a probability function."

According to the above trend of thought, it becomes apparent that in the final analysis of stability of a system it is necessary to choose arbitrarily a suitable probability of failure. However, this in effect means really choosing a factor of safety. Let us recall that, figuratively speaking, the factor of safety aims at the protection of the structural system against extreme, yet reasonable, fluctuations which statistical methods indicate have the probability of occurring. Thus, according to Freudenthal, mathematics, i.e., probability, and statistics seem to be the answer to the question of what factor of safety to use. This may be realistic under complex loading conditions such as are encountered in aircraft design, but may be too unreasonably involved for the more static loading conditions such as occur in a soil-foundation-load system. Statistical analysis seeks to analyze human experience of random events. Ultimate strength design (as, for example, in concrete technology) presents some of the concepts of the statistical approach to the factor of safety.[42, 43] It has also been said that the ultimate design method results in a more realistic design and one with a more accurately provided factor of safety.

Realizing that the choice of a suitable probability of failure "can only be examined on economical grounds where one must balance the cost of failure against the cost of increased strength to reduce the incidence of failure,"[44] one would be inclined to think that probability and statistical methods definitely have their place in the determination of the factor of safety to use, especially with the advent of the electronic computer.

Conclusions

How much of the design procedures suggested by the proponents of probability calculus and statistical analysis will be adopted by foundation engineers is open to question. In the words of A. Chibaro,[45] who discussed Freudenthal's paper,[40] "... it will probably be a long time before the use of the 'factor of safety' concept will be supplanted by the use of the 'probability of failure' concept. ... In order to use the concept of probability of failure a great mass of statistical data pertaining to load effects and strengths must be obtained and correlated."

Things become even worse when working with soil as a construction material. It appears that probability and statistics are applicable only to controlled, man-made engineering materials, such as steel and probably concrete. One field where the probability calculus and statistical analysis approach to the safety factor is clearly not applicable as yet is soil and foundation engineering. This is because of the nature of soil. There are no two soils in the world exactly the same, nor will they react in the same fashion under various conditions. Due to the nonuniformity of soils, the mathematical method cannot be applied to soils either directly or indirectly. In foundation engineering, according to J. M. Corso,[46] "Seldom, if ever, are enough test data available to determine the frequency distribution of the soil properties. Furthermore, in many cases the loading effect and resistance cannot be considered as independent variables, and interrelation between the load and the resistance, including variations with time, is often so complicated as to defy rigorous analysis."

It appears as though the question of applying probability calculus and statistical analysis to soil and foundation engineering problems is still premature, mainly because of our still relatively meager knowledge of the nature of deformation of soil subject to shear stresses.

Besides, Freudenthal himself[40] says that "the foregoing does not imply that the use of the probability theory is, in itself, sufficient to make design procedures more adequate and reliable. Probability concepts and statistical methods based thereon can be used effectively only in conjunction with a thorough knowledge of the operating conditions of the structure and of its structural action...."

Thus it seems that the expression "factor of safety" has become and will probably remain in the engineering lexicon as an integral term in soil and foundation engineering design, and that in this field experience and common sense cannot be dispensed with as yet in favor of probability and statistical analysis.

Until science and engineering contribute to the correct evaluation of the complex properties of soil as an engineering material and to the correct determination of its shear strength, thus enabling one to apply more sophisticated methods for determining a proper factor of safety than is now possible, it seems that we are still compelled to continue to resort to the present conventional methods in selecting a proper factor of safety to use.

The factor of safety must be chosen realistically and with deep awareness of what is occurring or what may occur. Then it could be used with confidence in any deliberations and decisions. The selection of a factor of safety for foundation and earthwork design purposes still requires sound experience, common sense, and careful engineering judgment.

REFERENCES

1. A. R. Jumikis, *Introduction to Soil Mechanics*, Princeton, N.J.: Van Nostrand, 1967, pp. 303-321.
2. A. R. Jumikis, *Soil Mechanics*, Princeton, N.J.: Van Nostrand, 1962, pp. 549-596.
3. A. R. Jumikis, *Mechanics of Soils: Fundamentals for Advanced Study*. Princeton, N.J.: Van Nostrand, 1964, pp. 219-329.
4. A. R. Jumikis, *Active and Passive Earth Pressure Coefficient Tables*, New Brunswick, N.J.: Bureau of Engineering Research, College of Engineering, Rutgers University Engineering Research Publication No. 43, 1962.
5. A. L. Bell, "Lateral Pressure and Resistance of Clay and the Supporting Power of Clay Foundations," *Proceedings of the Institution of Civil Engineers*, London, 1915, Vol. 199.
6. W. P. Crieger and B. O. McCoy. "Dams," in R. W. Abbett (ed.), *American Civil Engineering Practice*, Vol. II. New York: Wiley, 1956, pp. 14-19.
7. ASCE Commitee Report on "Uplift in Masonry Dams," *Trans. ASCE*, Vol. 117 (1952), p. 1218.
8. L. F. Harza, "Uplift and Seepage Under Dams on Sand," *Trans. ASCE*, Vol. 61, No. 8, Part 2 (1953), pp. 1352-1385. Discussions on pp. 1386-1406.
9. R. D. Chellis, *Pile Foundations*, New York: McGraw-Hill, 1951, pp. 104.
10. G. E. Monfore (ed.), "Ice Pressure Against Dams," a symposium, *Trans. ASCE*, Vol. 119 (1954), p. 26.
11. Anon., "Mackinac Bridge—Incredible but True." *Engineering News-Record*, Jan. 27, 1955.
12. N. C. Raab and H. C. Wood, "Earthquake Stresses in the San Francisco-Oakland Bay Bridge," *Trans., ASCE*, Vol. 106, (1944), pp. 1363-1384.
13. N. C. Raab and H. C. Wood, *Proc., ASCE*, October 1940, p. 1447.
14. R. Briske, *Die Erdbebensicherheit von Bauwerken*, Berlin: Ernst, 1927.
15. Discussion on "The Relationship between Earthquakes and Engineering Substructures," *Proc. ASCE*, Vol. 55 (1929), pp. 219-227.
16. N. H. Heck, *Earthquakes*, Princeton, N.J.: Princeton U.P., 1936, p. 202.
17. J. E. Rinne, "Building Code Provisions for Aseismic Design," *Symposium on Earthquake and Blast Effects on Structures* (Los Angeles, June 26-28, 1952),

Earthquake Engineering Research Institute and University of California, Berkeley, Calif., 1952, pp. 291-308.
18. H. W. Bolni, "The Field Act of the State of California," *Symposium on Earthquake and Blast Effects on Structures* (Los Angeles, June 26-28, 1952), Earthquake Engineering Research Institute and University of California, Berkeley, Calif., 1952, pp. 309-313.
19. F. M. Andrus, "Earthquake Design Requirements of the Uniform Building Code," *Symposium on Earthquake and Blast Effects on Structures* (Los Angeles, June 26-28, 1952), Earthquake Research Institute and University of California, Berkeley, Calif., 1952, pp. 314-316.
20. H. M. Westergaard, "Water Pressure on Dams during Earthquakes," Paper no. 1835, *Trans. ASCE*, Vol. 97-98 (1933), pp. 418-433.
21. J. M. Trefethen, *Geology for Engineers*, Princeton, N.J.: Van Nostrand, 1949, p. 384.
22. K. Terzaghi, *Theoretical Soil Mechanics*, New York: Wiley, 1943, p. 124.
23. A. R. Jumikis, *Soil Mechanics*, Princeton, N.J.: Van Nostrand, 1962, p. 627.
24. J. Brinch Hansen, "A General Formula for Bearing Capacity," Geoteknisk Institut, Copenhagen, Bulletin No. 11, 1961.
25. Swiss Union of Highway Specialist, *Stützmauern* (Murs de soutènement). Zürich: Vereinigung Schweizerischer Strassenfachmänner (Union Suisse des professionnels de la route), Vol. 1, 1966, pp. 673-692.
26. A. W. Skempton, "The Bearing Capacity of Clays," *Proceedings of the British Building Research Congress (London)*, Vol. 1 (1951), pp. 180-189.
27. 1968 *Book of ASTM Standards*: Part 11, "Standard Method Penetration Test and Split-Barrel Sampling of Soils." ASTM Designation D 1586-67. Philadelphia, Pa.: American Society for Testing and Materials, 1968, pp. 496-498.
28. A. R. Jumikis, "Stability Analyses of Soil-Foundation Systems," Engineering Research Publication No. 44, New Brunswick, N.J.: College of Engineering, Rutgers University, 1965.
29. R. B. Peck, W. E. Hanson, and T. H. Thornburn, *Foundation Engineering*, New York: Wiley, 1963, p. 224.
30. K. Terzaghi and R. B. Peck, *Soil Mechanics in Engineering Practice*, 2nd ed., New York: Wiley, 1967, p. 516.
31. A. R. Jumikis, *Theoretical Soil Mechanics*, Princeton, N.J.: Van Nostrand, 1969, p. 383.
32. R. L. Handy and N. S. Fox, "A Soil Bore-Hole Direct-Shear Test Device," Special Report, Contribution No. 66-8a, Soil Research Laboratory, Engineering Research Institute, Iowa State University, Ames, Iowa, 1966.
33. K. Terzaghi and R. B. Peck, *Soil Mechanics in Engineering Practice*, 2nd ed., New York: Wiley, 1967, p. 563.
34. A. R. Jumikis, "Engineering Aspects of Glacial Soils of the Newark Metropolitan Area of New Jersey," Engineering Research Bulletin No. 42, New Brunswick, N.J.: College of Engineering, Rutgers University, 1959.

35. A. R. Jumikis, "The Factor of Safety in Foundation Engineering," Highway Research Board Record No. 156 on Classification, Safety Factor and Bearing. Washington, D.C.: Highway Research Board, National Academy of Sciences, National Academy of Engineering, Publication 1433, 1967, pp. 23-32.
36. A. R. Jumikis, *Soil Mechanics*, Princeton, N.J.: Van Nostrand, 1962, pp. 640-642.
37. O. K. Fröhlich, *Druckverteilung im Baugrunde*. Berlin: Springer, 1934.
38. A. M. Freudenthal, "The Safety of Structures," *Proc. ASCE*, Vol. 71 (October 1945), pp. 1147-1193; also, *Trans. ASCE.*, Paper No. 2296, Vol. 112 (1947), pp. 125-180.
39. N. L. Svensson, "Factor of Safety Based on Probability," *Engineering* (London), Vol. 191 (1961), pp. 154-155.
40. A. M. Freudenthal, "Safety and the Probability of Structural Failure," *Trans. ASCE*, Vol. 121 (1956), pp. 1337-1372.
41. A. M. Freudenthal, "Safety, Reliability and Structural Design," *Trans. ASCE*, Paper No. 3372, Vol. 127, Part II (1962), pp. 304-319.
42. A. L. L. Baker, *The Ultimate-Load Theory Applied to the Design of Reinforced and Prestressed Concrete Frames*. London: Concrete Publications, 1956.
43. P. M. Ferguson, "Recent Trends in Ultimate Strength Design," *J. Struct. Div., Trans. ASCE*, Paper No. 3373, Vol. 127, Part II (1962), pp. 324-338.
44. A. I. Johnson, "Strength, Safety and Economical Dimensions of Structures," Report No. 12, Institute of Building Statistics, Royal Institute of Technology, Stockholm, 1953.
45. A. Chibaro, Discussion of paper "Safety and the Probability of Structural Failure," by A. M. Freudenthal, *Trans. ASCE*, Vol. 121 (1956), p. 1376.
46. J. M. Corso, Discussion of paper "Safety and the Probability of Structural Failure," by A. M. Freudenthal, *Trans. ASCE*, Vol. 121 (1956), p. 1383.

part II

EXCAVATIONS AND COFFERDAMS

chapter **5**

The Excavation

5-1. EXCAVATION–IMPORTANT ELEMENT IN LAYING FOUNDATIONS

The foundation and various subsurface structures require the excavation of soil and/or rock in order to establish a foundation pit. The foundation pit, or simply excavation, is one of the most important elements in laying the foundations.

Excavation work comprehends not only all kinds of difficulties, but it also requires most of the means and time for erecting a structure. Therefore the engineer should learn all that he can about the condition at a given construction site.

All excavations for laying of foundations may be classed as follows:
1. Shallow excavations (i.e., structures without basements, retaining walls, low dams)
2. Deep excavations (i.e., caissons, piers for bridges, heavy structures).

Excavations are made as follows:
1. In dry soil or rock requiring no coping with water. The banks of the excavation may be free, unsupported, or there may be a need for braced (supported) banks
2. In soils that contain groundwater
3. In and through open water (wet or subaqueous excavations as for docks, locks, sea walls, bridge piers, excavations in cofferdams, caissons)

In water-saturated soils a dry excavation can also be procured by dewatering it by
1. Open pumping
2. Lowering the groundwater table by the well-point method
3. Freezing water-logged sand soils

Size of Excavation

The delineation, form, and size of the excavation usually depend upon the plan of the structure, the properties of the soil and/or rock, and the groundwater conditions at the site.

When the excavation must be enclosed by sheet piling or cofferdam, or when the foundation pit must be excavated to a great depth, sharp angles in the outline of the excavation are omitted. One chooses either a possibly short-perimetered outline of the excavation consisting of straight sections or a round-perimeter outline of the excavation.

Several anticipated small excavations are sometimes arranged to merge in a common large excavation. Where there is a large influx of water, large excavations are sometimes subdivided into small sections in order to cope successfully with groundwater. This usually requires the use of sheet pilings and cofferdams. The site thus encompassed must therefore be chosen somewhat larger in extent in order to support, store and operate construction machinery and to store construction materials.

Excavation Slopes

In open excavations with unsupported slopes, good maintenance of slopes, repair activities, and coping with water and snow must be practiced. It is not wise to store excavated soil masses and various other materials on top of excavation slopes. These may impound water and cause slope failures brought about by seepage (Fig. 5-1). Also, during rainy periods steep clay slopes may slough down.

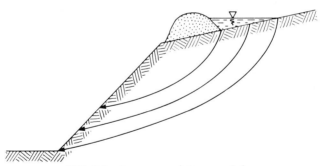

FIG. 5-1. Improper maintenance of slope.

Slopes exceeding 5 to 6 ft in height are formed with berms (Fig. 5-2) 2 to 3 ft wide. The angle of slope α should not exceed 60°.

To permit walking around the excavation, and to protect the top of slopes from mechanical damage by walking, it is well to place planks on the berms, and wooden or metal stairs on the slopes. At the very top of the slope there should be a load-free strip around the excavation. The excavated soil material may be placed beyond the load-free strip on the land side.

Sometimes the slopes are protected from erosion by water by covering them with tarpaulin sheets or other waterproof fabric or plastic sheets.

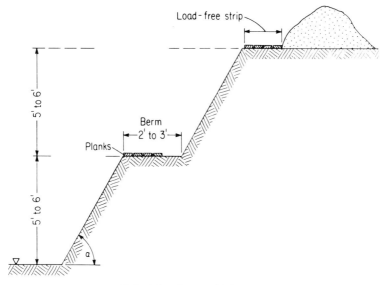

FIG. 5-2. Slopes with berms.

Slopes of large, deep excavations, which by necessity must be kept open for a long period of time as for the construction of locks, docks, and other large engineering structures, for example, must be checked for their stability against rupture and consequent sliding. Likewise, free earth slopes must be checked for their stability where heavy construction machinery operates and various materials are stored on top of them. For stability analyses of slopes the reader is referred to Ref. 1. Also, the stability of excavation slopes must be checked against slope failure from vibration brought about by vehicles, pile-driving machinery, and vibrations transmitted from nearby passing rolling stock, as well as those caused by blasting (where this applies).

5-2. MAXIMUM UNSUPPORTED DEPTH OF AN OPEN, STEPPED EXCAVATION

The maximum unsupported depth z_1 of an open, stepped excavation shown in Fig. 5-3 as section 1 may be calculated from equilibrium condition by means of the theory of major and minor principal stresses, σ_1 and σ_3, respectively, as follows.

At point A the lateral stress is the minor principal stress $\sigma_3 = 0$ (because of no lateral restraint). The corresponding major principal stress is $\sigma_1 = \gamma z_1$. Here $\gamma = 120$ lb/ft^3 = 0.06 ton/ft^3 is the unit weight of the given soil. Its cohesion is given as $c = 350$ lb/ft^2 = 0.175 ton/ft^2, and the angle of internal friction of this soil is given as $\phi = 18°$. Then, by Eq. 2-68,[2]

$$\sigma_3 = 0 = \sigma_1 \tan^2(45° - 18°/2) - 2c \tan(45° - 18°/2)$$
$$= (0.06)(z_1)(0.726)^2 - (2)(0.175)(0.726)$$
$$= (0.032)(z_1) - 0.254$$
$$\therefore z_1 = 0.254/0.032 = \underline{7.94 \text{ (ft)}}$$

As a check point for a conventional construction procedure when pressure on shoring is not calculated, the depth z_2 below the ground surface of the supported walls, section 2, is calculated as set forth.

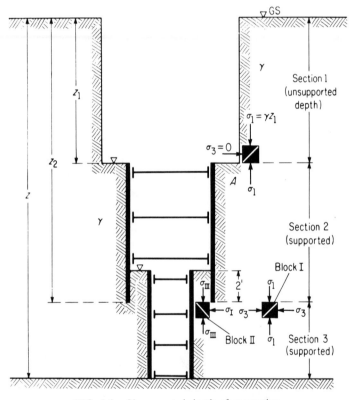

FIG. 5-3. Unsupported depth of excavation.

Block I

Major principal stress: $\sigma_1 = \gamma z_2$
Minor principal stress: $\sigma_3 = \gamma z_2 \tan^2 36° - (0.35) \tan 36$
$= (0.032)z_2 - 0.254$

Block II

Soil overburden = 2 ft thick:

$$\sigma_{III} = (2.00)(\gamma) = \underline{0.12} \quad (ton/ft^2)$$

is here the minor principal stress. Expressed as a function of its major principal stress σ_1, obtain:

$$\sigma_{III} = 0.12 = \sigma_1 \tan^2 36° - 2c \tan 36°$$
$$= [(0.032)(z_2) - 0.254](0.527) - (0.254)$$

or

$$= (0.0168)(z_2) = 0.508$$

$$\therefore z_2 = 0.5080/0.0168 = \underline{30.2 \text{ (ft)}}$$

Critical Depth of Excavation

Excavating a trench in a cohesive soil to a critical depth H_{crit} below the ground surface causes plastic zones at the corners at the bottom of the excavation (Fig. 5-4). The approximate critical excavation depth H_{crit} at which the plastic flow of soil begins is, after Fröhlich:[2,3]

$$H_{crit} = \frac{\pi p_i}{\gamma \left[\cot \phi - \left(\frac{\pi}{2} - \phi\right)\right]} \quad (cm) \quad (5\text{-}1)$$

where $p_i = c \cot \phi$ = initial stress of cohesive soil, kg/cm^2
c = cohesion of soil, kg/cm^2
γ = unit weight of soil, kg/cm^3
ϕ = angle of internal friction of soil (or, rather, a test parameter)

When $p_i = 0$, the $H_{crit} = 0$ (dry sand flows into excavation). When $\phi = 0$, then

$$H_{crit} = \pi(c/\gamma) \quad (5\text{-}2)$$

Example

If $p_i = 0.20 \text{ kg/cm}^2$, $\gamma = 0.0016 \text{ kg/cm}^3$, and $\phi = 25°$, then the critical depth of the excavation calculates as

$$H_{crit} = \frac{(0.20)\pi}{0.0016(2.145 - 1.134)} = \underline{382 \text{ (cm)}}$$

When $\phi = 0°$, $\gamma = 0.0016 \text{ kg/cm}^3$, and $c = 0.09326 \text{ kg/cm}^2$, then

$$H_{crit} = (3.14)(0.09326/0.0016) = \underline{183 \text{ (cm)}}$$

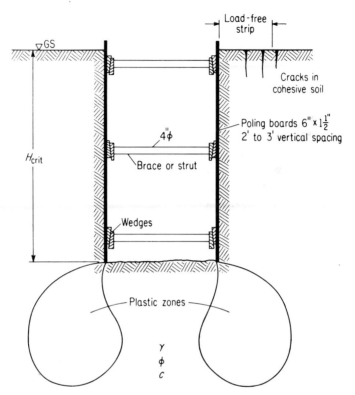

FIG. 5-4. Timbering of a trench in a firm soil.

5-3. PROTECTION OF EXCAVATION BANKS

In congested places such as are frequently encountered in cities, and for laying of deep-lying parts of structures (elevator shafts, boiler rooms, for example), excavations in soil must be made with vertical banks. Vertical banks must be protected against caving in. Likewise, foundation trenches and excavations made in weak soils and weak rocks, as well as excavations made to a considerable depth, must be protected against collapse of their banks. The protection is usually accomplished by sheeting, shoring, bracing and strutting—temporary structures which are usually dismantled after completion of the earth—and foundation work.

When important, complicated, and expensive structures—viz., foundations—are built, their excavations as well as those where lives or adjacent structures may be endangered, the bracing and strutting, as well as the stability of the excavation and the slope system must be analyzed and designed as for any other important structure.

Timbering of Trenches

As the excavation proceeds and the trench becomes deeper, appropriate bracing must be installed from side to opposite side of the excavation to protect the banks of the excavation. The method of bracing to use depends upon the nature of the soil and depth of the excavation. Firm soil would require less strutting than a weak soil, which may have to be sheet-piled. Figure 5-4 illustrates a method of strutting a trench in a firm, stiff soil. The struts, or braces, are wedged and clamped horizontally against the vertical side poling boards. The struts prevent the development of tension cracks in the clayey soil.

At a certain critical depth H_{cr} of the trench or excavation, dangerous plastic zones may develop in the soil at the bottom edge corners of the trench (Fig. 5-4)—a zone where soil particles begin to translocate or flow because of the critical overburden pressure, thus bringing about a dangerous soil instability at the bottom of the excavation.

For widths of trenches less than 5 ft metal extensible struts with turnbuckles and jackable telescoping steel pipes are used (see Figs. 5-5a and 5-5b, respectively).

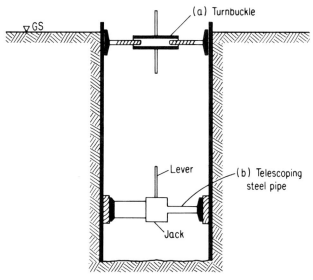

FIG. 5-5. Metal struts.

Sheeting Walls

Sheeting walls, or simply sheeting, are distinguished from sheet piling. When an excavation or trench is in progress and an inward movement of the soil can be anticipated, the excavation walls should be sheeted. Sheeting supports the sides of an open excavation against caving inward. The sheeting is placed or driven in position as the excavation progresses. Sheet piling, however, is usually driven before the excavation is started. Moreover, sheet piling is usually driven deeper than the final depth of the excavation.

The sheeting of timber planks or sheet-piling material may be placed vertically or horizontally.

In practice, vertical sheeting is practiced in a loose, running soil and in rupture-prone soils with heavy influx of water. Horizontal sheeting is practiced in a dense soil with little influx of water.

It should be kept in mind that it is difficult to pour concrete for foundations through a forest of wales, struts, pumping equipment, drains, drainage ditches, and all kinds of conduits.

In wide excavations the use of long, horizontal struts across the entire width may become impractical. Therefore inclined strutting of the sheeting is practiced.

In wide excavations where there is not enough room, inclined struts are anchored backwards away from the excavation. In a situation where there are structures adjacent to the excavation, and if there is room enough in the excavation itself, the bank-supporting inclined strutting system is installed in the excavation (Figs. 5-6a and 5-6b).

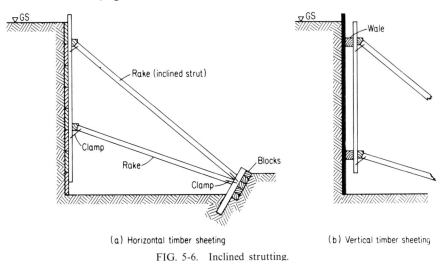

(a) Horizontal timber sheeting   (b) Vertical timber sheeting

FIG. 5-6. Inclined strutting.

Frequently horizontal timber sheeting is placed against the flanges of vertical H-piles, called the "soldier" piles (Fig. 5-7). Cleats hold timber in position. Soldier piles are used for protecting vertical banks of deep building excavations. Figure 5-8 illustrates a framed bracing of the excava-

The Excavation

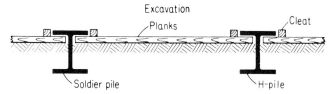

FIG. 5-7. Timber sheeting placed against flanges of vertical H-piles.

tion. This kind of bracing is practiced where slides and caving in of soil can be anticipated upon advancing the excavation in depth.

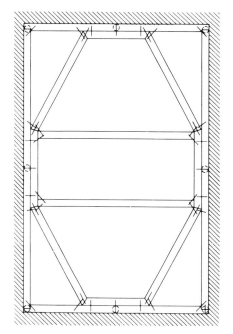

FIG. 5-8. A horizontal bracing frame.

5-4. SHEET PILING

Timber Sheet Piling

In foundation engineering, sheet piling is used mainly to restrain the influx of soil and water into the excavation. Sheet piling is made of timber, of steel, as well as of reinforced concrete. Sheet piling is driven into the soil below the bottom of the excavation.

Timber sheet piling is used for enclosing shallow foundation excavations to depths of about 15 ft in one tier. For greater depths than 15 ft and using timber sheet piling, the sheet piling enclosure is made in several tiers,

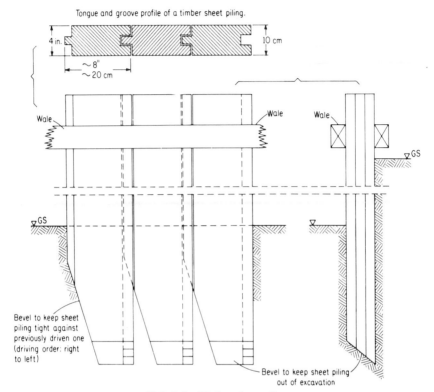

FIG. 5-9. Timber sheet piling.

whereby each successive and higher tier is set inside the wales of the previous tier. Some elements of a timber sheet piling are shown in Fig. 5-9.

Because timber is a material relatively weak in bending as compared with steel, the driving depths of timber sheet piling are shorter than those attained with steel sheet piling. This necessitates closer spacing of wales and struts, frequently obstructing construction work in trenches. Because of its considerable thickness, timber sheet piling, upon driving, displaces a considerable amount of soil, thus densifying it, and causes hard driving. Also, upon encountering boulders and other solid obstacles during the driving operations, timber sheet piling may be easily forced out of alignment or its tip may even become damaged, bringing about the splintering of the timber sheet piling. Also, timber sheet piling swells when taking up moisture, and warps.

Steel Sheet Piling

When the depth of the excavation exceeds about 15 ft, preferably steel sheet piling instead of timber sheet piling is used.

The Excavation

The use of steel sheet piling in trenches excavated in waterlogged sand is shown in Fig. 5-10. Here the sheet piling is designed for pressure of saturated soil. The penetration depth D of sheet piling in the soil may be taken from approximately $D=(0.8)H$ to $D=H$ for all b/H-ratios. This means that the sheet pile must be driven deep enough to resist cantilever moment, or else, the driving depth D can be calculated. Bracing is sometimes advisable.

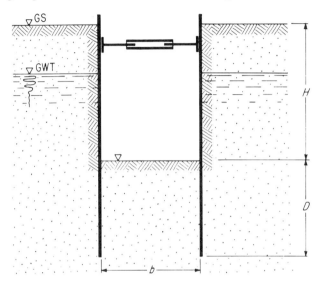

FIG. 5-10. Sheet piling in waterlogged soil.

Water is usually pumped out of the excavation to maintain a dry trench for laying of foundations or pipes, because of seepage and because the sheet piling is not 100 percent tight.

The use of sheet piling for cutting off the water-bearing strata sandwiched between clay layers is shown in Fig. 5-11. The penetration depth in clay is usually taken as approximately $D=(0.2)H$ or about 3 ft, whichever is greater. Bracing may be needed. Clay seals the toe of sheet piling, preventing water influx into the trench.

Because the sheet piling ordinarily is not watertight, the seepage of water through it into the excavation is coped with by introducing ashes or sawdust or fine sand in the seeping water. These materials, carried by the seeping water through the interlocked joints of the sheet piling, seal the joints. In the case of timber piles, they swell when wet. Dry running sand is kept out of the excavation by caulking the joints of the sheet piling.

Sometimes it is impractical to support vertical excavation walls by means of interior shoring and bracing. Instead, exterior anchorage of the sheet

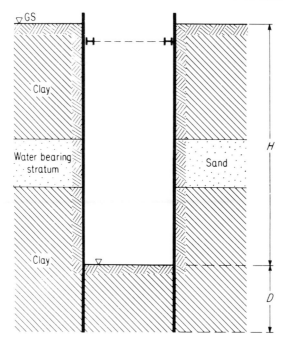

FIG. 5-11. Sheet piling driven into clay.

piling walls *AB* is used. The exterior anchorage is a tie rod *T-T* (or *t-T*) (see Fig. 5-12) tied to a dead man (a massive concrete block) or to an inclined steel sheet-piling anchorage placed outside the rupture surface *AC* of the active earth pressure soil rupture wedge *ABC*. Here angle ρ is the angle

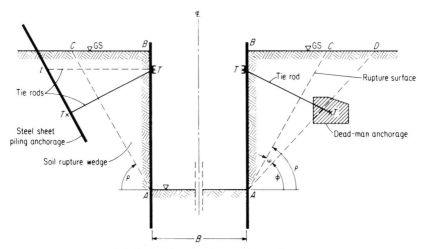

FIG. 5-12. Exterior sheet piling anchorage.

The Excavation

System Nos. and Signs of Angles	Active Earth Pressure System	Tangents of Rupture Angles, $\tan \omega$
1	2	3
① $+\alpha$ $+\delta$ $(+)\phi$ $(+)\phi_1$		$\tan \omega = \tan(\rho - \phi) = \dfrac{-\tan(\phi-\delta) + \sqrt{\tan(\phi-\delta)\,[\tan(\phi-\delta) + \cot(\phi-\alpha)]\,[1 + \tan(\phi_1+\alpha)\cdot\cot(\phi-\alpha)]}}{1 + \tan(\phi_1+\alpha)\,[\tan(\phi-\delta) + \cot(\phi-\alpha)]}$
② $+\alpha$ $\delta = 0$ $(+)\phi$ $(+)\phi_1$		$\tan \omega = \tan(\rho - \phi) = \dfrac{-\tan\phi + \sqrt{\tan\phi\,[\tan\phi + \cot(\phi-\alpha)]\,[1 + \tan(\phi_1+\alpha)\cdot\cot(\phi-\alpha)]}}{1 + \tan(\phi_1+\alpha)\,[\tan\phi + \cot(\phi-\alpha)]}$
③ $+\alpha$ $-\delta$ $(+)\phi$ $(+)\phi_1$		$\tan \omega = \tan(\rho - \phi) = \dfrac{-\tan(\phi+\delta) + \sqrt{\tan(\phi+\delta)\,[\tan(\phi+\delta) + \cot(\phi-\alpha)]\,[1 + \tan(\phi_1+\alpha)\cdot\cot(\phi-\alpha)]}}{1 + \tan(\phi_1+\alpha)\,[\tan(\phi+\delta) + \cot(\phi-\alpha)]}$
④ $\alpha = 0$ $+\delta$ $(+)\phi$ $(+)\phi_1$		$\tan \omega = \tan(\rho - \phi) = \dfrac{-\tan(\phi-\delta) + \sqrt{\tan(\phi-\delta)\,[\tan(\phi-\delta) + \cot\phi]\,[1 + \tan\phi_1\cdot\cot\phi]}}{1 + \tan\phi_1\,[\tan(\phi-\delta) + \cot\phi]}$

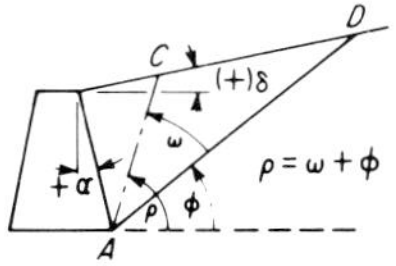

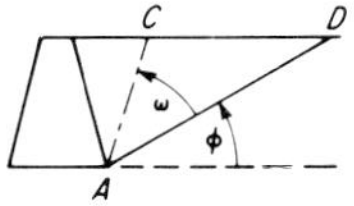

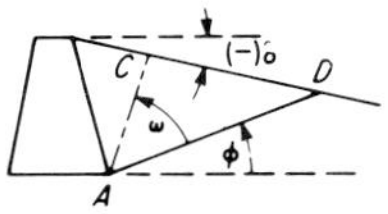

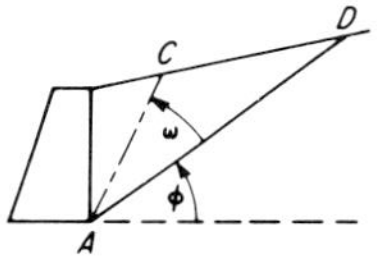

FIG. 5-13a. $\tan \omega$ equations (Ref. 4).

Case	Diagram	Equation
⑤ $\alpha=0$ $\delta=0$ $(+)\phi$ $(+)\phi_i$		$\tan\omega = \tan(\rho-\phi) = \dfrac{-\tan\phi + \sqrt{\tan\phi\,[\tan\phi + \cot\phi]\,[1+\tan\phi_i\cdot\cot\phi]}}{1+\tan\phi_i\,[\tan\phi + \cot\phi]}$
⑥ $\alpha=0$ $-\delta$ $(+)\phi$ $(+)\phi_i$		$\tan\omega = \tan(\rho-\phi) = \dfrac{-\tan(\phi+\delta) + \sqrt{\tan(\phi+\delta)\,[\tan(\phi+\delta)+\cot\phi]\,[1+\tan\phi_i\cdot\cot\phi]}}{1+\tan\phi_i\,[\tan(\phi+\delta)+\cot\phi]}$
⑦ $-\alpha$ $+\delta$ $(+)\phi$ $(+)\phi_i$		$\tan\omega = \tan(\rho-\phi) = \dfrac{-\tan(\phi-\delta) + \sqrt{\tan(\phi-\delta)\,[\tan(\phi-\delta)+\cot(\phi+\alpha)]\,[1+\tan(\phi_i-\alpha)\cdot\cot(\phi+\alpha)]}}{1+\tan(\phi_i-\alpha)\,[\tan(\phi-\delta)+\cot(\phi+\alpha)]}$
⑧ $-\alpha$ $\delta=0$ $(+)\phi$ $(+)\phi_i$		$\tan\omega = \tan(\rho-\phi) = \dfrac{-\tan\phi + \sqrt{\tan\phi\,[\tan\phi + \cot(\phi+\alpha)]\,[1+\tan(\phi_i-\alpha)\cdot\cot(\phi+\alpha)]}}{1+\tan(\phi_i-\alpha)\,[\tan\phi + \cot(\phi+\alpha)]}$
⑨ $-\alpha$ $-\delta$ $(+)\phi$ $(+)\phi_i$		$\tan\omega = \tan(\rho-\phi) = \dfrac{-\tan(\phi+\delta) + \sqrt{\tan(\phi+\delta)\,[\tan(\phi+\delta)+\cot(\phi+\alpha)]\,[1+\tan(\phi_i-\alpha)\cdot\cot(\phi+\alpha)]}}{1+\tan(\phi_i-\alpha)\,[\tan(\phi+\delta)+\cot(\phi+\alpha)]}$

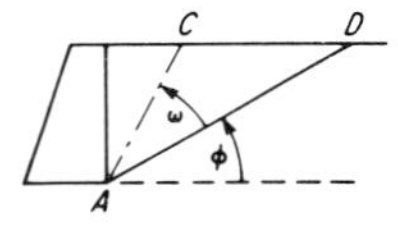

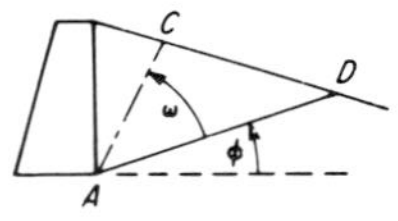

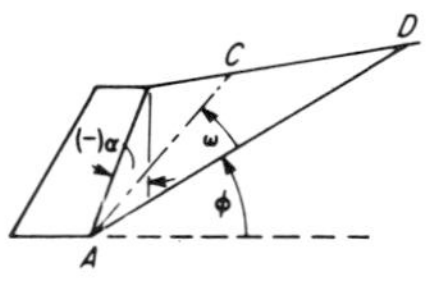

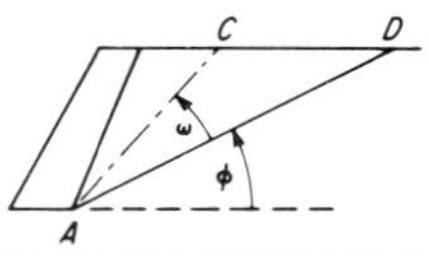

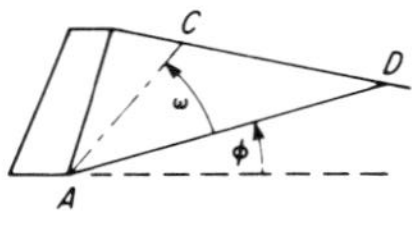

FIG. 5-13a *(continued)*

The Excavation

System Nos. and Signs of Angles	Passive Earth Pressure System	Tangents of Rupture Angles, $\tan \Omega$
1	2	3
① $+\alpha$, $+\delta$, $(-)\phi$, $(-)\phi_1$	(diagram with $+\delta$, $+\alpha$, Ω, ϕ, $\rho = \Omega - \phi$)	$\tan \Omega = \dfrac{\tan(\phi+\delta) + \sqrt{\tan(\phi+\delta)[\tan(\phi+\delta)+\cot(\phi+\alpha)][1+\tan(\phi_1-\alpha)\cdot\cot(\phi+\alpha)]}}{1+\tan(\phi_1-\alpha)[\tan(\phi+\delta)+\cot(\phi+\alpha)]}$
② $+\alpha$, $+\delta=0$, $(-)\phi$, $(-)\phi_1$	(diagram)	$\tan \Omega = \dfrac{\tan\phi + \sqrt{\tan\phi\,[\tan\phi+\cot(\phi+\alpha)][1+\tan(\phi_1-\alpha)\cdot\cot(\phi+\alpha)]}}{1+\tan(\phi_1-\alpha)[\tan\phi+\cot(\phi+\alpha)]}$
③ $+\alpha$, $-\delta$, $(-)\phi$, $(-)\phi_1$	(diagram with $-\delta$)	$\tan \Omega = \dfrac{\tan(\phi-\delta) + \sqrt{\tan(\phi-\delta)[\tan(\phi-\delta)+\cot(\phi+\alpha)][1+\tan(\phi_1-\alpha)\cdot\cot(\phi+\alpha)]}}{1+\tan(\phi_1-\alpha)[\tan(\phi-\delta)+\cot(\phi+\alpha)]}$
④ $+\alpha=0$, $+\delta$, $(-)\phi$, $(-)\phi_1$	(diagram)	$\tan \Omega = \dfrac{\tan(\phi+\delta) + \sqrt{\tan(\phi+\delta)[\tan(\phi+\delta)+\cot\phi][1+\tan\phi_1\cdot\cot\phi]}}{1+\tan\phi_1\,[\tan(\phi+\delta)+\cot\phi]}$

FIG. 5-13b. Tan Ω equations (Ref. 4).

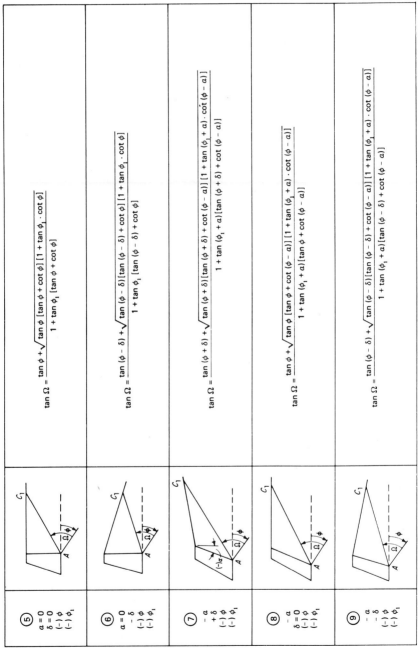

FIG. 5-13b (continued)

of rupture. For equations, graphical construction and numerical values of the rupture surface angles $\rho = \phi + \omega$ for various angles of internal friction of soil ϕ refer to Refs. 1, 2, and 4. Here ϕ is the angle of internal friction of soil, and ω is the analytical angle for active earth pressure. The equations for calculating angles of soil rupture ρ, viz., ω and Ω for active and passive earth pressures, respectively, are given in Figs. 5-13a and 5-13b.

5-5. LATERAL EARTH PRESSURE ON SHEETING

The problem of stability of excavation walls and in construction operations requires a careful investigation of the forces involved in the earth retaining structures. This is also required by the insurance companies in order to keep injuries and loss of life to the construction force, loss of equipment and material to a minimum.

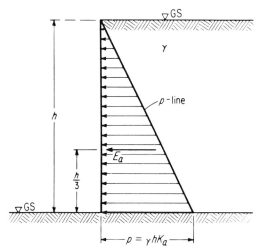

FIG. 5-14. Coulomb's linear pressure distribution diagram.

For rigid, nonyielding retaining structures, the lateral earth pressure E_a on sheeting, bracing and strutting systems may be ascertained, among other methods, by means of Coulomb's classical earth-pressure theory (Fig. 5-14): $E_a = (1/2)\gamma h^2 K_a$. However, because the sheeting and bracing system in reality is somewhat flexible, there takes place a pressure redistribution behind the wall. Hence the earth pressure distribution in reality deviates from Coulomb's straight-line pressure distribution.

Figure 5-14 shows comparative curvilinear earth-pressure distribution diagrams plotted from measurements in situ and the straight-line earth pressure diagram as obtained from analysis by means of Coulomb's active earth pressure theory. These curvilinear diagrams, due to Klenner,[5] result from measurements made during the construction of the subway system (U-Bahn) in Berlin, and from measurements performed with earthwork sheetings in Munich. This comparison gave Klenner reason to put forward a rectangular, lateral earth pressure diagram for flexible walls as shown in Fig. 5-15. The curvilinear lateral earth-pressure diagrams in this figure, when compared with Coulomb's ideal, triangular active earth-pressure diagram, reveal that the upper part of the measured actual pressure is usually ignored in calculations, and the lower part of the ideal, triangular lateral earth-pressure diagram is needlessly included in computations.

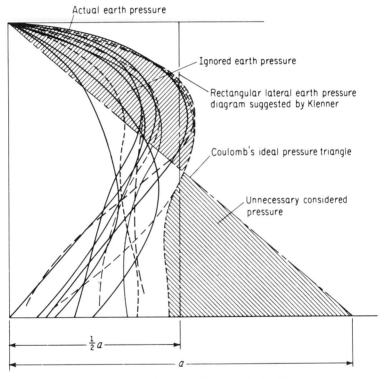

FIG. 5-15. Comparative earth pressure diagrams obtained from measurements (curvilinear) and from calculations (linear) (after Klenner, Ref. 5).

For open excavations in dense and medium dense sand, Terzaghi and Peck[6] suggested using their modified active earth pressure diagram as shown in Fig. 5-16. For loose sand, these authors recommended use of the active earth pressure diagram as shown in Fig. 5-17. For soft and medium clay, the Terzaghi and Peck diagram appears as shown in Fig. 5-18. The pressure

The Excavation

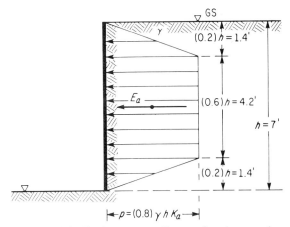

FIG. 5-16. Earth pressure diagram for dense and medium dense sand (after K. Terzaghi and R. B. Peck, Ref. 6).

diagram is based on results of measurements of lateral earth pressure against bracing of an open cut in medium clay of Chicago.

The pressure ordinate p of the trapezoidal pressure distribution diagram contains the quantity $c = \text{cohesion} = \tau = q_u/2$, or

$$p = \gamma h - 2q_u \tag{5-3}$$

where q_u = ultimate, unconfined, compressive strength of the clay.

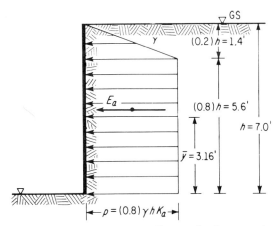

FIG. 5-17. Earth pressure diagram for loose sand (Ref. 6).

No plausible or even satisfactory theory has yet been put forward for computing earth pressures against strutting and bracing of a cut in soil.

Suggestions for such calculations vary from author to author, and thereby so frequently and in such a great amount that it is impossible to include all of them in this book. For "apparent" pressure diagrams for design of struts and braces in cuts excavated in sand and clay see Ref. 7, Art. 18, pp. 394-413.

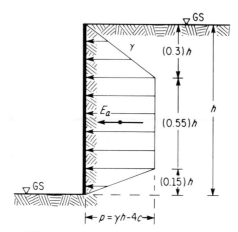

FIG. 5-18. Earth pressure diagram for soft and medium soft clay (Ref. 6).

5-6. SPACING OF BRACES FOR EQUAL AXIAL COMPRESSION

If the sheeting does not deflect and if Coulomb's law of triangular (linear) earth-pressure distribution is applicable, then Coulomb's pressure distribution diagram may be used for spacing horizontal braces in a trench so that each brace is subjected to the same compressive force. This is best performed graphically, as shown in Fig. 5-19, by the principle of the geometric division of a triangle into n equal parts. The graphical construction is shown for three rows of braces. The position of the centroids C_1, C_2, and C_3 determines the position, viz., spacing, of the three braces. The magnitude of the compressive axial force $P_1 = P_2 = P_3$ is any one of the three physical (and equal) areas in Coulomb's diagram, where $A_1 = A_2 = A_3$.

For constructive purposes, a fourth brace at the ground surface level may be installed.

5-7. STATICS OF STRUTTING

With inclined strutting of excavation walls the struts should be checked for buckling, and the whole wall and strutting system should be analyzed for their overall stability. The resultant earth pressure E_a should form

The Excavation

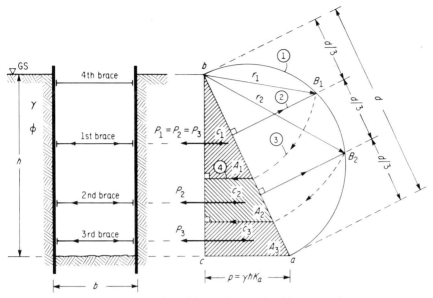

FIG. 5-19. Spacing of braces for equal axial compression.

with the forces in the struts a closed force polygon, viz., triangle, which should not contain any tensile force (see Fig. 5-20). If E_a does not pass through point J (*rule of statics*: in the case of equilibrium three forces meet in a common point), then the strutting system is not in equilibrium.

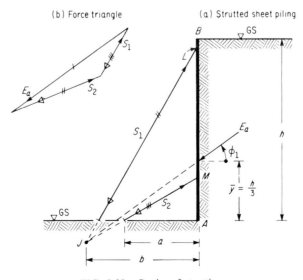

FIG. 5-20. Statics of strutting.

Depending upon the angle of wall friction ϕ_1 and the height of application of E_a above the bottom of the excavation, the struts may or may not fulfill their purpose. With too steep an arrangement of struts there form static moments which may cause the wall to pull upward and thus the studs or soldier beams or the sheeting may be forced out, or else, with a small angle of wall friction, the struts do not fulfill their purpose and the strutting may collapse.

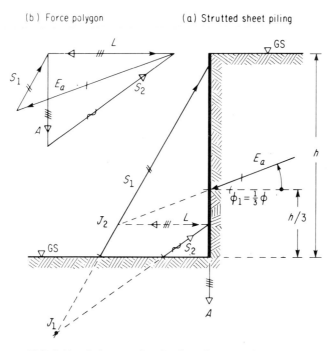

FIG. 5-21. A downward acting force in a strutting system.

In Fig. 5-21 a downward acting force is needed. This tensile force at A can be sustained only by a driven-in soldier beam holding the wall of the horizontal sheeting.

Particular attention must be given to these matters of strutting of excavation walls in cohesive soils, because the cohesion of the soil reduces by precipitation, and so do the angle of internal friction ϕ and that of the wall friction ϕ_1.

The Excavation

REFERENCES

1. A. R. Jumikis, *Soil Mechanics*, Princeton, N.J.: Van Nostrand, 1962, p. 681.
2. A. R. Jumikis, *Mechanics of Soils: Fundamentals for Advanced Study*, Princeton, N.J.: Van Nostrand, 1964, p. 32.
3. O. K. Fröhlich, *Druckverteilung im Baugrunde*. Vienna: Springer, 1934, p. 161.
4. A. R. Jumikis, *Active and Passive Earth Pressure Coefficient Tables*. New Brunswick, N.J.: College of Engineering, Bureau of Engineering Research, Rutgers University, Engineering Research Publication No. 43, 1962.
5. C. Klenner, "Versuche über die Verteilung des Erddruckes über die Wände ausgesteifter Baugruben," *Die Bautechnik*, No. 29 (July 4, 1941), pp. 316-319.
6. K. Terzaghi and R. B. Peck, *Soil Mechanics in Engineering Practice*. 1st ed., New York: Wiley, 1948, p. 348.
7. K. Terzaghi and R. B. Peck, *Soil Mechanics in Engineering Practice*. 2nd ed., New York: Wiley, 1967, pp. 394-413.

QUESTIONS

5-1. What is a foundation pit (excavation)?

5-2. Why is there a need for a dry excavation for laying foundations?

5-3. How does one cope with the influx of water in an excavation?

5-4. How do soil physical properties affect excavation work?

5-5. What is understood by "maintenance of excavation slopes"?

PROBLEMS

5-1. Given a braced trench as shown in Fig. 5-19. There is no surcharge. (a) Determine the earth pressures each lateral strut receives. (b) Space three struts in a vertical plane so that each strut receives equal pressure if the sheeting may be considered as being rigid. *Hint*: use graphoanalytical method for subdivision of pressure triangle into n equal parts (n = number of rows of struts). $\gamma = 100$ lb/ft^3; $\phi = 25°$; $\phi_1 = 0$; $c = 0$; $h = 12'$.

5-2. Construct lateral earth pressure distribution diagram for a layered system. Refer to the accompanying figure.

Active earth-pressure ordinates:

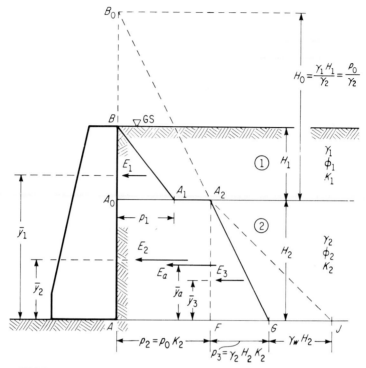

PROB. 5-2. If layer 2 is in water then $\gamma_2 = \gamma_{\text{sub}}$, and water pressure ΔGA_2J must be added to obtain the total lateral pressure diagram.

p_1-ordinate:

$$p_1 = \overline{A_o A_1} = \gamma_1 H_1 K_1 = \gamma_1 H_1 \tan^2\left(45° - \frac{\phi_1}{2}\right) \quad (1)$$

p_2-ordinate:

Consider thickness H_1 of layer 1 as a surcharge over layer 2. The intensity p_0 of the vertical surcharge pressure is

$$p_0 = \gamma_1 H_1 \quad (2)$$

This surcharge p_0 may be converted into an equivalent height H_0 of soil 2 whose unit weight is γ_2, and angle of internal friction is ϕ_2. Then the equivalent surcharge height H_0 over layer 2 is

$$H_0 = \overline{A_o B_o} = p_0/\gamma_2 = \frac{\gamma_1 H_1}{\gamma_2} \quad (3)$$

and the lateral earth-pressure ordinate $p_2 = \overline{A_o A_2} = p_0 K_2$ due to the vertical surcharge H_0 is calculated as

$$p_2 = p_0 K_2 = \gamma_2 H_0 K_2 = \gamma_2 \frac{\gamma_1 H_1}{\gamma_2} K_2 = \gamma_1 H_1 K_2$$

$$= \gamma_1 H_1 \tan^2\left(45^\circ - \frac{\phi_2}{2}\right) \quad (4)$$

p_3-ordinate:

The pressure diagram, $\Delta(FA_2G)$, represents physically the total active earth pressure E_3 brought about by soil layer 2, on the lower part $AA_0 = H_2$ of the retaining wall. The active earth pressure ordinate $p_3 = FG$ is

$$p_3 = \gamma_2 H_2 K_2 \quad (5)$$

Discussion

In this problem, the wall friction is assumed equal to zero.
(1) When $\phi_2 = \phi_1$, then $K_2 = K_1$ and $p_2 = p_1$
(2) When $\phi_2 < \phi_1$, then $K_2 > K_1$ and $p_2 > p_1$
(3) When $\phi_2 > \phi_1$, then $K_2 < K_1$ and $p_2 < p_1$

The total active earth pressure E_a on the wall is the physical area of the total lateral pressure-distribution diagram:

$$E_a = E_1 + E_2 + E_3$$

The resultant active earth pressure acts distance $\bar{y}_a$ from the base of the wall:

$$\bar{y}_a = \frac{E_1 \bar{y}_1 + E_2 \bar{y}_2 + E_3 \bar{y}_3}{E_a}$$

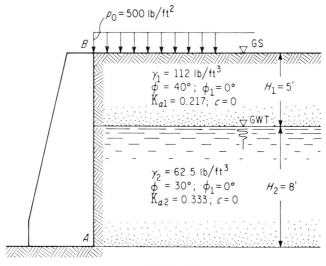

PROB. 5-3

5-3. Construct active earth pressure distribution diagram for the earth pressure-retaining wall system as shown in the figure. Also, add lateral water pressure diagram where it applies.

5-4. Given a soil-earth retaining wall system as shown in Prob. 5-3.

$$\text{Layer 1: } \gamma_1 = 112 \text{ lb/ft}^3; \ \phi = 30°; \ K_{a1} = 0.333$$
$$\text{Layer 2: } \gamma_2 = 62.5 \text{ lb/ft}^3; \ \phi = 40°; \ K_{a2} = 0.217$$

The wall friction is zero. Construct the effective and total active earth pressure distribution diagrams. Calculate the magnitude of the total active earth pressure E_a and its position $\bar{y}_a$.

5-5. Refer to Fig. 5-20. Strut the vertical wall of an excavation with two inclined struts S_1 and S_2 as indicated on Fig. 5-20. Determine the loads on each strut and the position a of the lower strut S_2. What is the best inclination of strut S_2?

Given: $h = 15$ ft; $BL = 3$ ft; $LM = 9$ ft; $AM = 3$ ft; $b = 12$ ft; $\phi = 30°$; $\phi_1 = 20°$; $\gamma = 110$ lb/ft^3

5-6. Strut the sheet piling as shown in Fig. 5-21, determine the position and inclination of struts and the magnitude of forces S_1, S_2 and A. What are the best inclinations of struts?

Given: $h = 12$ ft; $\phi = 30°$; $\phi_1 = 10°$; $\gamma = 110$ lb/ft^3

5-7. Given an anchored sheeting as shown. Determine the position of the strut, the magnitude and inclination of force S_1, and the magnitude of the anchor pull A.

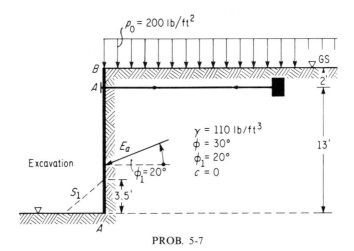

PROB. 5-7

5-8. Solve Probs. 5-5 through 5-7 by Klenner's method.

5-9. Solve Probs. 5-5 through 5-7 by methods suggested by Terzaghi and Peck.

chapter 6

Sheet Piling

6-1. INTRODUCTION

Sheet piling, or a bulkhead, is a special type of earth-retaining structure. Sheet piling consists of special shapes of interlocking piles (Fig. 6-1). Sheet piles may be fabricated from timber, steel, or reinforced concrete. They are driven, in close contact with each other, into the ground to form a continuous wall or sheet.

Sheet piling is used in foundation engineering, for waterfront structures (for example, in building wharves, quays, and piers), canal locks, dams, cofferdams, for river-bank protection, and as earth-retaining walls. Sheet piling is required (a) to resist lateral pressure caused by combinations of earth, water, and applied horizontal and vertical loads, and/or (b) to prevent or reduce leakage of water and of saturated, soft, plastic soil or dry sand into an excavation.

Sheet piling must therefore, be continuous and reasonably tight (achieved by interlocks). Furthermore, it must be driven to a satisfactory depth into the soil, and it must possess sufficient strength to resist bending.

Figure 6-2 shows a stage in the construction of a large cofferdam at the new $575 million World Trade Center in New York City. Here three templates 64 feet in diameter are positioned for removal after serving to guide the driving of 160 lengths of interlocking steel sheet piling contained in each cell of the cofferdam wall. Earlier, barge-mounted pile drivers could be seen installing up to three cells simultaneously, driving piles through hardpan to bedrock under 20 to 40 feet of water. The completed cofferdam was built to provide a receptacle for the excavation of 1,180,000 cubic yards of landfill from the site. Approximately 4,120 tons of flat sheet and Z-shaped piling, about half of the piling used in the project, were supplied by the Bethlehem Steel Corporation. The 23.5 acres of new land created by the landfill project will be turned over to New York City by The Port of New York Authority, builder of the twin-tower World Trade Center.

BETHLEHEM STEEL SHEET PILING

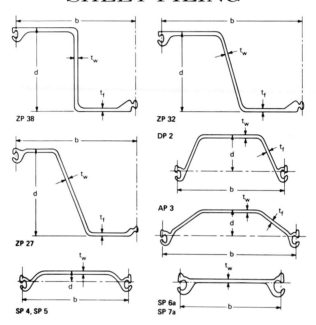

Dimensions and Properties

Section Number	Nominal Width b	Web Thickness t_w	Flange Thickness t_f	Nominal Depth d	Single Section				Per Foot of Wall	
					Weight per Foot	Area	I	Z	Weight per sq ft	Z
	in.	in.	in.	in.	lb	in.²	in.⁴	in.³	lb	in.³
ZP 38	18	3/8	1/2	12	57.0	16.77	421.2	70.2	38.0	46.8
ZP 32	21	3/8	1/2	11 1/2	56.0	16.47	385.7	67.0	32.0	38.3
ZP 27	18	3/8	3/8	12	40.5	11.91	276.3	45.3	27.0	30.2
DP 2	16	3/8	3/8	5	36.0	10.59	53.0	14.3	27.0	10.7
AP 3	19 5/8	3/8	3/8	3 1/2	36.0	10.59	26.0	8.8	22.0	5.4
SP 4	16	3/8	—	1 11/32	30.7	8.99	5.5	3.2	23.0	2.4
SP 5	16	1/2	—	1 11/32	37.3	10.98	6.0	3.3	28.0	2.5
SP 6a	15	3/8	—	—	35.0	10.29	4.6	3.0	28.0	2.4
SP 7a	15	1/2	—	—	40.0	11.76	4.6	3.0	32.0	2.4

FIG. 6-1. Bethlehem Steel sheet piling from Catalog 2331, 1967 ed. Courtesy and permission of the Bethlehem Steel Sorporation, Bethlehem, Pa.

Sheet Piling   131

FIG. 6-2. Sheet piling. Courtesy Bethlehem Steel Corporation, New York.

6-2. CONCERNING ANALYSES OF SHEET PILING

For static analysis of sheet piling, it is necessary to determine the horizontal soil reactions on the sheet piling. These in turn depend upon the lateral pressure distribution in the soil.

Depending upon the character of the soil into which sheet piling is driven, one distinguishes between (a) sheet piling in noncohesive soil, and (b) sheet piling in cohesive soil.

Cohesion of soil is a very sensitive physical property, especially relative to water. Because soil in the field is very difficult to protect against climatic variations which occur during the various seasons of the year, particularly in respect to changes in moisture content in soil, cohesion is very seldom considered in sheet-piling calculations.

Because of these and other reasons, most of the methods for calculating sheet piling are developed for noncohesive soils, and simplifications in theories usually consist in assuming that soil lateral stress distribution from active and passive earth pressures is probably linear.

In general, the calculation and proportioning of a sheet-piling wall is performed by satisfying the requirements of statics and strength of materials. In stability calculations of sheet-piling systems the self-weight of the sheet piling is usually neglected because of its insignificance in comparison with the magnitudes of the lateral loads acting on the sheet piling.

The problem of sheet-piling wall design requires
1. The necessary depth of driving or penetration of sheet piling to attain its stability with a satisfactory factor of safety.
2. Suitable size or cross section or profile of the sheet piling to resist bending stresses induced by lateral loads (earth pressure and water pressure, for example), as well as from surcharge loads on the ground surface. The cross section of sheet piling, of course, depends also upon the driving conditions in soil, which upon hard driving may require a heavier profile than the calculated one.
3. Correct size of the anchor (if any); hence the term anchored sheet piling or anchored bulkhead.

Based on its structural type and loading scheme, sheet-piling systems may be classed as (1) cantilever sheet piling, and (2) anchored sheet piling

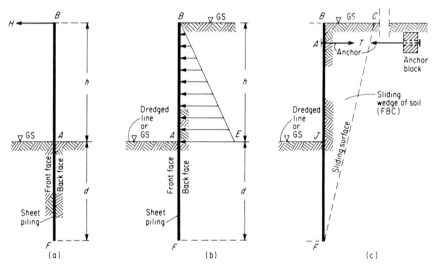

FIG. 6-3. Sheet-piling systems. (a) Free cantilever sheet piling loaded with concentrated horizontal load H, (b) Free cantilever sheet piling loaded with active earth preessure. (c) Anchored sheet piling.

Sheet Piling

(Fig. 6-3). Cantilever sheet piling, in turn, can be classed as (a) a free-cantilever sheet piling (Fig. 6-3a), and (b) a free-cantilever sheet piling loaded laterally with active earth pressure from the backfill material. Note that the ground surface on each side of the sheet piling is at different elevation (Fig. 6-3b).

Sheet piling is considered free if it derives its stability entirely from the lateral resistance of the soil into which the sheet piling is driven. Free-cantilever sheet piling is considered the simplest type of sheet piling.

An anchored bulkhead is one which is held above its driven depth by one or more tie rods or anchors (Fig. 6-3c).

The calculation and sizing of a sheet piling is performed by satisfying the requirements of statics and strength of materials. A method of calculation, in order to be successful, must be simple. The results of a survey through the technical literature reveal that several theories have been put forward for sheet-piling calculations, and that there is no unified opinion available as yet as to the best method of performing these calculations to arrive at a safe design. Cumbersome theories and complex equations are usually not appreciated by the practicing engineer, hence such difficult tools are seldom used. The method of calculation must be clear and easy to perform.

In practice there usually exist certain conventions relative to assumptions underlying a given theory. These conventions generally are based on, and are more or less justified by, the philosophy that the end results of a theory are no better or more precise than the data available on the soil physical properties in front of, behind, and along the length of the wall of the sheet piling. The same pertains to the mode of establishing the earth-retaining system: whether it is to be backfilled behind the driven wall or the soil dredged in front of the sheet-piling wall. The method of backfilling (by rolling, vibration, tamping, or by backfilling hydraulically) would also influence the pressure distribution (and redistribution) in soil and upon the wall. The temporarily created hydrostatic condition should not be forgotten in the stability analysis of sheet-piling system. Also, rapid drawdown of water from behind the wall may cause statically unfavorable or even undesirable conditions in the earth-retaining system. If groundwater is present in the soil behind the wall or if there is water in front of the wall, then the earth-retaining system must be treated accordingly.

In stability calculations, the weight of the sheet piling is usually neglected because of its relative insignificance in comparison with horizontal loads. Sometimes sheet piling also has to carry vertical loads.

6-3. ANCHORED SHEET PILING

An anchored sheet piling is one which is held above its driven depth by one or more rows of tie rods or anchors (Fig. 6-3). The anchors are attached or anchored to and held by anchor plates, or anchor walls, or anchor piles, or deadmen, commonly known as *anchorage*. Such anchorage should be constructed behind the sliding surface of the soil rupture wedge. By means of anchoring, the driving depth, or the total length of the sheet piling (and thus its weight, which costs money), are reduced considerably.

The stability of anchored sheet piling depends more upon the anchor than on deep embedment of the sheet piling. Thus anchored sheet piling permits

1. Smaller driving depth than that computed in the analysis of free cantilever sheet piling.
2. Greater earth-retaining heights for a given section modulus of sheet piling.

In other words, under comparable earth-retaining heights, the use of an anchor reduces the length, cross section, and thus weight of the sheet piling.

For a systematic treatment of rigid and flexible free-cantilever sheet piling loaded with a horizontal concentrated load and active earth pressure under assumption of straight line and parabolic earth-pressure distribution on sheet piling, see Ref. 1.

6-4. CALCULATIONS OF FREE-CANTILEVER SHEET PILING (CONVENTIONAL METHOD)

A loading system of a free-cantilever sheet-piling system[1,2] subjected to a single, concentrated, horizontal load H is shown in Fig. 6-4. The necessary equations for calculating the ordinates of the resulting lateral earth pressure-distribution diagram for this system are as set forth.

Minimum Driving Depth d:

$$d^4 - \frac{8H}{\gamma(K_p - K_a)b}d^2 - \frac{12Hh}{\gamma(K_p - K_a)b}d - \left[\frac{2H}{\gamma(K_p - K_a)b}\right]^2 = 0 \quad (6\text{-}1)$$

where H = horizontal load
h = length of free cantilever
γ = unit weight of soil
K_p = passive earth-pressure coefficient

Sheet Piling

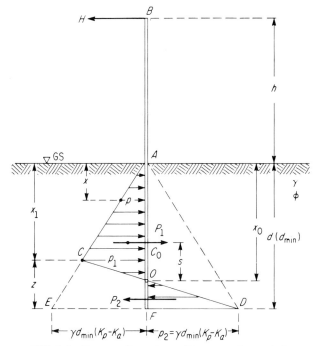

FIG. 6-4. Free-cantilever sheet-piling system for analysis.

K_a = active earth-pressure coefficient
b = width of sheet piling

Position of Pivot Point O:

$$x_o = \frac{2d^5 + 3hd^4 - 8ad^3 - 12ahd^2 + 8a^2d + 12a^2h}{3d^4 + 6hd^3 - 12ad^2 - 36ahd - 18ah^2 - 4a^2} \tag{6-2}$$

where

$$a = \frac{H}{Bb} \tag{6-3}$$

$$B = \gamma (K_p - K_a) \tag{6-4}$$

and

b = width of sheet piling along the run of the wall

Position x_1 of Ordinate p_1:

$$x_1 = \frac{2H(2d + 3h)}{\gamma b (K_p - K_a) d^2 - 2H} \tag{6-5}$$

Position of z:

$$z = \frac{\gamma b (K_p - K_a) d^2 - 2H}{2\gamma bd (K_p - K_a)} \qquad (6\text{-}6)$$

Lateral Pressure Ordinate p_1:

$$p_1 = \gamma x_1 (K_p - K_a) = Bx_1 \qquad (6\text{-}7)$$

Lateral Pressure Ordinate p_2:

$$p_2 = \frac{Bx_1 (d - x_o)}{x_o - x_1} \qquad (6\text{-}8)$$

At minimum driving depth the pressure ordinate p_2 is

$$p_2 = \gamma d_{min}(K_p - K_a) \qquad (6\text{-}9)$$

For detailed derivation of these equations consult Ref. 1.

To allow for full development of the passive earth pressure below the pivot point O on the right-hand side of the sheet piling as shown in Fig. 6-4, engineers increase the calculated theoretical minimum driving depth by approximately 20 percent.

Knowing x_o, x_1 (or z), and d, the resultant lateral earth-pressure diagram can be constructed, and the maximum bending moment M_{max} in the sheet piling calculated. The sizing of the necessary profile of the sheet piling is then made according to the allowable bending stress σ_{all} of the sheet-pile material:

$$Z = \frac{M_{max}}{\sigma_{all}} \qquad (\text{in.}^3 \text{ or cm}^3) \qquad (6\text{-}10)$$

where $Z = I/c$ is section modulus, in.^{3} or cm^3

I = moment of inertia, in.^{4} or cm,^{4} of cross section of sheet-piling profile

c = distance from the neutral axis of section to farthest-spaced fiber of cross section

However, sometimes deflection governs sizing of the sheet piling.

Relative to the minimum driving depth, the foregoing calculations pertain to a factor of safety of $\eta = 1.0$. For an n-fold factor of safety against the displacement of sheet-pile wall, stability analyses are performed with a force nH instead of H, where $n > 1$.

6-5. CALCULATIONS OF FREE-CANTILEVER SHEET PILING (H. BLUM METHOD)

Blum[3,4] presupposes a homogeneous soil at the fixed end of the sheet piling, and that part of the loading areas on the lateral soil pressure-distribution diagram due to passive resistance of the soil can be idealized as a right-angle triangle *AOE* down to the theoretical driving depth *d*, where the triangle terminates as shown in Fig. 6-5. A further simplification in Blum's theory is that the distributed load P_2 (acting to the left as shown in Fig. 6-4) below the pivot point *O* is replaced by a single, concentrated, equivalent reactive force *C*. This force *C* acts at point *O*. In calculations, the

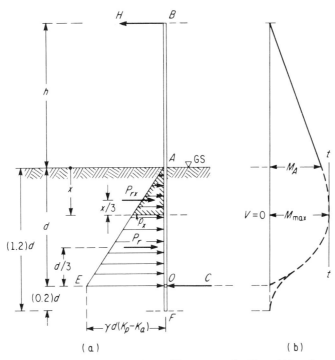

FIG. 6-5. Free-cantilever sheet-piling system, after Blum. (a) Loading diagram. (b) Bending-moment diagram.

magnitude of *C* is of no significance. However, because in reality this force *C* is the resultant of the acting resultant earth resistance below point *O*, and because the real force P_2 distributes over the lower part of the sheet-piling area below point *O*, and in order to allow to develop the resultant passive earth resistance fully, the sheet piling must be driven deeper than the calculated theoretical depth *d* by about 20 percent.

Referring to Fig. 6-5, the bending moment M_x at depth *x* below the ground surface is per 1 ft of run of sheet-piling wall:

$$M_x = H(h+x) - (1/6)\gamma(K_p - K_a)x^3 \tag{6-11}$$

where all symbols are same as before.

The position of the maximum bending moment of the sheet piling is calculated by differentiating M_x with respect to x, and setting this derivative equal to zero, i.e.,

$$\frac{dM_x}{dx} = 0 \qquad (6\text{-}12)$$

Differentiation renders

$$H - (1/2)\,\gamma\,(K_p - K_a)\,x^2 = 0 \qquad (6\text{-}13)$$

or

$$x = \sqrt{\frac{2H}{\gamma\,(K_p - K_a)}} \qquad (6\text{-}14)$$

Because the second derivative, $d^2 M_x/dx^2$, is less than zero, the M_x-function has a maximum value M_{max} at x (Eq. 6-14).

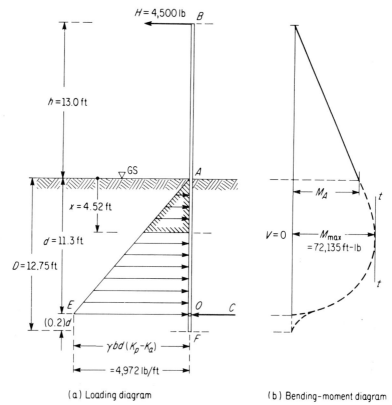

(a) Loading diagram (b) Bending-moment diagram

FIG. 6-6. Example of analysis of free-cantilever sheet-piling system.

Sheet Piling

The sizing of the sheet piling goes through calculation of the necessary section modulus of the sheet-piling profile: $Z = M_{max}/\sigma_{all}$, where σ_{all} = allowable bending stress.

Example

Given a sheet-piling system as shown in Fig. 6-6. The $b = 18$ in. wide sheet piling is loaded with a horizontal load $H = Lb = (3,000)(18/12) = 4,500$ lb; $h = 13$ ft; $\gamma = 110$ lb/ft^3; $\phi = 30°$.

$$K_p = 3.000$$
$$K_a = 0.333$$
$$\overline{K_p - K_a = 2.667}$$

$$\gamma(K_p - K_a) = (110)(2.667) = 293.4 \text{ (lb/ft}^3)$$

$$\gamma b(K_p - K_a) = (293.4)(18/12) = 440.0 \text{ (lb/ft}^2)$$

Solution

Position x of M_{max} (by Eq. 6-14):

$$x = \sqrt{(2H)/\gamma b(K_p - K_a)} = \sqrt{\frac{(2)(4,500)}{440}} = \underline{4.52 \text{ (ft)}}$$

Magnitude of Maximum Moment:

$$M_{max} = H[h + (2/3)\sqrt{(2)(H)/\gamma b(K_p - K_a)}]$$
$$= (4500)[(13) + (0.670)(4.52)] = 72,135 \text{ (ft-lb)} = \underline{865,620 \text{ in.-lb.}}$$

Check of Bending Stresses:

$$\sigma = M_{max}/Z = 865,534/45.3 = \underline{19,100 \text{ lb/in.}^2} < 20,000 \text{ lb/in.}^2$$

For a $b = 18$ in. wide sheet-piling profile ZP 27, $Z = 45.3$ in.^{3} (Bethlehem Steel, Fig. 6-1).

The driving depth d is calculated by taking moments about point O:

$$M_o = H(h + d) - (1/6)\gamma b(K_p - K_a)d^3 = 0 \tag{6-15}$$

resulting in the following cubic equation:

$$d^3 - 61.3d - 735.6 = 0 \tag{6-16}$$

and rendering a minimum theoretical driving depth of $d = 11.3$ ft. Increasing this value by 20 percent, obtain the practical driving depth as $D = (1.20)(11.3) = 12.65$ (ft). Assume 12.75 ft.

6-6. CANTILEVER SHEET PILING LOADED WITH ACTIVE EARTH PRESSURE

A cantilever sheet-piling system loaded with active earth pressure is sketched in Fig. 6-7. Note that the zero loading point O_1 is at distance m below the lower ground surface.

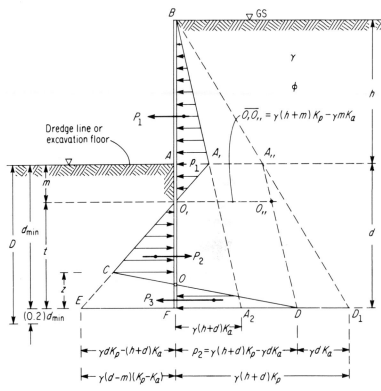

FIG. 6-7. Resultant earth-pressure diagram.

Minimum Theoretical Driving Depth d:

$$d^4 + h\frac{(K_p - 3K_a)}{(K_p - K_a)} d^3 - K_a h^2 \frac{(7K_p - 3K_a)}{(K_p - K_a)^2} d^2$$

$$- K_a h^3 \frac{5(K_p - K_a)}{(K_p - K_a)^2} d - \frac{K_a K_p}{(K_p - K_a)^2} h^4 = 0 \quad (6\text{-}17)$$

Ordinate z:

$$z = \frac{K_p d^2 - K_a(h + d)^2}{(K_p - K_a)(h + 2d)} \quad (6\text{-}18)$$

Sheet Piling

Lateral Pressure Ordinate p_2:

$$p_2 = \gamma(h+d)K_p - \gamma d K_a \tag{6-19}$$

Distance m:

$$m = \frac{p_1}{\gamma(K_p - K_a)} \tag{6-20}$$

Lateral Pressure Ordinate p_1:

$$p_1 = \gamma h K_a \tag{6-21}$$

With these quantities known, the resultant soil lateral pressure-distribution diagram (Fig. 6-7) can be constructed and the sheet-piling profile sized.

6-7. CALCULATION OF CANTILEVER SHEET PILING LOADED WITH ACTIVE EARTH PRESSURE ACCORDING TO BLUM

Blum's idealized sheet-piling system loaded with active earth pressure and fixed in a uniform soil[3,4] is shown in Fig. 6-8. The resultant earth resistance triangular diagram starts at point O_1 and terminates at point O. The equivalent reactive force C is shown to act at the pivot point O. The loading zero point O_1 is at a distance m below the lower ground surface.

The lateral pressure ordinates are

$$p_1 = \gamma h K_a \tag{6-22}$$
$$p_2 = p_1$$
$$p_r = \gamma(K_p - K_a)t \tag{6-23}$$

where $t = \overline{O_1 O}$ = minimum theoretical driving depth of the sheet piling below the zero loading point O_1.

The distance m is calculated as

$$\frac{m}{t} = \frac{p_1}{\gamma(K_p - K_a)t} \tag{6-24}$$

or

$$m = \frac{p_1}{\gamma(K_p - K_a)} \tag{6-24a}$$

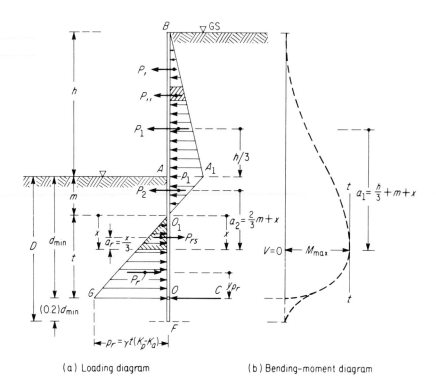

(a) Loading diagram (b) Bending-moment diagram

FIG. 6-8. Blum's free-cantilever sheet-piling system loaded with active earth pressure.

Driving Depth

To calculate the minimum depth t below point O_1, take static moments about point O and set the moment sum, ΣM_o, equal to zero. This results in a cubic equation in t. The solution of the cubic equation renders the quantity t. The minimum theoretical driving depth $d_{\min}$ is

$$d_{\min} = t + m \qquad (6\text{-}25)$$

to which about 20 percent is usually added to obtain the necessary driving depth D:

$$D = (1.20)(d_{\min}) \qquad (6\text{-}26)$$

The load P_r can now be calculated as

$$P_r = (1/2)\gamma(K_p - K_a)t^2 \qquad (6\text{-}27)$$

Sheet Piling

Maximum Bending Moment M_{max} in Sheet Piling

The maximum bending moment in sheet piling occurs where the shear force V is zero:

$$V = \sum_1^n P - (1/2)\gamma(K_p - K_a)x^2 = 0 \qquad (6\text{-}28)$$

$$\therefore x = \sqrt{\frac{2\sum_1^n P}{\gamma(K_p - K_a)}} \qquad (6\text{-}29)$$

where n = number of the active P-loads
$\sum_1^n P$ = the sum of the active P-loads

The maximum bending moment M_{max} is computed relative to the sheet-piling section at x:

$$M_{max} = (P_1 y_1 + P_2 y_2 + \cdots + P_n y_n) - (1/6)\gamma(K_p - K_a)x^3 \qquad (6\text{-}30)$$

where $y_1, y_2, \cdots, y_n$ are moment arms.

The necessary sheet-piling profile can now be determined.

Because the active pressure above the zero loading point O_1 may be variously distributed because of surcharge, groundwater, several layers of soil with different γ, ϕ, and c, the solution of such a sheet-piling system may be facilitated if the soil lateral pressure-distribution diagram is subdivided in several parts according to the soil and loading conditions, and each part of the distributed loads is substituted by corresponding concentrated loads.

Example

Determine the driving depth and the maximum bending moment for the sheet-piling system shown in Fig. 6-9.

Given: $h = 12.0$ ft; $\gamma = 110$ lb/ft^3; $\phi = 30°$
$K_p = 3.000$
$K_a = 0.333$
$K_p - K_a = 2.667$; $\gamma(K_p - K_a) = (110)(2.667) = 293$ (lb/ft^3)

Active Earth Pressure Ordinate p_1:

$$p_1 = \gamma h K_a = (110)(12)(0.333) = \underline{440 \text{ (lb/ft}^2)} \text{ per 1 ft of run of wall}$$

Excavation and Cofferdams

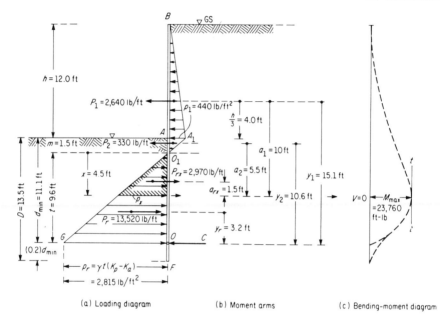

(a) Loading diagram (b) Moment arms (c) Bending-moment diagram

FIG. 6-9. Example of analysis of free-cantilever sheet-piling loaded with active earth pressure.

Passive Earth Resistance Ordinate p_r:

$$p_r = \gamma(K_p - K_a)\, t = (110)(2.667)\, t = (293)t$$

Position m of Zero Loading Point O_1:

$$\frac{p_1}{p_r} = \frac{m}{t}; \quad m = \frac{p_1}{p_r} t = \frac{440}{293} = \underline{1.5 \text{ (ft)}}$$

Total Active Earth Pressure:

$P_1 = (1/2)p_1 h = (0.5)(440)(12) = $ 2,640 (lb per 1 ft of run of wall)
$P_2 = (1/2)p_1 m = (0.5)(440)(1.5) = $ 330 (lb per 1 ft of run of wall)

$$\sum_1^2 P = 2{,}970 \text{ (lb per 1 ft of run of wall)}$$

Resultant Resisting Pressure P_r:

$$P_r = (1/2)\gamma(K_p - K_a)\, t^2 = (0.5)(293)\, t^2 = (146.68)\, t^2$$

Sheet Piling

Moment Arms:

$$y_1 = h/3 + m + t = (12.0/3) + 1.5 + t = 5.5 + t$$
$$y_2 = (2/3)m + t = (2/3)(1.5) + t = 1.0 + t$$
$$y_r = t/3$$
$$y_c = 0$$

Calculation of x:

$$\Sigma M_o = 0$$

$$P_1 y_1 + P_2 y_2 - P_r y_r = 0$$

or

$$(2{,}640)(5.5 + t) + (330)(1.0 + t) - (146.5)t^2(t/3) = 0$$
$$t^3 - 60.6t - 303.6 = 0$$
$$t = 9.6 \text{(ft)}$$

Minimum Theoretical Driving Depth d_{min}:

The theoretical minimum depth t is usually increased by 20 percent:

$$d_{min} = m + t = 1.5 + 9.6 = 11.1 \text{ (ft)} \qquad (6\text{-}25)$$
$$D = (1.20)d_{min} = (1.20)(11.1) = 13.32 \text{ (ft)} \sim \underline{13.5 \text{ ft}}$$

Position x of Zero Shear:

By Eq. 6-28, the position x of zero shear is

$$2{,}970 - (1/2)\gamma(K_p - K_a)x^2 = 0$$

$$x = \sqrt{2{,}970/146.5} = \underline{4.5 \text{ (ft)}} \text{ (below zero loading point } O_1 \text{ or 6 ft below ground surface)}$$

Maximum Bending Moment M_{max}:

Maximum bending moment M_{max} in the sheet piling is computed relative to the point where shear $V = 0$.

Moment Arms:

$$a_1 = (h/3) + m + x = 4.0 + 1.5 + 4.5 = 10 \text{ (ft)}$$
$$a_2 = (2/3)m + x = (1.0) + 4.5 \qquad\qquad = 5.5 \text{ (ft)}$$
$$a_{rx} = (1/3)x = (1/3)(4.5) \qquad\qquad\quad = 1.5 \text{ (ft)}$$

Load P_{rx}:

$$P_{rx} = (1/2)\gamma(K_p - K_a)x^2 = (146.5)x^2 = (146.5)(4.5^2) = 2{,}970 \text{ (lb/ft)}$$

This checks out with $P_1 + P_2 = 2{,}970$ lb/ft.

Maximum Moment:

$$\begin{aligned}M_{max} &= \Sigma M_x = P_1 a_1 + P_2 a_2 - P_{rx} a_{rx} \\ &= (2{,}640)(10) + (330)(5.5) - (2{,}970)(1.5) = \underline{23{,}760 \text{ (ft-lb)}}\end{aligned}$$

The size of the sheet-piling profile can now be determined from manufacturers' sheet-piling catalogs.

Resisting Pressure P_r:

$$P_r = (146.68)\ t^2 = (146.68)(9.6^2) = \underline{13{,}518}\ \text{(lb/ft)} = \underline{13{,}520}\ \text{lb/ft}$$

Moment equation about O gives $P_r = \underline{13{,}515}$ (lb/ft) $\approx \underline{13{,}520}$ (lb/ft)

6-8. ANCHORED SHEET PILING

The cantilever sheet piling, loaded with active earth pressure which is too large, can be anchored by tie rods at, or near, its top to anchor plates, anchor walls, anchor piles or deadmen. Such anchors should be constructed far enough behind the surface of rupture of the sliding soil wedge (see Fig. 6-3c) so that it acts independently of the bulkhead and the soil behind it. By this method of construction, the driving depth, or the total length of the sheet piling and its weight, are reduced considerably.

The stability of this kind of sheet piling depends more upon the anchor than on the inducement of large resultant earth pressures. Thus this method of design permits (1) a smaller driving depth than that computed in the analysis of free cantilever sheet piling, and (2) greater retaining heights for a given section modulus of sheet piling.

In other words, under comparable retaining heights, the use of an anchor reduces the length, cross section, and weight of the sheet piling.

There are two methods customarily used as a basis for the stability analysis of anchored sheet piling, namely, (1) the method of "free earth support," and (2) the method of "fixed earth support" (Fig. 6-10).

6-9. METHOD OF FREE EARTH SUPPORT

By this method, also known as the method of minimum penetration depth, the sheet piling is assumed to be inflexible. No pivot point exists in this statical system below the excavation floor. The earth pressures are computed by the classical earth-pressure theory (Rankine's case).

Sheet Piling

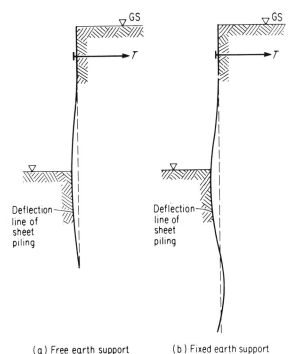

(a) Free earth support   (b) Fixed earth support

FIG. 6-10. Two methods for anchored sheet-piling analysis.

The stability analysis of an anchored sheet piling by the free earth-support method is based on the assumption that the soil into which the sheet piling is driven does not produce effective restraint to the piling, at least not to the degree necessary to induce negative bending moments at its support (point of anchorage and the lower part of the sheet piling in the soil). This type of sheet piling is used where no leakage of soil and/or water is expected. This design gives the smallest driving depth, just long and strong enough to prevent the toe of the sheet piling from yielding and displacing laterally. The passive earth pressure on the front face of the sheet piling resists this lateral displacement of the sheet piling at its toe.

The force system in this type of sheet piling includes (a) the tension T in the tie rods, (b) the active earth pressure P_1 on the backface of the sheet piling, and (c) the full passive earth pressure P_{2f} along the front face of the driven depth of the sheet piling. This pressure, together with the tensile force in the tie rod, counteracts the active pressure P_1 (see Fig. 6-11). This pressure system is here adapted after Krey[5], who, for reasons of simplicity, uses only part of the passive earth pressure diagram, namely part $12FA$.

148 Excavation and Cofferdams

The problem in analyzing the stability of a free earth-supported sheet piling is (a) to determine a minimum driving depth d which would ensure a sufficient lateral support, and (b) to determine the magnitude of the tensile force T in the tie rod.

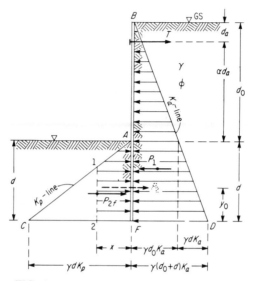

FIG. 6-11. Anchored sheet-piling system with free earth support.

6-10. PRESSURES

Active Earth Pressure

The magnitude of the active earth pressure is

$$P_1 = \frac{b}{2}\gamma(d_o + d)^2 K_a \qquad (6\text{-}31)$$

where b = width of the sheet pile
  d_o = the retained height
The quantity b can also be one unit of length along the wall of the sheet piling.

Passive Earth Pressure

In practice, for stability calculations of such a system of sheet piling as treated in this section, only a fraction of the possible maximum passive earth pressure is taken into account, namely,

$$P_2 = \frac{P_{2f}}{n} = \frac{b\gamma}{2n} d^2 K_p \qquad (6\text{-}32)$$

Sheet Piling

where P_{2f} = full pressure
n = factor of safety

Usually n is assumed to be 2 for temporary structures.

The reduced ordinate x of the passive earth pressure which would give $(1/n)$th of the passive earth pressure P_{2f} is calculated by setting up and solving the following equation (see Fig. 6-12):

$$\frac{b\gamma}{2n} d^2 K_p = \frac{y+d}{2} bx \qquad (6\text{-}33)$$

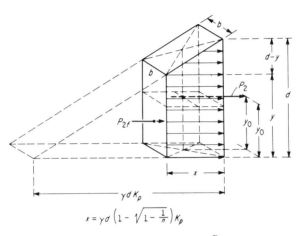

FIG. 6-12. Illustrating $P_2 = \frac{P_{2f}}{n}$.

From Eq. 6-33, the required ordinate is

$$x = \gamma(1 - \sqrt{1 - 1/n})dK_p \qquad (6\text{-}34)$$

The length y of the smaller base of the trapezoidal area is

$$y = d\sqrt{1 - 1/n} \qquad (6\text{-}35)$$

A graphical method for dividing a triangular area into n equal parts, for example, into two equal parts, is shown in Fig. 6-13.

The height, y_o of the application of force P_2 is

$$y_o = \frac{d^2 + y^2 + yd}{3(d+y)} \qquad (6\text{-}36)$$

or, substituting Eq. 6-35 into Eq. 6-36,

$$y_o = \frac{d}{3}\left[1 + \frac{n-1}{n(1 + \sqrt{1 - 1/n})}\right] \qquad (6\text{-}36a)$$

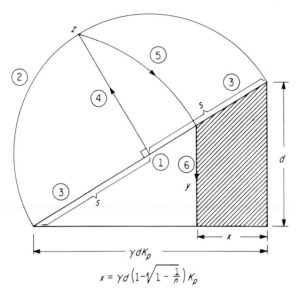

FIG. 6-13. Dividing a triangular area into two equal parts graphically.

or

$$y_o = \frac{Nd}{3} \tag{6-37}$$

where

$$N = 1 + \frac{n-1}{n(1+\sqrt{1-1/n})} \tag{6-38}$$

The driving depth d is now calculated from static equilibrium.

1. $\Sigma F_H = 0$

$$T - P_1 + P_2 = 0$$

or

$$T = P_1 - P_2 \tag{6-39}$$

where T = tensile force in the tie rod. Substitution of the P_1 and P_2-values from Eqs. 6-31 and 6-32, respectively, into Eq. 6-39 gives

$$T = \gamma b \left[d_o \left(d + \frac{d_o}{2} \right) K_a - \tfrac{1}{2} d^2 \left(\frac{K_p}{n} - K_a \right) \right] \tag{6-40}$$

Sheet Piling

2. $\Sigma M_T = 0$

$$P_1[(2/3)(d_o + d) - d_a] - P_2(d_o + d - d_a - y_o) = 0 \quad (6\text{-}41)$$

or

$$\frac{b\gamma}{2}(d_o + d)^2 K_a[(2/3)(d_o + d) - d_a]$$

$$-\frac{b\gamma}{2n} d^2 K_p(d_o + d - d_a - y_o) = 0 \quad (6\text{-}42)$$

where d_a = distance between the tie rod and the top of the sheet piling.

Grouping terms according to d gives the following general cubic equation:

$$\left[\frac{(3-N)}{2}\frac{K_p}{n} - K_a\right]d^3 - \frac{3}{2}\left[(d_o - d_a)\frac{K_p}{n} - (2d_o - d_a)K_a\right]d^2$$

$$- 3d_o(d_o - d_a)K_a d - (1/2)d_o^2(2d_o - 3d_a)K_a = 0 \quad (6\text{-}43)$$

or, rearranging,

$$d^3 + \frac{(3/2)[(d_o - d_a)(K_p/n) - (2d_o - d_a)K_a]}{\frac{(3-N)}{2}\frac{K_p}{n} - K_a} d^2 - \frac{3d_o(d_o - d_a)K_a}{\frac{(3-N)}{2}\frac{K_p}{n} - K_a} d$$

$$- \frac{(1/2)d_o^2(2d_o - 3d_a)K_a}{\frac{(3-N)}{2}\frac{K_p}{n} - K_a} = 0 \quad (6\text{-}44)$$

The solution of this cubic equation gives the minimum driving depth d for the assumed conditions.

When $d_a = 0$, i.e., the tie rod is attached at the top of the sheet piling, Eq. 6-44 can be rearranged according to (d/d_o)-ratios and rewritten as follows:

$$\left(\frac{d}{d_o}\right)^3 + \frac{(3/2)[(K_p/n) - 2K_a]}{\frac{(3-N)}{2}\frac{K_p}{n} - K_a}\left(\frac{d}{d_o}\right)^2 - \frac{3K_a}{\frac{(3-N)}{2}\frac{K_p}{n} - K_a}\frac{d}{d_o}$$

$$- \frac{K_a}{\frac{(3-N)}{2}\frac{K_p}{n} - K_a} = 0 \quad (6\text{-}45)$$

After the driving depth d has been found, the tensile force T in the anchor can be calculated by Eq. 6-40. Minimum driving depths result in shorter total lengths of sheet piling and greater bending moments, viz., heavier profiles of sheet piling (greater section modulus).

Excavation and Cofferdams

Example

Given an anchored sheet-piling system with free earth support as shown in Fig. 6-14. The unit weight of the dry soil is given as $\gamma = 120$ lb/ft³. The specific gravity of the soil is $G = 2.65$; $\phi = 30°$; $c = 0$; $\phi_1 = 0$.

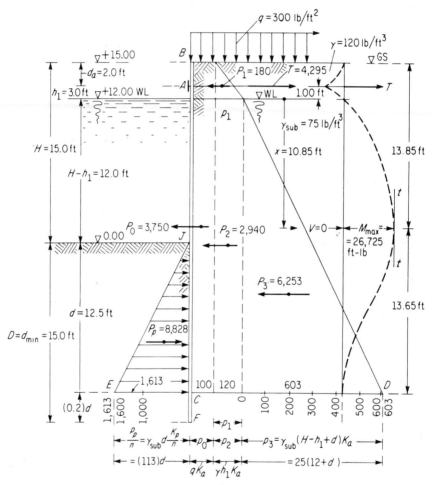

FIG. 6-14. Example of anchored sheet-piling system.

Determine

1. The minimum driving depth d_{min} of the sheet piling for a factor of safety $n = 2.0$ which would ensure a sufficient lateral support to prevent the toe of the sheet piling from displacing laterally and against groundbreak of the soil.

Sheet Piling

2. The tensile force T in the tie rod of the anchorage.
3. The maximum bending moment M_{max} in the sheet piling wall per one linear foot of run of wall.

The factor of safety for coarse-particled soils is usually taken as $n = 2$ to 3. For fine-particled soils $n = 1.5$ to 2.0 is used.

Auxiliary Calculations

Submerged Unit Weight of Soil:

$$\gamma_{sub} = \gamma - \frac{\gamma}{G} = 120 - \frac{120}{2.65} = 75 \ (lb/ft^3) \tag{6-46}$$

Active Earth Pressure Coefficient K_a:

$$K_a = \tan^2(45° - \phi/2) = \tan^2 30° = 0.333$$

Passive Earth Pressure Coefficient K_p:

$$K_p = \frac{1}{K_a} = \tan^2(45° + \phi/2) = \tan^2 60° = 3.000 \tag{6-47}$$

$$\frac{K_p}{n} = 3.000/2.0 = 1.5$$

Lateral Earth Pressure Ordinates:

$$p_o = qK_a = \quad (300)(0.333) = \quad 100 \ (lb/ft^2)$$
$$p_1 = \gamma h_1 K_a = (120)(3.0)(0.333) = 120 \ (lb/ft^2)$$
$$p_2 = p_1 = \quad\quad\quad\quad\quad\quad\quad\quad 120 \ (lb/ft^2)$$
$$p_3 = \gamma_{sub}(H - h_1 + d)K_a = (75)(15 - 3.0 + d)(0.333) = (25)(12 + d)$$
$$p_p/n = \gamma_{sub}d(K_p/n) = (75)(d)(1.5) = (113)d$$

Lateral Loads on Sheet Piling in lb per 1 ft or Run of Wall:

$$P_o = p_o(H + d) = (100)(15.0 + d)$$
$$P_1 = (1/2)p_1 h_1 = (0.5)(120)(3.0) = 180$$
$$P_2 = p_2(H - h_1 + d) = (120)(15.0 - 3.0 + d) = (120)(12.0 + d)$$
$$P_3 = (1/2)p_3(H - h_1 + d) = (0.50)(25)(12.0 + d)(12.0 + d) = (12.5)(12.0 + d)^2$$
$$P_p = (1/2)p_{p/n}d = (1/2)\gamma_{sub}d^2(K_p/n) = (1/2)(113)d^2 = (56.5)d^2$$

To find d, take moments about the tie-rod point A.

Moment Arms:

$$y_o = (H + d)/2 - d_a = (0.5)(15.0 - 4.0 + d) = (0.5)(11.0 + d)$$
$$y_1 = 0$$

$$y_T = 0$$
$$y_2 = (H - h_1 + d)/2 + (h_1 - d_a) = (0.5)(15.0 - 3.0 + d) + (3.0 - 2.0)$$
$$= (0.5)(14.0 + d)$$
$$y_3 = (2/3)(H - h_1 + d) + (h_1 - d_a) = (2/3)(12.0 + d) + 1.0$$
$$= (1/3)(27.0 + 2d)$$
$$y_{P_p} = H - d_a + (2/3)d = (13) + (2/3)d = (1/3)(39) + 2d)$$

The taking of moments about point A renders the following cubic equation:

$$\Sigma M_A = 0:$$

$$P_o y_o + P_2 y_2 + P_3 y_3 - P_p y_{P_p} = 0$$
$$d^3 + 12d^2 - 220d - 1090 = 0$$
$$\therefore d = 12.5 \text{ (ft)}$$

To allow for dredging and to safeguard against soil failure by soil particles under the toe of sheet piling, as well as against groundbreak of soil, and scour, it is the engineering practice to increase the theoretically computed minimum driving depth by 20 percent. Thus the driving depth of the sheet piling d_{min} is

$$d_{min} = (1.20)(12.5) = 15.0 \text{ (ft)}$$

Loads

$$P_o = (100)(15.0 + 12.5) = \quad 3,750$$
$$P_1 = \quad 180$$
$$P_2 = (120)(12.0 + 12.5) = \quad 2,940$$
$$P_3 = (0.5)(75)(24.5^2)(0.333) = \quad 6,253$$
$$\Sigma P = 13,123 \text{ (lb)}$$
$$P_p = (0.5)\gamma_{sub}(12.5^2)(1.5) = \quad 8,828 \text{ (lb)}$$

Anchor Pull

$$T = \Sigma P - P_p = 13,123 - 8,828 = 4,295 \text{ (lb)}$$

Position x of Maximum Bending Moment Below Ground Surface:

The maximum bending moment in the sheet piling is where the shear force V is equal to zero, i.e., $V = T - \Sigma P_x = 0$:

$$P_o + P_1 - T + P_2 + P_{3x} = p_o(h_1 + x) + (0.5)p_1 h_1$$
$$- 4,295 + p_2 x + (0.5)\gamma_{sub} x^2 K_a = 0$$
$$(12.5)x^2 + 17.6x - 306 = 0$$

Sheet Piling

$\therefore x = \underline{10.85 \text{ (ft)}}$ below the groundwater table

or

$10.85 \text{ (ft)} + 3.0 \text{ (ft)} = \underline{13.85 \text{ (ft)}}$ below ground surface

The maximum bending moment M_{max} in the sheet piling at point $x = 10.85$ (ft) where the shear V is zero $= 26{,}725$ ft-lb.

With the maximum bending moment known, the profile size of the sheet piling can now be chosen from manufacturers' catalogs.

The necessary section modulus Z is given by:

$$Z = \frac{M_{max}}{\sigma_{all}} = \frac{(26{,}725)(12)}{20{,}000} = \underline{16.03 \text{ (in.}^3)}$$

where $\sigma_{all} = 20{,}000$ lb/in.^{2} = allowable stress in tension (bending) of structural steel.

6-11. MOMENT REDUCTION

Conventional design methods of sheet-piling walls were used extensively for many years, and still are used today. So far as is known no failures have been attributed on the basis of these methods. However, it has been shown[6-11] theoretically and experimentally that the maximum bending moment on an anchored sheet piling is influenced, among other things, by the flexibility of the sheet piling, as well as by the density of the soil for a given driving depth D and the height of the soil to be retained. It was found that sheet-piling flexibility reduces the calculated bending moments. Therefore, for design purposes, some engineers reduce the calculated maximum bending moment M_{max} by a certain percentage, thus arriving at a "design moment," M_{des}. For practical application of moment reduction Rowe[6-8] developed, among other things, a theoretical relationship between maximum bending moment and flexibility of sheet piling and gave appropriate empirical moment-reduction curves. Such curves were prepared also for single-anchored sheet-piling walls with free earth support. The moment-reduction curve is applied to the maximum bending moment in the form of a ratio M_{des}/M_{max}. The bending-moment reduction factor (or number) ρ is computed from Eq. 6-48 and the moment ratio $M_{des}/M_{max}' = r_M$ as a function of ρ is read off the moment-reduction graph. Here

$$\rho = \frac{(H+D)^4}{EI} \quad \text{(in.}^2/\text{lb)} \tag{6-48}$$

where H = retained height of soil, ft
D = minimum driving depth, ft
$H + D$ = length of sheet piling, ft

E = Young's modulus of elasticity of sheet-piling material, lb/in.²
I = moment of inertia of cross section of sheet-piling profile, in.⁴

Also,

$$\frac{M_{des}}{M_{max}} = f(\rho) = r_M \tag{6-49}$$

and the design bending moment is

$$M_{des} = (r_M)(M_{max}) \tag{6-49a}$$

Thus the flexibility number governs the fixity of the free-earth-supported single-anchored sheet-piling wall at a given state of density of sand and/or gravel.

According to Rowe, his moment-reduction method is an improvement in design of such sheet-piling systems. Because of the reduced moments, his flexibility factor method is claimed by Rowe to be a saving over the conventional method.

A simple, easy-to-use moment-reduction graph as a function of flexibility factor ρ is shown in Fig. 6-15 as given by the *Design Manual* of the Bureau of Yards and Docks, Department of the Navy.[12] This graph may be used for sheet piling driven into very compact, coarse-particled soils or for medium, medium-compact, and compact, coarse-particled soils. According to this *Design Manual*, no reduction in M_{max} is permitted for driving of sheet piling in fine-particled soils or loose or very loose coarse-particled soils.

The flexibility factors ρ in Fig. 6-15 are computed on the basis of lubricated sheet-piling interlocks.

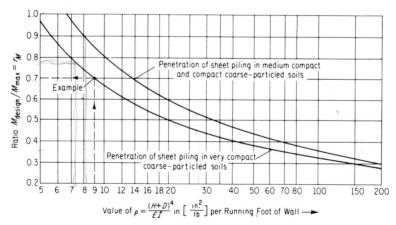

FIG. 6-15. Moment reduction ratio as a function of flexibility number ρ (after Ref. 12).

Sheet Piling

The routine of calculating the design moment M_{des} and the use of the moment-reduction graph consists of successive trials of assumed sheet-piling profile size until the design bending-moment stress in sheet piling equals the allowable bending stress of the sheet-piling material.

Suppose that the length $L = H + D$ of the given free-earth-supported sheet piling is $H + D = 47.35$ ft, and the theoretical calculated maximum bending movement is $M_{max} = 1,072,500$ (in.-lb) ft. The soil in question is given as a medium compact, coarse-particled soil. The next step to take is to select and try a sheet-piling profile, say Bethlehem Steel sheet-piling profile ZP 32, the properties of which for 1 ft of run of wall are: $Z = 38.3$ in.^{3}; $I = 385.7$ in.^{4}; $E = 30 \times 10^6$ lb/in.^{2}, and the allowable bending stress of structural steel is $\sigma_{all} = 20,000$ lb/in.^{2} With this information, the flexibility factor ρ is calculated as

$$\rho = \frac{(H+D)^4}{EI} = \frac{(47.35^4) \times 12^4 \times 1.75}{(30 \times 10^6)(3.857 \times 10^2)} = 15.75 \text{ (in.}^2/\text{lb)}$$

Now apply moment reduction for flexibility. To do this, enter the moment-reduction graph for flexibility, Fig. 6-15, at $\rho = 15.75$, intersect the "medium compact" curve, and read off the ordinate axis the moment ratio r_M:

$$r_M = \frac{M_{des}}{M_{max}} = 0.66 \approx 0.70$$

Thus the reduced (= design) bending moment calculates as

$$M_{des} = (0.70) M_{max} = (0.70)(1,072,500) = 750,750 \text{ [(in.-lb)/ft]}$$

Compare the maximum bending moments for design and from theoretical calculations in Fig. 6-16.

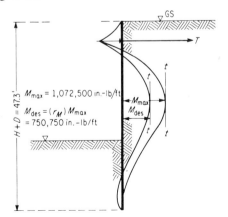

FIG. 6-16. Comparison between M_{max} and M_{des}.

Now check the bending stress f_s in sheet piling:

$$f_s = M/Z = 750{,}750/38.3 = 19{,}600 \text{ (lb/in.}^2) < 20{,}000 \text{ lb/in.}^2 = \sigma_{\text{all}}$$

If the bending stress is considerably less than 19,600 lb/in.², try a lighter profile. If the allowable bending stress of the sheet-piling material is, say, $\sigma_{\text{all}} = 25{,}000$ lb/in.² then try a smaller profile.

6-12. METHOD OF FIXED EARTH SUPPORT

In this method it is assumed that the sheet piling is flexible.[2] This analysis is used when the sheet piling must be driven deeper than usually in order to prevent leakage.

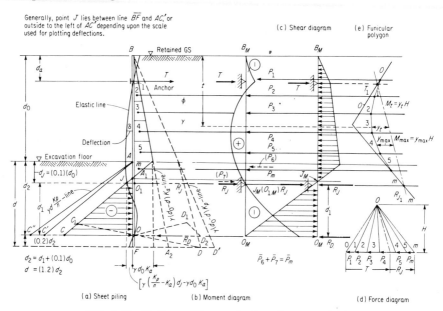

FIG. 6-17. Static conditions at fixed earth support (Ref. 2).

The stability analysis of sheet piling, applying the fixed-earth-support method, is based on the assumption that the deflections of the sheet piling form an elastic line (see Fig. 6-17a). The elastic line changes its curvature at the inflection point J. This implies that the soil into which the sheet piling is driven exerts an effective restraint on the sheet piling below the point of inflection. Also, it is quite reasonable to assume that a slight deflection of the sheet piling may take place below the pivot point O, inducing thereby a negligible resultant earth pressure along part OF on the back face of the sheet piling.

Sheet Piling

However, for reasons of simplicity, the pressure diagram $\triangle(A_1C_1O)$ is transformed into a $\triangle(A_1CO)$, thus increasing its area by the amount $\triangle(C_1CO)$. This increase, however, is counterbalanced by adding an equivalent pressure area $\triangle(OD_1D_2)$ onto the backface of the sheet piling. The resultant, R_D, of $\triangle(OD_1D_2)$ and $\triangle(ODF)$ is applied as a concentrated load at point O, acting to the left. The magnitude of R_D is unknown. However, it is automatically excluded from consideration when point O is chosen as the point of rotation for writing the moment equation (in the case of equilibrium, the sum of the moments must be equal to zero). Reaction R_D can be found when R_J and d_1 are known.

For practical purposes the location of the inflection point J can be assumed to be $d_J = (0.1)(d_o)$ units of length below the excavation floor.[5]

6-13. CONCEPT OF EQUIVALENT BEAM

To simplify the solution of the complex fixed-earth-support sheet-piling problem, it is assumed that the sheet piling BAF is a beam. This beam is supported freely at one end by the reaction T (the tensile force in the tie rod), and fixed in the soil at its other end (lower end of the sheet piling) (see Fig. 6-17). The character of the moment diagram for such a beam is shown in Fig. 6-17b. From this diagram it can be seen that point J_M on the beam, which corresponds to the inflection point J on the elastic sheet piling, does not transmit any moments, i.e., the moment at this point is zero. Hence, there is no bending. Therefore, theoretically, the lower end O_1O of the sheet piling, i.e., the part $O_{1M}O_M$ of the original beam B_MO_M, can be cut off at point J_M (see Figs. 6-17, b and c), and the shear force here replaced by a reaction R_J, to be determined.

By this operation, an equivalent, freely supported beam B_MJ_M of the original beam B_MO_M is obtained. The cut-off part J_MO_M, of the original beam, in turn, is also a simple, freely supported beam, the span of which is d_1 (Fig. 6-17c). It is loaded with a trapezoidally distributed load. If the reaction R_J of this short beam is known, a moment equation at equilibrium for this beam about point O_M gives the span d_1.

6-14. DETERMINATION OF REACTION R_J

The most convenient way to determine the reaction R_J of the equivalent beam B_MJ_M is by graphical solution. For this solution proceed as follows:

1. Divide part BO_1 of the sheet piling and its pressure diagram into a number of equal panels (see Fig. 6-17a).
2. Determine the concentrated loads $P_1, P_2, P_3 ... P_m$ of these pressure panels. These concentrated loads should be so large, viz., the panels so

small that the error thus introduced is reasonably small when compared with distributed loads. Then draw an equivalent beam $B_M J_M$ (see Fig. 6-17b), show reactions T and R_J, and apply to the beam the previously computed concentrated loads $P_1, P_2, P_3 \ldots P_m$.

3. Construct a force diagram by setting horizontally the forces ΣP (see Fig. 6-17d) head to tail.

4. Choose a pole, O, draw rays 0, 1, 2... m to it from head and tail of each of the forces in the force diagram (Fig. 6-17d).

5. Construct the funicular polygon (Fig. 6-17e).

6. Intersect that ray in the funicular polygon designated by O, with the line of action of the reaction T, at point T_1. At this point there is no deflection. Intersect that ray, designated by m, with the line of action of the reaction R_J, at point R_{J1}.

7. Connect these two points with a closing line $T_1 R_{J1}$.

8. Shift the closing line to the force diagram through pole O and parallel to itself in the funicular polygon. The closing line in the force diagram cuts the resultant force $\overset{m}{\underset{1}{\Sigma}} P$ into two parts, namely, into reactions T (the tensile force in the tie rod) and R_J, respectively (see Fig. 6-17d). R_J, again, is the reaction of the equivalent beam.

6-15. DETERMINATION OF d_1

As previously cited, part $O_1 O$ of the sheet piling can be regarded as a simple beam, the span of which is

$$d_1 = d_2 - d_J \tag{6-50}$$

This beam is loaded with a trapezoidally (or in special cases triangularly) distributed load as represented by the shaded pressure areas $CJO_1 O$, in Fig. 6-17a.

Figure 6-18 shows the various pressure ordinates necessary for the stability calculations of the sheet-piling system. The symbol n means the coefficient of safety. The passive pressure should not exceed the value of

$$p_n = \frac{\gamma d K_p}{n} \tag{6-51}$$

where d is the variable driving depth.

To achieve stability, the moments about any point of rotation – for example, about point O (this eliminates the unknown reaction R_D from our consideration) – should be equal to zero:

Sheet Piling

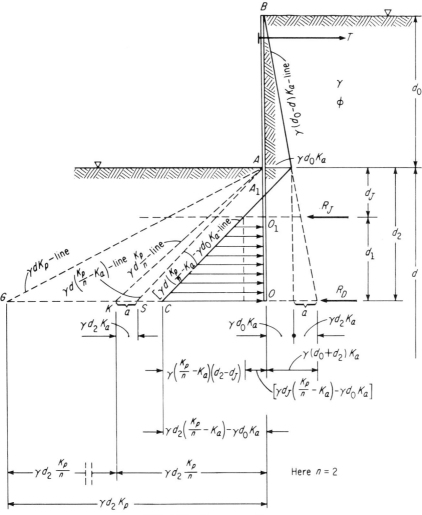

FIG. 6-18. Pressure ordinates.

$$R_J(d_2 - d_J) - \frac{\gamma}{2}(d_2 - d_J)^2 \left(\frac{K_p}{n} - K_a\right) \frac{(d_2 - d_J)^2}{3}$$

$$- \left[\gamma d_J \left(\frac{K_p}{n} - K_a\right) - \gamma d_o K_a\right] \frac{d_2 - d_J}{2} = 0 \quad (6\text{-}52)$$

Because $d_2 - d_J = d_1$, Eq. 6-52 can be rewritten as follows:

$$R_J d_1 - \frac{\gamma}{6} d_1^3 \left(\frac{K_p}{n} - K_a\right) - \left[\gamma d_J \left(\frac{K_p}{n} - K_a\right) - \gamma d_o K_a\right] \frac{d_1^2}{2} = 0 \quad (6\text{-}53)$$

or

$$d_1^2 + 3\left[d_J - \frac{d_o K_a}{(K_p/n) - K_a}\right]d_1 - \frac{6R_J}{\gamma[(K_p/n) - K_a]} = 0 \qquad (6\text{-}54)$$

and

$$d_1 = \frac{3}{2}\frac{d_o K_a}{(K_p/n) - K_a} - \frac{3}{2}d_J \pm \frac{1}{2}\sqrt{9\left[d_J - \frac{d_o K_a}{(K_p/n) - K_a}\right]^2 + \frac{24R_J}{\gamma[(K_p/n) - K_a]}} \qquad (6\text{-}55)$$

The first term under the square root is relatively small as compared with the second term. Therefore the first term can be omitted, and the equation can be rewritten as follows:

$$d_1 = \frac{3}{2}\frac{d_o K_a}{(K_p/n) - K_a} - \frac{3}{2}d_J + \sqrt{\frac{6R_J}{\gamma[(K_p/n) - K_a]}} \qquad (6\text{-}56)$$

Then

$$d_2 = d_1 + d_J \qquad (6\text{-}57)$$

or

$$d_2 = \frac{3}{2}\frac{d_o K_a}{(K_p/n) - K_a} - \frac{d_J}{2} + \sqrt{\frac{6R_J}{\gamma[(K_p/n) - K_a]}} \qquad (6\text{-}58)$$

To allow full development of the relatively small resultant pressures on the sheet piling below the pivot point O, it is sufficient for practical purposes to increase the calculated depth d_2 by an amount of 20 percent.[13] Hence, the practical driving depth of the sheet piling is

$$d = (1.2)(d_2) \qquad (6\text{-}59)$$

6-16. BENDING MOMENT

The bending moment at any point on the beam $B_M J_M$ is equal to the horizontal ordinate y_t, between the funicular line and the closing line (Fig. 6-17d) at that point, multiplied by the pole distance H from the force diagram:

$$M_t = y_t H \qquad (6\text{-}60)$$

Usually the ordinate y_t is measured to a linear scale to which the beam is drawn, but H is measured to the force scale to which the force diagram is

Sheet Piling

drawn. When $H = 1$, then the bending moment is numerically equal to the ordinate y_t.

Let us, for example, assume that the linear scale of the beam is 1 in. = 5 ft, and that the force scale is 1 in. = 800 lb. If the pole distance H happens to be 1.25 in., then

$$H = (800)(1.25) = 1,000 \text{(lb)}$$

The moment scale (M.S.) is

(linear scale times pole distance)

or

$$(\text{M.S.}) = 5(\text{ft/in.}) \times 1,000(\text{lb}) = 5,000(\text{lb-ft/in.})$$

Thus, if the ordinate y_t at a certain point measures 0.7 in., the bending moment at that point is

$$M = y_t(\text{M.S.}) = (0.7)(5,000) = 3,500 \text{ (lb/ft)}$$

The same result can be obtained by measuring y_t to a linear scale in feet, which in our example corresponds to 3.5 ft. Then

$$M = y_t H = (3.5)(1,000) = 3,500 (\text{lb/ft})$$

The maximum bending moment is obtained by finding $y_{\max}$ in the funicular polygon and multiplying it by the pole distance (H):

$$M_{\max} = y_{\max} H \tag{6-61}$$

If the allowable stress, σ_{all}, of the sheet-piling material is known, then the section modulus of the profile of the sheet piling can be calculated. From static tables of sheet piling the size of the sheet piling can be determined:

$$Z = \frac{12M}{\sigma_{all}} \text{ (in.}^3\text{)} \tag{6-62}$$

where M = bending moment in lb-ft
Z = section modulus of sheet piling

6-17. TSCHEBOTARIOFF'S ANCHORED SHEET-PILING SYSTEM

One of the methods for designing a single-anchored sheet-piling wall is that proposed by G. S. Tschebotarioff.[2] His method is known as the simplified equivalent beam method. The working system is illustrated in Fig. 6-19. In this design, Tschebotarioff recommends that for a factor

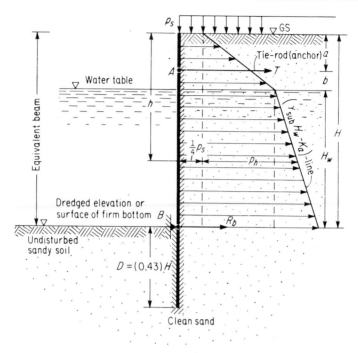

FIG. 6-19. Tschebotarioff's working system for designing single-anchored flexible bulkhead in non-cohesive soil. The sheet piling is loaded by earth pressure of the backfill. (By consultation with Dr. G. P. Tschebotarioff.)

of safety of $\eta = 2.0$ against toe kick-out the driving depth be $D = (0.43)H$. At this ratio, $\dfrac{D}{H} = 0.43$, the point of effective restraint of the lower part of the sheet piling is at point B. The concept of the "equivalent beam" is used for assessing the magnitude of the tension T in the tie rod.

ANCHORAGE

6-18. ANCHOR TENSION

According to American Civil Engineering Practice,[14] "in all cases where the solution for anchor tension T is based on the assumption of linear-pressure diagrams based on the hydrostatic principle of the Rankine-Coulomb theory, there may be an error on the side of danger. The exact location of resultant earth pressure has not been clearly established. For the purpose of designing the anchorage, the theoretical value of T should be generally increased by 20-30 percent, and if the anchor is in unyielding soil, the increase may be as much as 50 percent."

6-19. PLACING OF ANCHOR BLOCKS

The anchor block, anchor wall, and anchor plate or slab must be placed far enough from the sheet-piling wall that the sliding wedge $CF_1 D_1$ of passive earth resistance brought about by the anchor wall will not intercept the sliding wedge FBC of the active earth pressure behind the sheet-piling wall FB (Fig. 6-20).

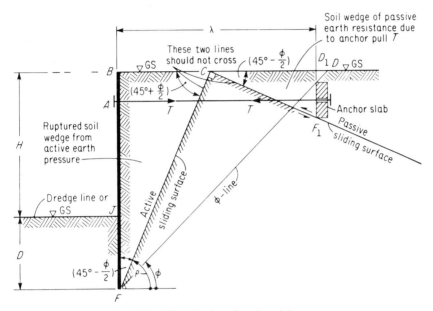

FIG. 6-20. Placing of anchor slab.

The anchored sheet piling wall must be anchored securely. The anchorage systems may be:
- Individual anchor blocks (deadmen)
- Continuous anchor walls
- Anchor slabs.
- Piled anchorage (see Fig. 16-21)

The most effective type of reinforced concrete anchorage (deadman) is the continuous wall rather than a multiplicity of individual blocks. For proper function the anchor wall should parallel the wall of sheet piling.

The position of anchor blocks, walls, and slabs is best determined graphically, as shown in Fig. 6-22. In this figure

Zone ① is an unsafe one for placing of anchorage. Upon failure of the active wedge, the anchor placed in zone ① provides no resistance against the failure of the sheet-piling wall.

166　Excavation and Cofferdams

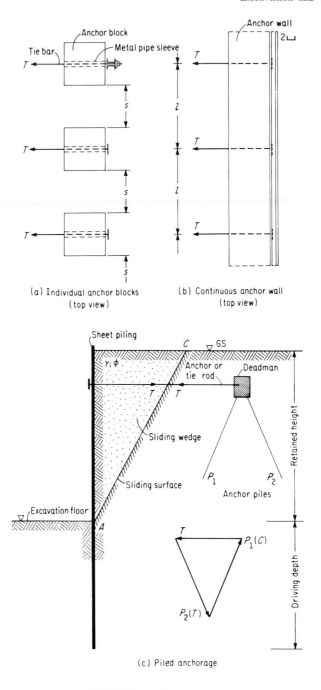

FIG. 6-21. Anchorage systems.

Sheet Piling

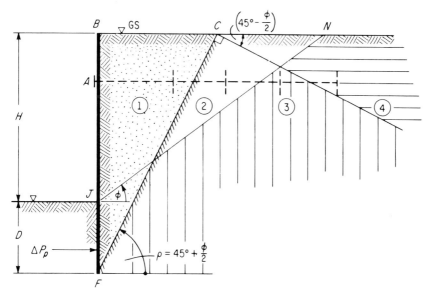

FIG. 6-22. Anchorage zones of various degree of safety.

Zone ② is an uncertain one relative to the placing of anchor block.
Zone ③ provides partial safety for anchorage: the anchor block provides partial resistance, and transfers lateral load P_a from the block to the base of the wall (ΔP_p is the reaction to P_a, Fig. 6-23). The

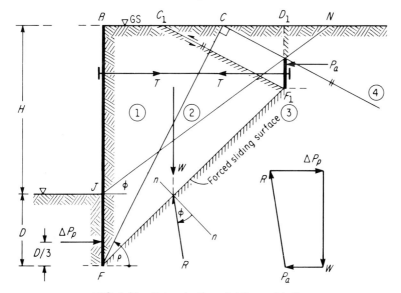

FIG. 6-23. Determination of ΔP_p graphically.

magnitude of ΔP_p may be ascertained from the force polygon of the free body FBD_1F_1F, where P_a is the active earth pressure on the face D_1F_1 at the anchor block.

Zone ④ is a completely safe one for the placing of anchorage. Here the block provides full resistance to anchor pull. No lateral load is transferred by the block to the sheet-piling wall.

Anchor blocks and walls should be placed deep enough below the ground surface to avoid shearing off and expelling a passive soil wedge in front of the anchor wall.

6-20. FORCES ACTING ON ANCHOR WALL

The forces acting on the anchor wall (block) are: (1) the anchor pull T, (2) the active earth pressure P_a on the back of the anchor wall, and (3) the passive earth pressure P_p in front of the anchor wall.

The frictional force in the bottom contact plane between anchor wall and soil is usually neglected.

At equilibrium,

$$T + P_a - P_p = 0 \tag{6-63}$$

as seen in Fig. 6-24b. Here

$$P_p = (1/2)\gamma d^2 (K_p/n)$$

γ = unit weight of soil
d = height of anchor wall = depth below ground surface if anchor wall extends up to ground surface
$K_p = \tan^2(45° + \phi/2)$
ϕ = angle of internal friction of soil
n = factor of safety, between 2.0 and 3.0

Thus Eq. 6-63 may be rewritten as

$$T + (1/2)\gamma d^2 K_a - (1/2)\gamma d^2 (K_p/n) = 0 \tag{6-63a}$$

From Eq. 6-63a, the height d of the anchor wall calculates as

$$d = \sqrt{\frac{2T}{\gamma[(K_p/n) - K_a]}} \tag{6-64}$$

The resultant of the earth pressure, $P_R = P_p - P_a$, acts at the lower third of the height d of the anchor wall. Therefore the tie rod must be purposely placed collinearly with the resultant pressure P_R if rotation of the wall due to eccentric loading of the anchor wall is to be avoided. Or else, if

Sheet Piling

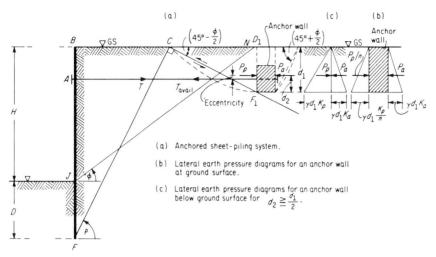

FIG. 6-24. Forces on anchor wall.

eccentric loading is tolerated, the anchor wall must be laid into the soil below ground surface to provide for a surcharge of soil on top of the anchor wall, thereby preventing rotation of the block (wall).

When $d_2 \geq d_1/2$ (Fig. 6-24c), then compute

$$P_a = (1/2)\gamma d_1^2 K_a \qquad (6\text{-}65)$$
$$P_p = (1/2)\gamma d_1^2 K_p \qquad (6\text{-}66)$$

and the available resistance T_{avail} to the anchor pull T is

$$T_{\text{avail}} = P_p - P_a = (1/2)\gamma d_1^2 (K_p - K_a) \qquad (6\text{-}67)$$

The available horizontal resistance T_{avail} of the anchorage should be $\eta = 1.5$ to 2.0 times as large as the anchor pull T. Thus with a factor of safety η, the anchor pull as a function of T_{avail} is

$$T = \frac{T_{\text{avail}}}{\eta} L = \frac{P_p - P_a}{\eta} L \qquad (\text{lb}) \qquad (6\text{-}68)$$

where L is the length of the anchor wall or anchor block, whichever is the case.

When there is a surcharge q on the ground surface (Fig. 6-25), and $d_2 \geq (d_1/2)$, the active earth pressure P_a may be reckoned with the surcharge, whereas for reasons of extra safety, the passive earth resistance P_p is reckoned without surcharge even if it is there (Fig. 6-25b).

When $d_2 < (d_1/2)$, then the required factor of safety η is expressed as

$$\eta = \frac{P_{ptr}}{T + P_{atr}} \geq 1.5 \tag{6-69}$$

(Fig. 6-25c).

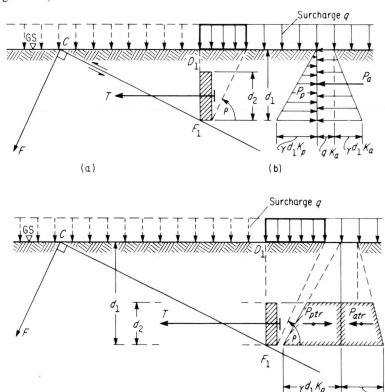

FIG. 6-25. Anchor wall below surcharged ground surface. (a) Surcharged ground surface, (b) Lateral earth preessure diagrams for $d_2 \geq d_1/2$, (c) Lateral earth pressure diagrams for $d_2 < d_1/2$.

6-21. FRICTIONAL FORCE R_L ON VERTICAL SIDE OF PASSIVE WEDGE

In the case of individual anchor blocks or anchor slabs, at limit equilibrium of the passive soil wedge F_1CD_1 there acts a frictional resisting force R_L on each face ($\triangle C_1 D_{11} F_{11}$ and $\triangle C_2 D_{12} F_{12}$) of the passive wedge. These resisting forces $2R_L$ may be calculated as set forth in writing a differential active earth pressure dE_a on a smooth wall ($\phi_1 = 0$) (Fig. 6-26).

Sheet Piling

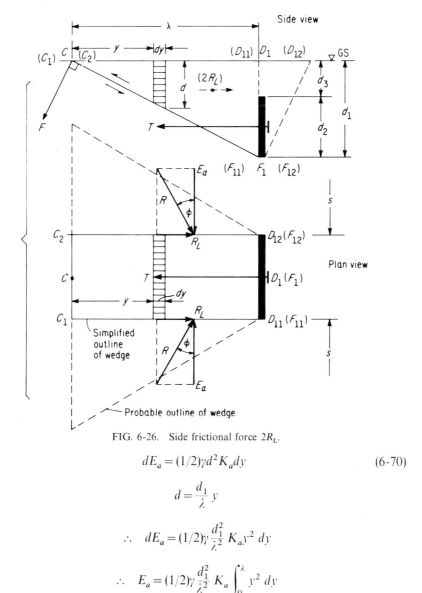

FIG. 6-26. Side frictional force $2R_L$.

$$dE_a = (1/2)\gamma d^2 K_a dy \tag{6-70}$$

$$d = \frac{d_1}{\lambda} y$$

$$\therefore \quad dE_a = (1/2)\gamma \frac{d_1^2}{\lambda^2} K_a y^2 \, dy$$

$$\therefore \quad E_a = (1/2)\gamma \frac{d_1^2}{\lambda^2} K_a \int_0^\lambda y^2 \, dy$$

$$= (1/6)\gamma d_1^2 \lambda K_a \tag{6-71}$$

$$\therefore \quad 2R_L = 2E_a \tan\phi = (1/3)\gamma d_1^2 \lambda K_a \tan\phi \tag{6-72}$$

When blocks or slabs are spaced too closely, then Eq. 6-72 is valid only as long as the friction in the side faces of the passive soil wedge is less than the soil resistance pertaining to the interspace s between the blocks, i.e., the criterion for the validity is

$$(1/3)\gamma \lambda d_1^2 \tan \phi \, K_a \leqslant (1/2)\gamma(d_1^2 K_p - d_3^2 K_a)s \qquad (6\text{-}73)$$

When blocks and slabs are too closely spaced, the anchorage must be treated as a continuous wall of constant height.

When anchor walls, blocks, and slabs cannot be accommodated to utilize full passive earth resistance, the tie rod can be anchored to piles.

6-22. STABILITY OF SHEET PILING-ANCHOR WALL-SOIL SYSTEM

The system comprising sheet piling, anchor wall, and soil must be safe relative to

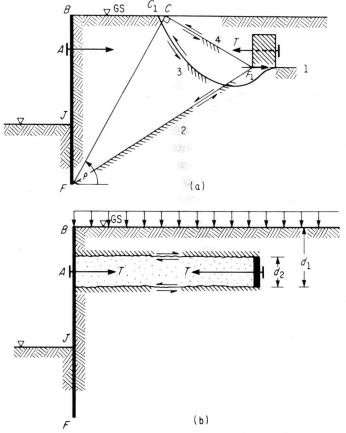

FIG. 6-27. Possible failures of anchorage. (a) Possible modes of failures of soil, (b) piercing of anchor slab through the soil.

1. Sliding of anchor wall on its own base.
2. Sliding along a forced sliding surface FF_1 through the toe of the sheet piling F wall and the base of the anchor wall (Fig. 6-27a).
3. Groundbreak of soil due to deep-seated shear failure of soil.
4. Shear failure of soil in front of anchor wall (passive earth-resistance wedge).
5. Simply piercing through the soil of the anchor wall laterally (Fig. 6-27b) in the case of deep-seated anchor walls. If the anchor wall or slab lies deep in the soil $(d_2 \ll d_1)$ then it may pierce through the soil laterally without bringing about the passive rupture surface in front of the anchor wall.

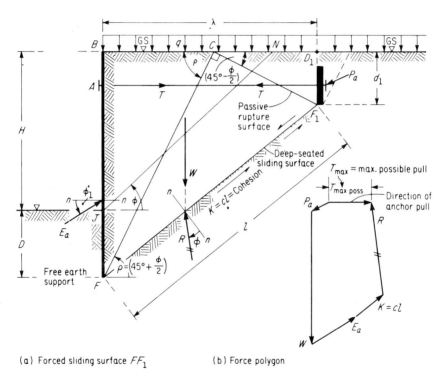

(a) Forced sliding surface FF_1 (b) Force polygon

FIG. 6-28. Deep-seated failure of soil.

The anchor wall may cause not only a passive rupture surface CF_1 with respect to the upper ground surface (Fig. 6-28), but it may also bring about a deep-seated forced rupture surface FF_1 passing through the toepoints of the sheet piling and anchor wall. The maximum possible tensile force $T_{\text{max poss}}$ for the rupture body FBD_1F_1F may be readily determined graphically as indicated in Fig. 6-28b. The anchorage is safe if the resultant of the forces acting in the direction of the anchor (the maximum possible anchor pull $T_{\text{max poss}}$) is greater than the actual anchor pull T. The factor of safety η against this deep-seated failure of soil is then calculated as

$$\eta = \frac{T_{\text{max poss}}}{T} \geq 1.5 \qquad (6\text{-}74)$$

or else the anchor length λ must be increased, viz., it must be placed deeper into the soil.

6-23. LENGTH OF ANCHOR

The necessary length λ of the anchor is determined by the condition as shown in Fig. 6-28. From geometry, the length λ calculates as

$$\lambda = (H + D) \cot (45° + \phi/2) + d_1 \cot (45° - \phi/2) \qquad (6\text{-}75)$$

6-24. WALE

The strength of the wale (at A, Fig. 6-28) along the sheet-piling wall is determined by means of the maximum midspan, viz., support point bending moment M of a continuous beam supported at the anchor points:

$$M = q_l \frac{l^2}{10} \qquad (6\text{-}76)$$

where l = span between two adjacent anchor points (Fig. 6-21b)

q_l = the distributed load from anchor pull T per foot of run of wall
Usually anchors are spaced from 2.5 to 8 ft center-to-center.

REFERENCES

1. R. W. Abbett (ed.), *American Civil Engineering Practice*, New York: Wiley, Vol. 2, 1956, pp. 21-67 to 21-68.
2. A. R. Jumikis, *Mechanics of Soils*, Princeton, N.J.: Van Nostrand, 1964, pp. 333-378.
3. H. Blum, *Beitrag zur Berechnung von Bohlwerken mit Berücksichtigung der Wandverformung*, Berlin: Ernst, 1951.
4. H. Blum, "Berechnung einfach oder nicht verankerter Spundwände," in *Grundbau Taschenbuch*, Berlin: Ernst, Vol. 1, 1955, pp. 328-333.
5. K. Terzaghi, *Theoretical Soil Mechanics*, New York: Wiley, 1943, p. 348.
6. P. W. Rowe, "Anchored Sheet-Pile Walls," Paper No. 5788, *Proceedings of the Institution of Civil Engineers* (*London*), Part 1, Vol. 1, No. 1 (January 1952), pp. 22-70.
7. P. W. Rowe, "A Theoretical and Experimental Analysis of Sheet-Pile Walls,"

Sheet Piling

Paper No. 5989, *Proceedings of the Institution of Civil Engineers (London)*, Part 1, Vol. 4, No. 1 (January 1955), pp. 32-69.
8. P. W. Rowe, "Sheet-Pile Walls at Failure," Paper No. 6107, *Proceedings of the Institution of Civil Engineers (London)*, Part 1, Vol. 5, No. 1 (January 1956), pp. 276-315.
9. R. Briske, *Erddruckverlagerung bei Spundwandbauwerken*. Berlin: Ernst, 1953.
10. J. B. Hansen, *Earth Pressure Calculations*, Copenhagen: The Danish Press (Institution of Danish Civil Engineers), 1953, p. 43.
11. G. P. Tschebotarioff, "Retaining Structures," G. A. Leonards (ed.), in *Foundation Engineering*, New York: McGraw-Hill, 1962, pp. 438-524.
12. Bureau of Yards and Docks, *Design Manual* (Soil Mechanics, Foundations and Earth Structures) Navdocks-7. Washington, D.C.: Department of the Navy, Bureau of Yards and Docks, 1962, pp. 7-10-18.
13. K. Terzaghi, *Theoretical Soil Mechanics*, New York: Wiley, 1943, p. 225.
14. R. W. Abbett (ed.), *American Civil Engineering Practice*, Vol. 2. New York: Wiley, 1956, pp. 21-69.

PROBLEMS

6-1. Determine the driving depth of the sheet-piling system as shown in the figure and select the size of the sheet-piling profile. Use conventional and Blum methods, compare results, and report.

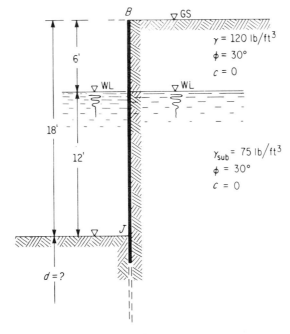

PROB. 6-1

6-2. In Prob. 6-1, use a factor of safety of $\eta = 2.0$ on the passive earth-resistance coefficient K_p. Use conventional and Blum methods and report.

6-3. Use Prob. 6-1 and perform sheet-piling analysis without water in front of sheet piling and without groundwater.

6-4. Refer to the accompanying figure and perform sheet-piling analysis using moment-reduction method.

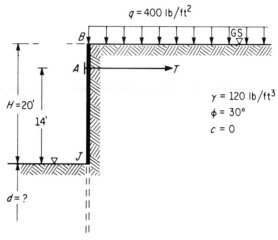

PROB. 6-4

6-5. Refer to the accompanying figure and perform sheet-piling analysis using moment-reduction method.

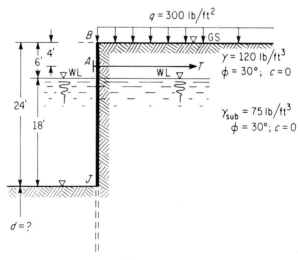

PROB. 6-5

6-6. Perform Prob. 6-5 by Tschebotarioff's method.

chapter 7

Cofferdams

7-1. DEFINITION. CLASSIFICATION

According to Webster's dictionary, the word "coffer" (from Greek *kophinos*) means (1) a casket, chest, or trunk; (2) the chamber of a canal lock; (3) a caisson; (4) a cofferdam.

In civil engineering and hydraulic structures, a cofferdam is a dam, a temporary structure built to enclose a foundation excavation with relatively dense walls to exclude the influx of water and/or soil from the excavation so that a foundation may be laid there in the dry. Thus, a cofferdam is only an aid to the work of the construction operations associated with the laying of foundations, namely: to allow a temporary dewatering of the excavation.

Cofferdams are applied when the size of the foundation excavation makes bracing and strutting impossible or uneconomical, or when strutting hinders construction operations.

Cofferdams are usually required for foundations of engineering structures (bridge piers, locks, docks, dams, power plants) which are built in open water (rivers, lakes, sea) or on land where there is a high groundwater table present.

Cofferdams may be broadly classified into two principal groups, namely:
- (A) Cofferdams built in open waters
- (B) Land cofferdams
- (A) Cofferdams built in open water, in their turn, may be classified further as
    - (a) Cofferdams on permeable (and impermeable) soil
    - (b) Cofferdams supported on bedrock

Among the cofferdams built in open waters and supported on soil are
1. Earth dikes and embankments and rock-fill cofferdams
2. Single-wall sheet piling
    - (i) with no bracing
    - (ii) with bracing
    - (iii) with earth fill on one or both sides of the sheet piling

3. Double-wall sheet piling filled with soil
   (i) Timber sheet piling
   (ii) Timber or steel sheet piling with wales and tie rods
4. Steel sheet piling
   (i) Semicircular cells with straight diaphragms
   (ii) Circular cells with connecting arcs
  (b) Cofferdams on bedrock in open water
   1. All as under A
   2. Cribs
   3. Concrete walls
(B) Land cofferdams
   1. Timber sheet piling with bracing
   2. Steel sheet piling with bracing

Land cofferdams enclose excavations made below the groundwater table.

7-2. EARTH COFFERDAMS. ROCK-FILL COFFERDAMS

The simplest type of a cofferdam is an earth bank or a dike with a clay core or vertically driven sheet piling or clay blanket facing the upstream slope of the bank, sometimes covered with a riprap of rocks. In the excavation thus enclosed construction is done in the dry (Fig. 7-1).

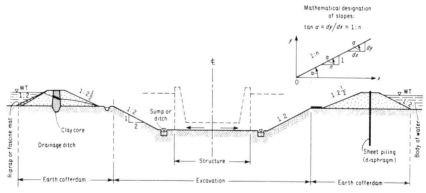

FIG. 7-1. Earth cofferdam.

In coarse sand or sand and gravel the slopes of the excavation are usually made 1:2; in fine sand, 1:2.5 to 1:3; in silt, 1:3.5; in clay, variable. However, depending upon the water regimen present and the height of the slopes, the design of the slopes must be based on stability analyses.

The designation of a slope $1:n$ means mathematically $\tan \alpha$, where α is the angle of the slope. For example, $\tan \alpha = dy/dx = 1:n$. Hence in this book slopes are designated as is done in analytical geometry, and as generally

Cofferdams

practiced by engineers in all countries where the mathematical slope designation is observed.

A successful cofferdam need not necessarily be completely watertight. For reasons of economy of design, some influx of water into the excavation is usually tolerated. This water is then usually pumped out of the excavation. The only requirement relative to water is that the systems of the earth and other cofferdams should be such that soil erosion of the cofferdam would not occur.

Because soil has little resistance to erosion by water and wave action, earth cofferdams are used in shallow and relatively calm waters. The wet slopes of an earth cofferdam are protected against damage by an impervious clay blanket or by riprap. Overtopping of the dam must be guarded against.

The soil materials to use for building earth cofferdams are sandy clay and clayey sand. Pure cohesive soil is unsuitable for this purpose because it does not lend itself to being worked into the dam under water and to becoming well compacted.

Reinforced concrete foundations can be successfully laid only if water is excluded from the excavation. This is accomplished by pumping, by installing a drainage system inside the excavation from which water is pumped out of the excavation, and by constructing drainage galleries and water-collector galleries along the toe of the earth cofferdam.

A rock-fill cofferdam is one made of a rock fill to enclose the site of a foundation to be dewatered. A rock-fill dam may be built by placing around the rock fill a course of earth or a rock and earth mixture. Or the upstream slope of the rock fill may be covered with an impervious blanket of soil, and the crest and the upper part of the impervious slope may be protected against wave action by riprap (see Fig. 7-2). Overtopping of a rock-fill cofferdam usually does not bring about serious damage to it.

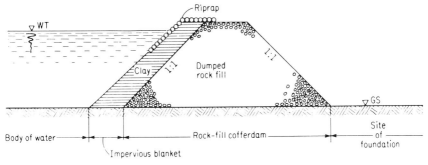

FIG. 7-2. Rock-fill cofferdam.

7-3. STABILITY REQUIREMENTS OF EARTH COFFERDAMS

Every cofferdam, including earth cofferdams, should satisfy requirements for its stability, strength, and reasonable degree of impermeability to water.

Because a cofferdam is a temporary structure, the requirement for its durability is usually dropped. For the same reason as above, the strength and impermeability requirements for temporary structures are also usually somewhat relaxed: the cofferdams may be allowed to permeate a quantity of seepage such that it would not erode the soil and the earth fill. Besides, even as yet soil mechanics does not have any analytical tool for designing absolutely water-impermeable, strong, and stable earth structures. Therefore in actual practice the calculations of a cofferdam are usually performed in each particular case by means of simplified methods accepted by one or another scientific-technical organization or by the owner of the structure.

The forces acting on a seeping earth cofferdam are
1. Weight of dam, G
2. Lateral pressure of water, W (hydrostatic and hydrodynamic)
3. Seepage pressure, D
4. Seismic forces (where this applies), $F = ma$, where m is mass of dam and a = seismic acceleration of the propagating earthquake wave
5. Vertical reaction, $R = G$
6. Frictional force, acting in the contact plane between the dam and its supporting soil (or rock) in the direction opposite to the water pressure W.

An earth cofferdam should be designed to resist the combined effect of all forces that will act on it. Usually an earth cofferdam is subjected to two kinds of stability analyses—stability against lateral sliding, and stability of slopes of the dam. Where settlement is critical, settlement analysis too must be performed.

7-4. SINGLE-WALL SHEET-PILING COFFERDAMS

Sheet-piling cofferdams are one of the most frequently used temporary-type systems. The sheet piling is penetrated into the soil by driving or vibration. Steel sheet piling can be driven to a greater depth and into harder soil than a piling of timber, and its strength requires less bracing than timber sheet piling. Single-wall sheet-piling cofferdams are usually used to enclose small foundation sites in water for bridges at a relatively shallow depth. These are also used as land cofferdams for laying building foundations.

Depending upon the length of the free cantilever of the single sheet piling,

Cofferdams

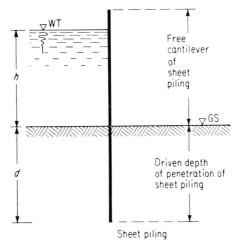

FIG. 7-3. Single-wall sheet piling.

the latter may have at times to be heavily braced. The sheet pile acts as a fixed-end cantilever beam loaded with a horizontal load (lateral earth pressure and hydrostatic pressure of water). A single-wall sheet piling is illustrated in Fig. 7-3. A strutted single-wall sheet piling is shown in Fig. 7-4.

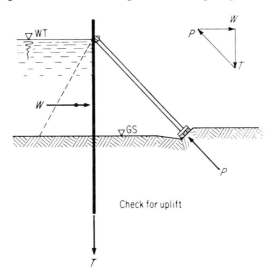

FIG. 7-4. Strutted single-wall sheet piling.

For moderate-flow velocities of water, the single-wall timber sheet piling cofferdams are adequate for a depth of water to about 4.5 ft to 5 ft, and for steel sheet piling from 6 ft to about 12 ft (depending on the section modulus,

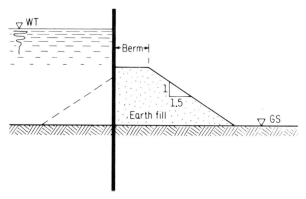

FIG. 7-5. Cofferdam supported by earth fill.

of course). For greater depths and great flow velocities, box-type double-walled cofferdams become necessary.

Leaky joints in single-wall timber cofferdams are sealed from the excavation side by means of oakum, cotton waste, or hemp packing. In the case of a single-wall steel sheet piling, leaks in joints are sealed by means of fast-setting mortar or cement, or by floating fine sand, sawdust, clay, or ashes from behind the wall of sheet piling to clog the leaks.

The toe penetration or driving depth of sheet piling is usually assessed from experience as a fraction of the hydrostatic head h as follows:

In coarse sand, or sand and gravel ...... $d = (0.25)h$
In fine sand. . . . . $d = (0.50)h$
In silt. . . . . $d = (0.85)h$

Figure 7-5 shows a single-wall sheet-piling cofferdam supported on one or two sides by earth fill. The purpose of the earth fill or berm is to aid in

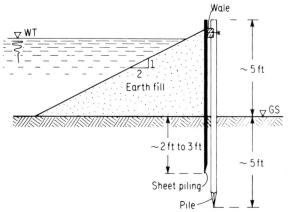

FIG. 7-6. Single-wall earth-supported timber cofferdam.

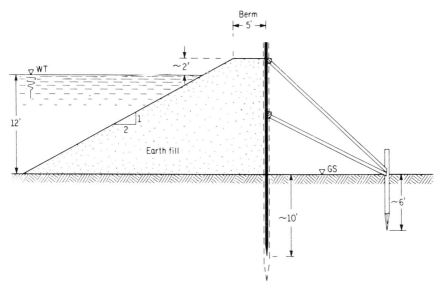

FIG. 7-7. Combined strutted and earth-supported cofferdam.

the lateral stability of the cofferdam, or to cope with leakage (see Fig. 7-6). Figure 7-7 illustrates a combined strutted single-wall timber sheet-piling cofferdam, earth-supported. Figure 7-8 shows an inclined decked-frame cofferdam. The deck of plank sheeting is covered with a bituminous roofing

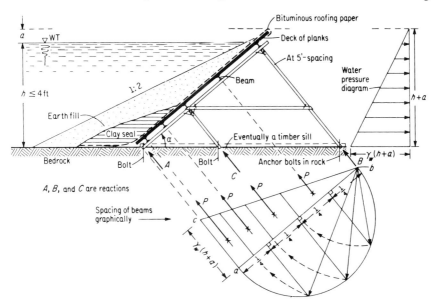

FIG. 7-8. Inclined-frame cofferdam on bedrock.

paper for sealing joints of the planking. This type of cofferdam is suitable where the bottom of the water basin is bare bedrock at shallow depth (up to 4 ft). The frame may be supported directly on the bedrock, or on a sill on bedrock. Whichever the case is, the frame must be tied to the bedrock by means of anchor bolts. The frame may be made of timber or of steel.

After completion of the construction work all cofferdams are removed.

7-5. LAND COFFERDAMS

Single-wall sheet piling is also used as a cofferdam on land for supporting excavation banks from caving in, and when excavation in water requires tight walls of sheet piling (see Fig. 7-9). Here the sheet piling is to be designed

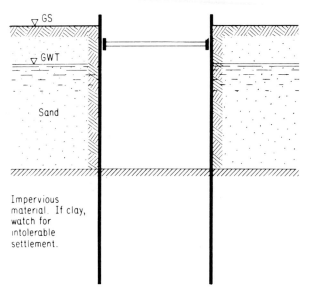

FIG. 7-9. A sheet-piling land cofferdam for excavation in water.

for lateral pressure of saturated soil. The impervious material seals the excavation against water entering into it from below. Because the sheet piling is not absolutely watertight, the excavation is pumped to keep it dry for laying the foundation. Before pumping is commenced and as pumping progresses, the sheet piling must be braced. This is because a single-wall sheet-piling cofferdam is relatively weak in resisting bending stresses from horizontal loads such as soil and water. Excavations of large extent are strutted or anchored on the outside of the excavation (in the case of a land cofferdam).

Cofferdams

Land cofferdams are also used for laying of large spread footings and mat foundations, as well as for foundations of piers and caissons in built-up areas. A land cofferdam is shown in Figs. 7-10a and 7-10b. To keep the excavation dry, it is provided with a drainage ditch, sumps and pumps for the removal of water. The ditches may be filled with crushed rock or gravel.

FIG. 7-10a. Deep land cofferdam, Columbus Circle, New York. Note to the right tie-backs in the pavement. By permission of the Thomas Crimmins Contracting Company, New York, N. Y.

For bracing wide cofferdams horizontal frames and trusses are used. Figure 7-11 shows a braced land cofferdam. Frame bracing is used when there exists a danger of rupture of soil, and when upon advancing the foundation pit the stability of adjacent structures would become impaired.

7-6. DOUBLE-WALL SHEET-PILING COFFERDAMS

The principal types of double-wall sheet-piling cofferdams are
1. Box type, with two straight, flush, parallel vertical walls (double row) of sheet piling (Figs. 7-12 to 7-17), tied to each other and filled between walls with soil

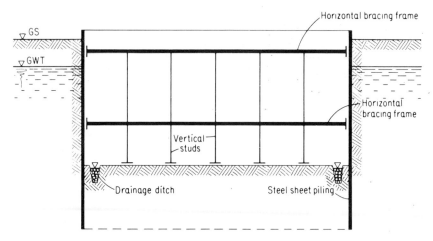

FIG. 7-10b. Land cofferdam.

2. A line of semicircular cells connected by diaphragms (Fig. 8-1)
3. A line of circular cells with tie rods or diaphragm walls (Fig. 8-2).

The double-wall sheet-piling cofferdams are very popular with engineers for enclosing large foundation sites. These types of cofferdams are usually

FIG. 7-11. Braced land cofferdam. Courtesy of U.S. Steel.

Cofferdams

constructed to resist an appreciable head of water. However, they involve some uncertainties such as overturning, sliding, bursting or collapse of the cells, sometimes excessive leakage, and nonuniform pressure distribution at their bases. Where sheet-piling walls in sand hold back an appreciable head of water on one side, sand boiling may set in on the other side.

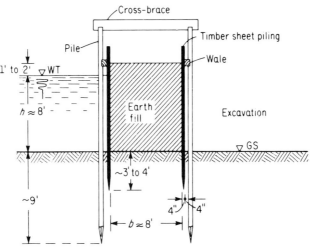

FIG. 7-12. Double-wall box-type timber sheet-piling cofferdam.

A box-type double-wall timber sheet-piling cofferdam is shown in Fig. 7-12. Its width b is empirically assessed as $b = h/2 + 3$ ft to 5 ft. Double-wall sheet-piling cofferdams which are higher than 8 ft should be strutted (Fig. 7-13).

The space between the two rows of sheet-piling walls is purposely filled with soil material (or concrete) having great weight and high coefficient of internal friction, to be determined in the laboratory. The fill material may be gravel, sand, crushed rock, or stones and is introduced by floating, tamping, or vibration to attain the desired density of the fill. Frequently also clay has been puddled into the cofferdam. The fill replaces lateral strutting, and takes care of lateral pressures. Because of its weight, the fill between the double-wall sheet piling performs as a massive body, giving the cofferdam stability against sliding and overturning. Because the soil is subjected to vertical and horizontal loads by the cofferdam, a thorough soil investigation is in order here.

The double-wall sheet-piling cofferdam has the advantage of being less leaky than the single-wall cofferdam. Also, there is freedom from the necessity for cross-bracing in the enclosed excavation.

In essence, steel sheet-piling cofferdams are built in a manner similar to the timber sheet-piling cofferdams. The spacing of the sheet piling is usually taken as $b = (0.3)h$ to $b = (1.0)h$, where b is the spacing between two rows of the sheet piling, and h is the hydrostatic head of water pressure.

The sheet-piling rows are tied with ties spaced 6 ft to 10 ft vertically. Besides, to stiffen the cofferdam, there are installed at horizontal spacings of $3b$ to $4b$ vertical, transverse sheet-piling diaphragms between the two parallel, longitudinal sheet-piling walls.

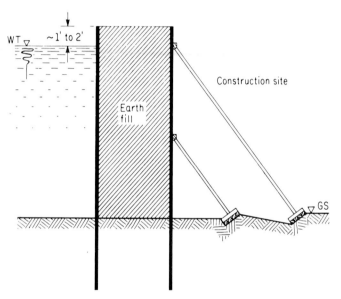

FIG. 7-13. Double-wall, box-type, strutted sheet-piling cofferdam.

The depth of water in front of the cofferdam (outside the excavation) may be designed from 30 to 36 ft and even more, depending upon the length, shape, and section modulus of the sheet-piling profile. The profiles of steel sheet piles may have straight webs, arch webs and also Z-shaped webs.

Figure 7-14 shows a double-wall, box-type, braced sheet-piling cofferdam.

For reasons of economy, uplift in cofferdams should be avoided if at all possible. This is achieved by
1. Driving the sheet piling on the upstream side as deep as is feasible; or, in the presence of an impermeable stratum, the sheet piling should be driven into it in order to seal off the excavation (Figs. 7-15a and b).
2. Dewatering of the cofferdam fill (Fig. 7-16).
3. Lowering the groundwater table of the site enclosed by the cofferdam.

Cofferdams

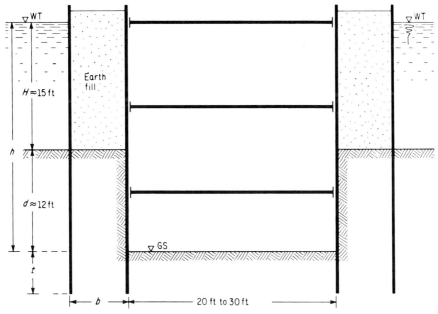

FIG. 7-14. Double-wall, box-type, braced sheet-piling cofferdam.

Effective dewatering requires that the cofferdam fill material be clean, permeable coarse sand or the like. Clay is not suitable for this purpose. In order to reduce the seepage flow velocity, the length of the seepage

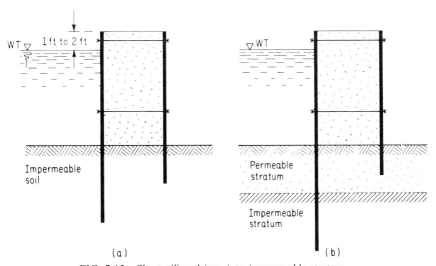

FIG. 7-15. Sheet piling driven into impermeable stratum.

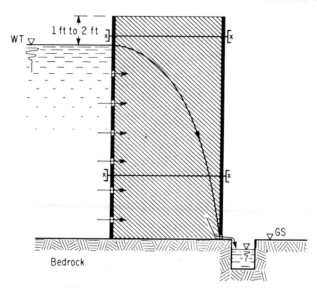

FIG. 7-16. Dewatering of cofferdam fill.

path may be increased and thus the hydraulic gradient reduced by driving the front row (1) of the sheet piling deeper into the soil than the inner row (2) (Fig. 7-17). The berm against row (2) increases the stability of the cofferdam.

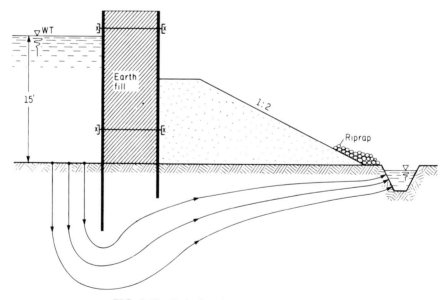

FIG. 7-17. Reducing the seepage flow velocity.

Cofferdams

7-7. UNDERWATER CONCRETING

Underwater concreting is frequently practiced in laying bridge foundations in water. Laying of such a foundation is done in the following sequence (Fig. 7-18):

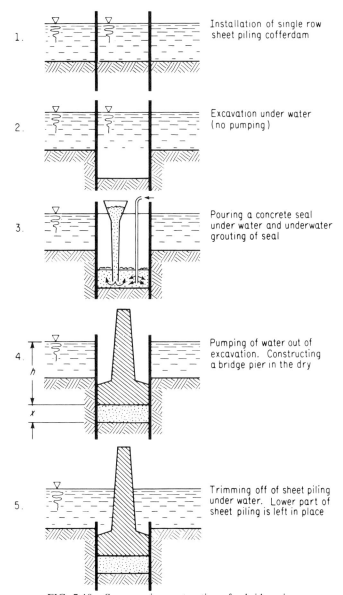

FIG. 7-18. Sequence in construction of a bridge pier.

1. Installation of a single-wall sheet-piling cofferdam.
2. Excavation under water (no pumping).
3. Pouring a concrete seal under water (tremie method) and grouting of the seal.
4. Pumping water out of cofferdam (excavation) and constructing a bridge pier in the dry.
5. Trimming off sheet piling under water. The lower part of the sheet piling is left in place.

The concrete seal must be thick enough to counteract uplift pressure. The approximate thickness x of the seal is calculated from the following equilibrium condition (see sequence 4 in Fig. 7-18):

$$\gamma_c x = \gamma_w (h + x)$$

From this,

$$x = h \frac{\gamma_w}{\gamma_c - \gamma_w} \tag{7-1}$$

where γ_c = unit weight of concrete seal
γ_w = unit weight of water
$h + x$ = uplift pressure head

In this method no effect of the constraint of the concrete seal between the sheet piling as forms is considered. As a matter of fact, this effect is very difficult to comprehend. Hence, this method gives a relatively thick concrete seal. To be sure that the seal does not break because of the uplift pressure, one usually makes the seal somewhat thicker ($> x$) than calculated.

In pouring an underwater concrete seal two important requirements must be met, namely:
1. Upon being placed, the concrete should not move in the water;
2. In the foundation excavation no influx of water should take place (for example through leaky interlock joints of the sheet piling).

To achieve a good underwater concrete seal, the seal is poured uninterruptedly through a stationary, waterproof, funneled pipe approximately 1 ft (~30 cm) in diameter, by introducing the plastic concrete at the bottom of the excavation—"up from below," so to speak (Fig. 7-19). In this method, only the upper surface of the concrete is in contact with the water above in the sheet piling enclosure.

The lower end of the concreting pipe is maintained approximately 3 ft (~0.90 m to 1.00 m) below the surface of the seal concrete. Thus, under water, the plastic concrete spreads laterally about 3 m in radius, and also upward with no segregation of the aggregate. This method is known in the United States as the *tremie method*.

Cofferdams

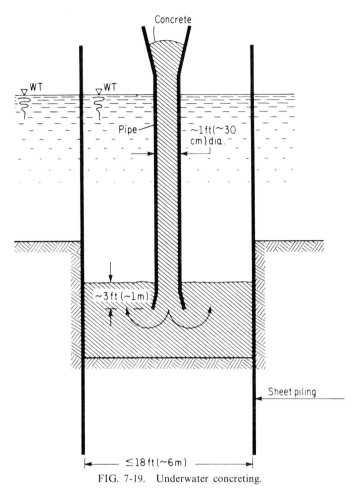

FIG. 7-19. Underwater concreting.

The thickness of the concrete seal to pour is achieved by raising the stationary concreting pipe up and maintaining its lower end always approximately 1 m below the surface of the concrete. In essence, this is a "vertical" method of underwater concreting, and known in Sweden by the term "contractor method." This method is suitable for concreting of small areas (up to about 6.0 m × 6.0 m). Larger areas must be subdivided into smaller sections and several concrete discharge pipes may be used, one in each section. In this method very little cement slurry develops.

Another method, the so-called *horizontal method* of underwater concreting, is the one where the seal is poured in courses through a vertical, mobile concreting pipe on top of each previous course. The lower end of the pipe is at or above the surface of the concrete. The direction of the pour of the concrete of each consecutive course is perpendicular to the direction in the previous, lower course. This method of pouring the concrete may be visualized in Fig. 7-20. This is done to avoid jointing upon jointing. The pouring of concrete is commenced as soon as the pipe and the funnel are filled to a height above the water level. This concrete level in the funnel is kept the same during the entire operation to avoid filling the pipe under the water table. This work can be hampered considerably if the foundation pit contains many struts and braces. A disadvantage of this method lies in the possibility of segregation of the aggregate of the freshly poured concrete,

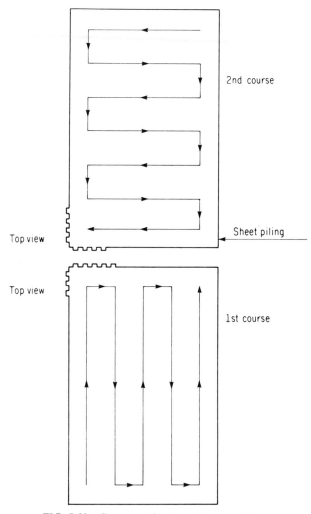

FIG. 7-20. Sequence of underwater concreting.

Cofferdams

and the development of a large amount of cement slurry. This method, in contradistinction to the vertical method, is good for concreting seals of large areal extent.

To make the seal tight, it is sometimes grouted under water before pumping water out of the foundation excavation.

After the seal (and the grout) has set, the water is pumped out of the sheet-piling enclosure, the surface of the seal is cleaned, and the footing and shaft of the pier is constructed in the dry state.

DOUBLE-WALL BOX-TYPE COFFERDAM

7-8. COFFERDAM ON ROCK

Because of the great static indeterminateness and the many uncertainties involved, the double-wall sheet-piling cofferdam does not lend itself to rigorous analysis. Hence, the stability and strength of such a cofferdam can be calculated only approximately.

In general, the height h of the cofferdam depends upon the hydrostatic pressure head plus an allowance for freeboard. However, for a given constant width b of the cofferdam, based on stability conditions, the maximum hydrostatic head, viz., height h, of a cofferdam placed on bedrock may be calculated, with reference to Fig. 7-21 from stability conditions against overturning as follows. Assume that the stage of the water level is at the top of the cofferdam. Allowing a maximum eccentricity $e = b/6$ for the resultant force R acting on the contact plane D-B at point S (no tensile force allowed at heel D of cofferdam), write the moment equilibrium equation with respect to the toe-point B:

$$W(h/3) + G(b/3) - G(b/2) = 0 \qquad (7\text{-}2)$$

where W = hydrostatic pressure of water per one unit of length of run of wall

G = self-weight of cofferdam, including shell and ballast.

Although the condition of $\sigma_{min} = 0$ at the heel when $e = b/6$ cannot exist in a cofferdam the cells of which are filled with a loose, granular pouring body such as sand, many engineers still use this method of calculation for the approximate sizing of the width b of the cofferdam as a function of the height for the purpose of orientation, at least, if for nothing else. With $W = (1/2)\gamma_w h^2$, and $G = N = \gamma b h(1)$, Eq. 7-2 is written as

$$\frac{\gamma_w}{6} h^2 + \frac{\gamma}{3} b^2 - \frac{\gamma}{2} b^2 = 0 \qquad (7\text{-}3)$$

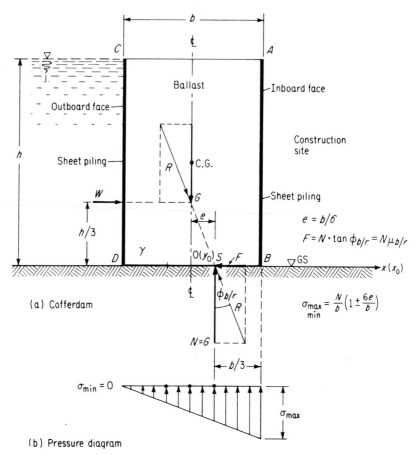

FIG. 7-21. Cofferdam on rock.

where γ_w = unit weight of water
γ = unit weight of ballast
b = width of cofferdam

The solution of this quadratic equation is

$$h = b \sqrt{\frac{\gamma}{\gamma_w}} \tag{7-4}$$

With $\gamma = 2\ t/m^3$, $\gamma_w = 1\ t/m^3$

$$h = (1.41)b \tag{7-5}$$

or

$$b = (0.71)h \tag{7-6}$$

Cofferdams

The relationship $b = (0.85)h$ is used in calculations of cellular-type cofferdams.

The check of the cofferdam against overturning is made only with cofferdams to which horizontal loads are applied. The stability is calculated as a ratio η (Fig. 7-22):

$$\eta = \frac{M_R}{M_D} = \frac{Ga}{Hz} \tag{7-7}$$

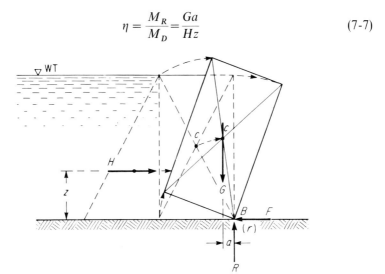

FIG. 7-22. Overturning of a cofferdam.

where $M_R = Ga$ = resultant resisting moment relative to a horizontal axis of rotation (r) through toepoint B
$M_D = Hz$ = resultant driving moment about point B
a = moment arm for vertical force G
z = moment arm for horizontal force H

Here η is the factor of safety. Its value is usually taken as equal to 1.5 or greater.

The stability of a cofferdam against lateral translation over the rock is characterized by the following equation:

$$H \leqslant \frac{F}{\eta} = \frac{G\mu_1}{\eta} \tag{7-8}$$

where H = horizontal force on cofferdam
$F = G\mu$ = frictional force resisting sliding (H)
G = weight of cofferdam
μ_1 = coefficient of friction between base of cofferdam and rock (or soil)
η = factor of safety = 1.5 against sliding

The edge pressures on rock (soil) exerted by the fully backfilled box-type cofferdam or crib wall are calculated as

$$\sigma_{\substack{max \\ min}} = \frac{G}{A} \pm \frac{M_\mathfrak{C}}{I_o} = \frac{G}{A}\left(1 \pm \frac{6e}{b}\right) \tag{7-9}$$

where σ_{max} and σ_{min} are maximum and minimum stresses under the edges of the contact area A of the cofferdam or crib wall on the rock or soil
$A = bd$ = contact area of the cofferdam on rock (soil)
b = width of box-type cofferdam
d = length of a section of cofferdam wall
$M_\mathfrak{C}$ = the resultant moment about centerline $\mathfrak{C}$, including buoyancy if present, as well as seepage forces, earthquake and other forces (if present and/or pertinent)
$I_o = bd^3/12$ = moment of inertia of horizontal base area with respect to central axis y_o perpendicular to the drawing plane
$c = d/2$ = maximum distance of edge (farthest spaced fiber) from centroidal axis $y_o - y_o$ (see Fig. 7-21)
e = eccentricity

The maximum edge pressure σ_{max} should not exceed the allowable bearing capacity of the rock (or soil).

7-9. COFFERDAM ON SOIL

According to Hager,[1] the requirements for the stability of the cofferdam are
1. The pressure line should lie always within the middle third of the width of the cofferdam in order that only compressive stresses should prevail in the ballast of the soil in the cofferdam,
2. The shear stresses induced in the soil are taken up completely by the internal friction of the soil or by the friction between wall and soil so that no displacement whatsoever of the cofferdam would occur.

From the first condition, and referring to Fig. 7-23, obtain the following equations:

Water pressure W_z:

$$W_z = \gamma_w(z^2/2), \tag{7-10}$$

and its moment about S is

$$M_z = W_z(z/3) = \gamma_w(z^3/6) \tag{7-11}$$

Static moment M_D with respect to any point D is

Cofferdams

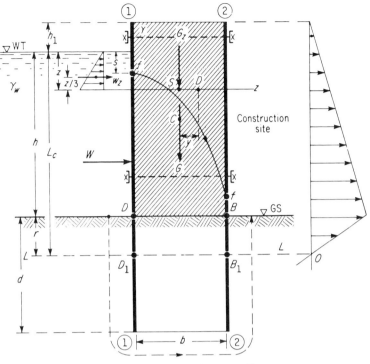

FIG. 7-23. Cofferdam on soil.

$$M_D = M_z - G_z y = \gamma_w(z^3/6) - \gamma(z+h_1)by \qquad (7\text{-}12)$$

This moment becomes zero when point D lies on the pressure line; then

$$y = \frac{\gamma_w z^3}{6\gamma b(z+h_1)} \qquad (7\text{-}13)$$

When $z = h$, then it should be that $y = b/6$. From this requirement the necessary width b of the cofferdam is calculated as

$$b_{min} \geqslant \sqrt{\frac{\gamma_w}{\gamma} \frac{h^3}{(h+h_1)}} \qquad (7\text{-}14)$$

Effective Height of Cofferdam

The most unfavorably stressed horizontal cross-sectional joint of a loose cofferdam sitting on rock is the base joint $D\text{-}B$ (Fig. 7-21). In the case of cofferdams with sheet piling driven into the soil (Fig. 7-23) the question as to the location of the most unfavorably stressed cross section cannot be plausibly resolved. By fixing the sheet piling in the soil, the forces externally applied to the cofferdam are partly transmitted to the soil because of the bending of the sheet piling and by wall friction. Nevertheless the most unfavorable point $D_1 - B_1$, it may be assumed, would be located at some distance r below the ground surface. Its location may be assumed to be the height $L_c = h + r$, where the quantity r is usually taken as 2 to 3 ft; or else, for reasons of safety, r may be taken even larger than 3 ft. The effective or computation height of the cofferdam is then to be taken for calculations as being this height, L_c.

Shear Stress

The maximum tangential forces in the soil ballast brought about by the lateral pressure of water occur in the cofferdam where the maximum shear force acts. This is at the base of the dam. Thus from the second condition:

$$W = \gamma_w (h^2/2)$$

and the maximum shear stress in a rectangular cross-section $A = b(1)$ or a vertical cross section is

$$\tau_{max} = (3/2)(W/A) = (3/2)(\gamma_w/2A)h^2 = (3/4)\gamma_w(h^2/b) \quad (7\text{-}15)$$

where $A = b(1)$ is the horizontal cross-sectional area of the cofferdam.

This shear stress must be kept in equilibrium by the frictional stress τ_μ:

$$\tau_\mu = \mu(G/A) = \mu(G/b) = \mu\gamma \frac{(h+h_1)b}{b} = \mu\gamma(h+h_1)$$

or

$$\tau_{max} = \tau_\mu = (3/4)\gamma_w(h^2/b) = \mu\gamma(h+h_1) \quad (7\text{-}16)$$

Thus for safety against shear the necessary width b of the cofferdam is

$$b \geqslant \frac{3}{4} \frac{\gamma_w}{\gamma} \frac{1}{\mu} \frac{h^2}{(h+h_1)} \quad (7\text{-}17)$$

If the water level is as high as the cofferdam ($h_1 = 0$), and if instead of h the effective or computation height $L_c = h + r$ is used, then Eq. 7-17 is written as

$$b \geqslant (0.75) \frac{\gamma_w}{\gamma} \frac{1}{\mu} L_c \quad (7\text{-}17a)$$

The latter equation gives a somewhat greater value of b than Eq. 7-17.

Of the two values of b (Eqs. 7-14 and 7-17) one chooses the larger one for the design.

The width b of a double-wall sheet-piling cofferdam is usually taken as not greater than the head h of water.

If the weight of the ballast in the upper part (h_1) of the cofferdam is $\gamma_1 h_1$ (drier than below), and the lower part h is saturated, having a saturated

Cofferdams

weight $\gamma_{sat} \cdot h$, the quantity $(h \cdot \gamma_{sat} + \gamma_1 \cdot h_1)$ must be used because of different self-weights γ_1 and γ_{sat} of the ballast.

On the vertical plane (joint) inside of the cofferdam the only lateral force is the earth pressure E exerted by the ballast. Its magnitude is considerably less than the vertical soil pressure, because the earth pressure coefficient K is less than 1.

For narrow cofferdams with confined ballast between walls of sheet piling, the value of K is, according to Krynine:[2,3]

$$K = \frac{\cos^2 \phi}{2 - \cos^2 \phi} < 1.0 \qquad (7\text{-}18)$$

For wide cofferdams the K-values may be obtained from Coulomb's earth-pressure theory,[4,5] or directly from *Active and Passive Earth Pressure Coefficient Tables*.[6]

Example

Calculate the minimum width b_{min} of a cofferdam for the following data:

$h = 10.0$ m $h_1 = 0$ m $r = 1.0$ m $L_c = 11.0$ m
$\gamma_w = 1.0$ ton/m³ $\gamma = 2.0$ ton/m³ $\mu = 0.5$

Solution

1. (i) By Eq. 7-14:

$$b_{min_h} = \sqrt{\left(\frac{1.0}{2.0}\right) \frac{(10.0)^3}{(10.0 + 0)}} = 7.05 \approx \underline{7.10 \text{ (m)}}$$

   (ii) By Eq. 7-14 for $L_c = 11.0$ m:

$$b_{min_{L_c}} = \sqrt{\left(\frac{1.0}{2.0}\right) \frac{(11.0)^3}{(11.0 + 0)}} = \underline{7.75 \text{ (m)}}$$

2. (i) By Eq. 7-17:

$$b_{min_h} = (0.75) \frac{(1.0)}{(2.0)} \frac{1}{(0.5)} \frac{(10.0)^2}{(10.0)} = \underline{7.50 \text{ (m)}}$$

   (ii) By Eq. 7-17a,

$$b_{min_{L_c}} = (0.75) \frac{(0.5)}{(0.5)} \frac{(11.0)^2}{(11.0)} = \underline{8.25 \text{ (m)}}$$

Ans. The minimum width of the cofferdam is $b_{min} = \underline{8.25 \text{ m}}$

Diaphragm Stress

The stress σ_d in the diaphragm of a parallel, straight double-wall box-type cellular cofferdam is calculated as

$$\sigma_d = p_E L \qquad (7\text{-}19)$$

where p_E is the largest earth pressure ordinate from the ballast in the cell

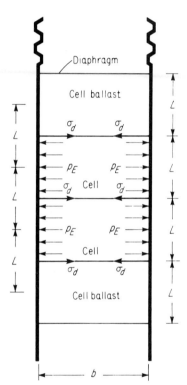

FIG. 7-24. Top view of cellular box-type cofferdam. Straight walls of sheet piling.

(see Fig. 7-24), and

$$L_{\max} = \sigma_{\text{all}}/p_E \qquad (7\text{-}20)$$

where L = length of cell
 σ_{all} = allowable stress in sheet piling (usually given in the catalogs of sheet piling manufacturers)

Cofferdams

Ballast Material

The ballast material for filling the cells of a cofferdam should be a clean, coarse-particled, free-draining soil. Fine-particled soil materials produce maximum bursting and minimum cell rigidity. Their use may necessitate interior berms and increased cell width b.

To increase the stability of the cofferdam, it is beneficial to place a berm against the inboard row 2 of the sheet piling (on the side of the construction site) (Fig. 7-25). This also adds to the economy of steel to use, because the inboard row of sheet piling can be made somewhat shorter than the one in the outboard row 1.

The stability check against lateral expulsion of soil from underneath the base (groundbreak) of an eccentrically loaded cofferdam (or a crib wall) may be performed by methods as shown in Refs. 4, 5, and 7.

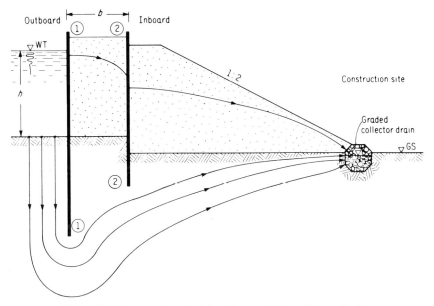

FIG. 7-25. Berm against the inboard row of sheet-piling cofferdam.

Hydraulic Gradient

If the permeability of the sheet-piling wall 2-2 is greater than that of wall 1-1, then the uppermost flow line of seepage f-f is steep, with a large drawdown s at the inside of the wall 1-1 (Fig. 7-23). If, however, the sheet-piling wall 2-2 is less permeable than the wall 1-1, then the position of the uppermost seepage line in the double-wall cofferdam may be close to the position of the water table outside the cofferdam.

To prevent the soil from becoming "quick" near the toeline (through point B) of the cofferdam on the construction side, the critical hydraulic gradient thus must be checked. The critical hydraulic gradient is calculated as

$$i_{crit} = \frac{G-1}{1+e} \tag{7-21}$$

where G = specific gravity of soil solid particles
e = void ratio of soil

This critical hydraulic gradient must be compared with the actual one, i. The actual hydraulic gradient should be less than the critical hydraulic gradient.

Stability of Soil Particles Against Erosion

The stability of soil particles against erosion at the bottom of the cell at point B may be assessed by requiring that the seepage pressure $D = \gamma_w i$ be less than the dry unit weight γ_{dry} of the soil. Here γ_w is the unit weight of water, and i is the actual or rather approximative hydraulic gradient at point B (Fig. 7-23):

$$i = \frac{h}{2d+b} \tag{7-22}$$

With a factor of safety $\eta > 1$, the stability condition of soil particles against erosion at point B is

$$\eta \cdot D = \gamma_{dry} \tag{7-23}$$

or

$$\eta \cdot \gamma_w \frac{h}{2d+b} = \gamma_{dry}$$

or

$$\eta \cdot h \frac{\gamma_w}{\gamma_{dry}} = 2d+b \tag{7-24}$$

7-10. DRIVING DEPTH OF SHEET PILING

The minimum driving depth d at the top tied sheet piling for a box-type cofferdam may be approximately calculated for the equilibrium condition when the active earth pressure E_a on wall 2-2 is set equal to the passive earth pressure E_p, i.e., when $E_a = E_p$ (Fig. 7-26). At such a lateral earth pressure equilibrium condition there exists a particular quantity d, which is then

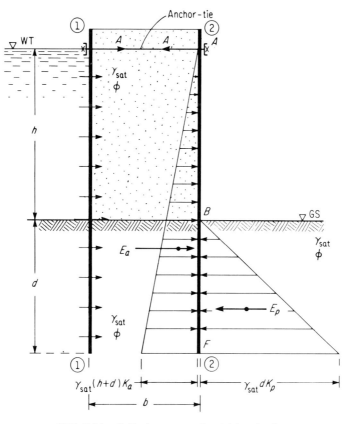

FIG. 7-26. Cofferdam system for driving depth.

calculated as the minimum driving depth d_{min}. In this method of calculation the hydrostatic pressure of water is disregarded, but in the earth pressure calculations the saturated unit weight of soil γ_{sat} is used. The equilibrium condition $E_a = E_p$ is now written for the given earth pressure system as

$$(1/2)\cdot\gamma_{sat}\cdot(h+d)^2 K_a = (1/2)\cdot\gamma_{sat} d^2 K_p \tag{7-25}$$

or

$$d = d_{min} = \frac{h}{K_p - 1} = \frac{h}{\tan^2\left(\frac{\pi}{4} + \frac{\phi}{2}\right) - 1}$$

$$= \frac{h \tan^2\left(\frac{\pi}{4} - \frac{\phi}{2}\right)}{1 - \tan^2\left(\frac{\pi}{4} - \frac{\phi}{2}\right)} \tag{7-26}$$

where $K_p = \tan^2\left(\dfrac{\pi}{4} + \dfrac{\phi}{2}\right)$ = passive earth-pressure coefficient

$K_a = \tan^2\left(\dfrac{\pi}{4} - \dfrac{\phi}{2}\right)$ = active earth-pressure coefficient

ϕ = angle of internal friction of soil on both sides of sheet piling 2-2

7-11. DEFLECTION OF SHEET PILING

To size approximately the sheet piling and the anchor or tie rods, assume the sheet piling to be loaded with triangularly distributed loads per one unit length of run of wall as shown in Fig. 7-27. Assume that the position L-L of fixity of the sheet piling so loaded is r units (from 2 to 3 ft) below the ground surface. Hence, the computation length of the cantilever is L_c. The ordinates of the pressure diagrams are shown in Fig. 7-27.

For the sake of simplicity, assume that the hydrostatic pressure of water is transmitted to the sheet piling wall 1-1 only, and that the unit weight

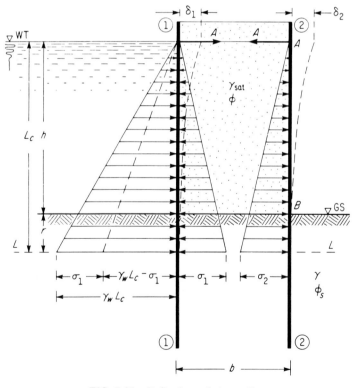

FIG. 7-27. Deflections of sheet piling.

Cofferdams

of the soil in the cell has a maximum value, i.e., the soil is in a saturated condition.

The stress ordinates σ_1 and σ_2 are then respectively:

$$\sigma_1 = \gamma_w L_c - \gamma_{sat} L_c K_a \tag{7-27}$$

$$\sigma_2 = \gamma_{sat} L_c K_a \tag{7-28}$$

where $K_a = \tan^2\left(\dfrac{\pi}{4} - \dfrac{\phi}{2}\right)$

Designating the force in the tie rod by A, the deflections δ_1 and δ_2 for the cantilever sheet piling 1-1 and 2-2, respectively, the deflections are[8]

$$\delta_1 = \frac{\sigma_1 L_c^4}{30 E_1 I_1} + \frac{A L_c^3}{3 E_1 I_1} \tag{7-29}$$

$$\delta_2 = \frac{\sigma_2 L_c^4}{30 E_2 I_2} - \frac{A L_c^3}{3 E_2 I_2} \tag{7-30}$$

Here E_1 and E_2 are the Young's moduli of elasticity of the sheet-piling material, and I_1 and I_2 are the moments of inertia of sheet-piling profiles.

But because the upper part of the sheet-piling wall 2-2 is tied by means of an anchor 1-2 to the wall 1-1, assume that both deflections are equal, i.e., $\delta_1 = \delta_2 (= \delta)$. With $E_1 = E_2$ and $I_1 = I_2$, obtain from Eqs. 7-29 and 7-30 the magnitude of the force A in the tie rod:

$$A = \frac{L_c}{20}(\sigma_2 - \sigma_1) \tag{7-31}$$

The maximum bending moments for the sheet piling are written with respect to the plane L-L of fixity of the cantilever beams:

$$M_{1_{max}} = \frac{\sigma_1 L_c^2(1)}{6} + A L_c \tag{7-32}$$

$$M_{2_{max}} = \frac{\sigma_2 L_c^2(1)}{6} - A L_c \tag{7-33}$$

With the bending moments known, the sheet piling of a proper strength to use may now be selected and ordered for cofferdam work.

Example

Given a double-wall box-type sheet-piling cofferdam system as follows (Fig. 7-27):

$h = 12$ ft $\quad b = 9$ ft

Saturated unit weight of sand in the cell: $\gamma_{sat} = 129.3$ lb/ft^3

Angle of internal friction: $\phi = 20°$
The angle of internal friction of soil below ground surface is $\phi_s = 30°$, porosity is $n = 35$ percent, and specific gravity of soil particles is $G = 2.65$.

Calculate:
1. Minimum driving depth of sheet piling
2. Minimum driving depth relative to stability of soil particles against erosion for $\eta = 3.0$
3.. Stresses for water and soil-pressure diagrams
4. Tension force in anchor
5. Maximum bending moments

Solution

1. Minimum driving depth (by Eq. 7-26):

$$d_{min} = h/(K_p - 1) = 12/(3.000 - 1) = \underline{6 \text{ (ft)}}$$

For system 5, Ref. 6, Table P-72, p. 278, and for $\phi = 30°$, $\phi_1 = 0$, $K_p = 3.000$

2. Minimum driving depth relative to stability of soil particles against erosion, by Eq. 7-24,

$$d = \eta \frac{h}{2} \frac{\gamma_w}{\gamma_{dry}} - \frac{b}{2} = (3.0) \frac{(12)}{(2)} \frac{(62.4)}{(107.5)} - \frac{9.0}{2}$$

$$= 5.9 \text{ (ft)} \approx 6.0 \text{ (ft)}$$

Here

$$\gamma_{dry} = (1 - n)G\gamma_w = (0.65)(2.65)(62.4) = \underline{107.5 \text{ (lb/ft}^3)}$$

3. Stresses at depth $L_c = h + r$.
The computation length L_c of cantilever sheet piling for $r = 3$ ft is

$$L_c = h + r = 12 + 3 = \underline{15 \text{ (ft)}}$$

$$\sigma_1 = \gamma_w L_c - \gamma_{sat} L_c K_a$$

$$= (62.4)(15) - (129.3)(15)(0.271) = 410.4 \text{ (lb/ft}^2) = \underline{0.205 \text{ ton/ft}^2}$$

$$\gamma_{sat} = \gamma_{dry} + n\gamma_w = (107.5) + (0.35)(62.4) = \underline{129.3 \text{ (lb/ft}^3)}$$

By Table A-72, for system 5, $\phi = 20°$, and $\phi_1 = 0$, $K_a = 0.271$ (see Ref. 6).

$$\sigma_2 = \gamma_{sat} L_c K_a = (129.3)(15)(0.271) = 525.6 \text{ (lb/ft2)} = \underline{0.263 \text{ ton/ft}^2}$$

Cofferdams 209

4. Tension force A in anchor (for 1 ft length of run of wall), by Eq. 7-31:

$$A = \frac{L_c}{20}(\sigma_2 - \sigma_1) = (15/20)(0.263 - 0.205)$$
$$= \underline{0.0435 \text{ (ton/ft)}} = \underline{87 \text{ lb/ft}}$$

5. Maximum bending moments: By Eqs. 7-32:

$$M_{1_{max}} = \frac{\sigma_1 L_c^2(1)}{6} + AL_c = \frac{(410.4)(15)^2}{6} + (87.0)(15)$$
$$= \underline{16{,}695.0 \text{ (ft-lb)}}$$

Section modulus Z:

$$Z = M/\sigma_{all}$$

where M = maximum bending moment
σ_{all} = allowable bending stress for sheet piling material

In the case when E_1 and $I_1 \neq E_2$ and I_2, respectively, the problem must be recalculated for each sheet-piling wall separately. If $E_1 = E_2 = E$, then

$$A = \frac{L_c}{20}\left(\frac{\sigma_2}{I_2} - \frac{\sigma_1}{I_1}\right)\left(\frac{I_1 I_2}{I_1 + I_2}\right) \tag{7-34}$$

REFERENCES

1. K. Hager, "Die Berechnung von Fangdämmen," *Wasserkraft und Wasserwirtschaft*, No. 14 (July 1931), pp. 165-166.
2. K. Terzaghi, "Stability and Stiffness of Cellular Cofferdams," *Trans., ASCE*, Paper No. 2253, Vol. 110 (1945), pp. 1083-1119, with discussion by D. P. Krynine on pp. 1175-1178.
3. A. R. Jumikis, *Theoretical Soil Mechanics*, Princeton, N.J.: Van Nostrand, 1969.
4. A. R. Jumikis, *Soil Mechanics*, Princeton, N.J.; Van Nostrand, 1962, pp. 549-596.
5. A. R. Jumikis, *Mechanics of Soils: Fundamentals for Advanced Study*, Princeton, N.J.; Van Nostrand, 1964, pp. 219-329.
6. A. R. Jumikis, *Active and Passive Earth Pressure Coefficient Tables*. New Bunswick, N.J.; Rutgers University, Bureau of Engineering Research, College of Engineering, Engineering Research Publication No. 43, 1962.
7. A. R. Jumikis, *Stability Analyses of Soil-Foundation Systems*. New Brunswick, N.J.; Rutgers University, Bureau of Engineering Research, College of Engineering, Engineering Research Publication No. 44, 1965.
8. L. Brennecke and E. Lohmeyer, *Der Grundbau*, 4th ed., Berlin; Ernst, 1927, Vol. 1, p. 226.

PROBLEMS

7-1. Given an open single-wall cofferdam as shown in Fig. 7-18, sequence 4.
   (a) Explain the preparation of the foundation pit (excavation) enclosed by the cofferdam.
   (b) What is the purpose of the concrete seal?
   (c) How is the concrete seal constructed?
   (d) What is the minimum thickness x of the seal if $h = 30$ ft?

chapter 8

Cellular-Type Cofferdams

8-1. TYPES AND USE OF CELLULAR COFFERDAMS

Cellular cofferdams are usually constructed of steel sheet piles, and are of two basic types, namely:

1. Segment-type (Fig. 8-1)
2. Circular-type (Fig. 8-2).

A third type may be mentioned, the so-called cloverleaf cofferdam, which is a succession of four merged cylindrical cells (Fig. 8-8). The four cells are subdivided by straight, flat diaphragms to form "cloverleaf" cells.

The *segment-type* cellular cofferdam consists of parallel walls of circular arcs with cross tie rods or straight connecting diaphragm walls.

The *circular-type* cellular cofferdam is a line of circular cells connected by diaphragms. The main characteristics of cellular cofferdams are the absence of bracing and durability.

Cellular-type sheet-piling cofferdams are used for a large variety of depths in water in foundation, hydraulic, and maritime-structures engineering for laying foundations and for building locks, docks, hydraulic power plants, and other pertinent structures. Such cofferdams are used as temporary as well as permanent structures, and are constructed on rock as well as on soil.

After the foundation work is completed, sheet piling is easily and quickly removed by pulling the sheet piling out or by cutting it off. Waste of sheet piling usually does not exceed 20 to 30 percent, and the recovered sheet piling may have a 3 to 5-fold turnover. Sheet piling can be easily vibrated into place in sand, hence is economical.

8-2. SEGMENT-TYPE CELLULAR COFFERDAMS

The segment-type cofferdam is a double-wall structure made of interconnected steel sheet piling. The inside and outside rows of this cofferdam are formed as segments of circular arcs. These walls are connected by straight, connecting, flat diaphragm walls or tie rods through the intersect-

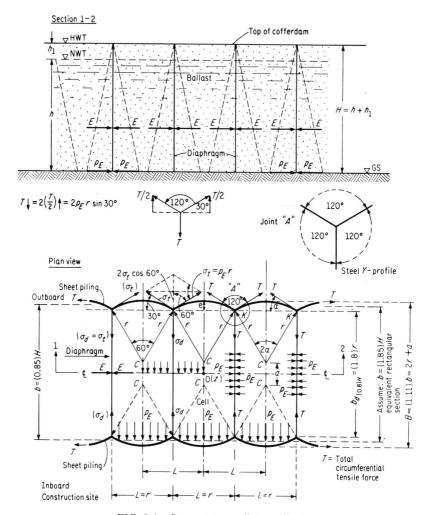

FIG. 8-1. Segment-type cellular cofferdam.

ing points of the arcs, thus forming cells of the cofferdam (Fig. 8-1). The connections of the curved parts and the diaphragm of the cofferdam are made by means of fabricated Y-elements. The cells are filled with coarse-particled, freely draining soil material. Thus the cofferdam is a gravity type structure.

If the Y-element is made so that its three legs form three equal angles of 120°, then the length of the chord $(k\text{-}k) = L$ (width of cell) is equal to the radius r of the arc. This means that the tensile forces in the arcs and diaphragms are of equal magnitude. The central angle of the arcs is $2\alpha = 60°$. In such a case the centers of the arcs lie on the longitudinal center line

Cellular-type Cofferdams

($\mathcal{C}$ – $\mathcal{C}$) of the cofferdam. With 120° leg angles, the width B of the cofferdam is about $B \approx (1.11)b$. Of course, the cofferdam may be designed for any angle α other than 60°. In such a case the centers of the arcs do not lie on the longitudinal center line of the cofferdam.

The height of the segment-type cofferdams is up to 45 to 55 ft (about 15 to 18 m). The width b of the equivalent rectangular cofferdam of the segmental cellular-type cofferdam is $b \approx (0.9)B$ and $b \approx (0.85)H$.

For the segmental-type of cofferdam the width L of its cells is assumed such that destructive (tensile) forces in interlocks of steel sheet piling do not exceed allowable stresses of the material. If $\alpha = 30°$, then $L = r$.

If $B = 2r = (1.11)b$, then the width L of the cell is approximately $L = r = (0.56)b$.

The width L of the cell is determined approximately as follows: if p_E is the largest earth-pressure ordinate, then the stress σ_t at the point of the junction of the curved and straight diaphragm walls is

$$\sigma_t = p_E r \tag{8-1}$$

and

$$r_{max} = \frac{\sigma_{all}}{p_E} \tag{8-2}$$

where σ_{all} pertains to the allowable stress in the sheet piling. Also,

$$\cos \alpha = 1 - \frac{e}{r_{max}} \tag{8-3}$$

and

$$L = 2 r_{max} \sin \alpha, \tag{8-4}$$

where e is the height of the segment (see Fig. 8-1). With $\sin \alpha = \sin 30°$, $L = r$, and also $\sigma_t = p_E$.

These equations are valid for cellular-type as well as circular-type cofferdams.

In order to increase the stability of the cofferdam, its cells are filled with clean, granular, well-draining material serving as ballast to increase the weight of the cofferdam. This also serves to decrease leakage, viz., seepage through the cofferdam.

The double-wall segmental cellular cofferdam is relatively insensitive to overtopping and erosion. These problems may be lessened by installing floodwater gates through the cofferdam.

The filling of the cells must be done very carefully. Because of the flat diaphragm walls, it is necessary to fill simultaneourly a large number of adjacent cells at approximately the same rate to prevent the diaphragms from being loaded by too large a one-sided lateral pressure. Otherwise, an unbalanced pressure from the filling ballast material may distort the diaphragm, thus causing a rupture of the interlocks of the sheet piling.

Rupture of interlocks in this type of cellular cofferdam occurs more often than in cylindrical cell cofferdams. This is one of the objections against the use of a segment-type cellular cofferdam. This type of cofferdam, however, requires less sheet piling as compared with other types of cellular cofferdams.

The cells are usually filled with ballast of soil up to about 6 ft (2 m) high along the front (outside) wall of the cofferdam.

8-3. CIRCULAR-TYPE CELLULAR COFFERDAM

For large depths of water and on bare rock the circular-type cellular cofferdam of steel sheet piling is usually used. The layout of this type of cofferdam consists of a linear group of individual, interconnected circular cylindrical cells of diameter D joined with smaller connecting cells of arcs of sheet piling, the diameter of which is approximately $(0.6)D$. Thus the walls of the connecting cells are perpendicular to the circular cylinder walls (Fig. 8-2). The practical limit of the diameter of the circular cells is about 70 to 75 ft; the average is 60 ft. The circular-type cofferdam cells are self-sustaining, independent of the adjacent circular cells, and each cell can be filled with ballast independently and without interruption. Their stability during construction and filling is greater as compared with the segmental cellular cofferdam. The circular cells can be filled in immediately after completion of construction, thus also affording support for construction equipment. However, the circular cells require more sheet piling than the segmental cells.

The diameter of the circular cells is chosen from $D = (1.0)H$ to $D = (1.2)H$, where $H = h + h_1$ is the height of the cofferdam. Cylindrical cells are constructed approximately 45 to 55 ft (about 15 m to 18 m) high. The circular cells are spaced empirically from $L = (1.05)D$ to $L = (1.0)D$ apart.

The circular cells are joined to the segmental part of the connecting cell by means of T-piles. These, four in number at each cylinder, are built into the circular sheet-piling wall, to which the arc piles are joined. The sketches of a circular cellular cofferdam are shown in Fig. 8-2.

The ballast in the cell brings about tensile stresses in the cylindrical wall of the cell. Both cells in this cofferdam, the circular and the segmental

Cellular-type Cofferdams

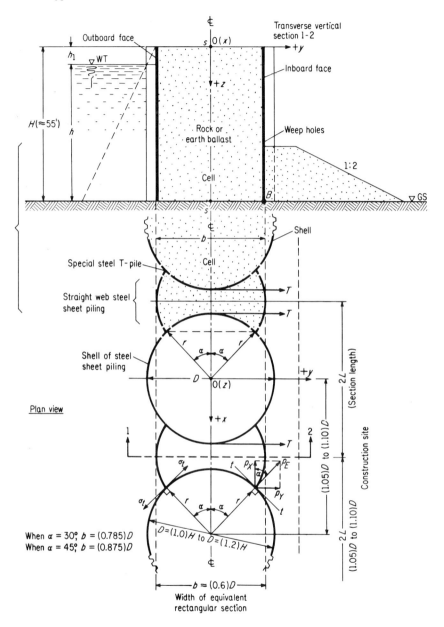

FIG. 8-2. Circular cellular cofferdam.

ones, are filled with clean, coarse-grained soil material. Possible rupture of the interlock in the circular cell cofferdam is usually localized in one or a few cells—a great advantage over the box or segmental types of cellular cofferdams. Also, the circular cofferdam cells and connecting arcs are relatively easy to install, the construction is simple, and its sheet piling can be salvaged after completion of the construction work and reused on subsequent projects. The diameter D of the cell shell is determined from stability conditions against sliding of the circular cell cylinder.

To resist the one-sided lateral pressure of water, the circular cell is designed as a gravity structure. To perform as a rigid unit, the ballast-filled cell must be able to resist vertical shear in the vertical, longitudinal plane through the longitudinal axis of the cofferdam wall.

Part of the strength of the circular cellular cofferdam, viz., cell, is derived from the resistance caused by friction induced by tension in the sheet-piling interlocks, but the major part comes from the internal shearing resistance of the ballast itself. The overall stability of the cofferdam is derived from the composite length of the ballast-filled steel shell. Thus the cell must be analyzed for tensile stresses in the interlocks, and for longitudinal shear.

For simplicity in calculating the stability of circular cell cofferdams the width b of the cofferdam is assumed as if the cofferdam were rectangular in horizontal as well as vertical cross section. The length of a repetitional section of the circular cofferdam for analysis is $2L$ (see Fig. 8-2).

The circular cellular cofferdam is thus analyzed for rupture in tension of the interlocks near the base of the shell, vertical shear, bearing capacity of soil at the base (contact) plane, stability against groundbreak, stability against sliding and overturning.

8-4. STATIC SYSTEMS AND NEEDED INFORMATION FOR DESIGN OF CELLULAR COFFERDAMS

The possible forces acting on a cellular cofferdam (save for earthquake forces) may be comprehended for the following two force cofferdam systems, namely:

1. For a cellular cofferdam with no hydraulic pressure head (Fig. 8-3a)
2. For a cellular cofferdam with a hydraulic head (Fig. 8-3b)

The symbols in Fig. 8-3 are:

p_0 = surcharge, lb/ft^2 (or in t/m^2)
$w_1 - w_0 - w_2$ = surface of wave
P_i = ice pressure, lb/ft (t/m)
P_w = wave pressure, lb/ft (t/m) (see pressure diagram 1-2-3-4-1)
W = water pressure (I-II-III-I), lb/ft (t/m)

Cellular-type Cofferdams

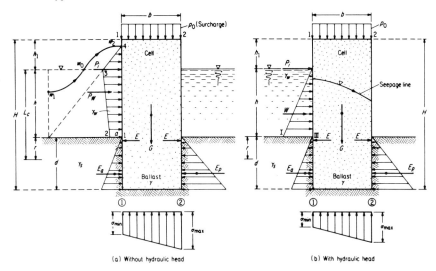

FIG. 8-3. Static systems. (a) Without hydraulic head, (b) With hydraulic head.

E_a = active earth pressure, lb/ft (t/m)
E_p = passive earth pressure or resistance, lb/ft (t/m)
ϕ_s = angle of internal friction of cofferdam-supporting soil
G = weight of cell, including self-weight, ballast of sand and uplift force, lb (or kg or tons)
$E = (1/2)\gamma h^2 K$ = lateral pressure against inside walls of cell from ballast
$K = \cos^2\phi/(2 - \cos^2\phi) = 1/(1 + 2\tan^2\phi)$
ϕ = angle of internal friction of ballast
γ = unit weight of ballast
γ_s = unit weight of soil
γ_w = unit weight of water
$1 - 1$ = outboard row of sheet piling
$2 - 2$ = inboard row of sheet piling
All other symbols are as before.

For the design and analysis of a cellular cofferdam the following minimum information is needed:

1. Engineering data on geological and hydrotechnical conditions pertaining to the construction site, such as geological formations, water regimen, flow velocity and stages, ice pressure, groundwater conditions, for example
2. Information on geotechnical exploration of the site

3. Test results on soil and/or rock physical, chemical and strength properties at the site
4. Data on maximum and minimum hydrostatic and/or hydrodynamic pressure heads (where they apply), and consideration of necessary allowable overtopping
5. Erosion and scour characteristics of a river bed at the proposed site of construction (where this applies)
6. Information on possible surcharge on cofferdams (type and load of equipment, materials, for example)

After this information is on hand the designer decides on the size of the cofferdam and the type and size of sheet piling to use, and checks the strength of the interlocks of the sheet piling against rupture. The cellular cofferdam dimensions are then checked against

1. Sliding of cell over its base support (soil or rock)
2. Overturning (occurs seldom, or is preceded by piping in soil)
3. Deformation in shear of soil in the central vertical plane in the middle of the width of cofferdam, brought about by an inward tilt of the cofferdam, or by a failure of the cofferdam-supporting soil
4. Groundbreak of cofferdam-supporting soil brought about by the ballast-filled cofferdam (a bearing capacity check)
5. Its strength, viz., rupture of sheet-piling interlocks and the Y- and T-elements, subjected to tensile forces brought about by the ballast in the cells and seepage pressure
6. Filtration through the base and the body of the cofferdam

In selecting the proper weight, viz., cross section, of the sheet piling for a permanent structure, an allowance for corrosion of about 0.1 to 0.2 mm/year should be considered.

Depending upon the geological conditions encountered at the construction site, cofferdams are supported on rock, on thick sand formations and on firm clay. Hence, methods of stability analyses of cellular cofferdams are classed accordingly.

For analysis of cellular-type cofferdams, transform these into equivalent parallel-wall cofferdams of width b. The analysis is good for cellular cofferdams of the segment-type as well as of the circular-type. The stability of a cellular cofferdam depends upon the ratio of width to height (b/h), presence of an inboard berm, type of cell-filling ballast material, and kind of drainage. For clean sand in the cell, the only drainage required is usually weep holes in the inboard wall of the sheet piling to drain the cell as much as possible. This reduces the hydrostatic pressure on the inside row of sheet piling and increases the shear strength of the ballast material in the cell.

The average width of a cellular cofferdam on rock is usually taken as 85 percent of the head of water to be resisted: $b = (0.85)h$.

If a cofferdam is placed on sand, its stability against sliding, overturning and shear can be increased by placing a berm against the inboard row (2) of the sheet piling. The berm also increases the length of the seepage path, thus decreasing the hydraulic gradient, and hence contributes to the stability against a quick condition of sand, piping or erosion, and subsequent collapse of the cofferdam.

To prevent pullout of the outboard row of sheet piling and piping at the interior row of the wall, the penetration of the sheet piling must be of adequate depth.

Relative to cofferdams on clay, the same requirements as above obtain, except that piping is probably not too critical here.

The cofferdam loading should not exceed the allowable bearing capacity of the soil, and the cofferdam soil system should be checked for groundbreak (lateral expulsion of soil from underneath the base of the obliquely loaded cofferdam).

8-5. CELLULAR COFFERDAM ON ROCK

The static calculations of the cellular cofferdam start out with assuming the equivalent width b of a rectangular cofferdam: $b = (0.85)H$ (Fig. 8-4).

Then the effective weight G_{eff} of the cell ballast is computed:

$$G_{\text{eff}} = \gamma_{\text{soil}} b(H - h_s) + \gamma_{\text{sub}} b h_s \tag{8-5}$$

Lateral pressures per one unit length of run on the outboard face of the cofferdam:

(i) Full possible hydrostatic pressure W of water:

$$W = (1/2)\gamma_w H^2 \tag{8-6}$$

here H = full height of cofferdam (Fig. 8-4).

(ii) Active earth pressure E_a for $\tan \phi_1 = \mu_1 = 0$:

$$E_a = (1/2)\gamma_{\text{sub}} d^2 K_a \tag{8-7}$$

where $K_a = \tan^2\left(\dfrac{\pi}{4} - \dfrac{\phi}{2}\right)$ = active earth-pressure coefficient

ϕ = angle of internal friction of soil

μ_1 = coefficient of wall friction

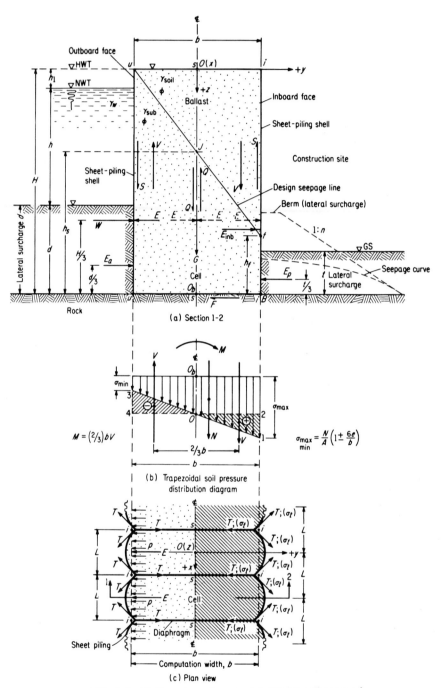

FIG. 8-4. Static system of segmental cellular cofferdam.

Cellular-type Cofferdams

d = driving depth of sheet piling

(iii) Passive earth pressure E_p:

$$E_p = (1/2)\gamma_{sub} t^2 K_p \tag{8-8}$$

where $K_p = \tan^2\left(\dfrac{\pi}{4} + \dfrac{\phi}{2}\right)$ = coefficient of passive earth pressure.

(iv) Lateral confined pressure E:

$$E = (1/2)\gamma h^2 K \tag{8-9}$$

where

$$K = \frac{\cos^2\phi}{2 - \cos^2\phi}$$

1. Stability Against Sliding on Base:

$$\eta = \frac{F_R}{F_D} = \frac{G \tan \phi_{b/r}}{W + E_a - E_p} \tag{8-10}$$

where η = factor of safety against sliding
 F_R = sum of all resisting forces in the cell system on rock
 F_D = sum of all driving forces on the cell system
 $G = G_{eff} + G_{sh}$ = total weight of cell resting on rock and comprising ballast in cell and its sheet-piling shell G_{sh}
 $\tan \phi_{b/r} = \mu_{b/r}$ = coefficient of friction between ballast and rock
 W = full hydrostatic pressure on cell per one unit of length of run of cofferdam
 E_a = active earth pressure on cell per one unit of length of run of cofferdam (where it applies)
 E_p = passive earth pressure on cell per one unit of length of run of cofferdam (where it applies)

This equation is valid if the cells are really drained (bottom filter course, weep holes in sheet piling).

When there is no berm or no surcharge on the tail side of the cell, then $E_p = 0$. When there is no surcharge on the upstream side of the cell, then $E_a = 0$.

For temporary structures it is required that $\eta = 1.25$.
For permanent structures it is required that $\eta = 1.50$.
For friction between sand and smooth rock $\mu_{b/r} \approx 0.50$. For all others $\mu_{b/r} = \mu = \tan \phi$ may be used.

2. Stability Against Overturning

If the cofferdam on bedrock can be regarded as a gravity wall, then the factor of safety η against overturning is calculated approximately as

$$\eta = \frac{M_R}{M_D} = \frac{(1/2)\gamma b^2 H}{M_D} \qquad (8\text{-}11)$$

where M_R = resultant resisting moment of cofferdam system relative to its toepoint B as a point of rotation (Figs. 7-22 and 8-2)
 M_D = resultant driving moment acting on cofferdam relative to its toepoint B
 γ = unit weight of ballast in cell
 b = width of cofferdam for calculations
 H = height (depth) of ballast in cell of cofferdam

Assuming a certain value for the factor of safety η, say, $\eta = 1.5$, the approximate width b of the cofferdam can be calculated, or else the height H of the filling of the ballast in the cell can be computed.

However, before failure by overturning occurs, the cofferdam, viz., cell, may fail in shear in a vertical plane s-s ($\mathsf{\xi} - \mathsf{\xi}$) (passed through the middle of the width b of the cell, i.e., failure in shear of the ballast material and also of the interlocks i-i of the sheet-piling walls in this vertical plane would take place) because of the tilt of the cell toward the construction side (Fig. 8-4).

3. Checking of Cellular Cofferdam Against Shear on a Vertical Plane Through Center of Cell

The static calculations as set forth here apply to both the segmental and the circular types of cellular cofferdams.

As shown by Hager[1] in 1931 and by Terzaghi[2] in 1945, the driving or overturning moment M acting on the cofferdam brings about stresses at the base of the cofferdam, the distribution of which is shown by the hatched $-$ and $+$ triangles, $\triangle$ 012 and $\triangle$ 034 [Fig. 8-4b, trapezoidal soil (rock) pressure diagram]. Each of these hatched triangles represents physically a vertical force V. These two V-forces form a couple, viz., static moment M:

$$M = V(2/3)b \qquad (8\text{-}12)$$

Note that in the case of equilibrium of the cofferdam system, moment M is the externally applied resultant moment, known as the driving moment M_D.

The vertical force V is a shear force of magnitude

$$V = (3/2)(M/b) \qquad (8\text{-}13)$$

Cellular-type Cofferdams

Further, the total vertical shear force in the vertical, neutral plane s-s ($\mathcal{C} - \mathcal{C}$) is of magnitude $Q = V + S$. To prevent failure in shear in this plane, the shear force must be resisted by a shear resistance Q_r of the ballast material (sand):

$$Q_r = E \tan \phi = E\mu = (1/2)\gamma H^2 K \tan \phi = (1/2)\gamma \mu H^2 K \tag{8-14}$$

where $E = (1/2)\gamma H^2 K =$ total lateral pressure on inside vertical wall from ballast

$K = \dfrac{\sigma_y}{\sigma_z} = \dfrac{\cos^2 \phi}{2 - \cos^2 \phi} =$ lateral confined earth pressure coefficient, and

$\mu = \tan \phi =$ coefficient of internal friction of ballast material

All other symbols are as before.

In the case of both segmental and circular cellular cofferdams, to this shear resistance Q_r there must be added the frictional resistance Q_i in the interlocks of the sheet piling comprising the metal shell of the cofferdam cell. This is so because no failure in shear can take place in the vertical plane s-s in the ballast without a simultaneous slip, viz., friction, in the steel sheet piling interlocks in this plane. The frictional force Q_i in the interlocks is calculated as

$$Q_i = Ef_i \tag{8-15}$$

where $E =$ total lateral pressure on one vertical interlock line

$f_i =$ coefficient of interlock friction

The ballast inside the cell of the cofferdam exerts an outward thrust or lateral pressure against the shell walls of the cofferdam. In these walls, thus, tensile forces T are brought about. In the sheet piles these tensile forces should be less than the allowable tension in the interlocks (about 8,000 to 12,000 lb/linear inch). This condition governs the radius and thus the equivalent computational width b of the cofferdam. This is so because from theory on inside pressure on a thin-walled cylinder the tensile stress σ_t in the cylindrical shell is proportional to its radius r. This calculation applies to the diaphragm and to tie rods of a segmental cellular cofferdam (one cross-diaphragm, spacing $= L$), as well as to a cylindrical cellular cofferdam (two cross-diaphragms, spacing $= 2L$). Recall that for $\alpha = 120°$, the magnitude of the tensile force in the plane diaphragm wall is the same as in the circularly curved part of the segmental cofferdam.

To continue the calculation of the interlock resistance Q_i, the shear resistance in the vertical plane s-s induced by the slipping interlock per one unit length of run of segmental or circular type cofferdam is

$$Q_i = \frac{T f_i}{L} = (1/2)\gamma f_i H^2 K(r/L) \tag{8-16}$$

If, as in the case of a segmental cell, $L = r$, then

$$Q_i = (1/2)\gamma f_i H^2 K \tag{8-17}$$

If, as in the case of a circular cell, $L = (1.1)r$, then

$$Q_i = (1/2)\gamma f_i H^2 K \frac{r}{(1.1)r} = (0.455)\gamma f_i H^2 K \tag{8-18}$$

The factor of safety to use with respect to a failure in tension in the interlocks is approximately 2.5.

The total frictional (viz., shear) force in the vertical plane s-s is

$$Q = Q_r + Q_i = E(\mu + f_i) \tag{8-19}$$

where all symbols are the same as before. Or, adding up Eqs. 8-14 and 8-16, we obtain

$$Q = (1/2)\gamma \mu H^2 K + (1/2)\gamma f_i H^2 K(r/L)$$

$$= (1/2)\gamma H^2 K(\mu + \frac{r}{L} f_i) = E(\mu + \frac{r}{L} f_i) \tag{8-20}$$

Because $r \approx L$,

$$Q = E(\mu + f_i) \tag{8-21}$$

Disregarding the friction between ballast and shell of steel sheet piling, the stability against failure in shear in the vertical plane s-s of the cell is written as

$$\eta = \frac{F_R}{F_D} = \frac{Q}{V} = \frac{Q_r + Q_i}{V} = \frac{E(\mu + f_i)}{V} = \frac{(1/2)\gamma H^2 K(\mu + f_i)}{(3/2)(M/b)}$$

$$= (1/3)\gamma b H^2 K \frac{(\mu + f_i)}{M_D} \tag{8-22}$$

where $M = M_D = W(H/3) + E_a(d/3) - E_p(t/3) = (1/3)(WH + E_a d - E_p t)$ (8-23)
= the resultant overturning (driving) moment on the cell system according to Fig. 8-4.

When $E_a = 0$ and $E_p = 0$, then

$$M = M_D = W(H/3) = (1/2)\gamma_w H^2(H/3) = (1/6)\gamma_w H^3 \tag{8-24}$$

In such a case the factor of safety η against shear in the vertical plane s-s is

$$\eta = (1/3)\gamma b H^2 K \frac{(\mu + f_i)}{(1/6)\gamma_w H^3} = 2\frac{\gamma}{\gamma_w}(\mu + f_i)\frac{b}{H}K \tag{8-25}$$

Cellular-type Cofferdams

Force S

Force S is a frictional force composed of a frictional resistance S_r between ballast material and sheet-piling material plus a frictional force S_i in the interlocks of the sheet piling:

$$S = S_r + S_i = E \tan \phi_1 + Ef_i = E(\mu_1 + f_i) \tag{8-26a}$$

or

$$S = (1/2)\gamma H^2 K(\mu_1 + f_i) \tag{8-26}$$

where $\tan \phi_1 = \mu_1 =$ coefficient of friction between ballast material (sand) and steel of sheet-piling material
$f_i =$ coefficient of interlock friction
$E =$ lateral pressure of ballast on vertical inside wall of sheet piling

Note here that $S_i = Q_i = Ef_i$ (see Eq. 8-17).

Force V

Now the vertical force V is calculated as

$$V = Q - S \tag{8-27}$$

Stability of Cofferdam Cell Against Vertical Shear when $V = Q - S$

If the vertical force V is computed as $V = Q - S$ (Fig. 8-4), then the stability of the cell of the cofferdam against rupture in vertical shear is expressed as the ratio of the system's resultant resisting moment M_R of all resisting shear forces F_R to the system's resulting driving moment M_D of all driving shear forces F_D as

$$\eta = \frac{M_R}{M_D} \tag{8-28}$$

The static moments M_R and M_D about the middle point O_b of the base of the cofferdam are calculated as set forth below.

Resisting Moment M_R

This moment is composed of the following three counterclockwise moments:

$$M_R = M_V + M_S + M_{E_p} \tag{8-29}$$

where $M_V = (2/3)bV =$ moment from soil reactive shear forces
$M_S = bS =$ moment formed by the couple S

M_{E_p} = moment of forces acting on the tail side of the cofferdam ($M_{E_p} = E_p(t/3)$, for example, where they apply) relative to point O_b.

Thus,

$$M_R = (2/3)bV + bS + M_{E_p} = (Q-S)(2/3)b + bS + M_{E_p}$$
$$= (2/3)bQ + (1/3)bS + M_{E_p} \qquad (8\text{-}30)$$
$$= (2/3)bE(\mu + f_i) + (1/3)bE(\mu_1 + f_i) + M_{E_p}$$
$$= (b/3)E(2\mu + \mu_1 + 3f_i) + M_{E_p} \qquad (8\text{-}31)$$

This is the total resisting moment of frictional force Q, shear resistance from the ballast in the vertical neutral plane s-s, plus the moment from the frictional force S between the ballast and the sheet-piling material, plus the moment of the frictional force Q_i in the interlocks, when the cell rests on bedrock.

Driving Moment M_D

The system's resultant driving moment M_D is composed of all clockwise moments $\Sigma M = M$ externally applied to the given cofferdam system causing rotation (failure) of the cofferdam about the middle point O_b in the base of the cell, as shown in Fig. 8-4. This moment is usually composed of the moment from external forces acting on the side of the hydrostatic head of the cofferdam. Thus the corresponding driving moment components are

From water pressure
$$M_W = W(H/3)$$
and from the active earth pressure
$$M_{E_a} = E_a(d/3) \qquad \text{(where it applies)}$$
as well as from the passive earth pressure (earth resistance) (where this applies):
$$M_{E_p} = E_p(t/3)$$
$$M_D = W(H/3) + E_a(d/3) - E_p(t/3) \qquad (8\text{-}32)$$

The moments pertain to one unit of length of run of cofferdam, viz., cell.

Factor of Safety η Against Vertical Shear of Ballast and Interlock

The factor of safety against vertical shear can now be written as

$$\eta = \frac{M_R}{M_D} = \frac{b}{3}\frac{E}{M_D}(2\mu + \mu_1 + 3f_i) + \frac{M_{E_p}}{M_D} \qquad (8\text{-}33)$$

Cellular-type Cofferdams

where in practice η is allowed from approximately 1.3 to 1.6. If there is no berm, then $E_p = 0$, and $M_{E_p} = 0$.

The foregoing calculations are given for the case when it is impossible either to penetrate the sheet piles into or pull them out from the soil. By means of this equation and for a factor of safety $\eta = 1$, the minimum width b of a cellular cofferdam, segmental or circular, can be calculated.

In the case of a loose, weak soil where resistance to the cell's pulling out from the soil is negligible, the quantity S may be regarded as a small, if not negligible, quantity and therefore it may be omitted from consideration (see Eq. 8-30). Hence, by setting in Eq. 8-30 $S = 0$, the factor of safety against vertical shear may then be calculated as

$$\eta = \frac{2}{3} b \frac{E}{M_D} (\mu + f_i) + \frac{M_{E_p}}{M_D} \qquad (8\text{-}33a)$$

If there is no berm (or no inside lateral surcharge), then $E_p = 0$, and $M_{E_p} = 0$, and Eq. 8-33a transforms into Eq. 8-22.

The *Design Manual* of the U.S. Navy's Bureau of Yards and Docks[3] for the design of foundations and earthworks provides that this stability calculation be made as

$$\eta = \frac{2b[\mu E + f_i(E_{\text{inb}} - E_p)]}{WH + E_a d - E_p t} \qquad (8\text{-}34)$$

When $E_p = 0$, then

$$\eta = \frac{2b(\mu E + f_i E_{\text{inb}})}{WH + E_a d} \qquad (8\text{-}34a)$$

where E_{inb} = resultant earth pressure on inboard wall of cell's sheet piling calculated by means of $K = \dfrac{\cos^2 \phi}{(2 - \cos^2 \phi)}$

For calculating stability against shear failure on the center line of the cell, this manual recommends a factor of safety of $\eta > 1.25$ for temporary construction and $\eta > 1.5$ for a permanent wall.

The quantity E_{inb}, according to this manual, is to be computed from part of the total pressure diagram, namely, from diagram 1-2-3.[3] The total lateral earth pressure is computed from the Navy's lateral stress diagram as shown in Fig. 8-5.

For calculating the interlock friction, a coefficient of interlock friction of $f_i = 0.3$ may be used.[2,3]

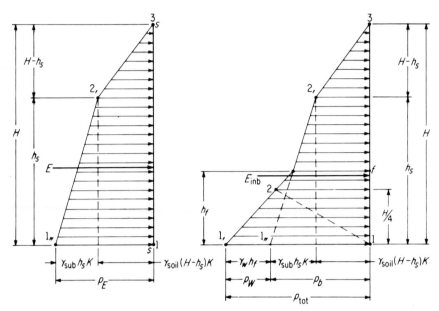

(a) Effective lateral pressure diagram p_E on vertical wall through s-s for calculating lateral pressure E

(b) Total compressive stress p_{tot} on inboard wall of cell of sheet piling by ballast and water

FIG. 8-5. Lateral pressure diagrams (after Ref. 3). For symbols in this figure refer to Fig. 8-4.

Equations 8-33 and 8-33a may be converted in terms of the radius r of the segmental cellular cofferdam, as well as a function of D of the diameter of the circular cellular cofferdam.

4. Stability Against Failure in Shear Between Ballast and Sheet Piling

This stability is calculated as

$$\eta = \frac{Sb}{M} = \frac{3bE(\mu_1 + f_i)}{WH + E_a d} \geqslant 1.3 \tag{8-35}$$

where $S = E(\mu_1 + f_i)$ by Eq. 8-26a
M = driving moment by Eq. 8-32
E is to be computed from lateral stress diagram in Fig. 8-5
Here the factor of safety is required to be $\geqslant 1.3$.

By the method of computation, of the Bureau of Yards and Docks, the factor of safety η is

$$\eta = \frac{3bS}{WH + E_a d - E_p t} = \frac{3b\mu E}{WH + E_a d - E_p t} \geqslant 1.3 \tag{8-36}$$

where $S = E \tan \phi = \mu E$.
When $E_p = 0$, then

$$\eta = \frac{3b\mu E}{WH + E_a d} \geqslant 1.3 \qquad (8\text{-}36a)$$

5. Stability of Cell Against Rupture in Tension of the Interlocks

Rupture of the interlocks of sheet piling occurs first before its web fails. Assuming empty space between the circular cells, the resistance or critical strength of the vertical sheet-piling interlocks against tension from ballast in the cell p_b and from hydrostatic pressure p_W inside the circular cell is calculated based on the theory of strength of the bottom of a thin cylindrical shell as

$$(p_b + p_W)(D/2) = p_{\text{tot}} \cdot r \leqslant \sigma_t \qquad (8\text{-}37)$$

where D = diameter of circular cellular cofferdam (L should be used for $D/2$ in a segmental cellular cofferdam, i.e., $p_{\text{tot}} \cdot L \leqslant \sigma_t$)
$p_{\text{tot}} = p_b + p_W$ = maximum pressure on inboard sheet piling from ballast and water
$p_b = \gamma_{\text{soil}}(H - h_s)K + \gamma_{\text{sub}} h_s K$ (Fig. 8-5b)
$p_W = \gamma_w h_f$ (Fig. 8-5b)
$\sigma_{t_{\text{all}}} = 1{,}400 - 2{,}260$ kg/cm $\approx 7{,}800 - 12{,}800$ lb/in. (or whatever the pertinent specifications provide for) = allowable tensile strength in interlock (in the horizontal direction along the length of wall of sheet piling)

The total horizontal thrust E from ballast and water is:

$$E = 2rH(1/2)p_{\text{tot}} = rHp_{\text{tot}} \qquad (8\text{-}38)$$

which is resisted by the tension in two lines of sheet piling, so that the total tensile force T across an axial vertical section on one sheet pile is

$$T = E/2 = (1/2)\, r\, Hp_{\text{tot}} \qquad (8\text{-}39)$$

This calculation is the same both with and without a berm (E_p) or lateral surcharge t units thick.

The determination of destructive tensile forces in interlocks is made by constructing tensile-force diagrams along the height of the sheet piling for various horizontal sections.

The magnitude of the tensile forces in the cells depends on
 1. Lateral earth pressure transmitted on the sheet-piling shell of the cell

2. Unbalanced hydrostatic pressure acting within the cell and transmitted to the inboard row (2) of the sheet piling
3. Surcharge loads on top of the cell of the cofferdam
4. Lateral concentrated loads transferred to the cell (ice pressure, lateral static and dynamic loads)

(i) *For segmental cells with* $L \geqslant r$, the tensile stress in the cross-wall diaphragm from ballast is (Fig. 8-1):

$$\sigma_{t_1} = \sigma_d = p_E L \qquad (8\text{-}40)$$

where $L=$ spacing of transverse diaphragm walls.

(ii) *For* $L \leqslant r$,

$$\sigma_{t_1} = p_E \, l \qquad (8\text{-}41)$$

Here $\sigma_{t_1} = \gamma_{\text{ballast}} z K$ is the lateral pressure-diagram ordinate of confined ballast in the shell of the cell of sheet piling as for a straight wall; z is the depth coordinate from the top of the ballasted cell, and γ_{ballast} is the weighted unit weight of the ballast material.

(iii) *For circular cells* the stress from ballast is

$$\sigma_{t_1} = p_E r$$

(iv) *For cloverleaf-type cells* with mutually intersecting transverse diaphragm walls, the stress σ_{t_1} from ballast is

$$\sigma_{t_1} = p_E L$$

where $L=$ computation length of the cloverleaf-type cofferdam.

6. Tensile Stress from Ice Pressure

This is given by

$$p_{\text{conc}} = \beta C p_{\text{ext}} \qquad (8\text{-}42)$$

where $p_{\text{conc}} =$ stress induced in interlock from an externally applied load p_{ext} per unit length of run of cell, kg/cm²
$\beta = 0.25$ to $0.33 =$ coefficient of load characteristic indicating nature of concentrated load, linear ice pressure
$C =$ coefficient indicating the form of cofferdam cell
 for a segmental cell $C = L_s$
 for a circular cell $C = L_c/2$, where $L_c =$ computation length of cylindrical cell;

for a cloverleaf-type cell with intersecting transverse diagrams $C = L_{c_1}$.

The total tensile stress σ_t in the interlock is thus made up of the four following stresses:

$$\sigma_t = \sigma_{t_1} + p_W + p_{p_o} + p_{\text{conc}} \tag{8-43}$$

where p_W = hydrostatic stress

p_o = intensity of surcharge on top of cell

To satisfy all design requirements one chooses and uses for the design the maximum value of b for width of cofferdam rendered by the foregoing calculations for the various conditions of the stability analyses.

8-6. STABILITY OF A CELLULAR COFFERDAM CONSTRUCTED ON THICK DEPOSIT OF SAND

Stability against sliding on the base of the cofferdam and stability against failure in shear between ballast and sheet-piling material does not control the design and thus stability of cofferdams built on deep (thick) sand deposits. It is the shear in the vertical plane through the ballast which governs the design and hence stability of this kind of cofferdam.

Calculations of stability against shear failure on the center line of ballast are performed in the same manner as for cells placed on rock. But for sand support, the total inboard pressure E_{inb} should be computed from the entire lateral pressure diagram $1\text{-}1_1\text{-}2\text{-}2_1\text{-}3\text{-}f\text{-}1$ (Fig. 8-5). However, cellular cofferdams on a sand deposit require adequate driving depth of sheet piling as a means of underseepage cutoff, as well as protection against the formation of a quick condition of sand. The safeguarding against a quick condition of sand is done either by installing a loaded, inverted filter in the cell, or by placing a sand berm with a broad base along the inboard side of the cofferdam. The seepage problem is evaluated by means of a hydrodynamic flow net.

To counteract erosion of soil by piping at the inboard toe, the driving depths d_1 and d_2 of both rows of sheet piling should be at least $d_1 = d_2 = (2/3)H$. If the groundwater table is lowered at least $H/6$ below the inboard ground surface, the driving depths should be $d_1 = d_2 = H/2$.

The tension in sheet piling interlocks is calculated in a way similar to that of cells on rock.

1. Stability Against Groundbreak of the Inboard Toe

Stability against bearing capacity failure (groundbreak) is not critical with the presence of a berm. Without the berm, the stability is investigated against the failure of bearing capacity of the inboard toe. The factor of safety η in this investigation is written as

$$\eta = \frac{\text{Ultimate bearing capacity}}{\dfrac{G}{A}\left(1 + \dfrac{6e}{b}\right)} \geqslant 2 \qquad (8\text{-}44)$$

where G = effective weight of cell
A = base contact area of cofferdam with soil
e = eccentricity of resultant vertical load on base of cell
b = computation width of cell of cofferdam

Eccentricity is calculated for the case of the presence of a berm as

$$e = \frac{b}{2} - \frac{W\dfrac{H}{3} + E_a\dfrac{d}{3} - G\dfrac{b}{2} - E_p\dfrac{t}{3}}{G} \qquad (8\text{-}45)$$

If there is no berm present in the static cofferdam system, then $E_p = 0$.

2. Stability Against Pull-Out of Sheet Piling

The factor of safety η against pull-out of the outboard sheet piling with a berm along the inboard wall:

$$\eta = \frac{\text{Ultimate pull-out capacity per unit of length of wall of sheet piling}}{\text{Pull-out force per unit length of wall of sheet piling}}$$

or

$$\eta = \frac{Z_u}{Z_p} \qquad (8\text{-}46)$$

where

$$Z_p = \frac{WH + E_a d - E_p t}{3b\left(1 + \dfrac{b}{4L}\right)} \qquad (8\text{-}47)$$

This check applies also to cofferdams without a berm. In the latter case, $E_p = 0$.

According to the *Design Manual*[3] of the U.S. Navy's Bureau of Yards and Docks Chapter 13, the ultimate pull-out capacity Z_u of a pile is calculated as

$$Z_u = 2\pi r C_A L \qquad (8\text{-}48)$$

where C_A = adhesion, pounds per square foot.
For medium stiff clay $C_A = 460\text{-}700$ psf
For stiff clay $C_A = 700\text{-}720$ psf

8-7. STABILITY OF A CELLULAR COFFERDAM CONSTRUCTED ON A THICK AND HOMOGENEOUS CLAY DEPOSIT

The stability of a cellular cofferdam on clay depends mainly upon the clay's bearing capacity, its compressibility from the ballast in the cell, and the driving depth of the sheet piling.

The main dimensions of a cellular cofferdam on a firm clay may be assessed by means of the same calculations as those used for the sizing of cellular cofferdams supported on rock.

1. *Stability against failure in bearing capacity.* For a cofferdam cell supported on clay the contact pressure from cell and ballast on the clay should not exceed its allowable bearing capacity σ_{all}:

$$\sigma_{\text{all}} = \frac{q}{\eta} = \frac{(5.7)c}{\eta} \tag{8-49}$$

where q = ultimate bearing capacity of clay
$\eta \geqslant 2.5$ = factor of safety
c = cohesion of clay

Note that the ultimate bearing capacity of a pure clay depends on its cohesion only.

Stability against bearing capacity failure of the inboard toe is evaluated in the same manner as for a cofferdam constructed on a thick deposit of sand, except that here the factor of safety η is required $\geqslant 3.0$.

2. *Stability against sliding.* Stability against sliding of a cofferdam on its base on clay can be increased by driving the sheet piling to a greater depth than ordinarily. This brings about a lateral surcharge of clay adjacent to both sides of the cofferdam to resist shear or rupture of soil, viz., groundbreak.

3. *Stability against piping.* For cofferdams on clay, the requirement for greater depths against piping is usually unimportant.

4. *Stability against pull-out of sheet piling.* The stability against pull-out of outboard sheet piling (1) driven in clay is evaluated by means of the same criterion as that applied to cofferdams constructed on sand.

5. *Strength of interlocks.* For a medium stiff clay the tension in and the strength of sheet-piling interlocks is evaluated by the same method as for cofferdams constructed on sand.

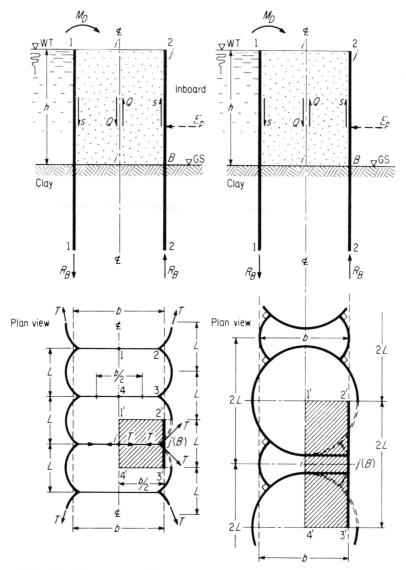

FIG. 8-6. Elements for calculating stability of cofferdam whose sheet piling is driven into clay.

6. *Stability against shear failure of cell in vertical central plane.* Stability against failure of the cell (ballast and interlocks) in shear in the vertical central plane s-s of a cofferdam built on clay is calculated by means of the same formulas as those used for cofferdams constructed on bedrock.

Cellular-type Cofferdams

If cofferdam sheet piling is driven into a clay soil, excessive shear stresses may bring about failure by slippage in the interlocks at points 1 and 2 in the neutral plane of the cell (see plan view in Fig. 8-6) and bring about sliding between the ballast at its vertical contact surfaces 1-2-3-4 with the sheet piling.

The shear strength of the shell (wall) of the cell along the interlocks is equal to the friction of the interlocks.

For stability calculations of shear in the vertical plane of the cofferdam cell constructed on clay, refer to Fig. 8-6 and consider an overturning moment applied to the cell system. This moment adds to the uniform pressure on clay from the cell filled with ballast an additional pressure R_B at the lower end of the inboard sheet-piling wall, row 2. Simultaneously, the moment M_D reduces the pressure by an amount R_B at the lower end of the outboard sheet piling, row 1.

The sum of all sheet-piling vertical reactions, ΣR_B, on one side of the vertical neutral plane (Fig. 8-6) of *one cell* is

For a segmental cell:

$$\Sigma R_B = R_B \left(L + \frac{b}{2}\right) \tag{8-50}$$

For a circular cell:

$$\Sigma R_B = R_B (2L + b) \tag{8-51}$$

In special cases, for segmental cells, $L = r$.

Because of the overturning moment M_D the interlocks of the sheet piling tend to slide past each other. The sliding or frictional force in the interlock (viz., two interlocks) is equal to:

For segmental cell:

$$Tf_i = (1/2)\gamma f_i r H^2 K \tag{8-52}$$

For circular cell:

$$2Tf_i = \gamma f_i r H^2 K \tag{8-53}$$

Here T = circumferential tensile force in interlock
f_i = coefficient of interlock friction
γ = unit weight of ballast
r = radius of circular arcs of cofferdams
$K = \dfrac{\cos^2\phi}{(2 - \cos^2\phi)}$ = coefficient of confined lateral pressure caused by weight of ballast

The equilibrium of the reactive force ΣR_B and the frictional force is

For segmental cell:

$$Tf_i = R_B \left(L + \frac{b}{2}\right)$$

For circular cell:

$$2Tf_i = R_B (2L + b)$$

or

$$(1/2)\gamma f_i rH^2 K = R_B \left(L + \frac{b}{2}\right) \quad (8\text{-}54)$$

$$\gamma f_i rH^2 K = R_B(2L + b) \quad (8\text{-}55)$$

giving the magnitude of the reactive force R_B induced by the overturning moment M_D:

For segmental cell:

$$R_B = \frac{Tf_i}{L + \frac{b}{2}}$$

For circular cell:

$$R_B = \frac{2Tf_i}{2L + b} \quad \begin{array}{c}(8\text{-}56)\\(8\text{-}57)\end{array}$$

or

or

$$R_B = \frac{\frac{1}{2}\gamma f_i rH^2 K}{L + \frac{b}{2}}$$

$$R_B = \frac{\gamma f_i rH^2 K}{2L + b} \quad \begin{array}{c}(8\text{-}56\text{a})\\(8\text{-}57\text{a})\end{array}$$

or

or

$$R_B = (1/2)\gamma f_i H^2 K \frac{r}{L + (0.5)b}$$

$$R_B = (1/2)\gamma f_i H^2 K \frac{r}{L + (0.5)b}$$

$$\begin{array}{c}(8\text{-}56\text{b})\\(8\text{-}57\text{b})\end{array}$$

It can be visualized that the reactive moment M_R to the overturning moment M_D consists of two components, namely

$$M_D = M_R = M_{f_i} + M_\mu \quad (8\text{-}58)$$

where M_R = reactive moment from the reactions $\downarrow R_B$ and $\uparrow R_B$ (as functions of interlock friction f_i only) with the moment arm b
M_μ = moment of the frictional resistance of the ballast only

The total interlock moment M_{f_i} per length of cell may be calculated from the pile reactions on both sides of the neutral plane s-s as set forth:

Cellular-type Cofferdams

For segmental cell: For circular cell:

$$LM_{f_i} = LR_B b + R_B \frac{b}{2}\frac{b}{2} \qquad 2LM_{f_i} = 2LR_B \cdot b \underbrace{}_{\substack{\text{end}\\\text{reaction}}} \underbrace{}_{\substack{\text{arm}}} + 2R_B \underbrace{\frac{b}{2}}_{\substack{\text{diaphragm}\\\text{reaction}}} \underbrace{\frac{b}{2}}_{\text{arm}}$$

(8-59)

(8-60)

Moment per unit length of run of cofferdam for segmental and circular cells is the same, namely

$$M_{f_i} = R_B b + R_B \frac{b}{4L} = R_B b \left(1 + \frac{b}{4L}\right) \qquad (8\text{-}61)$$

or

$$M_{f_i} = \frac{1}{2}\gamma f_i b H^2 K \frac{r}{L}\frac{L+(0.25)b}{L+(0.50)b} \qquad (8\text{-}62)$$

Being a function of the interlock friction only, this moment is independent of the compressibility of the cofferdam-bearing clay.

If the cofferdam is constructed on a firm clay, then the pressure on the clay from the ballast-filled cell may be assumed to be uniformly distributed, i.e., the soil pressure distribution diagram is a rectangle, but not a trapezoid or a triangle.

Upon applying the overturning moment M_D to a cell system on firm clay shear forces S_{μ_1} are induced along the inside walls 1 and 2 as shown in Fig. 8-7. By Eq. 8-26, the magnitude of this shear, viz., frictional force, is

$$S''_{\mu_1} = \mu_1 E = (1/2)\gamma \mu_1 H^2 K \qquad (8\text{-}63)$$

This is the greatest value S''_{μ_1} can attain.

The total frictional force S along the perimeter of the inside shell of the cell from ballast on one side of the neutral plane s-s is, per cell of a cofferdam,

For segmental cell: For circular cell:

$$S = S_{\mu_1}\left(L + \frac{b}{2}\right) \qquad S = S_{\mu_1}(2L + b) \qquad (8\text{-}64)$$

(8-65)

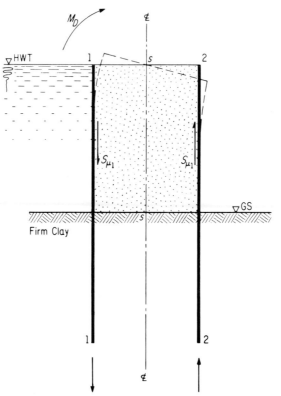

FIG. 8-7. Shear forces S_{μ_1} along inside surfaces of cell shell walls.

This force is equal to the total shear resistance in the vertical, neutral plane of the ballast in one cell:

$$Q = LQ_\mu \qquad\qquad Q = 2LQ_\mu \qquad (8\text{-}66)$$
$$(8\text{-}67)$$

where

$$Q_\mu = \mu Q = (1/2)\gamma\mu H^2 K \qquad (8\text{-}14)$$

$$\therefore S = Q \qquad\qquad\qquad \therefore S = Q$$

or or

$$S_{\mu_1}[L + (0.5)b] = LQ_\mu \qquad S_{\mu_1}(2L + b) = 2LQ_\mu \qquad (8\text{-}68)$$
$$(8\text{-}69)$$

and and

$$S_{\mu_1} = \frac{LQ_\mu}{L + (0.5)b} \qquad S_{\mu_1} = \frac{2LQ_\mu}{2L + b} = \frac{2L(1/2)\gamma\mu H^2 K}{2L + b}$$

Cellular-type Cofferdams

$$= (1/2)\gamma\mu H^2 K \frac{L}{L+(0.5)b} < S'_{\mu_1} \qquad = \frac{\gamma\mu L H^2 K}{2L+(0.5)b}$$

$$= (1/2)\gamma\mu H^2 K \frac{L}{L+(0.50)b} < S'_{\mu_1}$$

(8-70)
(8-71)

Therefore the cofferdam will fail in vertical shear before the wall friction (ballast and steel) is fully developed.

The part of the overturning moment M_μ for one cell at which failure occurs is

$$LM_\mu = (LS_{\mu_1})b + \left(2S_{\mu_1}\frac{b}{2}\right)\left(\frac{b}{2}\right) \qquad 2LM_\mu = 2LS_{\mu_1}b + 2S_{\mu_1}\frac{b}{2}\frac{b}{2}$$

or

or

$$LM_\mu = LS_{\mu_1}b\left(1+\frac{b}{2L}\right) \qquad LM_\mu = S_{\mu_1}bL\left(1+\frac{b}{4L}\right)$$

(8-72)
(8-73)

Therefore, the part of the overturning moment M_μ per unit length of run of cofferdam is

$$M_\mu = S_{\mu_1}b\left(1+\frac{b}{2L}\right) \qquad M_\mu = S_{\mu_1}b\left(1+\frac{b}{4L}\right)$$

$$= (1/2)\gamma\mu H^2 Kb \frac{1+0.5\frac{b}{L}}{1+0.5\frac{b}{L}} \qquad = (1/2)\gamma\mu H^2 Kb \frac{1+0.25\frac{b}{L}}{1+0.50\frac{b}{L}}$$

$$= (1/2)\gamma\mu b H^2 K$$

(8-74)
(8-75)

The total overturning moment M_D is the sum of M_{f_i} and M_μ, i.e.,

$$M_D = M_{f_i} + M_\mu \qquad M_D = M_{f_i} + M_\mu$$

$$= (1/2)\gamma b H^2 K\left(f_i\frac{r}{L}\frac{L+0.25b}{L+0.50b}+\mu\right) \qquad = (1/2)\gamma b H^2 K\frac{L+0.25b}{L+0.50b}\left(f_i\frac{r}{L}+\mu\right)$$

(8-76)
(8-77)

With $r/L = 1$

$$M_D = (1/2)\gamma bH^2 K \left(\frac{L+0.25b}{L+0.50b} f_i + \mu\right) \quad M_D = (1/2)\gamma bH^2 K \frac{L+0.25b}{L+0.50b}\left(f_i + \mu\right)$$

(8-76a)
(8-77a)

By means of these equations the minimum width b of a segmental or circular cellular cofferdam for a factor of safety of $\eta = 1$ (limit equilibrium) can be calculated. Of course, the necessary b can be calculated for any $\eta > 1$:

$$\eta = \frac{M_{fi} + M_\mu}{M_D} = \frac{M_{fi} + M_\mu}{W\frac{H}{3} + E_a\frac{d}{3} - E_p\frac{t}{3}} \quad (8\text{-}78)$$

According to the *Design Manual* of the Navy's Bureau of Yards and Docks,[3] the stability against failure in shear on the center line of a cell for a cofferdam built on clay is calculated after Terzaghi[2] as

$$\eta = \frac{(E_{\text{inb}} - E_p)rf_i\left(\frac{b}{L}\frac{L+0.25b}{L+0.50b}\right)}{WH + E_a d - E_p t} \quad (8\text{-}79)$$

where $\eta \geqslant 1.25$ = factor of safety required for temporary structures
$\eta \geqslant 1.50$ = factor of safety required for permanent structures
E_{inb} = lateral earth pressure to be assessed from Fig. 8-5
E_p = passive earth resistance (where applicable)

8-8. CLOVERLEAF CELLULAR COFFERDAM

The cloverleaf cellular cofferdam with intersecting diaphragm walls is more complex than the parallel-wall type cofferdams. However, with dimensions as shown in Fig. 8-8, the calculation of strength and the various stabilities of the cloverleaf cofferdams are calculated by the same methods and formulas as those applied to segmental and circular cellular cofferdams, namely: a computation section of a cloverleaf cell is regarded as a ranctangular-prismatic box with a length $2L$ and width b.

8-9. EXAMPLE OF ANALYSIS OF A CIRCULAR CELLULAR COFFERDAM

Refer to Figs. 8-4 and 8-9. Design a circular cellular cofferdam to rest on bedrock. Calculate the stability of the cofferdam against sliding,

Cellular-type Cofferdams

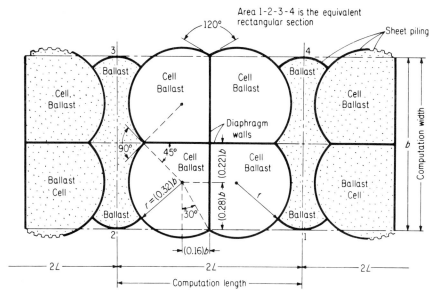

FIG. 8-8. Cloverleaf-type cellular cofferdam.

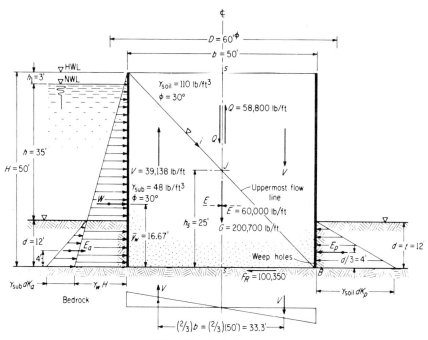

FIG. 8-9. Cofferdam on rock.

overturning, shear in the central vertical plane s-s in the ballast, and check failure in tension of the interlocks of the sheet piling near the base of the cell.

In order to safeguard against uplift, fissured bedrock below the base of the cofferdam should be grouted.

Free drainage of the ballast in cofferdam cells may be facilitated by providing weep holes in the inboard row of sheet piling. Assume that the uppermost flow line in the cell is a straight line, and that the water level at the outboard face coincides with the height H of the cofferdam. To prevent the reduction in strength and stiffness of the cofferdam, no lubrication of interlocks should be permitted for easy pulling out of the sheet piling after the completion of work.

Given:
$H = 50$ ft; $h_f = 0$; $d = 12$ ft; $t = d = 12$ ft; $h_1 = 3$ ft
$\gamma_w = 62.4$ lb/ft^3
$\gamma_{soil} = 110$ lb/ft^3 (for ballast and soil)
$\gamma_{sub} = 48$ lb/ft^3
$\mu = \tan\phi = 0.58 =$ coefficient of internal friction of soil (ballast) for dry as well as for submerged conditions
$\mu_1 = \tan\phi_1 = 0.40 =$ coefficient of friction between ballast material (soil) and steel
$\mu_{b/r} = 0.500 =$ coefficient of friction between ballast material and rock
$f_i = 0.40 =$ coefficient of interlock friction
$K_a = 0.333 =$ active earth-pressure coefficient
$K_p = 3.000 =$ coefficient of passive earth resistance
$K = 0.600 =$ Krynine's lateral earth pressure coefficient for ballast pressure E on central, vertical shear plane s-s in ballast (Fig. 8-10)

Solution. The diameter of the cell is governed by the allowable interlock strength in tension σ_i expressed in lb/in. This strength varies from 6,000 lb/in. to 12,000 lb/in. For example, for the U.S. Steel sheet-piling section *MP* 110 the allowable interlock tension is given by that manufacturer as $\sigma_i = 8,000$ lb/in.

The relationship between inside pressure $p = \gamma_{ave} HK$ on the sheet piling at the bottom of the cell and the radius $D/2$ of the cell is

$$\sigma_i = \frac{p}{12}\frac{D}{2} \qquad (8\text{-}80)$$

If the allowable interlock strength to use is assumed to be $\sigma_i = 6,000$ lb/in., then the diameter D of the cell calculates as

Cellular-type Cofferdams

FIG. 8-10. Vertical shear plane in ballast.

$$D = \frac{24\sigma_i}{24} = \frac{(24)(6,000)}{24} = 60 \text{ (ft)}$$

By empirical design,

$$D = (1.2)(H) = (1.2)(50) = \underline{60 \text{ (ft)}}$$

∴ try $D = 60$ ft

Here $p = (80)(50)(0.600) = 2,400$ lb/in²., where (80) is the average unit weight of the ballast.

Design width b:

$$b = (1.0)(H) = (1.0)(50) = \underline{50.0 \text{ (ft)}}$$

Length $2L$ of repetitional section (panel):

$$2L = (1.05)(D) = (1.05)(60) = \underline{63.0 \text{ (ft)}}$$

Effective weight G_{eff} of cell ballast:

$$G_{eff} = \gamma_{ball} b(H - h_s) + \gamma_{sub} b h_s$$

$$= (110)(50.0)(50.0 - 25.0) + (48)(50.0)(25.0)$$

$$= 197,500 \text{ (lb/ft)}$$

Making weep holes near the bottom, $h_f = 0$.

$$h_s = \frac{H - h_f}{2} + h_f = \frac{50.0 - 0}{2} + 0 = \underline{25.0 \text{ (ft)}}$$

Weight of shell:
$$G_{\text{shell}} = (2)(\gamma_{\text{steel}})(H)(1.0)$$
$$= (2)(32)(50.0)(1.0) = \underline{3{,}200 \text{ (lb/ft)}}$$

where $\gamma_{\text{steel}} = 32 \text{ lb/ft}^2$ of the sheet piling used.

Total effective weight G of ballast and shell:
$$G = G_{\text{eff}} + G_{\text{shell}} = 197{,}500 \text{ lb/ft} + 3{,}200 \text{ lb/ft}$$
$$= \underline{200{,}700 \text{ lb/ft}}$$

Full possible hydrostatic pressure W of water on outboard face of the cofferdam:
$$W = (1/2)\gamma_w H^2 = (0.5)(62.4)(50.0^2) = \underline{78{,}000 \text{ (lb/ft)}}$$

Active earth pressure E_a on outboard face of cofferdam for $\phi_1 = 0$:
$$E_a = (1/2)\gamma_{\text{sub}} d^2 K_a = (0.5)(48)(12.0^2)(0.333)$$
$$= \underline{1{,}150 \text{ (lb/ft)}}$$

Passive earth resistance E_p on inboard face of cofferdam for $\phi_1 = 0$:
$$E_p = (1/2)\gamma_{\text{soil}} d^2 K_p = (0.5)(110)(12.0^2)(3.000)$$
$$= \underline{23{,}760 \text{ (lb/ft)}}$$

Stability against sliding on base:
$$\eta_{\text{sliding}} = \frac{F_R}{F_D} = \frac{G \tan \phi_{b/r}}{W + E_a - E_p} = \frac{(200{,}700)(0.500)}{55{,}390} = \underline{1.8 > 1.5}$$

Disregarding E_p, the factor of safety η against sliding is
$$\eta_{\text{sliding}} = (100{,}350)/79{,}150 = 1.26 < 1.5$$

Stability against overturning:

Resisting moment M_R (without the stabilizing effect of passive earth resistance)
$$M_R = (1/2)(\gamma_{\text{ave}}) b^2 (H)(1.0) = (0.5)(80.0)(50.0^2)(50.0)(1.0)$$
$$= \underline{5{,}000{,}000 \text{ (ft-lb)}}$$

where

Cellular-type Cofferdams

$$\gamma_{ave} = G/(b)(H)(1.0) = 200{,}700/(50.0)(50.0)(1.0)$$
$$= \underline{80.0 \ (lb/ft^3)}$$

Driving moment M_D about point B:

$$M_D = W\frac{H}{3} + E_a\frac{d}{3} = (78{,}000)(50.0/3) + (1{,}150)(12.0/3)$$
$$= \underline{1{,}304{,}600 \ (ft\text{-}lb)}$$

Factor of safety against overturning:

$$\eta_{overturning} = \frac{M_R}{M_D} = \frac{5{,}000{,}000}{1{,}304{,}600} = \underline{3.8 > 1.5}$$

Stability against shear of ballast in central, vertical plane s-s:

$$\eta_{s\text{-}s} = \frac{F_R}{F_D} = \frac{Q}{V} = \frac{Q_r + Q_i}{V} = \frac{E\mu + Ef_i}{V} = E\frac{\mu + f_i}{V}$$

Shear force V caused by moment $M = M_D$:

$$V = (3/2)\frac{M}{b} = \frac{(1.5)(1{,}304{,}600)}{50.0} = 39{,}138 \ (lb/ft)$$

Shear resistance Q_r (assume $r/L = 1.0$):

$$Q_r = E \tan \phi = (1/2)(\gamma_{ave})(H^2)(K) \tan \phi$$

where $(1/2)(\gamma_{ave})(H^2)(K)$ is the lateral confined pressure E on the vertical shear plane s-s in the ballast.

$$E = (0.50)(80.0)(50.0^2)(0.600) = \underline{60{,}000 \ (lb/ft)}$$
$$E(\mu + f_i) = (60{,}000)(0.58 + 0.40) = \underline{58{,}800 \ (lb/ft)}$$

Factor of safety $\eta_{s\text{-}s}$:

$$\eta_{s\text{-}s} = 58{,}800/39{,}138 = \underline{1.5 = 1.5 \ (allowable)}$$

Interlock tension, σ_t:

$$\sigma_t = \frac{p}{12}\frac{D}{2} = \frac{(2{,}400)(60.0)}{24} = \underline{6{,}000 \ (lb/in.) < \sigma_i = 8{,}000 \ (lb/in.)}$$

where $p = \gamma_{ave}HK = (80.0)(50.0)(0.600) = 2{,}400 \ (lb/ft^2)$ is pressure on sheet piling near base of cell

REFERENCES

1. K. Hager, "Die Berechnung von Fangdämmen," *Wasserkraft und Wasserwirtschaft*, No. 14 (July, 1931).
2. K. Terzaghi, "Stability and Stiffness of Cellular Cofferdams," *Trans., ASCE*, Paper No. 2253, Vol. 110 (1945), p. 1083.
3. Department of the Navy, Bureau of Yards and Docks, *Design Manual* (Soil Mechanics, Foundations and Earth Structures), Navdocks DM-7. Washington, D.C.: Department of the Navy, Bureau of Yards and Docks, 1962, pp. 7-10-7.

QUESTIONS

8-1. What is a cofferdam?
8-2. What is the purpose of a cofferdam?
8-3. Systematize cofferdam types as to the form of cells (in plan), and the mode of their support.
8-4. What is a diaphragm in cofferdam design?
8-5. What is a segmental cellular cofferdam?
8-6. What is a circular cellular cofferdam?
8-7. Why is it necessary to fill cofferdam cells with ballast?
8-8. Why is it necessary to fill a large number of adjacent segmental cells at approximately the same rate simultaneously?
8-9. What is Krynine's lateral earth-pressure coefficient, and how is it used?
8-10. What is one of the major objections against the use of a segmental cellular cofferdam?
8-11. What is the advantage of the segmental cellular cofferdam as compared with other types of cellular cofferdams?
8-12. What kind of cellular cofferdam is usually used for large depth of water and on bare bedrock?

PROBLEMS

8-1. All conditions are given the same as in the Example on page 240, except that the bedrock is not overlaid by a layer of sand. The cofferdam is to rest on bedrock. Determine factors of safety of the cofferdam against sliding, overturning and shear in the vertical, central plane s-s.

8-2. All conditions are given the same as in the Example on page 240, and in Prob. 8-1, except that a berm must be used on the dry side of the cofferdam in order to increase its stability. The cofferdam rests on bedrock. Determine factors of safety, and compare them with those obtained in the Example, and in Prob. 8-1.

Cellular-type Cofferdams

8-3. Referring to Fig. 8-5, recalculate the Example on page 240 for a segmental cellular cofferdam on bedrock. Use the same conditions as given in Prob. 8-1.

8-4. Calculate stabilities for a circular cellular cofferdam whose sheet piling is driven into sand. Use same size cofferdam above the ground as in Prob. 8-1. Also, all physical data are the same as in the Example on page 240.

8-5. Determine stability of a circular cellular cofferdam whose sheet piling is driven into clay. Refer to Figs. 8-6 and 8-7. Use same size cofferdam as in Prob. 8-4, and assume all engineering data needed for clay.

chapter 9

Dewatering of Excavation

9-1. METHODS OF DEWATERING

To facilitate the laying of foundations in the dry, it is necessary to dewater the excavation, or to seal the bottom of the excavation against a quick condition of soil by means of underwater concreting.

Excavations may be dewatered by open pumping, by lowering the groundwater table, and by artificial and natural freezing of soil. Figure 9-1 shows dewatering systems applicable to different soils.

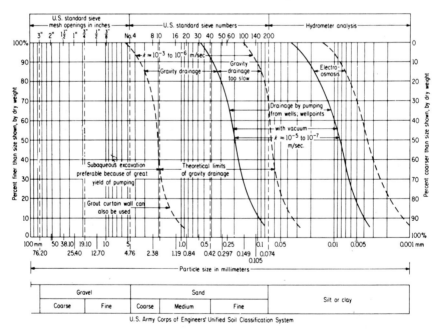

FIG. 9-1. Dewatering systems applicable to different soils. (Courtesy of the Moretrench Corp., Rockaway, N.J.)

Open Pumping

Open pumping is carried out by various kinds of pumping systems. This method is most effective in rocky and gravelly soils. In sandy soils and silts, the method of dewatering by open pumping frequently brings about sloughing, erosion and the flowing down of unsupported excavation slopes, loosening of the soil, quick condition, and other consequent difficulties in executing earthworks. Therefore in such troublesome soils excavation walls are made as vertical banks which are protected against caving in by sheeting and sheet piling, whichever method is most appropriate to use.

To cope with water in the excavation, sumps are installed, from which the water is removed by pumping.

The sump must be dug before each lift of the excavation is made. The size of the sump should be assessed large enough to cope for a short time with emergencies such as a breakdown of a pump. The depth of a sump is about 3 ft for small structures, and about 5 ft for large structures.

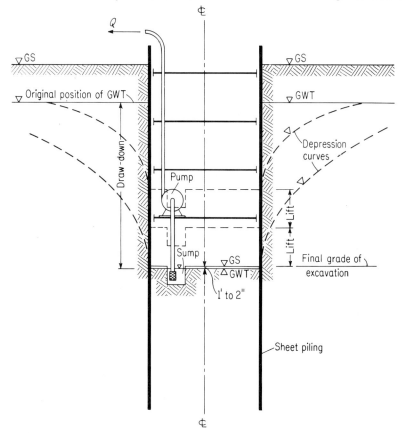

FIG. 9-2. Dewatering of excavation.

Sometimes several sumps are dug. The collection of water in the excavation takes place by ditches.

If the sheet piling is driven into a permeable material with groundwater present then the excavation is advanced by predrainage. The excavation is dug in several lifts, thus lowering the groundwater table in and around the excavation by pumping water out from a sump (Fig. 9-2).

Open excavation must be done with great caution and slowly, or else a quick condition of sand may set in, resulting in "boiling" of the toes of slopes and bottom of the excavation. A sudden drawdown may bring about an unbalanced pressure and thus a hydraulic groundbreak.

9-2. AMOUNT OF INFLUX OF WATER

The amount of influx of water, viz., discharge Q in volume per unit of time t into an open excavation through its bottom for permeable sand and/or gravel may be calculated with a sufficient, practical degree of accuracy by means of Darcy's law of filtration as

$$Q = kiA \qquad (9\text{-}1)$$

where k = coefficient of permeability of soil, cm/sec
$i = h/L$ = hydraulic gradient
h = pressure head of water, cm
L = length of filtration, cm
A = cross-sectional area of flow, cm^2

Of course, any other consistent units of measurements than those as indicated above may be worked with, for example: discharge Q in m^3/sec, or in ft^3/sec, or ft^3/min.

When filtration takes place in the vertical direction perpendicular to the horizontal bedding planes of the soil formation, then the coefficient of vertical permeability k_v for that soil must be used. For horizontal flow, the coefficient of horizontal permeability of flow, k_h, must be used.[1]

In the case of layered soil systems, the corresponding weighted coefficients of permeability must be used in discharge calculations.

Sometimes the average coefficient of permeability $k_{ave} = \sqrt{k_v k_h}$ is to be used.

It is necessary to know the discharge quantity in order to assess the proper capacity of pump to use for dewatering the excavation.

Values of Coefficients of Permeability

For orientation purposes some ranges of coefficients of permeabilities for textural fractions of soil are given in Table 9-1. The classification of

Dewatering of Excavation

the ranges of permeability by various degrees is here adopted after Terzaghi and Peck.[2] This table shows that a composite soil consisting of various textural fractions and in various proportions may have k-values which can vary within a very wide range. Therefore these k-values are not suitable for use in design of hydraulic structures, earthworks, and foundations. For design work actual values of coefficients of permeability of soil are to be obtained.

TABLE 9-1

Range of Order of Magnitude of Coefficients of Permeability of Soil Textural Fractions

Degree of Permeability	Range of Coefficient of Permeability k, cm/sec	Approximate Textural Soil Fraction
High	$> 10^{-1}$	Medium and coarse gravel
Medium	10^{-1}-10^{-3}	Fine gravel; coarse, medium and fine sand; dune sand
Low	10^{-3}-10^{-5}	Very fine sand; silty sand
Very low	10^{-5}-10^{-7}	Dense silt; dense loess; clayey silt; clay
Impervious	$< 10^{-7}$	Homogeneous clay

A graphical representation of values of coefficients of permeabilities of different types of soils, the application of k's to earth dams, and the method of determination of k's is shown after A. Casagrande and R.E. Fadum[3] in Fig. 9-3.

Influx of Water Through Permeable Bottom of Excavation

For excavations through water in water-logged soil with permeable bottom of the foundation pit there are as yet no reliable methods for calculating the amount of influx of water Q available. However, with a satisfactory approximation, one may calculate Q as follows:

$$Q = k_v A \frac{h}{d_1 + d_2} \tag{9-2}$$

where the symbols are shown in Fig. 9-4, and A is the horizontal area of the bottom of the excavation. The flow net theory [1,5] may also be used for ascertaining Q.

When the steel sheet piling is driven into an impermeable layer of soil, then only that amount of water seeping into the excavation through the

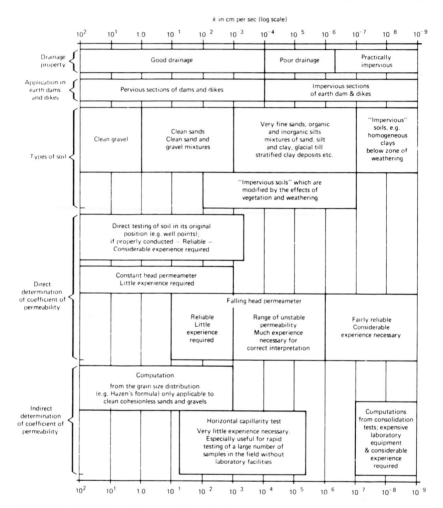

FIG. 9-3. Coefficient of permeability (after A. Casagrande and R. E. Fadum, Ref. 3).

leaky seams of the interlocks of the sheet piling must be removed from the excavation.

The influx, viz., discharge quantity Q, may be calculated approximately for sandy soils by means of the Hazen[4] empirical flow-velocity equation:

$$v = c\, C\, d_{10}^2 \frac{h}{L} [0.70 + (0.03)T] \quad \text{(m/day)} \qquad (9\text{-}3)$$

where v = velocity of flow of water in meters per day in a solid column of the same area as that of the sand

$c \approx 1{,}000$ — an empirical constant, experimental coefficient depending

Dewatering of Excavation

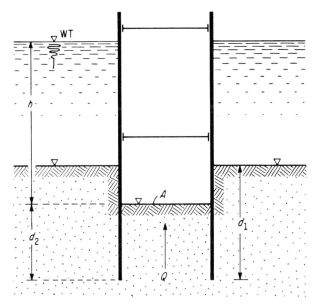

FIG. 9-4. Influx Q of water in excavation through its bottom.

upon the physical nature of the soil: the value depending upon "cleanliness" of the soil: for clean sands $c \approx 700$ to $1,000$; for less clean sands $c \approx 500$ to 700; generally, c varies from 400 to 1,200

C = a numerical coefficient, depending upon units in which the flow velocity is expressed; for example, for v in meters/day, $C = 1$; for v in feet per day, $C = 3.28$

d_{10} = effective particle size diameter in millimeters, determined from the soil particle size accumulation curve

h = pressure head

L = thickness of soil layer used in test, and through which water permeates

h/L = hydraulic gradient

T = temperature in °C. When $T = 10$°C, the temperature coefficient in the brackets of Eq. 9-3 becomes equal to unity.

Thus, by inspection, Eq. 9-3 appears to be Darcy's law, where the coefficient of permeability k may be seen to be

$$k = cd_{10}^2 \left[0.70 + (0.03)T\right] \quad \text{(m/day)} \tag{9-4}$$

or

$$k = 116 d_{10}^2 \left[0.70 + (0.03)T\right] \quad \text{(cm/sec)} \tag{9-4a}$$

The coefficient of permeability k is a function of the effective size of the soil particles, as well as of temperature. Hazen's experiments pertained to filter sand, and his equation gives satisfactory results for soils with 0.1 mm $< d_{10} <$ 3 mm under the condition that the coefficient of uniformity U of the soil particles is $d_{60}/d_{10} \leqslant 5$.

This equation is satisfactory for approximative calculations for fine, clean, graded sand. It is not applicable to clayey silts and clays.

"As a provisional basis which best agrees with the known facts," writes Allen Hazen,[4] "the size of grain where the curve cuts the ten percent line is considered to be the 'effective size' of the material. This size is such that 10 percent of the material is of smaller grain and 90 percent is of larger grain than the size given." Thus Hazen defined the effective size d_e as the diameter of soil particles such that 90 percent of the particles by weight is coarser and 10 percent is finer than the grain size of a sample of the material. It is sometimes also called the 10 percent size.

Sometimes Hazen's k-equation is written in the form

$$k = cCd_e^2 \tag{9-5}$$

which is nothing else than the coefficient of permeability at 10°C. Expressed in meters per day ($C = 1$), Eq. 9-5 transforms to

$$k_{10} = cd_e^2 \tag{9-5a}$$

where $d_e = d_{10} =$ effective size of soil particles.

Hazen's experimental results indicate that the finer 10 percent has as much influence upon the action of a material in filtration as the coarser 90 percent. This is explained by the fact that in a mixed material, containing particles of various sizes, the water is forced to go around the larger particles and through the finer parts which occupy the intervening spaces, and so it is the fines' portion which mainly determines the frictional resistance, the capillary attraction, and, in fact, the action of the sand in almost every way.

Example

Determine by means of the Hazen formula the coefficient of permeability k_{20} at $T = 20$°C of a clean sand the effective size of which has been determined from the soil particle size accumulation curve as $d_{10} = 1.20$ mm. The coefficient of uniformity of this material is $U = d_{60}/d_{10} = 4 < 5$.

Solution. By Eq. 9-3,

$$k_{20} = cC\,[0.7 + (0.03)T]d_{10}^2$$
$$= (1{,}000)(1)[0.7 + (0.03)(20)](1.2)^2 = \underline{1872.0 \quad \text{(m/day)}}$$

Dewatering of Excavation

or
$$k_{20} = (3.28)(1872.0) = \underline{6140 \quad (\text{ft/day})}$$

With 1 m = 100 cm, and 1 day = 86,400 sec, Eq. 9-3 converts into

$$k_{20} = 2.166 \text{ (cm/sec)}$$

or
$$k_{20} = 0.852 (\text{in./sec})$$

Problem

Calculate discharge Q through the bottom of the excavation shown in Fig. 9-5.

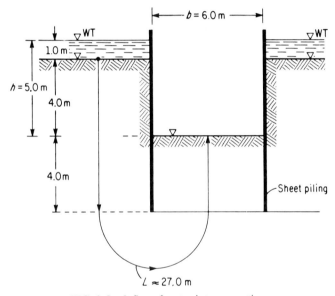

FIG. 9-5. Influx of water into excavation.

Solution. If a single sheet-piling cofferdam encloses a construction site the bottom area of which is $A = 6.0 \text{ m} \times 15.0 \text{ m} = 600 \text{ cm} \times 1500 \text{ cm} = 900{,}000 \text{ cm}^2 = 90 \text{ m}^2$, and if the excavation depth is $h = 5 \text{ m} = 500 \text{ cm}$ below the outside water table, and if the average length of travel path of a water particle to the center of the bottom of the excavation is about $L = 27 \text{ m} = 2{,}700 \text{ cm}$, then the hydraulic gradient is $i = h/L = 5.0/27.0 = 0.185$, the velocity of flow is $v = k_{20} i = (2.166)(0.185) = 0.400$ (cm/sec), where k_{20} is here given as equal to 2.166 cm/sec, and the influx (discharge) Q is $Q = vA = (0.400)(900{,}000) \approx (3.60)(10^5) \text{ (cm}^3/\text{sec)} = (3.60)(10^{-1}) \text{ m}^3/\text{sec}$ through the entire bottom of the excavation.

9-3. LOWERING THE GROUNDWATER TABLE

One of the greatest opponents and hazards in foundation engineering operations is the groundwater. Unfavorable groundwater conditions may threaten the entire construction program.

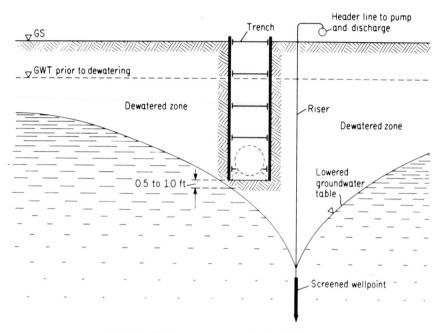

FIG. 9-6. Single-stage asymmetrical wellpoint system.

Sometimes the site of the excavation for laying foundations is temporarily dewatered by means of a wellpoint system. This method is expedient and successful in sandy soils. Figures 9-6 through 9-9 illustrate dewatering by the wellpoint method in single and multiple tiers. Upon dewatering of foundation pits and trenches by means of wellpoints the position of the groundwater table is lowered to about 0.50 m (≈ 1.5 ft) below the bottom of the excavation. This permits laying of the foundation in the dry.

The groundwater table can be lowered rapidly by means of wellpoints. This method is effective for flow velocities from $v = 1$ m/day to 100 m/day. When the flow velocity of the water is $v < 1$ m/day, then wellpoints are ineffective for dewatering work.

For the theory of wellpoints the reader is referred to Refs. 1 and 5.

Dewatering of Excavation

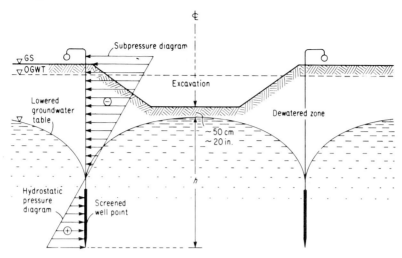

FIG. 9-7. Single-stage symmetrical wellpoint system.

Pumping from Multiple Wells

The general equation of the depression curve of the lowered groundwater table when pumping from a single, ordinary, perfect well is (see Fig. 9-10):[1,5]

$$y_2^2 - y_1^2 = \frac{q}{\pi k}(\ln x_2 - \ln x_1) \tag{9-6}$$

where q = rate of water pumped from a single well
k = system's coefficient of permeability
All other symbols are shown in Fig. 9-10.

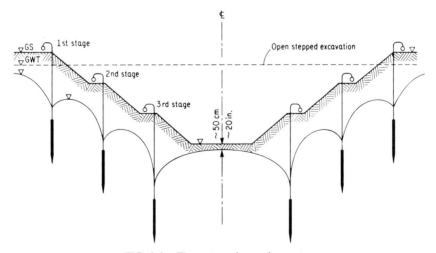

FIG. 9-8. Three-stage dewatering system.

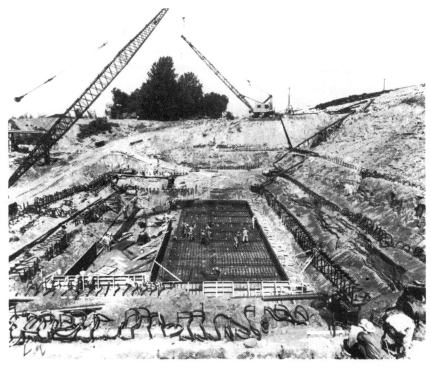

FIG. 9-9 Multiple-stage dewatering system. (Courtesy of Griffin Wellpoint Corporation, New York.)

When $x_2 = R$ and $x_1 = r_o$, and with r_o and k known, the quantity q of water pumped per unit of time (= discharge) is

$$q = \pi k \frac{H^2 - h^2}{\ln R - \ln r_0} \tag{9-7}$$

where R = radius of influence of the soil-well system
r_0 = radius of well
H = thickness of water-bearing stratum

It must now be tested whether the well can handle the necessary quantity of water. According to Sichardt,[5,6] the tapping capacity f in m³/sec of a single well may be calculated as

$$f = 2\pi r_0 h_0 \sqrt{k/15} \tag{9-8}$$

where r_0 = radius of well in meters
h_0 = height of wetted filter area of well, in meters
k = coefficient of system's permeability, m/sec

Dewatering of Excavation

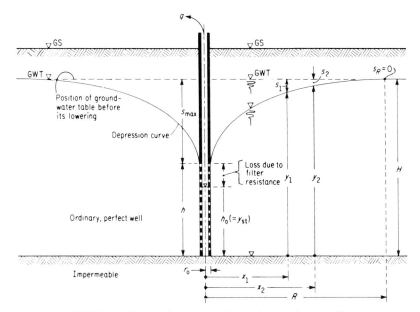

FIG. 9-10. Depression curve of lowered groundwater table.

For each k this capacity is attained at a practical maximum hydraulic gradient i_{max} at the wall of the well. With $q = kiA$, where $A =$ filter area,

$$i_{max} = q/kA = \frac{2\pi r_0 h_0 \sqrt{k}/15}{2\pi r_0 h_0 k} = \frac{1}{15\sqrt{k}} \qquad (9\text{-}9)$$

The maximum drawdown s_{max} at the outside surface of the well casing upon pumping is

$$s_{max} = H - h = H - \sqrt{H^2 - \frac{q}{\pi k}(\ln R - \ln r_0)} \qquad (9\text{-}10)$$

When $y_2 = H$, $y_1 = y$, $x_2 = R$ and $x_1 = x$, Eq. 9-6 transforms to

$$H^2 - y^2 = \frac{q}{\pi k}(\ln R - \ln x) \qquad (9\text{-}11)$$

With a circular arrangement of n wells in number, encompassing a round foundation pit (Fig. 9-11), and each well yielding an amount of water q, the total quantity Q of water pumped, in liters per second, is $Q = nq$, and Eq. 9-11 is now written as

$$H^2 - y^2 = \frac{Q}{\pi k}(\ln R - \ln \rho) \qquad (9\text{-}12)$$

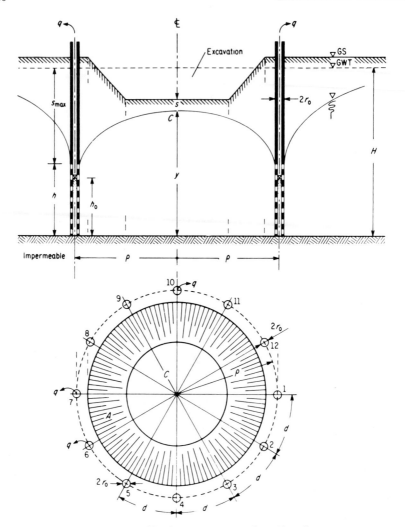

FIG. 9-11. Circular arrangement of $n = 12$ wells.

where y = ordinate of a point on the depression cone at center C of circle
ρ = radius of circular area to be dewatered

If instead of a circular arrangement of wells one must deal with a rectangular arrangement of wells (Fig. 9-12), then in Eq. 9-12 substitute $\ln \rho$ by the arithmetical mean of the algebraic sum of natural logarithms of the individual distances $x_1, x_2, x_3, \ldots, x_n$ from a reference point C of the excavation to be dewatered to the individual wellpoints Nos. 1, 2, 3, ..., n:

$$\ln \rho = \sqrt[n]{\ln x_1 + \ln x_2 + \ln x_3 + \cdots + \ln x_n} \qquad (9\text{-}13)$$

Dewatering of Excavation

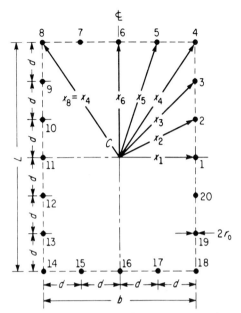

FIG. 9-12. Rectangular arrangement of $n = 20$ wells.

$$= (\ln x_1 + \ln x_2 + \ln x_3 + \cdots + \ln x_n)^{1/n} \qquad (9\text{-}13a)$$

$$= \frac{1}{n} \ln (x_1 x_2 x_3 \cdots x_n) \qquad (9\text{-}13b)$$

Thus for an excavation rectangular in plan, Eq. 9-12 becomes

$$H^2 - y^2 = \frac{Q}{\pi k} \left[\ln R - \frac{1}{n} \ln (x_1 x_2 x_3 \cdots x_n) \right] \qquad (9\text{-}14)$$

Hence, if the point of reference C of an excavation rectangular in plan is located at its geometric center, then from Eq. 9-13b

$$\sqrt[n]{x_1 x_2 x_3 \cdots x_n} = \rho \qquad (9\text{-}15)$$

may be considered as the radius ρ of a circle (whose area is A) equivalent to the given rectangular system of the same area A in plan. Thus

$$H^2 - y^2 = \frac{Q}{\pi k} (\ln R - \ln \rho) = \frac{Q}{\pi k} \left[\ln R - \frac{1}{n} \ln (x_1 x_2 x_3 \cdots x_n) \right] \qquad (9\text{-}16)$$

This method of calculation can be used when the number n of wells is known. When $x_1 = x_2 = x_3 = \cdots = x_n = x = \rho$ as for a circle, then

$$\frac{1}{n} \ln (x x x \cdots x) = \frac{1}{n} \ln (x^n) = \ln x = \ln \rho \qquad (9\text{-}17)$$

To work for a rectangular excavation system with an equivalent circular system is an approximation, of course, and is practically justified when the rectangular excavation is neither very narrow nor too long. For a closed, concentrated plan of a rectangular foundation excavation, such as would pertain to many small structures, the rectangular area encompassed by wells may be assumed with sufficient precision as an equivalent circle, and its radius ρ determined by setting the known area of the rectangular foundation pit to be dewatered equal to the area A of the equivalent circular pit, so that $x = \rho$. In such case Eq. 9-12 may be used for calculating total discharge Q (Fig. 9-13).

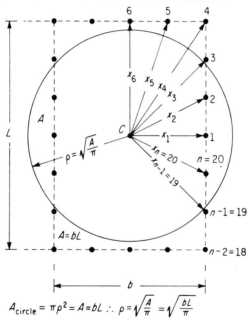

FIG. 9-13. Substitution of a rectangular area by an equivalent circular one.

By Eq. 9-16 the total yield Q of water of the circular multiple well system is now computed as

$$Q = \frac{\pi k (H^2 - y^2)}{\ln R - (1/n) \ln (x_1 x_2 x_3 \cdots x_n)} = \frac{\pi k (H^2 - y^2)}{\ln R - \ln \rho} \qquad (9\text{-}16a)$$

The discharge Q from the system must be known in order to assess the necessary pump capacity.

Dewatering of Excavation

When pumping is done from imperfect wells one usually performs Q calculations as for ordinary, perfect wells and increases this discharge by approximately 20 percent to obtain discharge Q_i from imperfect wells, i.e., $Q_i = (1.20)Q$.

Drawdown. At the center C of the circle the drawdown s of the groundwater table is

$$s = H - y \qquad (9\text{-}18)$$

In order to attain this desired s, the necessary discharge quantity of water Q to be pumped off the given excavation system is, by Eq. 9-16a:

$$Q = \pi k \frac{(2H-s)s}{\ln R - \ln \rho} = \pi k \frac{(2H-s)s}{\ln R - (1/n) \ln (x_1 x_2 x_3 \cdots x_n)} \qquad (9\text{-}19)$$

The lowering of the groundwater table by an amount of s is, by Eq. 9-10:

$$s = H - \sqrt{H^2 - \frac{Q}{\pi k} \ln (R/\rho)} \qquad (9\text{-}20)$$

The necessary number of wells n is calculated as $n \approx Q/f$, where f = tapping capacity of a single well.

Stage in a Well. If it is desirable to know the stage of the water in the individual wells, one makes use of Eq. 9-16. In this equation, substitute each time for $x_1, x_2, x_3, \cdots x_n$ the distance (center to center) of the individual wells from that well for which the stage is tested, its own (test well) distance which is $x = r_0 = r_{st}$, or, r_{st} = radius of well for which stage is checked. The stage ordinate is $y = y_{st}$. Hence, Eq. 9-16 is written as

$$H^2 - y_{st}^2 = \frac{Q}{\pi k} \left[\ln R - (1/n) \ln (r_{st}\, x_1 x_2 x_3 \cdots x_{n-1}) \right] \qquad (9\text{-}21)$$

(see Fig. 9-14).

Spacing of Wells

According to Sichardt,[6] the spacing d of individual wells at which the well-tapping capacity can be fully utilized may be calculated (Fig. 9-11) as

$$d \geq (5)(2\pi r_0) \qquad (9\text{-}22)$$

If the site to be dewatered is of a rectangular form whose area is A, with the number of wells n not yet known but to be determined, then the magnitude of the equivalent radius ρ of the equivalent circular area $A = \pi \rho^2$ is calculated for a rectangular area as

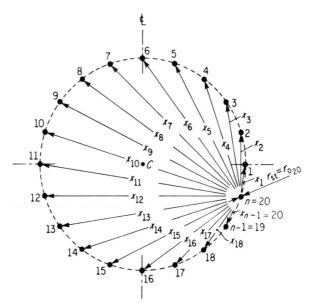

FIG. 9-14. Scheme for testing stage y_{st} in well No. 20.

$$\rho = \sqrt{A/\pi} \qquad (9\text{-}23)$$

The circular spacing d of the wells is calculated empirically after Sichardt[6,7] in consistent units of measurement as $d \geqslant (5)(2\pi r_0)$.

The spacing of wells depends upon the type of soil and its permeability k. In clean sand, with a depth of water up to about 15 ft, wellpoints are installed in a row along the header line from 4 to 5 ft apart.

Coefficient of Permeability

The coefficient of system's permeability k must be determined from pumping tests. It may also be assessed for preliminary calculations of small dewatering installations from the following compilation:

Soil Type	Coefficient of Permeability k, cm/sec (order of magnitude)
Very fine sand	1×10^{-4} to 2×10^{-4}
Fine sand	1×10^{-3} to 2×10^{-3}
Coarse sand	5×10^{-3} to 2×10^{-2}

Radius of Influence

The radius of influence R, the magnitude of which is of very little significance in the results of calculating Q as compared with k (because R enters into the Q-equation under the sign of logarithm, but logarithms of the R/r_0- or R/x-ratios vary very slowly), is calculated by an empirical equation after Sichardt[1.8] as

$$R = (3000)(s)\sqrt{k} \qquad (9\text{-}24)$$

where R = radius of influence in meters
  s = drawdown in meters
  k = coefficient of permeability, in m/sec

According to Schoklitsch,[9] the values of the radius of influence R of the depression cone for various soils are about as follows:

Dune sand.	$R =$	5 m to 10 m ≈	15 ft to ≈ 30 ft
Sand.	$R =$	50 m ≈	165 ft
Fine gravel	$R =$	100 m to 5000 m ≈	330 ft to ≈ 1640 ft
Coarse gravel	$R \leqslant$	500 m ≈	1640 ft

For lack of any better data for R, in preliminary calculations the R-values can be used as indicated in Table 9-2. The R-values were collected from many sources; they were evaluated and adjusted for the soil classification system according to the Bureau of Soils, U.S. Department of Agriculture.

TABLE 9-2

Radius of Dewatering Influence in Various Soils

Description of soil		Radius of Influence R, meters
Type	Particle Size d, mm	
Gravel,		
Coarse	10	1,500
Medium and fine	1 -10	400-1,500
Sand,		
Coarse and medium.	0.25 - 1	100- 400
Fine	0.05 - 0.25	10- 100
Silty	0.025- 0.05	5- 10

Lift

The maximum lift of water that can be attained by a single-stage dewatering system is about 18 to 20 ft (practically). The so-called "suction height" of the pump is obtained from $(H - h)$, plus (1) the height of the horizontal axis of the pump located above the unlowered groundwater table, and (2) plus the resistance (losses) of the length of the filter of the well. The height of the lift of a pump is the sum of the suction height and the pressure height. The line resistances $\Sigma \zeta_L$ may be included in the coefficient of efficiency, η, of the dewatering system.

For lifting water from a depth greater than 20 ft, a multiple stage dewatering system is installed (Figs. 9-8 and 9-9). A multiple-stage dewatering system consists of two or more tiers of single-stage dewatering systems. With n stages, a lowering of the groundwater table of $(n \times 18)$ ft can be attained.

To lift water at heights greater than 18 to 20 ft, deep-well pumps (termed *submersible pumps*), in one tier are used. Such pumps are submerged into the casing of the well. The water can then be pushed up to a considerable height at a small suction height.

9-4. PUMP CAPACITY

The pump horsepower N_p in foot-pound units is calculated as

$$N_p = \frac{Q_{\text{eff}} H_L}{550 \eta_p} \quad \text{(hp)} \tag{9-25}$$

where $Q_{\text{eff}} = cQ\gamma_w$ = effective discharge in lb/sec
$c = 1.5\text{-}2.0$ = a safety factor for pumping quantity Q of water in ft^3/sec
$\gamma_w = 62.4$ lb/ft^3 = unit weight of water
H_L = total head (lift) against which a pump must pump water
$\eta_p = 0.20\text{-}0.50$ = coefficient of mechanical efficiency of pump
1 hp = 550 ft-lb per sec

Motor horsepower, N_m:

$$N_m = \frac{N_p}{\eta_m} \quad \text{(hp)} \tag{9-26}$$

where $\eta_m = 0.3$ to 0.9 = coefficient of mechanical efficiency of motor for driving the pump.

Dewatering of Excavation

Pump horsepower N_p in the metric system of units of measurement:

$$N_p = \frac{Q_{eff} H_L}{75 \eta_p} \quad (PS) \tag{9-27}$$

where $Q_{eff} = cQ\gamma_w$ = effective discharge, kg/sec
c = 1.5 to 2.0 = a safety factor for pumping quantity Q of water, liters per second
Q = discharge, l/sec
γ_w = 1 kg/liter approximately; 1 liter = 10cm × 10cm × 10cm = 1000 cm³
H_L = full manometric height of lift (for Q), in meters
η_p = 0.30-0.50 = coefficient of mechanical efficiency of pump; the efficiency of some centrifugal pumps is about η_p = 0.75-0.80
1 PS = metric or continental horsepower = 75 kg-m/sec, and is the most common mechanical unit of power

Motor horsepower, N_m. The capacity of the motor or any other machine to drive the pump is calculated as

$$N_m = \frac{N_p}{\eta_m} \quad (PS) \tag{9-28}$$

where η_m = 0.30-0.90 = coefficient of mechanical efficiency of motor for driving the pump.

Conversion factors:

1 hp = 1.014 PS = 0.7457 kw = 550 ft-lb/sec
1 PS = 542.5 ft-lb/sec = 0.986 hp = 0.7355 kw = 75 kg-m/sec
1 kw = 1.341 hp = 737.6 ft-lb/sec = 1.360 PS
1 kg-m = 7.233 ft-lb
1 ft-lb = 0.138254 kg-m

Example

Given large, ordinary, perfect wells, the diameter of which is $2r_0 = 6$ in. to be used for dewatering a foundation excavation as shown in Fig. 9-15. The dimensions of the rectangular area encompassed by the layout of the wells is 108 ft × 72 ft. The thickness of the aquifer is $H = 32$ ft, and the given dewatering system's coefficient of permeability was determined from pumping tests as being $k = 0.0055$ ft/sec = 0.0017 m/sec. It is required to calculate the dewatering system.

268 Excavation and Cofferdams

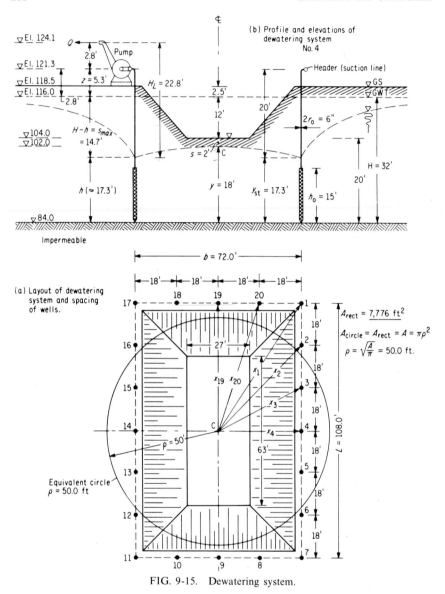

FIG. 9-15. Dewatering system.

Solution

Necessary amount of lowering s of groundwater table at center C of excavation:

$$s = \text{El. } 116.0 - \text{El. } 104.0 + 2 = \underline{14.0 \text{ (ft)}}$$

Dewatering of Excavation

Ordinate y of center C above lower end elevation of wells:

$$y = H - s = 32.0 - 14.0 = \underline{18.0 \text{ (ft)}}$$

Area A encompassed by wells:

$$A = 108.0 \times 72.0 = \underline{7{,}776 \text{ (ft}^2\text{)}}$$

Radius ρ of equivalent circle, by Eq. 9-23:

$$\rho = \sqrt{A/\pi} = \sqrt{7776/3.14} = \underline{50.0 \text{ (ft)}}$$
$$\ln \rho = \ln 50.0 = \underline{3.91202}$$

Radius of influence R, by Eq. 9-24:

$$R = (3000)(s)\sqrt{k} = (3000)(4.27)\sqrt{0.0017} = 525.2 \text{ (m)}$$

where $s = 14.0 \text{ ft} = (14.0)(\text{ft})(0.3048)(\text{m/ft}) = 4.27 \text{ m}$
$k = 0.0017 \text{ m/sec}$
$R = (525.2)(\text{m})(3.2808)(\text{ft/m}) = 1{,}723.07 \text{ ft} \approx \underline{1{,}723 \text{ ft}}$
$\ln R = \ln 1723 = \underline{7.44}$

Total discharge Q from the dewatering system, by Eqs. 9-12 or 9-16a:

$$Q = \pi k \frac{H^2 - y^2}{\ln R - \ln \rho} = (3.14)(0.0055)\frac{32^2 - 18^2}{7.44 - 3.91} = \underline{3.42 \text{ ft}^3/\text{sec}}$$

$$= (3.42)(0.028317) = 0.0968 \text{ m}^3/\text{sec} = 96.8 \text{ l/sec} \approx \underline{96.8 \text{ kg/sec}}$$

Tapping capacity f of a single well, by Eq. 9-8:

$$f = 2\pi r_0 h_0 \frac{\sqrt{k}}{15}$$

$$= (2)(3.14)(0.075)(4.56)\frac{\sqrt{0.0017}}{15} = 0.00586 \text{ m}^3/\text{sec}$$

$$= (0.00586)(35.314) = \underline{0.2069 \text{ (ft}^3/\text{sec})}$$

where $r_0 = 3.0 \text{ in.} \approx 0.075 \text{ m}$
$h_0 = 15.0 \text{ ft} = (15.0)(0.304) = 4.56 \text{ m}$ is the *assumed* wetted length of the well filter

Necessary number n of wells:

$$n = Q/f = 3.42/0.2069 = 0.0968/0.00586 = 16.5 \approx \underline{20}$$

Use 20 wells spaced at a distance of

$$d = \frac{(2)(108.0) + (2)(72.0)}{20} = \frac{360.0}{20} = \underline{18.0 \text{ ft center to center}}$$

or

$$d = (18.0)(0.3048) = \underline{5.49 \text{ m}}$$

Check of well spacing d, by Eq. 9-22:

$$d = (5)(2\pi r_0) = (5)(2)(3.14)(0.075) = 2.355 \text{ (m)} < 5.49 \text{ m}$$

or

$$d = (2.355)(3.28) = \underline{7.72 \text{ (ft)}} < 18.0 \text{ ft}$$

Hence the well can be fully exploited because $d = 7.72$ ft < 18.0 ft

Checking the y-value at center of excavation, by Eq. 9-12:

$$y = \sqrt{H^2 - \frac{Q}{\pi k}(\ln R - \ln \rho)} =$$

$$= \sqrt{32^2 - [(3.42)/(3.14)(0.0055)](7.44 - 3.91)} = 18.0 \text{ (ft)}$$

Thus, y checks out with that calculated at the outset of the problem, where $y = H - s = 18.0$ ft.

Checking the equivalent radius ρ, by Eq. 9-15:

$$\ln \rho = (1/n) \ln (x_1 \, x_2 \, x_3 \cdots x_4)$$
$$= (1/20)[\ln (64.9)^4 + \ln (50.9)^4 + \ln (40.3)^4 + \ln (36.0)^2$$
$$+ \ln (56.8)^4 + \ln (54.0)^2] = 78.51/20 = 3.93 \approx \ln \rho = 3.91$$

$$\therefore \rho = \underline{50.7 \text{ (ft)} \approx 50.0 \text{ ft}}$$

The radius ρ of the equivalent circle represents a very good approximation in this case with that ρ as calculated by Eq. 9-23, both methods of calculation thus rendering the same magnitude of Q. Here

$x_1 = x_7 = x_{11} = x_{17} = 64.9$ ft   4 ln 64.9 = (4)(4.17285) = 16.69
$x_2 = x_6 = x_{12} = x_{16} = 50.9$ ft   4 ln 50.9 = (4)(3.92986) = 15.72
$x_3 = x_5 = x_{13} = x_{15} = 40.3$ ft   4 ln 40.3 = (4)(3.69635) = 14.79
$x_4 = x_{14} = 36.0$ ft   2 ln 36.0 = (2)(3.58352) = 7.17
$x_8 = x_{10} = x_{18} = x_{20} = 56.8$ ft   4 ln 56.8 = (4)(4.03954) = 16.16
$x_9 = x_{19} = 54.0$ ft   2 ln 54.0 = (2)(3.98898) = 7.98

Checking the stage y_{st} of water level in wells (viz., drawdown in wells), by Eq. 9-21), in well No. 4 (Fig. 9-16),

Dewatering of Excavation

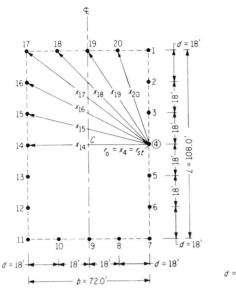

FIG. 9-16. Checking the stage y_{st} in well No. 4.

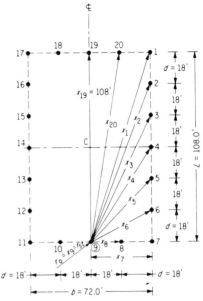

FIG. 9-17. Checking the stage y_{st} in well No. 9

$$y_{st} = \sqrt{H^2 - \frac{Q}{\pi k}\left[\ln R - \frac{1}{n}\ln(r_{st}\, x_1\, x_2\, x_3 \cdots x_{n-1})\right]}$$

$$= \sqrt{32^2 - \frac{3.42}{(3.14)(0.0055)}\left\{\ln 1723 - \frac{1}{20}\left[\ln(0.25) + \ln(54.00)^2\right.\right.}$$

$$\overline{+ \ln(36.00)^2 + \ln(18.00)^2 + \ln(56.92)^2 + \ln(64.90)^2 + \ln(76.36)^2}$$

$$\overline{\left.\left.+ \ln(90.00)^2 + \ln(80.50)^2 + \ln(74.21)^2 + \ln(72.00)\right]\right\}} = \underline{17.3 \text{ (ft)}}$$

$y_{st} = 17.3$ ft > 15.0 ft (approximately equal, as was assumed at the outset of the solution of the problem).

Here $r_{st} = 3$ in. $= 0.25$ ft; ln $0.25 = -1.38629$

Stage y_{st} in well No. 9 (Fig. 9-17):

$x_1 = x_{17} = 113.80$ ft	2 ln 113.8	$= (2)(4.73621) = 9.47$
$x_2 = x_{16} = 96.92$ ft	2 ln 96.9	$= (2)(4.57368) = 9.15$
$x_3 = x_{15} = 80.50$ ft	2 ln 80.5	$= (2)(4.38826) = 8.77$
$x_4 = x_{14} = 64.90$ ft	2 ln 64.9	$= (2)(4.17285) = 8.34$
$x_5 = x_{13} = 50.91$ ft	2 ln 50.9	$= (2)(3.92986) = 7.85$

$$\begin{aligned}
x_6 &= x_{12} = 40.50 \text{ ft} & 2\ln 40.5 &= (2)(3.70130) = 7.40\\
x_7 &= x_{11} = 36.00 \text{ ft} & 2\ln 36.0 &= (2)(3.58352) = 7.17\\
x_8 &= x_{10} = 18.00 \text{ ft} & 2\ln 18.0 &= (2)(2.89037) = 5.78\\
x_9 &= r_{st} = 0.25 \text{ ft} & \ln 0.25 &= -1.38629 = -1.39\\
x_{18} &= x_{20} = 109.50 \text{ ft} & 2\ln 109.5 &= (2)(4.69591) = 9.39\\
x_{19} &= 108.00 \text{ ft} & \ln 108.0 &= 4.68213 = 4.68
\end{aligned}$$

By Eq. 9-21, $y_{st} = 17.57$ (ft) > 15.00 ft (hence approximately equal, as was assumed at the outset of the solution of this problem).

Height of lift:
Position of horizontal axis of pump: $z = 5.3$ ft above GWT.

Suction height:

$$H - y_{st} + z = 32.0 - 17.3 + 5.3 = \underline{20.0 \text{ (ft)}}$$

Discharge point: 5.6 ft above ground surface, or 2.8 ft above horizontal axis of pump

Full manometric height H_L of Q:

$$H_L = 20.0 + 2.8 = 22.8 \text{ (ft)} = \underline{6.95 \text{ m}}$$

In order to keep the frictional resistances of the pipes within reasonable, practical limits, the flow velocities of water in suction lines should not exceed 1.5 m/sec ≈ 5 ft/sec, and in the pressure lines not to be more than 2 to 3 m/sec ≈ 6.5 ft/sec to ≈ 10 ft/sec.

Pump capacity:

$$Q_{eff} = cQ\gamma_w = (2.0)(3.42)(62.4) = 426.8 \approx \underline{430 \text{ (lb/sec)}}$$

where $c = 2.0$.

Pump horsepower N_p in foot-pound units:
Coefficient η_p of efficiency of pump: assume $\eta_p = 0.33$

$$N_p = \frac{Q_{eff} H_L}{(550)\eta_p} = \frac{(430)(22.8)}{(550)(0.33)} = \underline{54 \text{ (hp)}}$$

Pump capacity in the metric system (continental horsepower PS):

$$N_p = \frac{Q_{eff} H_L}{(75)\eta_p} = \frac{(2.0)(96.8)(6.95)}{(75)(0.33)} = 54.5 \approx \underline{55 \text{ (PS)}}$$

1 hp = 1.014 PS.

Dewatering of Excavation

Electric motors for driving the pumps:

Motor horsepower in British units of measurement:

$$N_m = \frac{N_p}{\eta_m} = 54/0.75 = 72 \approx \underline{75 \text{ (hp)}}$$

or in electric units,

$$N_m = (75)(0.7457) = 55.927 \approx \underline{56 \text{ kw}}$$

Motor horsepower in metric units:

$$N_m = \frac{N_p}{\eta_m} = 55/0.75 = 73.5 \approx \underline{75 \text{ (PS)}}$$

or $\underline{56 \text{ kw}}$.

9-5. ELECTRO-OSMOSIS

Fine-particled soils, such as silts and silty clays ($k = 10^{-3}$ to $k = 10^{-5}$ cm/sec) cannot be drained successfully by gravity (wells and wellpoints). They can be most effectively dewatered by the method of electrical drainage. In this method the underlying principle is the electrokinetic phenomenon of *electro-osmosis*.[1]

The term electro-osmosis is used to describe the electrokinetic phenomenon of liquid (water) moving through a system of a porous medium (soil) relative to a fixed solid (i.e., the colloidal soil particles are not moving) under the influence of a direct current electrical field. Also, electro-osmosis is possibly due to the existence of the electric diffuse double layer in a moist soil system when the water film is in motion past the solid soil particles.

Electro-osmosis is thus utilized for the dewatering of silty and clayey soils difficult to drain by gravity. These soils cannot be drained easily by gravity because the relatively large surface tension forces of water in such soils tend to retain water in their voids.

Electrical dewatering is pursued to facilitate laying of foundations in a dry pit or excavation, for the stabilization of soil in natural or artificial slopes, as well as for other purposes.

The principle of dewatering fine-particled soils electrically is illustrated in Fig. 9-18. In this electric field the water contained in the soil migrates through the voids of the soil from anode to cathode. Electrical dewatering brings about an increase in the shear strength of the soil and thus stability of the slope.

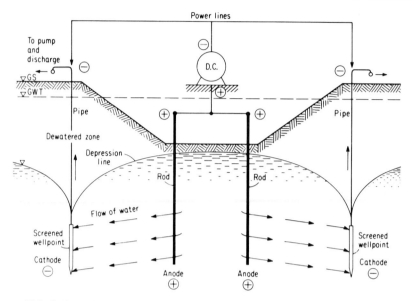

FIG. 9-18. Principle of dewatering of a fine-particled soil by electro-osmosis.

In this method of dewatering and stabilizing fine-particled soil such as silt, silty clay, and clayey silts, direct electric current is passed to flow through electrodes to the wells placed in the soil to be stabilized and encompassing the foundation site to be dewatered and thus stabilized.

To keep the excavation dry, positive electrodes (anodes in the form of rods) are installed in saturated soil near the toes of the slopes of the excavation (Fig. 9-18). The negative electrodes (cathodes) are installed in the mass of soil away from the slopes of the cut and are made in the form of perforated pipes resembling wellpoints. Their function is to collect the water flowing from the positive electrode when there exists in the soil an electric field (i.e., a direct current circuit) between the electrodes. The water collected in the negative electrode is pumped out and discharged.

The soil moisture migration in a direct current electric field in a fine-particled soil is based on the electrokinetic phenomenon that the dipole water molecules (water film, for example) align themselves in the electric field. Also, in the electric field the cations (+ charges) are set in motion, dragging with them the oriented free water molecules and also part of the water film surrounding the solid soil particles along the immobile part of the water film toward the negative electrodes where the water collects. On the surface of the solid soil particles there reside (upon motion of water past the solid surfaces) negative charges (−). Thus the water that migrates

Dewatering of Excavation

consists of water in bulk and of the less-stressed part of the water film adsorbed to the solid soil particles.

The resistance to motion of the water molecules is assumed to follow Newton's law of viscous flow. Hence the movement of the mobile part of the water film over the immobile part of the moisture film surrounding the soil particle is a tangential slip by overcoming the shearing resistance of the liquid (Fig. 9-19).

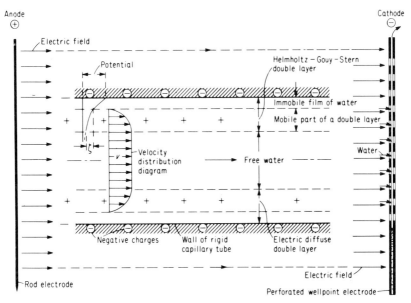

FIG. 9-19. Electro-osmotic flow of water.

According to Leo Casagrande,[10-13] the average electro-osmotic coefficient of permeability k_e for sand, silt, or clay may be assumed for practical purposes to be $k_e = 0.5 \times 10^{-4}$ cm/sec for an electric gradient of 1 volt/cm, or $k_e = 0.5 \times 10^{-4}$ (cm/sec)/(volt/cm).

The volume V of water moved per unit of time by electro-osmosis is directly proportional to the applied electric current I, the dielectric constant of water D, and inversely proportional to the dynamic viscosity η and electroconductivity λ of water:

$$V = \frac{\zeta D I}{4\pi \eta \lambda} \quad (\text{cm}^3/\text{sec}) \tag{9-29}$$

or in any other consistent units, such as the technical units m³/min, or ft³/min. Here ζ = coefficient of proportionality, also known as the *zeta potential*.

If the current is kept constant, the flow of water is independent of the length and area of the void passages.

For calculating the electro-osmotic dewatering installation of a soil, the following basic quantities are to be computed: (1) the spacing of electrodes, (2) the working voltage of the electrical installation, (3) the density of the electric current, (4) the specific ohmic resistance of the soil, and (5) the capacity of the electrical installation.

The distance L in meters between the rows of electrodes is computed as

$$L = \frac{100 V}{\rho \phi I} \quad \text{(m)} \quad (9\text{-}30)$$

where V = working electrical tension, volts
ρ = specific ohmic resistance, ohm/cm
ϕ = 2 to 3 = a coefficient, depending upon dimensions and spacing of electrodes
I = electric current density in amperes on an A_s = 1 m² cross-sectional area of soil to be dewatered in the plane of an electrode row

The most convenient working voltage of an electrical dewatering installation is from 60 volts d.c. to 100 volts.

The current density influences the intensity of the dewatering of the soil. With insufficient density the soil cannot be dewatered, and with excess intensity (strength) the soil will warm up and dry out. The optimum density of current is assumed within limits from 0.5 to 7 amp per 1 m² cross section of soil.

The specific ohmic resistance of soil depends on the nature of the soil mineral contents, the properties of the groundwater and principally on the moisture content of the soil. This resistance is calculated as

$$\rho = R(a/h) \quad \text{(ohm-cm)} \quad (9\text{-}31)$$

where ρ = specific resistance of soil in ohm-cm
R = ohmic resistance of the test sample, to be determined by means of a bridge, ohms
a = cross-sectional area of soil sample, cm²
h = length of soil sample, cm

The capacity N in kilowatts of the electrical installation is calculated as

$$N = \frac{VIA_s}{1{,}000} \quad \text{(kw)} \quad (9\text{-}32)$$

where A_s = cross-sectional area of a massive soil to be dewatered, in the plane of a row of electrodes perpendicular to current flow, m²

The electricity consumption amounts from 3 to 30 kwh/cu. yd of soil drained according to the nature of the job.[14]

Electrochemical Stabilization

In electrochemical stabilization aluminum anodes and copper cathodes are installed in the soil to be stabilized. When the electrical circuit is closed, the aluminum anode is consumed, the aluminum ions migrating to the soil surrounding the cathode. The insoluble aluminum compounds strengthen the soil and the soil does not soften on saturation with water.

9-6. SOIL DENSIFICATION BY HYDROVIBRATION

Loose, saturated sandy soils encountered in a cofferdam-enclosed foundation excavation often need to be densified if they are likely to be set in motion by superimposed loads such as construction equipment and/or structural loads. The necessity for densification of soil is established after a comprehensive soil exploration at the site and soil testing.

The choice of selecting and adapting a suitable method of densification of granular soil is based upon economic considerations, for example, on a comparison of the cost of the various methods of densification (surface vibration, deep vibration), upon the rapidity of the densification process, on the availability of necessary materials and special equipment. Also, the effectiveness of a method of densification and its duration (service life) must be given consideration.

Densification of loose, granular soil may be achieved by the vibration method. Soil vibration may be performed by surface vibration or by the method of deep vibration. Deep vibration may be performed with or without simultaneous introduction of water into the soil. The latter method is known as *hydrovibration*, a method of densification of noncohesive soil by deep vibration with simultaneous saturation of the soil from the vibration equipment. Hydrovibration is a fast, efficient and relatively successful method for densification of sand.

The *vibrator*, a cylindrical body approximately 2-3 m in length and 25 to 40 cm in diameter, is provided at both ends with water-jetting devices. It is connected to a pipe rod. The latter also accommodates electric cables to the vibration motor and pipes supplying saturation water to the jets. In essence, the vibrator is usually a two-mass oscillator. The frequency of hydrovibration equipment is from 3,000 to 5,000 c/min.

Process of Hydrovibration

Before any vibration is commenced, the vibrator is first jetted into the soil through the thickness of strata to be densified. For jetting the vibration

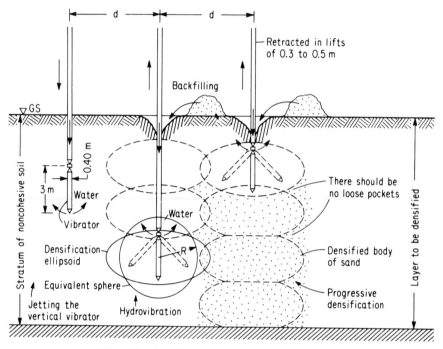

FIG. 9-20. J. Keller's method of hydrovibration (Ref. 15).

equipment, only the lower jet is opened (Fig. 9-20). Through this jet water is jetted into the soil under pressure in order to ease the penetration of the vibrator. The jetting action is so effective that, depending upon the efficiency of available vibration equipment, a 35-m penetration depth can be attained by its own weight (say 2,270 kg) within a few minutes without turning on the vibrator motor. After the desired depth of penetration, viz., thickness of strata to be densified, is attained, the lower jet is closed, the upper jet is opened, and the hydrovibration is commenced.

Densification brings about a decrease in porosity of the soil of an amount of 10 to 20 percent. Also, vibration forms around the shaft of the vibration equipment at the ground surface a crater or funnel which is continuously backfilled during vibration with sand supplied for this purpose.

Densification of soil by this method of hydrovibration is brought about, so to speak, "from below up" by pulling the vibrator up gradually. By this method, the sandy soil becomes uniformly densified in depth and also in extent, and densification is not affected by groundwater.

The density to be attained in a sand stratum by hydrovibration must be assessed by testing the soil for its relative density.

The time consumption for penetrating the vibrator by jetting is about 1 m/min. Densification of soil requires on the average 10 min/m.

Densification of soil brings about an increase in its shear strength and thus bearing capacity of the soil, hence reducing settlement. The bearing capacity of a sandy soil thus densified is to be determined from the results of soil tests with static loads.

Hydrovibration may be applied also to sandy soils containing not more than 20 percent of silt and clay particles.

The radius of influence of densification by hydrovibration, depending upon the properties of the soil, is 1-2 m.

The shape of the densified body of sand is an approximate ellipsoid. For practical calculations the ellipsoid may be assumed to be a sphere of radius R. The densified zone may be imagined to consist of a stack of merged densified ellipsoids, viz., spheres.

The spacing d between the positions or vertical axis of the vibrator shaft is calculated by considering the radius of the body of densification.

The thickness of one lift of densification of the soil should not exceed 3-4 m.

Hydrovibration can be used for densifying soil in the foundation pit, backfill of sheet-piling walls and massive earth-retaining structures, as well as granular materials in double-wall cofferdams. Hydrovibration reduces pressure on cofferdams and saves on the width of cofferdams, thus increasing their stability against overturning and sliding.

9-7. SOIL STABILIZATION

Soil stabilization has for its general purpose increasing its strength, decreasing its perviousness to water, and increasing its resistance to erosion or to wear and tear. These improved qualities of a soil or, as meant here, its chemical stabilization, are achieved by introducing into the soil special additives (cement, clay, or bitumen) for filling the voids and thus giving a more or less monolithic structure to the foundation soil.

Some of the commonly known methods of stabilization of soil are:
- Cement grouting
- Bitumen grouting
- Clay grouting
- Chemical injections

Some methods of improvement of some engineering properties of soils and rocks are compiled in Table 9-3 for the reader's orientation.

TABLE 9-3
Methods of Improvement of Some Engineering Properties of Soils

Soil Types	Method	Technological Process	Results	Limitations of Application of Method	Engineering Applications
Rock, semirock and gravel	Grouting (cementation)	Forcing (pumping) cement grout through bore holes	Imperviousness to water	Fissures: discharge flow 0.05-10 l/min	Curtains and increasing strength of soil (solidification of soil)
	Hot bituminization	Forcing bitumen through bore holes into the fissures	Impermeability to water	Flow 1-100 l/min	Curtains and increasing the strength of the soil
	Clay grouting	Forcing clay grout with lime additives through boreholes	Impermeability to water	Caverns: flow 0.1-100 l/min	Curtains and increasing the strength of the soil
Sand	Two-solution chemical injection	Forcing two chemical solutions of potassium silicate and calcium chloride through injectors	Impervious to water; strength, 10-50 kg/cm^2	Coefficient of permeability $k = 2$-80 m/day	Strengthening of soil underneath foundations; in tunneling, shaft-driving, and as impervious curtains
	One-solution chemical injection	Forcing potassium silicate solution with phosphorous acid through an injector	Impervious to water; strength; 3-5 kg/cm^2	$k = 0.5$-2 m/day	Curtains and aprons

TABLE 9-3 (*continued*)

Soil Types	Method	Technological Process	Results	Limitations of Application of Method	Engineering Applications
Sand	Lowering of groundwater table	Pumping water from wells and/or wellpoints	Stability of slopes of foundation excavations and their bottoms	$k = 1\text{-}150$ m/day	Providing for a dry foundation pit for laying of foundation
	Cold bituminization	Forcing bituminous emulsions through injectors	Imperviousness to water; bonding of soil	$k = 10$ m/day	Curtains and aprons
	Vibration; hydro-vibration; hydro-flotation	Densification with surface and depth vibrators of naturally deposited and filled soils	Increase in soil bearing capacity	Method limited to granular soil materials	Densification of fills and foundation-supporting soils
Loess (Silt)	Chemical injection	Forcing one-solution potassium silicate through a bore hole	Strength; 6-10 kg/cm^2; reduction of settlement	$k - 0.2\text{-}2.0$ m/day	Increasing strength of soil
	Firing (thermal treatment)	Forcing hot gases through bore holes	Strength; 10 kg/cm^2; reduction of settlement; stabilizing slopes	Method limited to cohesive soils	Increasing strength of soil.
	Sand piles (drains)	Introducing a column of sand in soil through a casing or mandrel	Decrease of porosity and reduction of settlement of soil		Densification of fill- and foundation-supporting soil

TABLE 9-3 (*continued*)

Soil Types	Method	Technological Process	Results	Limitations of Application of Method	Engineering Applications
Clays, silts	Soil drainage by electro-osmosis	Passing direct electric current through soil	Stability of slopes of foundation excavations	Saturated soils	Excavations for laying of foundations
	Electro-strengthening (solidification)	Same as before, but a slower process	Water-resistant increase in bearing capacity of foundation-supporting soil	Saturated soils	Densification (solidification)
	Sand piles (drains)	Introducing a vertical column of sand drain (pile)	Increase in bearing capacity of fill-and foundation-supporting soil		Expulsion of water from soil, densification of soil, reduction of settlement, acceleration of consolidation of soil
All types of sandy soils	Artificial freezing of soil	Circulation of cooling brine through freezer points installed in waterlogged soil	Formation of ice wall; impervious to water; great depth of ice wall	Waterlogged sandy soils	For laying foundations in the dry; curtains

9-8. GROUTING

Grouting is the corrective process of injecting under pressure a cement slurry or grout and other suitable materials into the voids of the soil and/or rock to solidify them. When the cement grout sets, the soil transforms into a solid, unerodable mass of improved bearing capacity, and the amount of seepage waters is cut down.

Grouting in soil is effective only when the sizes of the voids of the soil are considerably larger than the size of the particles of the cement. Therefore cement grouting is effective in gravelly soils, coarse and medium-coarse sand, as well as in fissured, faulty, and cavernous rocks. Cement grouting is less effective in fine-particled soil, and is ineffective in clayey soil.

Pressure grouting results in high strength of soil, thus affording important advantages in many types of foundation, soil and rock stabilization.

Grouting is also used for sealing porous foundation-supporting materials—such as quartzite, coral limestone, and in cavernous limestone, for example—to keep water out of open foundation excavations as well as to provide for water-impermeable and/or diversion aprons underneath hydraulic structures (such as hydraulic power plants, dams, tunnels, locks, docks, bridge foundations) or to reduce hydrostatic uplift under a massive dam.

The need for grouting is established by obtaining rock core samples from the site of the foundation work and studying their strength and permeable properties.

The rate of injecting a grout into a bore hole is determined by test. This test consist of forcing water under pressure into the sealed bore hole, and recording the rate of flow of water into the bore hole and the corresponding pressure.

A quick decrease in the rate of flow and a rapid increase in pressure may be interpreted as indicating few seams and fissures, hence easy grouting. A high rate of water flow with practically no increase in pressure would mean a very porous and fissured rock formation, hence extensive grouting may be anticipated.

A historical survey of grouting in engineering may be found in Ref. 16.

Cement Grout and Its Additives

The grout is commonly composed of (1) portland cement and water (usually 1 : 2, sometimes 1 : 10 by volume), (2) portland cement, sand (or rock flour) and water, or (3) portland cement, clay and water (1.5 : 7 : 6).

The best grout is the stiffest mix that can be grouted effectively, and should be ascertained experimentally.

To accelerate the setting time of the grout, additives such as calcium chloride or sodium silicate are introduced. The setting time of cement grout is retarded by gypsum.

Fine-particled bentonite clay increases plasticity of the grout. Most cement grout now also contains fly ash, which is cheap, and which by its spherical particles increases the ease of pumping and reacts pozzolanically with lime liberated by the cement.

Cement grout can be used to best advantage in gravelly sand and gravel the particle sizes of which are > 1.5 mm.

After forcing the grout into the soil or rock, the grout gradually solidifies, thus forming a strong, resistant, unerodable foundation support.

The pressure to use in grouting operations should be a safe one in order not to lift the rock up above the jointing.

The effectiveness of the grouting operation is usually verified by making check borings and examining the cores extracted from these bore holes. The presence of sufficient grout with all seams, cavities, and fissures filled may be construed as a successful operation.[17] Also, if the grouted bore holes for checking do not take water or grout under pressure, then this may be considered an indication that the results of the grouting operation are satisfactory.

Pressure grouting is applicable to igneous rocks, siliceous sandstones, and other noncavernous rocks through which the grout could not be washed away.

At Hoover Dam an extensive program of grouting was used to control seepage and leaks, and to reduce the hydrostatic uplift pressure on the base of the structure.[18]

The shortcoming of grouting is "working blind," because there is little control of where the grout is going. Therefore complete filling of all voids is impossible to insure.[19] A major problem occurs where a highly permeable layer in stratified soil may sap off the grout so intermediate layers are not grouted. This can be prevented by packers, or allowing to set, redrilling, and regrouting.

To insure continuous, uninterrupted grouting operations, two sets of grouting pumps are advisable: one for active work, and one spare to cope with breakdown and other emergencies.

Bitumen Grouting

The technology of bituminous grouting of soil, in essence, is similar to cement grouting. Bituminous grouting is applicable to strengthen and seal fissured rocks.

In jointed and fissured geological formations through which water flows the cement grout may be carried away by water before the grout can set. Therefore heated bituminous materials are injected into seams, fissures, cracks, and voids of rocks and soil to cut off infiltrating waters. The main function of a bituminous grout is to seal off the flow of water in the rocks

Dewatering of Excavation

or soil. After the flow has been stopped, cement grouting operations may follow.

Among bituminous grouting methods one distinguishes between hot bituminous grout and cold liquid bituminous emulsions. The sealing effect on fissures of a hot bituminous grout is explained as follows. The outside course of the injected bituminous (asphalt) "plug" cools off gradually and thus solidifies. Its central part, however, remains still liquid ($T \approx 100°C$), and continues to flow further under the inside pressure. This means that the bituminous grout solidifies under pressure and takes the shape of the fissures, thus cutting them off against permeating water.

The great relative ease of permeation of cold, liquid bituminous emulsions (particle size $\approx 2\mu$) as compared with hot bituminous grout permits utilizing the former for stabilization of sandy soils. Bituminous emulsions are introduced into soil at relatively low pressures. Upon filling the voids, the bituminous emulsion bonds the soil particles and thus strengthens the soil. However, bituminous emulsions have not been used extensively to reduce permeability of soil.

To retard seepage and to cut down future maintenance work a quick-setting chemical grout may be used in conjunction with cement and asphalt grouting.[20]

Clay Grouting

Clay grout is used to fill large seams and cavities where low hydrostatic pressure heads prevail. Clay grout is not suitable for sealing fissures through which water flows. Clay grout does not contribute much to the strength properties of the soil. However, clay suspensions offer great resistance to seepage.

Pressure grouting using materials other than cement has been practiced for a long time but generally has been regarded as a substitute for neat cement grouting. The use of soils as grouting materials in soil-water mixes, or with the addition of cement, is similar in many respects to the use of neat cement grout, since these are all basically suspension grouts in which solid material is suspended in water.

When using colloidal suspensions, consideration of the electrokinetic effects should not be forgotten.

9-9. CHEMICAL SOLIDIFICATION OF SOIL

Since 1925 one of the best known methods of chemical soil solidification is that by the process patented by Joosten.[22,23] In Joosten's method, two types of chemical solutions are consecutively injected into the soil under pressure: first, a solution of silicic acid (water glass, liquid glass),

$$Na_2 \cdot nSiO_2 + H_2O$$

and second, a strong solution of calcium chloride (an electrolyte). The silicate of sodium is introduced into the soil through injection pipes first. The injection pressure varies from 7 to 10 atm, or 103 to 147 psi. Silicic acid is a colloidal solution, coating the soil particles. Then, as the pipes are withdrawn, the second chemical is introduced. The latter—a salt solution—accelerates the transition from sol to gel.

When an electrolyte (calcium chloride, a salt) is added to silicic acid (sol), it coagulates, transforming immediately and permanently the sol into an irreversible gel. Because of the immediate and quick reaction of these two chemicals, forming instantly a precipitation of calcium silicate gel within the voids of the soil and thus binding the soil particles, this silicating method of soil solidification is applicable also in the presence of groundwater in the soil ("petrification of 'quick condition' of sand") where the coefficient of permeability of the groundwater is high (between $k = 1.5 \times 10^{-2}$ cm/sec and $k = 1.0 \times 10^{-1}$ cm/sec). This transforms a sand from its quick condition into an impervious solid.

The calcium silicate gel coats the soil particles with a thin film. This film has a high surface tension, cementing the soil mass.

There is no dilution of the chemicals by groundwater. Strength tests performed in Germany on soil materials chemically solidified by Joosten's method showed no decrease in density, strength or permeability after a period of ten years.

The injection operation may be visualized from Fig. 9-21.

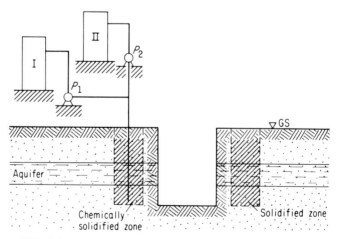

FIG. 9-21. Principle of injection. I, first chemical; II, second chemical.

The properties of the various brands of water glass vary like those of cements. However, all water glasses have one thing in common, namely: in a concentrated condition they possess a high amount of colloidal silicic acid which precipitates in the presence of suitable chemicals (metallic salt solutions) so that in noncohesive soils an instant petrification takes place, forming an artificial sandstone.

The radius of influence of the solidified mass of soil from a perforated injection pipe 35 mm in diameter is about 1.4 m. Injection pipes are usually spaced at 60 cm center to center. The perforation consists of holes 1 mm in diameter made at the lower end of the injection pipe over a length of about 50 cm.

The strength attained by the solidified soil material depends upon the type and properties of the soil. Under normal conditions, strengths of approximately 7.5 kg/cm^2 to 30.0 kg/cm^2 (140-425 psi) have been obtained in sands, approximately 40 to 50 kg/cm^2 (570-710 psi) in gravels, with strengths up to approximately 90 kg/cm^2 ($\sim$ 1,280 psi) under the most favorable conditions.[24]

Joosten's method for chemically solidifying loose sands and sealing geologic rock formations and soils that are permeable to water also has been applied for stabilizing foundation-supporting soils under existing structures, pile foundations, grouting of fissures and cracks in rocks and concrete, and sealing rocks in mines and tunnels against the influx of water.[25]

Figures 9-22 through 9-24 illustrate chemically solidified foundation-supporting soil, a chemically solidified cofferdam and chemically solidified soil for supporting a pile foundation, respectively.

American Cyanamid's AM-9 chemical grout is essentially a mixture of two organic monomers, acrylamide and methylenebisacrylamide, in proportions which produce very stiff gels from dilute, aqueous solutions when properly catalyzed.[26] The process by which gelation occurs is a polymerization-crosslinking reaction.

The AM-9 gel is a rubbery, elastic material which by itself has negligible load-supporting capacity. However, it is substantially impermeable to water ($k \approx 10^{-10}$ cm/sec). Hence, this type of grout is used primarily for seepage control of water and not as load-bearing soil grout because of its low strength.

The chrome-lignin process is also known as a means of soil grout. More recently, a variety of chemical injection processes have been studied by T. W. Lambe.[27]

Winsol resins and other suitable chemicals, too, are used in soil chemical stabilization and/or solidification.[28] For example, epoxy resins have been

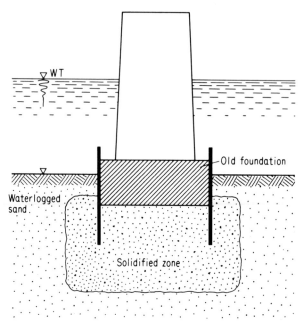

FIG. 9-22. Chemically solidified foundation-supporting soil.

used to bond cracks in concrete structures in order to restore the original strength.

Whereas successful results have been achieved with epoxy resin and polyester resin grouts, in his summary Erickson writes,[29] "The life of chemical grouts is not definitely known as they have been in use for only a relatively short time. It is possible that the chemical reaction may proceed until the grout becomes brittle, or the grout may eventually lose strength. A long-time study to determine the terminal conditions is needed".

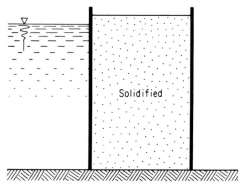

FIG. 9 23. Chemically solidified cofferdam.

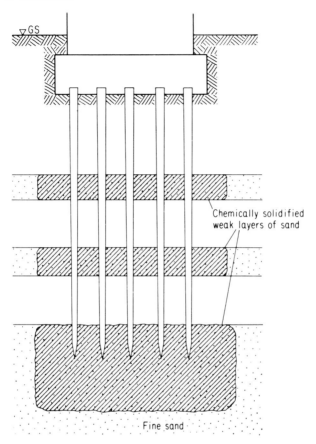

FIG. 9-24. Chemically solidified soil for supporting a pile foundation.

Figure 9-25 illustrates the idea of sealing the bottom of a cofferdam constructed in fine sand through harbor silt and clay against influx of water. The method of sealing is injection of a bituminous emulsion. After the seal is established the excavation is carried down to the level of the base of the foundation.

An adaptation of drilled-lime stabilization was tried on an active landslide on the north city limits of Des Moines by Handy and Williams.[30] The slide was in poorly compacted fill containing about 30 percent montmorillonitic clay, and resting on an eroded shale surface sloping about 15-20 percent. The project appears to have been successful, indicated by physical and chemical soil tests and measurements on the houses; meanwhile adjacent untreated or drained properties continued to fail.

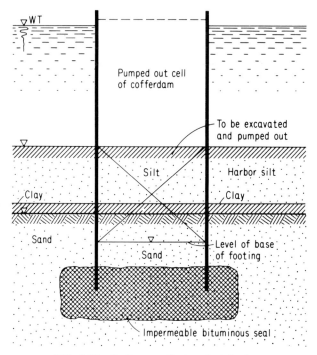

FIG. 9-25. Sealing a cofferdam by injection.

A summary (1957) of field applications of chemical grouting in foundation and earthwork engineering with annotated bibliography is contained in Ref. 31. A bibliography (1966) on chemical grouting is given in Ref. 32. For control of vibration of two gas compressors on separate concrete foundations over medium to dense sand by the Joosten process, see Ref. 33.

9-10. ARTIFICIAL FREEZING OF SOIL

Dry foundation excavations may also be attained by holding the water out of the excavation by freezing the soil surrounding the excavation.

Artificial freezing of soil in excavation operations is an expedient, efficient, and successful means of stabilizing temporarily the walls and bottom of foundation pits in fine-particled, permeable, saturated soils and in sandy, waterlogged soils subject to quick condition. In this method of thermal soil solidification, prior to excavation work the soil layers are temporarily converted into a solid wall of ice around the excavation. This is done by artificially freezing the soil at the construction site.

Upon freezing, water expands 9 percent in volume; ice cements the aggregate particles; a saturated soil becomes sufficiently impermeable to

Dewatering of Excavation

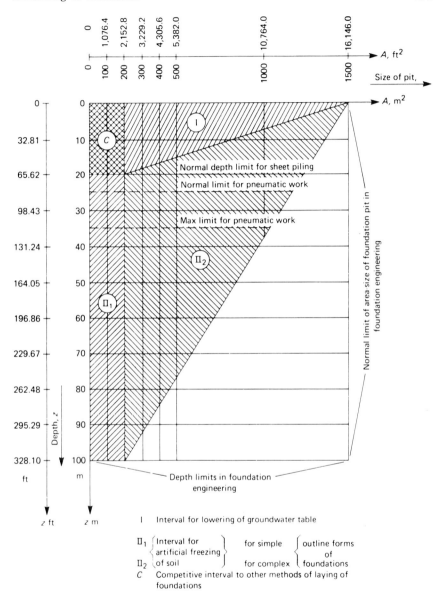

FIG. 9-26. Engineering and geometric limits of artificial soil freezing as compared with other methods of laying foundations (Ref. 34).

water for excavation, and the soil acquires sufficient mechanical strength for its intended purpose.

By freezing the soil around a foundation excavation, the influx of groundwater into the excavation can be checked. Also, artificial soil freezing for foundation engineering is no longer seasonal, but can be carried out in any season of the year and in any climate.

Artificial soil freezing can be used to great advantage in foundation systems that exceed the depth limit for sheet piling,[34] or where sheet piling cannot be applied because of unfavorable soil conditions (sheet piling is difficult to drive into gravelly soil, for example), or where open and/or pneumatic caissons cannot successfully be sunk because of the quick condition of fine sand occurring at great depth. Nor can it be used where the designed depth of the foundation exceeds the maximum economical drawdown of the groundwater table, or where pumping is difficult to perform if the velocity of the groundwater flow is greater than two meters per day.

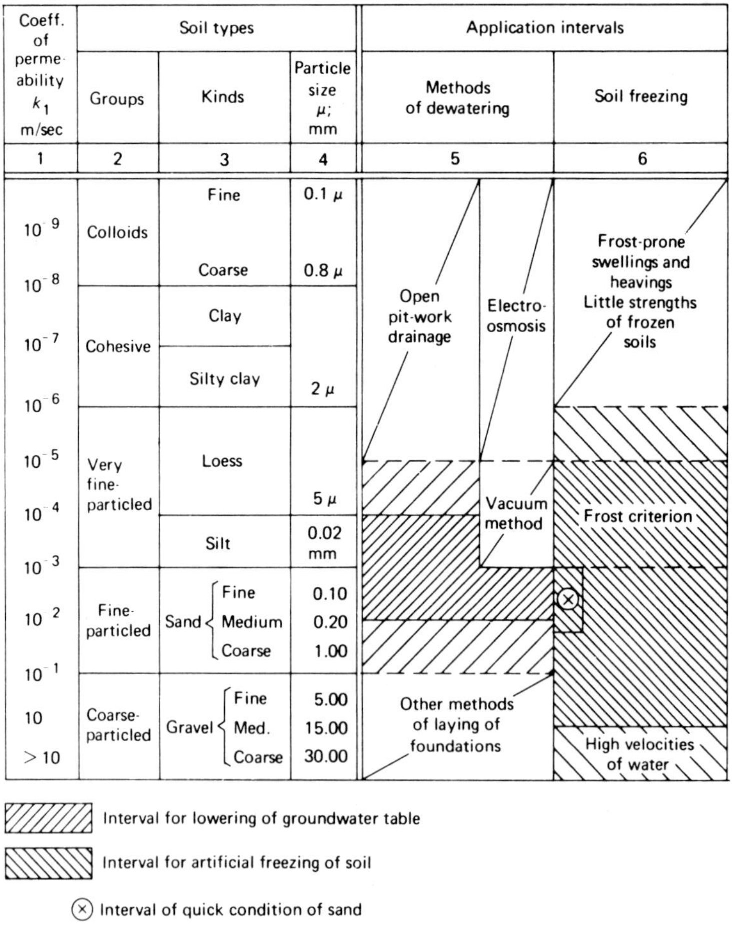

Coeff. of permeability k_1 m/sec	Soil types			Application intervals	
	Groups	Kinds	Particle size μ; mm	Methods of dewatering	Soil freezing
1	2	3	4	5	6
10^{-9}	Colloids	Fine	0.1 μ	Open pit-work drainage / Electro-osmosis / Vacuum method	Frost-prone swellings and heavings / Little strengths of frozen soils / Frost criterion
10^{-8}		Coarse	0.8 μ		
10^{-7}	Cohesive	Clay			
10^{-6}		Silty clay	2 μ		
10^{-5}	Very fine-particled	Loess	5 μ		
10^{-4}					
10^{-3}		Silt	0.02 mm		
10^{-2}	Fine-particled	Sand — Fine	0.10		
		Medium	0.20		
		Coarse	1.00		
10^{-1}					
10	Coarse-particled	Gravel — Fine	5.00	Other methods of laying of foundations	High velocities of water
> 10		Med.	15.00		
		Coarse	30.00		

▨ Interval for lowering of groundwater table

▧ Interval for artificial freezing of soil

⊗ Interval of quick condition of sand

FIG. 9-27. Soil physical limits for laying foundations (Ref. 34).

Dewatering of Excavation

Soil freezing is also applicable for tunnel work, for sealing leaky cofferdams, for underpinning structures, in soils where pile-driving vibrations cannot be tolerated, and for placing of water-impermeable curtains in soil underneath power plants, dams and other structures.

The soil physical limits for laying foundations after Ref. 34 are shown in Figs. 9-26 and 9-27.

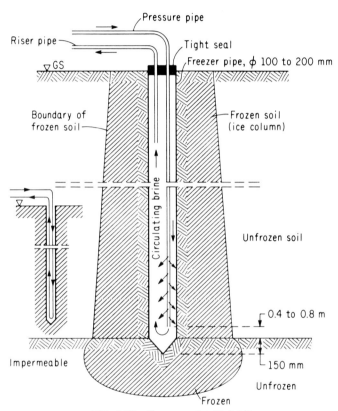

FIG. 9-28. Freezer point (Ref. 36).

A wall of frozen soil, the so-called "ice-wall," is created by installing freezer points vertically or obliquely along the perimeter of the excavation site at a predetermined spacing. A freezer point is sketched in Fig. 9-28. The cold freezer point removes heat from the adjacent soil and lowers soil water temperature to the freezing point.

For open excavations freezer points are spaced from about $s = 3$ ft to $s = 7.5$ ft (~ 1.0-2.5 m) center-to-center, depending upon the properties of the soil, soil and air temperatures, diameters of freezer pipes, and also the rate of freezing. Such a spacing forms an ice wall 2.5-3.0 m thick in

294   **Excavation and Cofferdams**

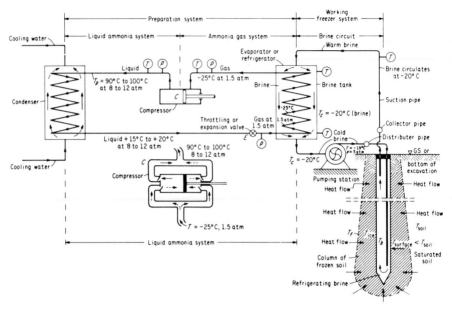

FIG. 9-29a. Schematics of an ammonia refrigeration system.

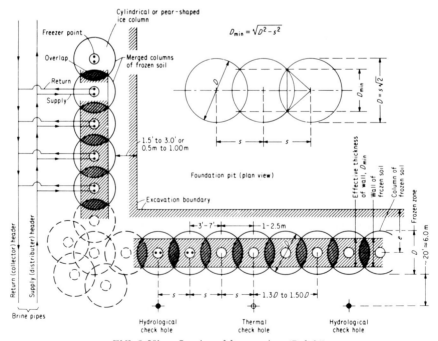

FIG. 9-29b. Spacing of freezer pipes (Ref. 36).

Dewatering of Excavation

about 45-60 days. An example of a refrigeration plant, the arrangement and spacing of freezer pipes around an excavation and some forms of ice walls are shown in Figs. 9-29a, 9-29b and 9-29c, respectively.

The thickness ξ of the ice wall is calculated as

$$\xi = m\sqrt{t} \qquad (9\text{-}33)$$

where t = time of freezing the soil, and

m = a coefficient to be determined from freezing theory as dealt with in Refs. 35 and 36.

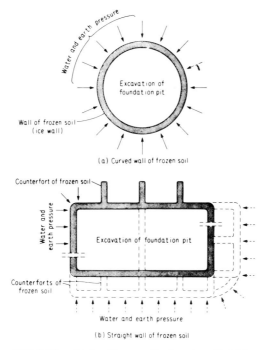

FIG. 9-29c. Walls of frozen soil: plan views (Ref. 36).

The artificial soil freezing method for making ice-walled cofferdams is especially useful in excavating deep shafts through water-logged soils. Figures 9-30a and 9-30b illustrate the freezing of the tunnel shaft for the Richmond, Staten Island, N.Y., water-supply system.

9-11. THE SLURRY-TRENCH METHOD

To avoid complicated bracing and framing systems and to protect foundation trench and excavation walls from caving in, the so-called slurry-trench method may be used. By this method the walls of the trench are

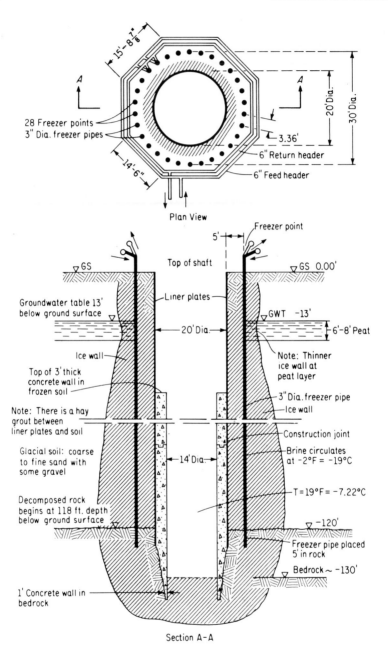

FIG. 9-30a. Artificial freezing of soil for a water-supply tunnel shaft.

Dewatering of Excavation

FIG. 9-30b

kept supported by means of a bentonite clay slurry. The slurry is kept in the trench while it is being excavated. After a certain length of the foundation trench has been excavated, the slurry is displaced by tremie concrete into the next trench section being excavated.

Besides its use in foundation engineering, the slurry-trench method* has been utilized also by the Jersey Central Power and Light Company in 1963 for the construction of a cutoff wall at the lower reservoir dam in connection

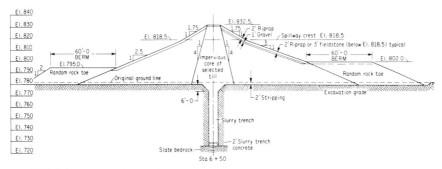

FIG. 9-31. Lower dam section at Sta. 6 + 50, Yards Creek Pumped Storage Project.

---

* The slurry trench method is patented (Patent No. 2.757.514 issued to Cronese, Inc., 12623 East Imperial Highway, Santa Fe Springs, Calif., 90670).

with the execution of its Yards Creek Pumped Storage Project in Warren County, New Jersey (see Fig. 9-31*), a 330,000 kw hydroelectric power plant.

The specifications for the construction of the slurry-trench cutoff wall, prepared by the consulting firm Ebasco Services, Inc., of New York, N.Y., called for the digging of an approximately vertical-walled trench 8 ft wide in the overburden soil along the longitudinal axis of the dam between Sta. 4 + 20 to Sta. 9 + 10. At the same time the trench had to be filled with a slurry of stable colloidal suspension of bentonite clay in water.

For a description of thixotropic fluids in engineering operations the reader is referred to the author's books on soil mechanics, Refs. 1 and 5, pp. 68-76 and 437-447, respectively.

The purpose of the slurry was to support the walls of the cutoff trench and to facilitate backfilling (the construction of the cutoff wall) so that a homogeneous cutoff wall could be secured.

The *overburden* at the dam site consisted of a deposit of clay and silt at the ground surface underlain by deposits of varved clay, sand, gravel, and boulders overlying the slate bedrock (see Fig. 9-32, varved clay).

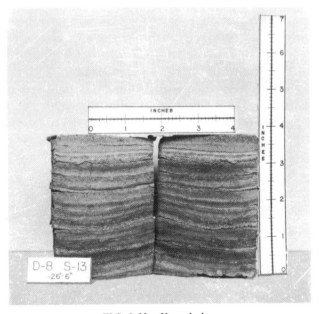

FIG. 9-32. Varved clay.

* By courtesy of Mr. S. B. Palmeter, Manager of Construction, Jersey Central Power and Light Company, Morristown, N.J., and by permission of the Central Jersey Power and Light Company, Morristown, N.J.

Dewatering of Excavation

The *slurry* was specified to have its properties as determined by standard tests described in American Petroleum Institute Code RP 29 dated May 1957—Standard Field Procedure for Testing Drilling Fluids.

The specifications provided
1. At the time of introduction of slurry into the excavation:
    (i) The viscosity of the slurry shall not be less than 15 centipoises at 20°C.
    (ii) The water loss shall not be greater than 20 cm^3 in 30 minutes.
2. At the time of placing backfill into the slurry-filled trench:
    (i) The viscosity shall not be greater than 30 centipoises at 20°C.
    (ii) The density of the slurry shall not be greater than 80 lb/ft^3 or as approved by the Engineer.
    (iii) The sand content of the slurry shall not exceed 20 percent by volume or as approved by the Engineer.

The specifications also provided for the use of the creek water for making the slurry. They also provided that admixtures of the types used in control of oil-field drilling muds might be used to alter the characteristics of the slurry in the trench as approved by the Engineer.

The slurry was introduced into the trench at the time of beginning the excavation and the level of the slurry in the trench was maintained above static water level throughout the excavating operations.

To facilitate *backfilling* operations, excavation of the slurry-filled trench was carried out in a single segment along the longitudinal axis of the trench.

Where the bottom of the trench extended to rock (slate), the latter was surfaced with a layer of concrete 2 ft thick. The concrete was lowered to the bottom of the trench through a tremie and/or buckets through the slurry-filled trench in such a manner as to prevent segregation or mixing of the concrete and slurry. No free fall of concrete through the slurry was allowed.

The backfill soil material was specified to conform with the following gradation requirements at the time of its placement:

Screen Size or Sieve Number (U.S. Standard)	Percent Passing by Dry Weight
3 in.	80-100
$\frac{3}{4}$ in.	70-100
No. 4	55- 85
No. 30	40- 70
No. 200	20- 50

The mixing, grading, and placing of the stockpiled material from borrow pits intended for backfill was performed by light earthmoving and/or grading equipment such as bulldozers and blade graders. The backfilling of the cutoff material into the trench was done by the use of crane and clamshell buckets, jet pipes, and probe rods.

The backfill material had to be thoroughly mixed with slurry into a stiff, homogeneous mass, free from lumps of silt or pockets of fines, sand and gravel.

The backfilling was done in two stages as set forth.

1. *Initial layer of backfill.* The initial backfill was placed upon the layer of rock-surfacing concrete to the lines and grades specified on the drawings before the initial set of the concrete. Initial backfill was placed by lowering it to the bottom of the trench with crane and bucket. The remainder of the backfill was placed only after the concrete in the area being filled had attained an initial set.

2. *Remaining backfill.* The remaining backfill was placed so as to avoid any pockets of slurry in the completed cutoff wall. Placing of backfill below the surface of the slurry was required to follow a reasonably smooth grade and to have no hollows which might trap pockets of slurry during subsequent backfilling. To avoid such pockets, the face of the backfill below the slurry was rodded. Free dropping of backfill material through the slurry was not allowed.

The backfill was placed by means of a crane and bucket starting at one end of the trench until the surface of the backfill rose above the slurry level. Additional backfill was then placed on top of the deposited backfill by a front end grader and bulldozer so that the backfill below the slurry surface pushed forward along the bottom of the trench by the superimposed load of backfill material above the slurry level. The bulldozer walked as near the end of the backfill as possible to assist the forward movement along the bottom of the trench. No material was allowed to be pushed into the slurry by the bulldozer blade, thus avoiding air pockets and segregation.

9-12. SAND PILES

Prior to the advent of concrete piles, "sand piles" were used for the purpose of densification of the soil. The sand piles were usually made from 3 ft to 4 ft in diameter to support foundations.

Today, the sand pile is a special type of pile, prepared directly in place by compacting sand (by self-weight and/or pressure) in a metal mandrel,

Dewatering of Excavation

while the mandrel is being withdrawn. When used for accelerating the consolidation process of soil, sand piles are known by the term *sand drains*. Sand drains act as wicks to facilitate the rapid drainage of water from the

FIG. 9-33. Sanddrain-installation rig. (By courtesy and permission of Hercules Concrete Pile Co., Ridgefield, N.J.)

soil.[1] Sand drains have been used in many projects, mostly in highway work on both the East Coast and the West Coast of the United States.

A sand-drain installation rig is shown in Fig. 9-33.

REFERENCES

1. A. R. Jumikis, *Soil Mechanics*, Princeton, N.J.: Van Nostrand, 1962, p. 261.
2. K. Terzaghi and R. B. Peck, *Soil Mechanics in Engineering Practice*, New York: Wiley, 1948, p. 331.
3. A. Casagrande and R. E. Fadum, *Notes on Soil Testing for Engineering Purposes*. Cambridge, Mass.: Harvard University, Publication No. 268, 1939-40, Fig. 11, p. 23.
4. A. Hazen, "Some Physical Properties of Sands and Gravels, with Special Reference to their Use in Filtration," 24th Annual Report of the State Board of Health of Massachusetts, 1892, Public Document No. 34. Boston, Mass.: Wright and Potter, 1893, p. 553.
5. A. R. Jumikis, *Mechanics of Soils: Fundamentals for Advanced Study*, Princeton, N.J.: Van Nostrand, 1964, pp. 391-414.
6. W. Sichardt, *Das Fassungsvermögen von Rohrbrunnen*, Berlin: Springer, 1928, p. 28.
7. Hütte, *Des Ingenieurs Taschenbuch*, 26th ed., Berlin: Ernst, 1936, Vol. 3, p. 129.
8. W. Kyrieleis and W. Sichardt, *Grundwasserabsenkung bei Fundierarbeiten*, Berlin: Springer, 1930, p. 30.
9. A. Schoklitsch, *Hydraulic Structures*, trans. by S. Shulits, New York: American Society of Mechanical Engineers, 1937, Vol. 1, p. 179.
10. Leo Casagrande, *The Application of Electro-osmosis to Practical Problems in Foundations and Earthworks*. Technical Paper No. 30, London: Department of Scientific and Industrial Research, H.M. Stationery Office, 1947.
11. Leo Casagrande, "Electro-osmosis." Proceedings, Second International Conference on Soil Mechanics and Foundation Engineering, Rotterdam, June 21-30, 1948, Vol. 1, pp. 218-223.
12. Leo Casagrande, "Electro-osmosis in Soils," *Géotechnique* (London), Vol. 1, No. 3 (June 1949), pp. 159-177.
13. Leo Casagrande, "Electro-osmotic Stabilization of Soils." *Journal of the Boston Society of Civil Engineers*. January 1952.(Also Harvard University's *Soil Mechanics Series*, No. 38.)
14. D. J. Maclean, "Soil Drainage by an Electrical Method," *Civil Engineering*, February 1945, pp. 34-37.
15. R. Hoffmann and M. Muss, "Die mechanische Verfestigung sandigen und kiesigen Baugrundes." *Bautechnik*, No. 33/36 (1944).
16. R. Glossop, "The Invention and Development of Injection Processes." *Géotechnique* (London), No. 3 (September 1960), pp. 91-100, No. 4 (1961). pp. 255-279.
17. A. W. Simmonds, "Cement and Clay Grouting of Foundations: Present Status of Pressure Grouting Foundations," *Proc., ASCE*, Paper No. 1544 (February 1958), pp. 1544-1/1545-11.
18. A. W. Simmonds, "Final Foundation Treatment at Hoover Dam," *Trans., ASCE*, Vol. 118 (1953), p. 78.

19. *Cement Grouting.* Progress Report of the Task Committee on Cement Grouting, Committee on Grouting, Soil Mechanics and Foundations Division. *J. Soil Mech. Foundations Div.*, (April 1962), *Proc., ASCE*, No. SM2, pp. 49-98.
20. J. T. Moore, Jr., "Controlling Leakage from Cowans Ford Dam," *Civil Engineering*, Vol. 35, No. 6 (June 1965), pp. 52-55.
21. S. J. Johnson, "Cement and Clay Grouting of Foundations: Grouting with Clay-Cement Grouts," *Proc. ASCE*, Paper 1545, February 1958, p. 1545.
22. Patented in 1927 by Dr.-Ing. H. Joosten, Deutschlands Patentschrift, No. 44166, Klasse 5c, Gruppe 1.
23. H. J. Joosten, *Das Joosten-Verfahren zur chemischen Bodenverfestigung und Abdichtungen seiner Entwicklung von 1925 bis heute*. The Netherlands: privately published, 1953.
24. *Highway Research Abstracts*, after Schütz, on "New Methods of Increasing the Stability and Tightness of Earth." Trans. by A. W. Johnson. Washington, D.C., Highway Research Board, March 1948.
25. R. O. Pynnonen and A. D. Look, "Chemical Solidification of Soil in Tunneling at a Minnesota Iron-Ore Mine," Bureau of Mines Circular No. 7846. Washington, D.C.: U.S. Department of the Interior, 1958.
26. R. H. Karol, "Chemical Grouting Technology." *J. Soil Mech. and Foundations Div., Proc. ASCE*, Vol. 94, No. SM1 (January 1968), pp. 175-204.
27. T. W. Lambe, "Chemical Injection Processes," *Proc. ASCE*, 1954.
28. H. F. Winterkorn, "Soil Stabilization," Proceedings of the Second International Conference on Soil Mechanics and Foundation Engineering, Rotterdam, Vol. 5 (1948), p. 209.
29. H. B. Erickson, "Strengthening Rock by Injection of Chemical Grout," *Journal of the Soil Mechanics and Foundations Division, Proceedings of the ASCE*, Vol. 94, No. SM1, January 1968, pp. 159-175.
30. R. L. Handy and W. W. Williams, "Chemical Stabilization of an Active Landslide," Special Report, Contribution No. 66-11 from the Soil Research Laboratory, Engineering Research Institute, Iowa State University, Ames, Iowa, 1968.
31. T. W. Lambe, "Chemical Grouting," Progress Report of the Task Committee on Chemical Grouting of the Soil Mechanics and Foundation Division, *Proc. ASCE*, Paper No. 1426-1, Vol. 83, No. SM4 (1957), pp. 1426-1/1426-106.
32. "Bibliography on Chemical Grouting," Third Progress Report, Committee on Grouting, ASCE, (J. P. Elston, Chairman), *J. Soil Mech. and Foundations Div., Proc. ASCE*, Vol. 92, No. SM6 (November 1966), pp. 39-66.
33. G. P. Tschebotarioff, "Vibration Controlled by Chemical Grouting," *Trans. ASCE*, Vol. 130 (1965), pp. 253-254.
34. K. H. Seydel, "Grundsätzliches über die Baugrundvereisung und deren Bedeutung für Grundbautechnische Zwecke," *Bautechnik*, No. 5 (1953), pp. 131-137; No. 7 (1955), pp. 199-202.
35. A. R. Jumikis, *The Frost Penetration Problem in Highway Engineering*, New Brunswick, N.J.: Rutgers U. P., 1955, pp. 77-93.
36. A. R. Jumikis, *Thermal Soil Mechanics*, New Brunswick, N.J.: Rutgers U. P., 1966, pp. 163-173, 236-252.

PROBLEMS

9-1. Discuss the three kinds of foundations as shown in the accompanying figure. Especially, observe the relative position of the water table and suggest methods of laying these foundations. Indicate methods of preparing dry excavations for laying the foundations. Assume, if necessary, factors to help to explain and to describe the problem. (*Partial answer*: Prob. 9-1 illustrates the increasing difficulties encountered such as depth of excavation, depth of water and type of soil in laying foundations from case *a* through *b* to *c*).

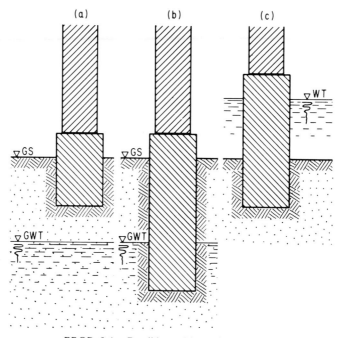

PROB. 9-1. Possible positions of water table.

9-2. Refer to Fig. 5-18. Given a braced trench the depth of which is $h = 10.0$ ft, the unit weight of soil $\gamma = 100$ lb/ft^3, and the angle of internal friction of the soil $\phi = 30°$. The width of the trench is $b = 6.0$ ft. There is no surcharge on either side of the trench. In doing this problem disregard the one foot or so of the driving depth of sheet piling shown in this figure. Also, assume that the sheet piling is rigid and does not deflect from earth pressure. In this problem, operate with the three lower braces only. Use the fourth (upper) brace merely for constructional purposes. *Required*: (1) determine the magnitude of earth pressure each of the three lower braces receives if the braces are spaced at uniform spacing; (2) space the three lower braces in a vertical plane so that each brace receives

Dewatering of Excavation

equal pressure. The calculations are to be performed for a 3-ft-long section of the trench (= a repeating section for placing braces).

9-3. Design an inclined frame cofferdam on bedrock for a 4 ft depth of water and $\alpha = 45°$. Assume all engineering factors and their values for use in this design. Refer to Fig. 7-8.

9-4. Given a box type of cofferdam on rock (refer to Fig. 7-21), where $h = 12.0$ ft and $b = 9.0$ ft. Calculate the factors of safety of this cofferdam against sliding and overturning, and calculate edge pressures on the rock.

9-5. Given a dewatering system 100 ft long as shown in the acompanying figure.

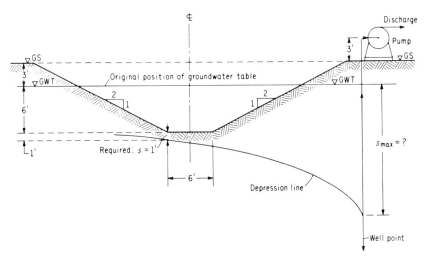

PROB. 9-5. Dewatering system.

(a) What should the maximum drawdown s_{max} be at the wellpoint and for the conditions shown in this figure to achieve a dry foundation trench so that the depression line is $s = 1$ ft below the toe point of the slope?
(b) Is it possible or impossible to achieve with the dewatering system as shown? Give reasons.
(c) If yes, calculate the pump horsepower to handle the discharge Q. The length of the trench to be dewatered is 100 ft.
(d) If no, what kind of solution can be suggested to have the trench dry?
(e) What is the equation of the depression line? Plot it. Soil: medium sand, the coefficient of the system's permeability being $k = 0.02$ ft/sec; radius of influence: $R = 1,000$ ft; outside diameter of wellpoint: $2r_o = 0.4$ ft; $h_o = 4$ ft.

9-6. Given a circular foundation in fine sand. Position of groundwater table: 3 ft below ground surface. Bottom radius of excavation to be dewatered: 12.5 ft; $k = (1.7) \times 10^{-3}$ m/sec: required depth of dewatering, 10 ft. Assume all other necessary engineering quantities.

Required: to lower the groundwater table for purpose of laying the foundation in the dry. Include the following:
(a) Determine the number of wellpoints, their spacing, and plot the depression curve.
(b) What kind of wells should be used—ordinary perfect or imperfect?
(c) What is the maximum drawdown?
(d) What is the discharge per single wellpoint and from the entire dewatering system?
(e) Calculate pump horsepower for the final dewatering system.

part **III**

SHALLOW FOUNDATIONS

chapter **10**

Shallow Foundations and Footings

10-1. CLASSIFICATION OF FOUNDATIONS AND FOOTINGS

Foundations may be classed according to their nature, their function, the materials they are constructed of, and according to the method of their analysis. All foundations, however, may be classed generally as (1) shallow foundations, (2) deep foundations, and (3) special foundations.

If the foundation is laid directly on a competent load-bearing soil at a minimum depth below the ground surface such that the foundation is safe against lateral expulsion of soil from underneath the base of the footing and is also safe against volume changes of the soil due to frost action, swelling for example, and/or shrinkage (settlement, heave), one then speaks of a *shallow foundation*. Shallow foundations apply mainly to buildings and certain engineering structures.

If, however, the load from the structure should be transmitted to a considerable depth by means of end-bearing piles, piers, and caissons through weak soil to a geologic stratum competent to support the structural load, or by means of friction piles where no underlying strong stratum is economically feasible, one then speaks of a *deep foundation*. Deep foundations apply mainly to engineering structures such as bridges and hydraulic structures, as well as to buildings if they have to be built on sites of poor soil. Wherever possible, deep foundations should be avoided because their cost increases rapidly with depth.

10-2. CLASSIFICATION OF SHALLOW FOUNDATIONS

Shallow foundations, in their turn, may be classed as (1) direct, (2) spread, and (3) floating.

1. *Direct or unenlarged foundations,* usually used for light loads and small buildings only (Fig. 10-1): the walls (columns) are built with but little spread, directly on sound rock or on firm soil. Unsound rock is considered a soil and foundations are designed as if founded on soil.

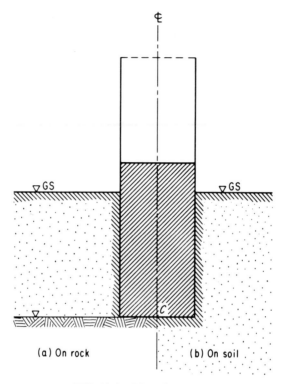

FIG. 10-1. Direct foundation.

2. *Spread foundations* are constructed with (a) footings, and (b) on a timber, steel, or reinforced concrete grillage.

A footing is the basis on which anything is established or supported. One distinguishes between column footings and wall footings.

A grillage is a kind of framework of sleepers and crossbeams of timber, steel I-beams, or reinforced concrete (Fig. 10-2) forming a foundation support on weak soil. The spaces between steel beams around the side, top and bottom of the steel grillage are filled with concrete. Grillage footings have been extensively used in the past.

As the word implies, a spread footing spreads or distributes structural loads over an adequately large contact area of bearing-capable soil greater than the base of the primary wall or column in order not to exceed the

Shallow Foundations and Footings

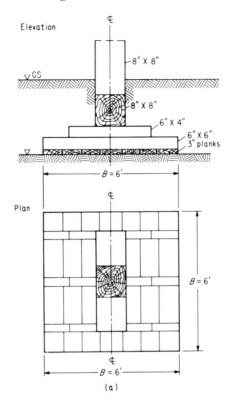

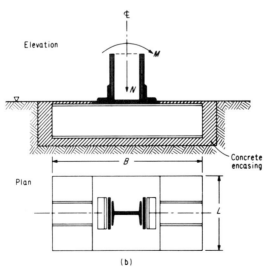

FIG 10-2. (a) Timber grillage. (b) Steel I-beam grillage.

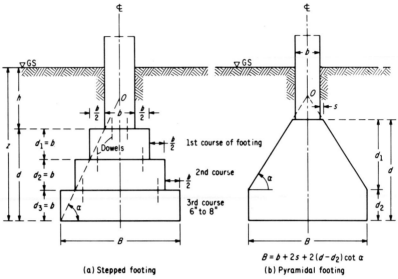

FIG. 10-3. Stepped and pyramidal footings.

allowable soil bearing capacity σ_{all} and/or tolerable settlements. There are four kinds of spread footings: (i) simple footings, (ii) stepped footings, (iii) combined footings (rectangular and trapezoidal), and (iv) continuous footings.

Simple and stepped footings are used as wall or strip and as column footings (Fig. 10-3).

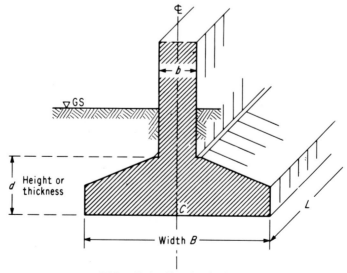

FIG. 10-4. Sloped strip footing.

Shallow Foundations and Footings

They are generally masonry units of concrete or reinforced concrete designed to distribute the column or wall load over a bed or contact area between the base of the footing and the soil (rock).

A wall or strip footing is one which supports a wall (Fig. 10-4). It is simply a strip of reinforced concrete, wider than the wall, to distribute its pressure on the soil.

A column footing supports a single column or other kind of concentrated load. Column footings are also known as single, isolated, or independent footings.

For light loads footings are made of plain concrete. To bridge soft spots in the soil, reinforced concrete footings are used. Reinforced concrete is a rational union of concrete and reinforcing steel.

In respect to cost, experience shows that a reinforced concrete footing with a sloping top is more expensive to construct than a single-course footing with no slopes. Various column footings are shown in Fig. 10-5.

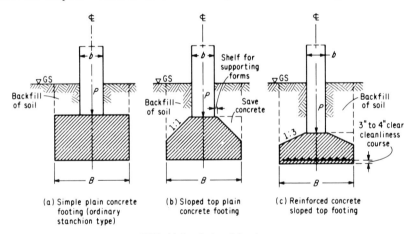

FIG. 10-5. Isolated footings.

To avoid making sloped forms which may turn out to be expensive in labor, contractors sometimes are willing to bear the cost of the extra concrete to cast a footing of rectangular cross section rather than a sloped one.

Reinforced concrete footings offer savings in their thickness and thus in weight of concrete. These footings are usually provided with a lateral reinforcement. If the soil is poor, then the footing also receives longitudinal reinforcement to bridge soft spots in the soil.

Simple Footing

In designing a wall or column foundation it is frequently necessary to widen the base of the foundation. This is accomplished by an appropriate footing. The widening depends upon the allowable bearing capacity of the soil, the compressive strength of the wall (column) and footing materials, and the nature and magnitude of the load to be transmitted to the soil.

A *widened foundation* is sketched in Fig. 10-6. However, sharp-angled rims and edges at the base of a footing such as shown at *a-a* in Fig. 10-6 are not permissible because sharp concrete edges spall thus reducing the contact area needed for transmission of the load to the ground. Besides, sloped-surface forms are difficult to make and foundations with sharp corners are difficult to cast. A widened pyramidal foundation such as that shown in Fig. 10-3b is more practical than that shown in Fig. 10-6. Otherwise the widening of footings may be made in steps.

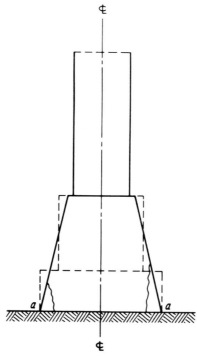

FIG. 10-6. Widened footing. Correct geometry is indicated by dashed lines.

Widening of a foundation may also be made by steps (several footing courses) or in a pyramidal form (Fig. 10-3). A cubical block of concrete would be wasteful. The changes in footing thickness are made in a few steps. To make many small steps results in little saving in concrete and great expense in labor and time on deferred progress of work (setting time

Shallow Foundations and Footings

of concrete in each footing step and in form-making for the stepped footing).

A combined or strap footing supports two column loads consisting of two footings connected or strapped together by a beam (the strap) or a slab (Fig. 10-7). A combined footing must be sized so that the centroid C of its contact area on the soil lies in the line of action of the resultant of the two column loads in order to avoid a tendency for the strap footing to rotate in a vertical plane. Such a design usually results in a trapezoidal footing—usually made of reinforced concrete.

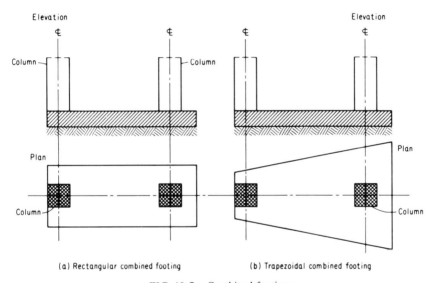

FIG. 10-7. Combined footings.

A *continuous footing* supports a row of three or more columns on a continuously extending reinforced concrete slab.

3. *Floating foundations.* A "floating foundation" is one in which the weight of soil removed for a depth h below the ground surface is assumed to increase the allowable soil pressure: $\sigma_o = \sigma_{all} + \gamma h$, where σ_o is the safe soil contact pressure; γ = unit weight of the soil, and σ_{all} = allowable soil bearing capacity at the ground surface, for example.

Floating foundations are (Fig. 10-8): (a) a mat, or raft, or slab foundations, and (b) inverted arch, or inverted T-beams and a slab.

Mats are used (i) to distribute loads from the periphery of the structure over the entire area of the structure, (ii) to reduce concentrations of high contact pressures on soil, and (iii) to resist hydrostatic head.

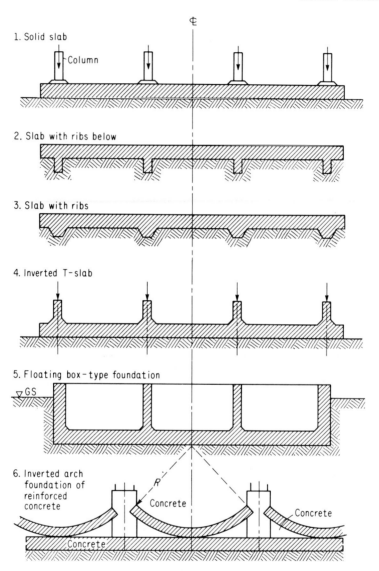

FIG. 10-8. Mat foundations.

Mats or rafts are designed as heavy, continuous, two-way reinforced concrete slabs 4-8 ft thick, extending under the entire structure. The slabs should be strong in order to distribute a uniform contact pressure on the soil, and to bridge weak spots in the soil. Large mats are reinforced by ribs or built with T-sections for stiffness and thus for a reasonably good pressure distribution on soil.

Shallow Foundations and Footings

10-3. DEPTH OF SHALLOW FOUNDATIONS

The depth to which shallow foundations are to be laid depends upon the following factors:

1. Technological significance of structure—for example, subsurface structures such as tunnels; monumental, or profane, or temporary structure and its safety requirements
2. Functional requirements of the structure (underground structures; need for cellars, basements; service utilities; adjoining and neighboring structures)
3. Kind of foundation material (timber, stone, bricks, plain concrete; reinforced concrete; steel, for example)
4. Kind and magnitude of loads transmitted to foundation-supporting soil (centric, eccentric, static, dynamic loads)
5. Geological conditions at the construction site (nature and types of soil and/or rock and their stratification and inclination; frost-proneness of soil; swelling and/or shrinkage of soil; quick condition of soil)
6. Hydrological conditions at the construction site (precipitation regimen; possibility of inundation of site and foundations; surface runoff; soil erosion by water; scouring of foundations; drainage conditions; presence, position, and fluctuation of groundwater table)
7. Seasonal variations in climatic conditions (frost penetration depth in soil and its associated heaving where applicable; swelling and shrinkage of soils brought about by seasonal changes in moisture content in soil)
8. Allowable bearing capacity of soil, viz., shear strength of soil
9. Consolidation properties and settlement of the soil
10. Tolerable settlement of structure
11. Distance and depth of foundations of existing adjacent structures where applicable (pressure overlap, settlements)
12. Cost of foundation.

These and possibly many other factors governing the choosing of the right depth of shallow foundations to use merely indicate the great responsibility of the design engineer and the great skill, wide experience, and knowledge he needs in soil mechanics and foundation engineering.

The safety of a structure is the most important and imperative requirement. It is to be strived for with the least expense possible.

The soil-foundation system must be safe against groundbreak. For reasons of durable stability, monumental structures are founded deeper than small buildings with lighter loads. Also, structures whose functions require basements or any other functional underground space are founded deeper than light and small structures. To avoid decaying of a timber grillage of a foundation, the grillage should be laid below the lowest elevation of the groundwater table.

The greater the load transmitted by the foundation, the deeper must the foundation be laid to provide for a lateral counterweight (overburden) of the adjacent soil against lateral expulsion of soil from underneath the base of the footing. Also, one remembers from soil mechanics that in a uniform soil the bearing capacity increases with increase in depth below ground surface.[1-3]

Relative to the pressure distribution in soil, in designing foundations one should strive to achieve stress distribution diagrams as uniform as possible underneath the base of the footing by suitable combinations of the dead and live loads.

Where there is a horizontal force component acting in the base area (contact plane) of the footing (such as occurs in earth-retaining wall systems, massive dams, vaults, frames, and arches), one should reckon with settlements from dead loads. With a trapezoidal stress distribution diagram, the originally horizontal base of the footing becomes inclined in the course of time. Therefore, here one should strive to attain in design the least nonuniform pressure distribution diagram on soil from constant (= dead) loads.

With heavy loading of shallow foundations large, sometimes intolerable, settlements are unavoidable. If the structure is sensitive to settlement, other types of foundations are to be considered.

Because of the danger of scour, where this is pertinent, structures to be founded in riverbeds require a relatively great depth of foundation, say several feet below the maximum anticipated depth of scour. Otherwise in this case the foundation requires protection against scour by encompassing the foundation with a sheet-piling enclosure, for example.

If a soil on which the foundation is to be laid flows out under the foundation load, such as a dry running sand, or as sand would run out upon seepage or during drainage (pumping) operations for an excavation, the desired results are usually attained by walls of sheet piling driven down to a depth below which any subsequent excavation is unlikely to be carried.

It should also be remembered that the fluctuation of the groundwater table near the loaded footing affects adversely the soil bearing capacity, bringing about settlement of the structure.

The depth of laying foundations depends also on the severity of the climatic region where the foundations are laid, namely, upon weather effects. In

Shallow Foundations and Footings

regions where seasonal frost occurs, the minimum depth of foundations is down below the ground surface at the lowest zero isotherm. This is to say that the base of the footing of exterior walls and all footings of unheated buildings must be laid at a depth somewhat greater, but never less, than the maximum frost-penetration depth in that region. A minimum depth of 1 ft below the frost line is commonly specified.

Relative to semiarid regions and to seasonal freezing zones, it can be said generally that the foundations should be laid below the depth of seasonal changes in the water regime, for example, the freezing and subsequent heaving of frost-prone soils, swelling of swelling-prone soils by increase in their moisture content and shrinking upon their drying out during dry seasons.

Structures with basements, naturally, require foundation depths greater than the frost-penetration depth, and so do heavy structures.

Under no conditions should a foundation be laid on frozen soil unless specially designed to cope with frost heaves and settlement by thawing. Structures whose foundations are laid in frost-prone soils at depths shallower than frost penetration may be heaved up, and cracking of their foundations and walls may occur upon thawing.

Depending upon the significance and service requirements of the structure whose foundations are laid in a frost-prone soil, the latter must be subjected to consolidation tests and settlement analysis.

Usually the minimum depth of laying of foundations with respect to frost-penetration depth is regulated by building codes. If there are no building codes for some municipalities, then for calculation purposes relative to maximum regional frost-penetration depths in the United States, one may resort to the Extreme Frost Penetration Chart (Fig. 10-9). Otherwise the frost-penetration depth may also be calculated using Stefan's or Neumann's formulas by means of climatologic freezing temperature records, and some thermal properties (see Refs. 1, 4, and 5).

The frost-penetration depth ξ in soil, thus, depends upon the frost-prone soil, freezing temperature, and proximity of the source of water supply (groundwater or perched groundwater).

In the northern part of the United States, the zero isotherm may be as low as 6 ft below the ground surface, whereas some parts of the southern United States never freeze. In New Jersey, one usually reckons with an average frost-penetration depth from $\xi = 30$ in. to $\xi = 36$ in., depending upon the altitude of the construction site.

In most cities the frost-penetration depth is fairly well established by the depth at which water mains must be placed to prevent freezing.

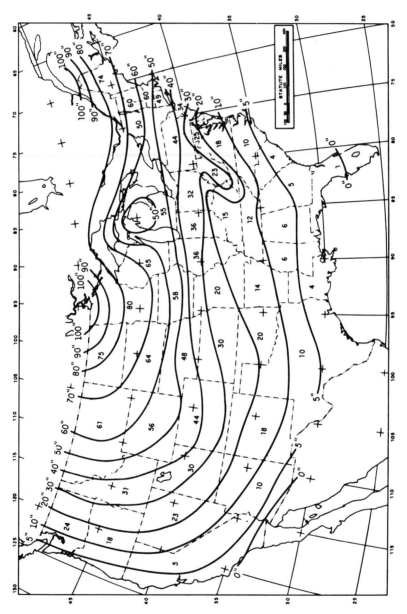

FIG. 10-9. Extreme frost penetration (in inches) based upon state averages. (Courtesy of U.S. National Weather Records Center.)

Shallow Foundations and Footings

For frost criteria of frost-susceptible soils refer to Ref. 4.

Today, one can say with a fair degree of confidence that the matter of frost penetration in soil and its adverse effects does not really lie merely in the climatological frost, but rather in the *heaving and thawing of frost-prone soils* and the proximity of a source of water supply for the growth of ice lenses in the soil. This would also mean that a foundation should not be laid unnecessarily deep where and when the soil does not heave, or swell, or shrink.

The depth of a foundation is measured from the ground elevation of the ground surface (GS).

Unless they are supported on rock, and if there are no frost problems in soil in the locality concerned, foundations for all but single-story buildings or other light structures should be founded at least 3 ft below the ground surface, because it is very unlikely that undisturbed soil is encountered at a depth less than about 3 ft.

Shallow foundations must be placed on clean, natural mineral soil. All kinds of artificial fill, garbage and trash, organic soil, remnants of organic matter and topsoil covering the good soil must be removed prior to laying of foundations.

10-4. CALCULATING THE MINIMUM (CRITICAL) DEPTH OF FOUNDATION

The minimum or critical depth z of foundations, considering equilibrium of the soil-foundation system against groundbreak, may be approximately calculated by Ref. 2, p. 33, as follows. Consider the strip footing as shown in Fig. 10-10. This strip footing exerts a uniform contact pressure $\sigma_1 = 5.0$ ton/ft^2 on the soil at the bottom of the excavation.

Solution

Minor principal stress σ_{III} from overburden soil of thickness z:

$$\sigma_{III} = \gamma z = (0.055)(z) \tag{10-1}$$

Lateral resistance σ_I:

$$\sigma_I = \sigma_{III} \tan^2 (45° + \phi/2) + 2c \tan (45° + \phi/2) \tag{10-2}$$

For equilibrium, $\sigma_I = \sigma_3$, and $\tag{10-3}$

$$\sigma_{III} = \gamma z = \sigma_1 \tan^4 (45° - \phi/2) - 2c \tan^3 (45° - \phi/2) - 2c \tan (45° - \phi/2) \tag{10-4}$$

From Eq. 10-4, the minimum or critical depth z calculates as

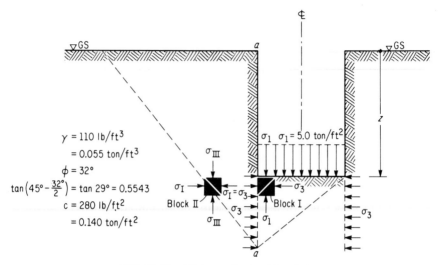

FIG. 10-10. Minimum depth of foundation.

$$z = \frac{1}{0.055} \left[(5.0)(0.554)^4 - (0.28)(0.554)^3 - (0.28)(0.554) \right] = \underline{4.89 \approx 5.00 \text{ (ft)}}$$

(10-5)

When $c = 0$, then

$$z = \frac{0.471}{0.055} = \underline{8.47 \approx 8.50 \text{ (ft)}}$$

(10-6)

10-5. LAYING OF SHALLOW FOUNDATIONS

Shallow foundations may be laid in dry or under water. If concrete is poured against or upon any rock surface, the surface must be freed from all loose material before the concrete is poured and thoroughly cleaned in order to assure a good bond between rock and footing concrete, and to avoid contamination of concrete.

For firm seating of reinforced concrete footings on the ground, the bottom of the excavation, should be covered with a course of lean concrete called the *cleanliness course* or *blinding layer*. Its purpose is to provide for a clean surface on which to place the bottom reinforcing bars for their protection. The thickness of the cleanliness course is recommended by the American Concrete Institute (ACI) Building Code to be not less than 3 in. To avoid crushing, the safe compressive strength of this concrete should be not less than the maximum bearing pressure on the ground.

Shallow Foundations and Footings

The thickness of the cleanliness layer does not enter into the static calculations and proportioning of the size of the footing, but is simply disregarded in strength analyses.

The plain concrete cleanliness course or carpet 3 to 4 in. thick may be poured over the bottom of the excavation immediately after the excavation of the foundation pit is completed, and allowed to set and harden.

The advantage of the cleanliness course is not only for placing the reinforcing steel bars, but also for protecting the soil from being disturbed by the laborers (walking over the bars), and against damage by precipitation, drought and to some degree by frost action.

Light Plain Concrete Wall Footings

Plain concrete footings are constructed when the loads are light and the soil (rock) conditions are excellent. When the loading and soil (rock) conditions are unfavorable, plain concrete footings become too bulky, heavy, and uneconomical. In such instances reinforced concrete footings are constructed.

Light, plain concrete wall footings, and footings for residence buildings and flats of bricks the walls of which are usually made from 8 to 12 in. thick (20 cm-30 cm), are laid with projections beyond the wall one half the thickness of the wall. The depth (thickness) of the footing is twice the projection, or the thickness of the wall. Thus for a wall 10 in. (~ 25.4 cm) thick, the footing course should be 10 in. thick and have a total width of 20 in. (≈ 51 cm). The footing thus designed should be checked to ascertain that the contact pressure σ_o on the soil does not exceed the allowable bearing capacity σ_{all} of the soil: $\sigma_o \leqslant \sigma_{all}$.

Heavy Wall Footings

The design of a heavy wall footing may be illustrated by the following problem.

Design a centrally loaded plain concrete footing for a wall ($b = 2$ ft thick) of a warehouse the wall of which transmits to the top course of the footing a linear load of $p_w = 10.5$ ton/ft. The allowable soil bearing capacity is given as $\sigma_{all} = 1.5$ ton/ft^2 (Fig. 10-11).

Solution. Assume that the weight of the footing is $p_f = 1.5$ ton/ft. Total load p on soil per linear foot:

$$p = p_w + p_f = 10.5 + 1.5 = 12.0 \text{ (ton/ft)}$$

Necessary width B of base of footing:

$$B = p/\sigma_{all} = 12.0/1.5 = 8.0 \text{ (ft)}$$

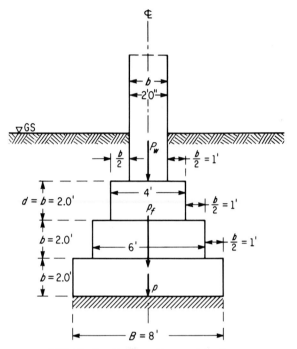

FIG. 10-11. Heavy wall stepped footing.

requiring three footing courses:

$$\frac{B-b}{2(b/2)} = \frac{8-2}{2(2/2)} = 3$$

each $d = b = 2$ ft thick with projections of 1 ft.

Note in this calculation that all allowable soil bearing capacity is utilized.

Column Footing

Design a centrally loaded plain concrete footing for a column load of $P_c = 260{,}000$ lb $= 260$ kips if the allowable soil bearing capacity is $\sigma_{\text{all}} = 3{,}500$ lb/ft² $= 3.5$ kips/ft² (Fig. 10-12).

Solution

Column load.................... $P_c = 260.0$ kips
Estimated weight of footing.......... $P_f = 90.0$ kips
Total load on soil.................. $P = 350.0$ kips

Required contact area A of base of footing:

$$A = P/\sigma_{\text{all}} = 350.0/3.5 = 100 \text{ (ft}^2)$$

Shallow Foundations and Footings

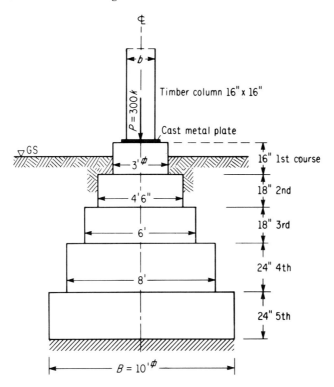

FIG. 10-12. Stepped column footing.

Use a square footing of $a = 10.0$ ft × 10.0 ft for bottom course of footing.

For the first (upper) course of footing use a concrete with a compressive strength of $\sigma_c = 800$ lb/in.^{2} This requires a steel billet plate $260{,}000/800 = 325 \sim 400$ in.^{2} (20 in. × 20 in) under the timber column.

Assume the upper (first) course of footing to be 3 ft square and 16 in. thick, with a projection of $b/2 = 8$ in. The footing courses are shown in Fig. 10-12.

In plain concrete footings, the minimum edge thickness d of the footing should not be less than 8 in. (≈ 20 cm) for footings on soil (see Fig. 10-13), nor less than 14 in. (≈ 35 cm) above the tops of the piles for footings on piles.

In reinforced concrete footings, the minimum edge thickness above the steel reinforcement should be not less than 6 in. (≈ 15 cm) for footings on soil, nor less than 12 in. (≈ 30 cm) for footings on piles (Fig. 10-14).

The transverse reinforcements as shown in Figs. 10-13b and 10-13c are meant to resist curling of the footing.

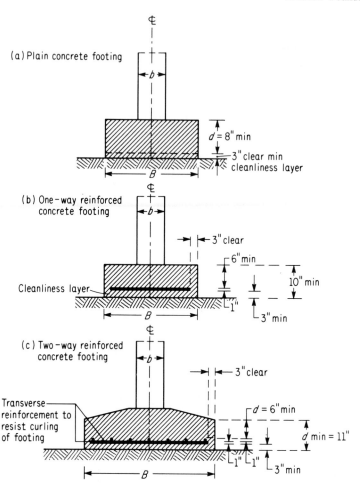

FIG. 10-13. Cleanliness layer and thickness of edges of footings.

For example, for a two-way reinforced concrete footing the total minimum edge thickness d_{min} is composed as follows:

Cleanliness layer.....................	3 in. min.
Thickness of two-way reinforcement bars, 1-in. dia. each way.................	2 in.
Net minimum edge thickness.......... $d =$	6 in. min.
Total thickness, $d_{min} =$	11 in.

The reinforcement of footings and other principal structural elements exposed to the weather must be protected with not less than 2 in. concrete cover for bars larger than No. 5 and 1.5 in. for No. 5 bars or smaller. In

Shallow Foundations and Footings

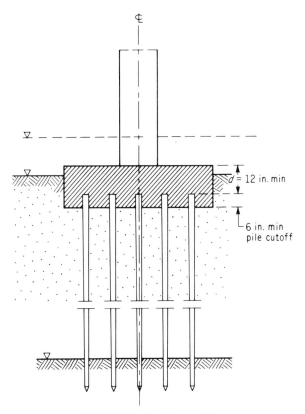

FIG. 10-14. Footing on piles.

extremely corrosive atmospheres or other severe exposures, the amount of protection should be suitably increased.

The concrete must be deposited as nearly as practicable in its final position to avoid segregation because of rehandling or flowing. No concrete that has partially hardened or been contaminated by foreign materials should be deposited in the structure.

Needless to say, good and proper curing of concrete should be practiced at all times, and all concrete materials and all reinforcement, forms, fillers and ground with which the concrete is to come in contact must be free from frost. No frozen materials or materials containing ice should be used in concrete foundation work. A good, reliable inspection of concrete work by a competent inspector is an absolute necessity.

If the depth of foundation is greater than about, say, 5 or 6 ft, a monolithic, cubical block of concrete would be wasteful. Therefore the so-called *pedestal* type of footing may be used. It is a slab, or courses of slabs, lying over the necessary bearing area surmounted by a short column or pedestal (Figs. 1-1 and 10-15). According to the definition given by the American Concrete Institute Building Code,[6] "a pedestal is an upright compression member whose height does not exceed three times its average least lateral dimension."

Thus, the length-to-width ratio of the pedestal is usually made $H/b \sim 3/1 = 3$.

The pedestal, or the steps and the footing (whichever the case is) that are designed as an integral, monolithic unit should be cast as a unit. The pedestal is tied to the footing by means of dowels or vertical reinforcements, whichever the case is. This applies also to tying columns and footings.

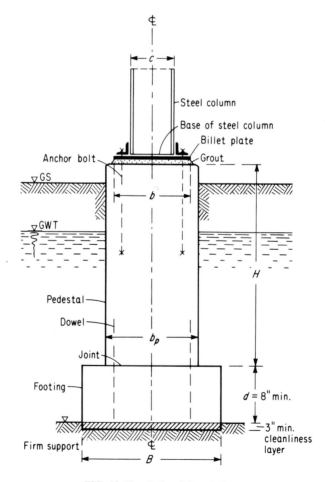

FIG. 10-15. Pedestal foundation.

Shallow Foundations and Footings

To safeguard against excessive corrosion, it is imperative that the bases of steel columns be placed on foundations above the ground surface and above the groundwater table.

The billet plate under a steel column distributes the column load on the foundation. If a steel column bears on a concrete footing, the footing is designed the same way as for a concrete column-footing system, using an average column dimension of $(b+c)/2$, where b and c are as in Fig. 10-15.

10-6. CALCULATION OF CENTRICALLY LOADED PLAIN CONCRETE SQUARE FOOTINGS

Vertical Resultant Load

One of the objectives of footing design is the correct determination of the size of the base of the footing for its contact area with the soil. The size of a foundation and its footing depends upon
1. The nature, magnitude and direction of action of the resultant structural load to be transmitted and distributed on the foundation-supporting soil.
2. The strength of the material of which the foundation, viz., footing, is made.
3. The safe, allowable soil bearing capacity σ_{all}.

Such footings are designed to resist safely bending and diagonal tension (shear) stresses.

One distinguishes between centrically and eccentrically loaded footings. When a footing is subjected to a vertical, centrically applied load and when the center of gravity of superimposed normal loads $N = P + W$ coincides with the centroid of the base area of the footing, then for practical calculations it may be assumed that the distribution of the contact pressure σ_o over the soil (rock) is reasonably uniform (Fig. 10-16). Its average intensity σ_o is then calculated as being equal to the total load $N = P + W$ divided by the contact area A of the footing, i.e.,

$$\sigma_o = \frac{N}{A} = \frac{P+W}{A} \qquad (10\text{-}7)$$

where $N = P + W =$ normal load transmitted by the foundation to soil (rock) over the contact area A
$P =$ load from structure on foundation
$W =$ weight of footing

Traditionally in foundation design the footing is assumed to be rigid and the soil supporting it elastic. This assumption also implies that a

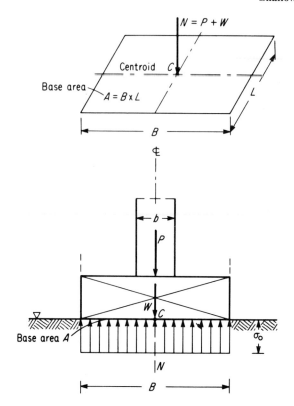

FIG. 10-16. Centrically loaded footing.

centrically loaded footing is supported by a uniformly distributed soil pressure. Relative to the assumption of uniform stress distribution on soil, soil mechanics has shown in most practical cases that uniform soil pressure does not necessarily bring about uniform settlement. However, based on experience of the safe performance over long periods of time of footings designed on the assumption of uniform distribution of contact pressure on soil, and for the sake of simplicity in design, this assumption is considered sound in practice. The uniform soil pressure distribution in design really means, though, "if possible."

Problem

Given a soil-foundation system as shown in Fig. 10-16. The resultant normal load applied centrically to a square footing is $N = 200$ short tons. The contact area of the footing is given as $A = 100$ ft². Calculate the average contact pressure σ_o on the soil.

Shallow Foundations and Footings

Solution. The average contact pressure is

$$\sigma_o = N/A = \frac{200}{100} = 2.0 \text{ (ton/ft}^2\text{)}$$

This contact pressure should be less than or equal to the safe allowable soil bearing capacity σ_{all}, i.e., it is required that $\sigma_o \leq \sigma_{all}$.

10-7. CALCULATION OF ECCENTRICALLY LOADED FOOTINGS

When the resultant of the loads acting on the footing does not pass through the centroid of the base area, the footing is said to be eccentrically loaded.

Eccentricity e on the base of the footing may be brought about either from placing the column or wall, viz., the resultant vertical normal load N, off the centroid of the base area of the footing, or because of an externally applied resultant static moment M to a centrically and/or eccentrically loaded footing system a moment which tends to overturn the footing (Fig. 10-17). Such eccentrically loaded footings are also termed footings subjected to a moment.

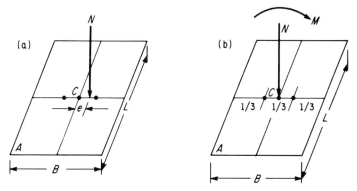

FIG. 10-17. Eccentrically loaded footings.

The eccentrically loaded footing with a vertical, resultant normal load N and its eccentricity e is statically equivalent to a footing system which is loaded centrically with a load N ($e = 0$) and a moment $M = Ne$. This moment is visualized if one adds to the footing system with eccentric normal load N at its centroid C of the base area two equal, collinear, oppositely directed forces N ($\downarrow$) and N ($\uparrow$). By doing so, no change in the statical footing system is made. Thus there now acts on the base a centrical load N and a couple (moment) $M = Ne$.

Conversely, if M and N are given, eccentricity e calculates as $e = M/N$. Among eccentrically loaded footings one distinguishes between
(a) eccentricity e less than one-sixth of the width B of the footing, i.e.,
$e < (B/6)$
(b) $e = B/6$
(c) $e > B/6$

(a) $e < B/6$. In this case the resultant normal load N is situated within the middle third of the width of the base of the footing (Fig. 10-18).

When the total normal load $N = P + W$ is applied on the base of the footing eccentrically, then the distribution of the contact pressure on the soil (rock) is not uniform, but varies from a minimum σ_{min} at the edge of the footing,

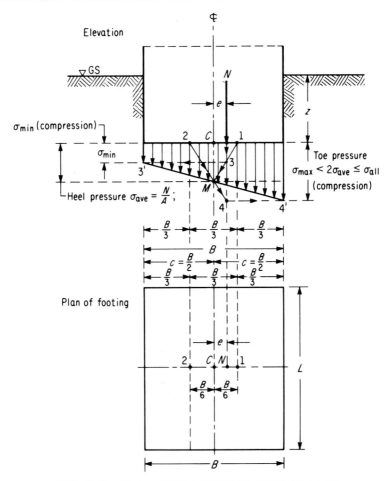

FIG. 10-18. Contact pressure distribution on soil for $e < B/6$.

Shallow Foundations and Footings

which is farther away from the resultant normal load N on the base, to a maximum σ_{max} at the edge of the footing at the side, which is nearer the resultant normal load N (Fig. 10-18). For practical purposes the variation in the soil contact pressure distribution σ may be assumed to be linear. The variation in contact pressure between σ_{min} and σ_{max} depends upon the position and magnitude of the eccentricity e.

For eccentrically loaded contact areas when $e < B/6$ the contact edge pressures on soil are calculated as

$$\sigma_{\substack{max \\ min}} = \frac{N}{A} \pm \frac{M}{I/c} \tag{10-8}$$

where σ_{max} = maximum compressive stress on soil at edge on side of eccentricity

σ_{min} = minimum compressive stress at lightly loaded edge (no tension on soil is allowed)

$M = Ne$ = moment caused by eccentricity e

I = moment of inertia of the base area

$c = B/2$ = distance from neutral axis n-n to farthest spaced fiber in horizontal cross-sectional area

$N/A = \sigma_{ave}$ = average contact pressure on soil from centrically applied load

$M/(I/c) = M/Z$ = pressure on soil brought about by resultant moment M

Z = section modulus of base area of footing

When $e < B/6$, then $\sigma_{max} < 2\sigma_{ave}$, $\sigma_{min} > 0$, and the pressure distribution diagram is a trapezoid (Fig. 10-18).

For rectangular footings the edge pressures can also be expressed in terms of eccentricity e:

$$\sigma_{\substack{max \\ min}} = \frac{N}{A}\left(1 \pm \frac{6e}{B}\right) \tag{10-9}$$

Based on this equation, the edge pressures σ_{max} and σ_{min} can be obtained graphically as follows (Fig. 10-18):

1. Calculate and plot to a stress scale the average contact pressure diagram $\sigma_{ave} = N/A$ (a rectangular diagram),
2. Connect midpoint M on the average pressure distribution diagram with the third points 1 and 2 on the base of the footing to obtain lines 1-M and 2-M,
3. Extend the line of action of the resultant vertical normal load N to intersect line 1-M at point 3 and to intersect the extension of line 2-M at point 4,

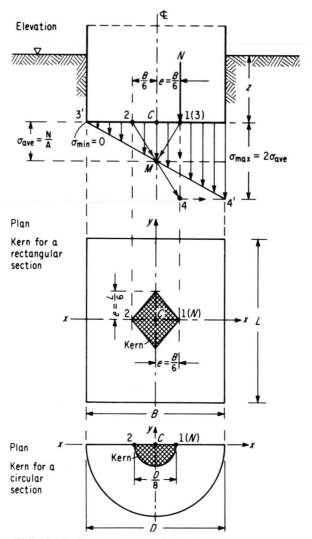

FIG. 10-19. Contact pressure distribution on soil for $e = B/6$.

4. Drawing from point 3 a horizontal line 3-3′, obtain graphically to stress scale the magnitude of σ_{min}.
5. Drawing from point 4 a horizontal line 4-4′, obtain graphically the magnitude of σ_{max}.

Note that this graphical method is a quick and expedient means of determining edge pressures under the bases of rectangular footings.

(b) $e = B/6$. When $e = B/6$, this means that the resultant normal load N on the base of the footing is applied at the third-point of the width B of the

Shallow Foundations and Footings

footing (see point 1 in Fig. 10-19). In such a case $\sigma_{max} = 2\sigma_{ave}$, $\sigma_{min} = 0$, and the pressure distribution diagram is a triangle. The latter is constructed graphically as shown in Fig. 10-19. Note in this case that point 3 coincides with point 1 and that the horizontal line 3-3, coincides with line 1-2-3. Within the kern of the base area the load may be applied anywhere without bringing about tensile stresses on any one of the four sides (edges) of the footing.

(c) $e > B/6$. When the resultant normal load is outside the middle third of the width of the footing ($e > B/6$), then $\sigma_{max} > 2\sigma_{min} < 0$, the light side of the footing loses its contact with the soil and is lifted up (Fig. 10-20). The moment $M = (2/3)BN$ tends to overturn the foundation. The stress distribution diagram consists of two stress triangles: one "plus" triangle (i.e., compression) and one "minus" triangle for tension.

When $\sigma_{min} < 0$ (tensile stress), this condition cannot be used in soil-foundation problems, because soil as a material is very weak in resisting tensile stresses. It is not advisable to anchor the footing to the soil with the idea of taking care of the uplifting tensile force. The foundation should be so designed that ordinarily the resultant normal load N should fall in the middle third ($1\text{-}2 = B/3$) of the width of the contact area of the footing (or, say, within the kern) so that there is an allowable compressive stress on the soil under ordinary performance of the foundation at all times.

The graphical construction of σ_{max} and σ_{min} is shown in Fig. 10-20.

When N lies in the outer one third of the width of the footing the linear contact pressure distribution on soil must be determined by another method. In the case when $e > B/6$, to perform calculations with compressive stresses only, one writes the following two equations for a footing x units wide which has the magnitude of eccentricity $e' = x/6$:

$$N = \underbrace{(1/2)\sigma_{max}Lx}_{\text{Average total pressure on area } Lx} = (1/2)\sigma_{max}(3)(B/2 - e)L \qquad (10\text{-}10)$$

where Lx = contact area for pressure only

$$x = 3(B/2 - e) \qquad (10\text{-}11)$$

$x/3 = B/2 - e$ = distance from edge to point where resultant N cuts base (Fig. 10-21)

Here L = length of footing perpendicular to drawing plane
         x = part of footing under compressive stress only

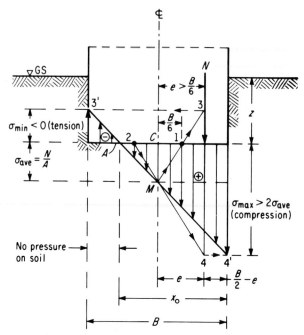

FIG. 10-20. Contact pressure distribution on soil for $e > B/6$.

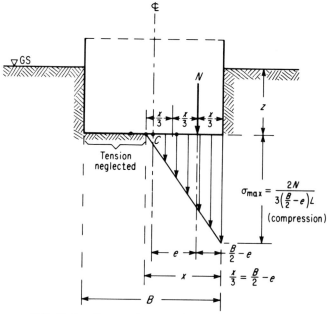

FIG. 10-21. Compressive stress diagram for $x = 3(B/2 - e)$.

Equations 10-10 and 10-11 permit one to calculate the maximum contact pressure σ_{max} on soil:

$$\sigma_{max} = \frac{2N}{3\left(\dfrac{B}{2} - e\right)L} \tag{10-12}$$

Note from Eq. 10-12 that the intensity of pressure at the toe increases as the distance $\left(\dfrac{B}{2} - e\right)$ decreases and that a condition of instability of the footing results when $\left(\dfrac{B}{2} - e\right) = 0$.

10-8. INCLINED RESULTANT LOAD

If the resultant load R of all superimposed loads applied at the centroid of the base of the footing is inclined, one resolves R into normal and horizontal components, N and H, respectively (Fig. 10-22). The force N produces an average contact pressure σ_o on the soil, namely: $\sigma_o = N/A$.

The horizontal component H ($\rightarrow$) in the plane of the base of the footing tends to bring about horizontal translation or sliding of the footing. If no other means against sliding are provided, friction against sliding between the base material of the footing and the soil (rock) supporting the footing must be reckoned with.

If the resultant inclined load R is applied on the base of the footing off its centroid, then one again resolves R into normal and horizontal

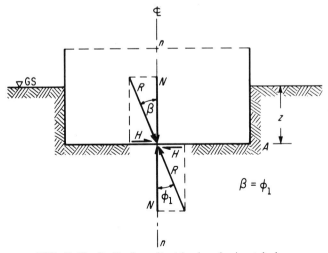

FIG. 10-22. Inclined resultant load on horizontal plane.

components, N and H, respectively, and checks pressures on soil for an eccentrically loaded soil-footing system for N, as was described in the foregoing sections, and checks the system against sliding for the horizontal force component H.

10-9. STABILITY AGAINST SLIDING

There is safety against sliding of the foundation if the angle β of inclination of the resultant force R (from the normal n-n) is less than the angle of base friction on soil (rock) ϕ_1, i.e., when $\beta \leqslant \phi_1$ (Fig. 10-23).

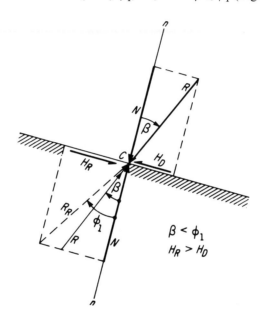

FIG. 10-23. Inclined resultant load on inclined plane.

The stability against sliding is expressed by means of a factor of safety η as

$$\eta = \frac{H_R}{H_D} = \frac{N \tan \phi_1}{N \tan \beta} = \frac{\tan \phi_1}{\tan \beta} \qquad (10\text{-}13)$$

where $H_R = N \tan \phi_1 =$ tangential frictional resistance in plane area (joint) of the base of footing

$N =$ normal component of resultant force R

$H_D =$ tangential driving force in plane of contact area of base of footing

$\tan \phi_1 =$ coefficient of friction between base material of footing and soil (rock)

Shallow Foundations and Footings

If, for example, $\phi_1 = 30°$, and $\beta = 20°$, then, by Eq. 10-13, the factor of safety against sliding is

$$\eta = \frac{\tan 30°}{\tan 20°} = \frac{0.577}{0.364} = 1.585 > 1.00$$

hence the foundation is safe against sliding.

To attain a possible and reasonably large stability against sliding and in order simultaneously to utilize the soil bearing capacity well, consistent with economy, the designer should strive to arrange that the resultant force R not only lies within the middle third of the width of the base of the footing, but also that there be as small an eccentricity as possible and that the resultant R be directed perpendicularly or nearly perpendicularly to the base area (joint) of the footing.

The factor of safety η against sliding should be equal to or greater than 1.5:

$$\eta = \frac{H_R}{H_D} \geq 1.5$$

The factor of safety against rotation of the foundation should be at least $\eta = 1.5$, or greater.

The factor of safety against ultimate failure of soil or groundbreak should be at least $\eta = 2.0$.

Problem

Given a mill bent column footing as shown in Fig. 10-24. The base of the column is hinged.

Load on footing from leeward column load. . . . . $P = 16,000$ lb $=$ 8 short tons
Weight of concrete footing . . . . $W = 4,000$ lb $=$ 2 short tons

Total vertical, normal load. $N = 20,000$ lb $= 10$ short tons

Contact area:

$$A = B^2 = 4.0 \times 4.0 = 16.0 \text{ (ft}^2)$$
$$\frac{B}{6} = \frac{4.0}{6} = 0.67$$

Coefficient of base friction: $\tan \phi_1 = 0.500$
Required factor of safety: $\eta = 2.0$
The allowable noncohesive soil bearing capacity is given as $\sigma_{all} = 1.5$ ton/ft^2.

Calculate contact edge pressures and stability against sliding of the footing.

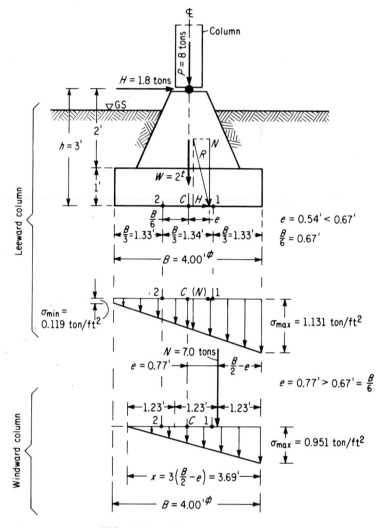

FIG. 10-24. Mill bent column footings.

Solution

Eccentricity, e, because of the horizontal load $H = 3{,}600$ lb $= 1.8$ tons:

$$e = \frac{Hh}{N} = \frac{(1.8)(3.0)}{10.0} = \frac{5.4}{10.0} = 0.54 \text{ (ft)} < (B/6) = 0.67$$

Contact pressures:

$$\sigma_{\substack{max \\ min}} = \frac{N}{A}\left(1 \pm \frac{6e}{B}\right) = \frac{10.0}{16.0}\left(1 \pm \frac{6 \times 0.54}{4.0}\right) = 0.625(1 \pm 0.81)$$

$$\sigma_{max} = (0.625)(1.81) = 1.131 \text{ (ton/ft}^2) < 1.50 \text{ ton/ft}^2 = \sigma_{all}$$

$$\sigma_{min} = (0.625)(0.190) = 0.119 \text{ (ton/ft}^2) < 1.50 \text{ ton/ft}^2$$

These soil pressures should not exceed the safe allowable bearing capacity of the given soil.

Stability against sliding:
Horizontal driving force: $H_D = 1.8$ ton.
Horizontal frictional resistance:

$$H_R = N \tan \phi_1 = (10.0)(0.500) = 5.0 \text{ (tons)}$$

$$\eta = \frac{H_R}{H_D} = \frac{N \tan \phi_1}{H_D} = \frac{(10.0)(0.500)}{1.8} = \frac{5.0}{1.8} = \underline{2.7 > 2.0 = \eta_{allow}}$$

Safe resistance to sliding, H_s, available for $\eta = 2.0$:

$$H_s \eta = N \tan \phi_1$$

or

$$H_s = \frac{N \tan \phi_1}{\eta} = \frac{(10.0)(0.500)}{2.0} = \underline{2.5 \text{ (ton)} > 1.8 \text{ ton}}$$

Problem

Given a hinged windward column on a footing as shown in Fig. 10-24.

Total vertical normal load on base of footing ...... $N = P + W = 14,000$ lb $= 7.0$ tons
Horizontal load at hinge.................. $H =$ 3,600 lb $= 1.8$ tons

Check factor of safety against sliding for $\tan \phi_1 = 0.500$, and calculate contact pressure on soil.

Solution

Eccentricity:

$$e = \frac{(1.8)(3.0)}{7.0} = \frac{5.4}{7.0} = \underline{0.77 > 0.67}$$

Hence the resultant vertical normal load is applied in the outer third of the width of the footing.

Distance $(B/2 - e)$ from toe:

$$B/2 - e = 4.0/2 - 0.77 = \underline{1.23 \text{ (ft)}}$$

Maximum pressure on soil, by Eq. 10-12:

$$\sigma_{max} = \frac{2N}{3\left(\frac{B}{2} - e\right)L} = \frac{(2)(7.0)}{(3)(1.23)(4.0)} = \frac{14.0}{14.76} = \underline{0.95 \text{ (ton/ft}^2) < 1.50 \text{ ton/ft}^2}$$

Factor of safety against sliding:

$$\eta = \frac{H_R}{H_D} = \frac{N \tan \phi_1}{H_D} = \frac{(7.0)(0.500)}{1.8} = \frac{3.5}{1.8} \approx 2.0 \approx \eta_{\text{allow}}$$

Problem

Given a mill bent column footing as shown in Fig. 10-25. The base of the column is fixed into the foundation. The load on the footing is given as set forth:

Load from column.............	$P = 14{,}000$ lb	$= 7.0$ tons
Weight of footing.............	$W = 25{,}000$ lb	$= 12.5$ tons
Total vertical, normal load	$N = 39{,}000$ lb	$= 19.5$ tons
Horizontal load...............	$H = 3{,}600$ lb	$= 1.8$ tons
Moment: given as	$M = 36{,}000$ lb-ft	$= 18.0$ ton-ft
Contact area..................	$A = 6.0$ ft $\times$ 9.0 ft	$= 54$ (ft^2)
Allowable soil pressure.........	$\sigma_{\text{all}} = 1.0$ ton/ft^2	

Coefficient of base friction:

$$\tan \phi_1 = 0.500$$

Required factor of safety against sliding:

$$\eta = 2.0$$

$$B/6 = 9.0/6 = 1.5$$

Calculate contact pressures and stability against sliding. If the soil has a unit weight $\gamma = 100$ lb/ft^3 and the angle of internal friction of this soil is ϕ, calculate also Fröhlich's critical edge pressure σ_{crit}. For this soil $p_i = 0$. The depth of the footing below the ground surface is given as $z = 4.0$ ft.

Solution

Eccentricity:

$$e = \frac{Hh + M}{N} = \frac{(1.8)(6.0) + 18.0}{19.5} = 1.476 \text{ (ft)} \approx 1.48 \text{ ft} < 1.5 = B/6$$

Contact pressure:

$$\sigma_{\substack{\max \\ \min}} = \frac{N}{A}\left(1 \pm \frac{6e}{B}\right) = \frac{19.5}{54.0}\left(1 \pm \frac{6 \times 1.48}{9}\right) = 0.361(1 \pm 0.986)$$

$$\sigma_{\max} = 0.361 + 0.356 = 0.717 \text{ (ton/ft}^2) < 1.00 \text{ ton/ft}^2$$

$$\sigma_{\min} = 0.361 - 0.356 = 0.005 \text{ (ton/ft}^2) < 1.00 \text{ ton/ft}^2$$

Shallow Foundations and Footings

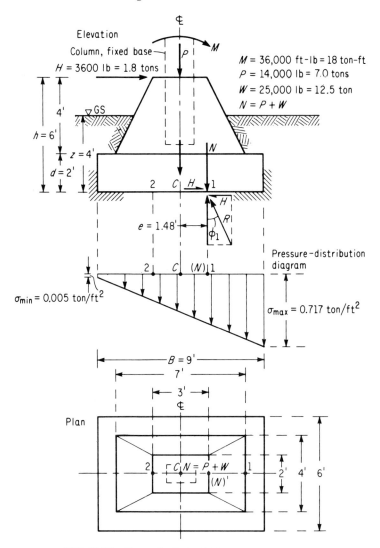

FIG. 10-25. Eccentrically loaded bent column footing.

Stability against sliding:

$$\eta = \frac{N \tan \phi_1}{H} = \frac{(19.5)(0.500)}{1.8} = \frac{9.75}{1.80} = \underline{5.4 > 2.0}$$

Safe available resistance to sliding:

$$H_s = \frac{N \tan \phi_1}{\eta} = \frac{9.75}{2.0} = \underline{4.375 \text{ (ton)} > 1.8 \text{ ton} = H}$$

Fröhlich's Critical Edge Pressure

The soil-foundation system should also be checked for critical edge pressure based on the theory of plastic flow of soil.[1-3,7-8] By critical edge pressure is understood that pressure at which soil particles just begin to flow out from underneath the base of the footing.

According to Fröhlich,[7] the critical edge pressure σ_{crit} on soil is calculated as

$$\sigma_{crit} = \frac{\pi(\gamma H + p_i)}{\cot\phi - \left(\frac{\pi}{2} - \phi\right)} \tag{10-14}$$

where σ_{crit} = allowable or critical limiting pressure on soil at edge of footing, a stress at which, when exceeded, soil will be expelled laterally from underneath base of footing

$\pi = 3.14...$
γ = unit weight of soil between ground surface and base of footing
H = depth of the base of footing below ground surface
$p_i = c \cot\phi$ = initial stress in soil
c = cohesion of soil
ϕ = angle of internal friction of soil in region underneath base of footing

To continue the problem, the critical edge pressure, by Eq. 10-14, for $p_i = 0$ is

$$\sigma_{crit} = \frac{\pi\gamma H}{\cot\phi - (\pi/2 - \phi)} = \frac{(3.14)(100)(4.0)}{1.732 - (1.570 - 0.523)} = \underline{0.917 \text{ (ton/ft}^2) < 1.0 \text{ ton/ft}^2}$$

Problem

If $p_i = 0$, and if the effective unit weight (buoyant weight) of the soil is

$$\gamma = (\gamma_s - \gamma_w)(1 - n) \tag{10-15}$$
$$= (0.00265 - 0.00100)(1 - 0.394) = 0.001 \text{ (kg/cm}^3)$$

where $\gamma_s = 0.00265$ kg/cm^3 = unit weight of the solid phase of the soil
$\gamma_w = 0.001$ kg/cm^3 = unit weight of water
$n = 0.394$ = porosity of the given soil

and if the depth of the footing is $H = H$ cm below the ground surface, and the angle of internal friction for the given soil is $\phi = 30°$ (cot $30° = 1.732$), then, by Eq. 10-14, the critical edge pressure σ_{crit} is calculated as

$$\sigma_{crit} = \frac{\pi(\gamma_s - \gamma_w)(1 - n)H}{\cot\phi - \left(\frac{\pi}{2} - \phi\right)} \tag{10-14a}$$

$$= \frac{(3.14)(0.001)H}{1.732 - \left(\frac{\pi}{2} - \frac{\pi}{6}\right)} = (0.00457)H \ (\text{kg/cm}^2)$$

If the depth of the foundation is given as $H = 200$ cm, then

$$\sigma_{\text{crit}} = (0.00457)(200) = 0.914 \approx 0.9 \ (\text{kg/cm}^2)$$

10-10. TWO-WAY ECCENTRICITY

When a footing is subjected to a static moment simultaneously about both axes of the base area of the footing, then the general edge stress equation for footing on soil is (Fig. 10-26):

$$\sigma_{\substack{\max \\ \min}} = \frac{N}{A} \pm \frac{M_x}{I_x/c_y} \pm \frac{M_y}{I_y/c_x} \tag{10-16}$$

$$= \frac{N}{A} \pm \frac{Ne_y}{I_x/c_y} \pm \frac{Ne_x}{I_y/c_x} = \frac{N}{A}\left(1 \pm \frac{6e_y}{L} \pm \frac{6e_x}{B}\right) \tag{10-16a}$$

Here $I_x/c_y = Z = \dfrac{BL^2}{6}$ = section modulus of soil contact area, viz., base area of footing

e_y = eccentricity along y-axis

All other symbols are the same as before.

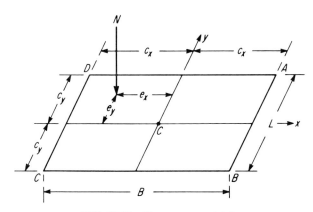

FIG. 10-26. Two-way eccentricity.

The corner stresses (Fig. 10-26) are:

$$\sigma_A = \frac{N}{A}\left(1 + \frac{6e_y}{L} - \frac{6e_x}{B}\right)$$

$$\sigma_B = \frac{N}{A}\left(1 - \frac{6e_y}{L} - \frac{6e_x}{B}\right)$$

$$\sigma_C = \frac{N}{A}\left(1 - \frac{6e_y}{L} + \frac{6e_x}{B}\right)$$

$$\sigma_D = \frac{N}{A}\left(1 + \frac{6e_y}{L} + \frac{6e_x}{B}\right)$$

Problem

Given a load-footing system as shown in Fig. 10-27. The total normal load is $N = P + W = 100.0 + 25.0 = 125.0$ tons, where $W = 25.0$ tons is the weight of the rectangular footing. The contact area is $A = B \times L = (12.0)(10.0) = 120.0$ ft^2; $B/6 = 12.0/6 = 2.0$ ft. Calculate stresses under corner points A, B, C, and D of the footing.

Solution

Moments of inertia of rectangular base area of footing
With respect to x-axis:

$$I_x = \frac{BL^3}{12} = \frac{(12.0)(10^3)}{12} = \underline{1{,}000 \ (\text{ft}^4)};$$

With respect to y-axis:

$$I_y = \frac{LB^3}{12} = \frac{(10)(12^3)}{12} = \underline{1{,}440 \ (\text{ft}^4)}$$

(i) The given problem may be transformed into that of a two-way eccentricity.

Eccentricities:

1. Along x-axis:

$$e_x = \frac{H_x d}{N} = \frac{(7.5)(5)}{125.0} = \underline{0.30 \ (\text{ft}) < 2.0 \ \text{ft}}$$

2. Along y-axis:

$$e_y = \frac{H_y d}{N} = \frac{(5.0)(5)}{125.0} = \underline{0.20 \ (\text{ft}) < 2.0 \ \text{ft}}$$

The normal force N lies within the kern of the cross section. Hence all stresses on the soil (rock) are compressive stresses.

The moment from $H_y = 5.0$ tons about the x-axis is

$$M_x = H_y d = (5.0)(5.0) = 25.0 \ \text{ton-ft}$$

Shallow Foundations and Footings

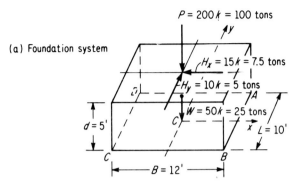

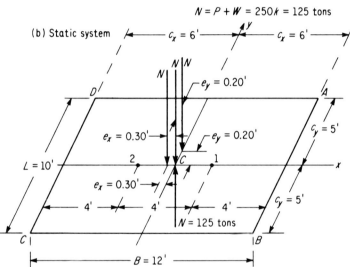

FIG. 10-27. Two-way eccentrically loaded footing.

This moment is shown as a couple made up of two oppositely directed forces of $N = 125.0$ tons with their moment arm $e_y = 0.20$ ft: $M_x = Ne_y = (125.0)(0.20) = 25.0$ ton-ft.

The same is shown about the y-axis: $M_y = H_x d = (7.5)(5.0) = 37.5$ ton-ft. Its couple has an arm of $e_x = 0.30$ ft (Fig. 10-27b): $M_y = Ne_x = (125.0)(0.30) = 37.5$ ton-ft.

Stresses:
By Eq. 10-16a, the stresses are:

$$\sigma_{\substack{max \\ min}} = \frac{N}{A}\left(1 \pm \frac{6e_y}{L} \pm \frac{6e_x}{B}\right) = \frac{125.0}{120.0}\left(1 \pm \frac{(6)(0.20)}{10.0} \pm \frac{(6)(0.30)}{12.0}\right)$$

$$= 1.042(1 \pm 0.120 \pm 0.150)$$

$\sigma_A = 1.042(1 + 0.120 - 0.150) = 1.011$ ton/ft^2 (compression)
$\sigma_B = 1.042(1 - 0.120 - 0.150) = 0.761$ ton/ft^2 (compression)
$\sigma_C = 1.042(1 - 0.120 + 0.150) = 1.073$ ton/ft^2 (compression)
$\sigma_D = 1.042(1 + 0.120 + 0.150) = 1.323$ ton/ft^2 (compression)

(ii) This problem can also be solved by means of moments, Eq. 10-16:

$$\sigma = \frac{N}{A} \pm \frac{M_x}{I_x/c_y} \pm \frac{M_y}{I_y/c_x} = 1.042 \pm \frac{(25.0)(5)}{1000} \pm \frac{(37.5)(6)}{1440}$$

$$= 1.042 \pm 0.125 \pm 0.156$$

$\sigma_A = 1.042 + 0.125 - 0.156 = 1.011$ ton/ft^2
$\sigma_B = 1.042 - 0.125 - 0.156 = 0.761$ ton/ft^2
$\sigma_C = 1.042 - 0.125 + 0.156 = 1.073$ ton/ft^2
$\sigma_D = 1.042 + 0.125 + 0.156 = 1.323$ ton/ft^2

One notes that these results are the same as those obtained by means of eccentricities.

REFERENCES

1. A. R. Jumikis, *Soil Mechanics*, Princeton, N.J.: Van Nostrand, 1962, pp. 624-627, 632-633.
2. A. R. Jumikis, *Mechanics of Soils: Fundamentals for Advanced Study*, Princeton, N.J.: Van Nostrand, 1964, pp. 147-151, 157-161, 169.
3. A. R. Jumikis, *Theoretical Soil Mechanics*, Princeton, N.J.: Van Nostrand, 1969.
4. A. R. Jumikis, *The Frost Penetration Problem in Highway Engineering*, New Brunswick, N.J.: Rutgers U. P., 1955, pp. 77-93.
5. A. R. Jumikis, *Thermal Soil Mechanics*, New Brunswick, N.J.: Rutgers U. P., 1966, pp. 107-114, 236-246, 247-252.
6. American Concrete Institute, *ACI Building Code, ACI Standard Building Code Requirements for Reinforced Concrete* (ACI-318-63). Detroit, Mich., 1963, p. 318.
7. O. K. Fröhlich, *Druckverteilung im Baugrunde*, Vienna: Springer, 1934, p. 142.
8. A. R. Jumikis, "Soil Mechanics," in James H. Potter (ed.), *Engineering Science Handbook*, Vol. 1, Princeton, N.J.: Van Nostrand, pp. 1120-1184.

PROBLEMS

10-1. Refer to Fig. 10-10. A foundation exerts a uniform contact pressure of $\sigma_o = 3$ ton/ft^2 on a cohesive soil. The properties of the cohesive soil are: $\gamma = 115$ lb/ft^3; $\phi = 20°$; cohesion $c = 300$ lb/ft^2.
   (a) What is the ultimate depth x for the laying of foundation?

Shallow Foundations and Footings

(b) Calculate the depth of foundation for a factor of safety of $\eta = 2$.

(c) What would be the conditions relative to x if the walls were supported by sheeting and strutting, rather than by outside anchorage?

10-2. A footing whose size is 5 ft × 9 ft in plan transmits a load of $N = 135$ kips to the soil. To how large a moment in the direction of the longitudinal axis of the footing can the footing be subjected before the contact pressure on the soil (at the edge of the footing) exceeds (a) 5 k/ft^2, and (b) 8 k/ft^2?

10-3. Given the same foundation system as shown in Fig. 10-24, except that the horizontal load H acts from right to the left. Analyze this foundation for its stabilities and plot the soil contact pressure distribution diagram.

10-4. Given the same foundation system as shown in Fig. 10-25, except that the horizontal load H acts from right to the left. Analyze this foundation for its stabilities and plot the soil contact pressure distribution diagram.

10-5. Given the same kind of foundation as shown in Fig. 10-25. $B = 10$ ft; $L = 6$ ft; contact area $A = 6 \times 10 = 60$ ft^2; $h = 6$ ft.

Loads from column.......................... $P = 6.0$ tons
Weight of footing.......................... $W = 10.0$ tons
Horizontal load; acting to the right.......... $H = 2.0$ tons
Moment M given as 18 ton-ft
Allowable soil pressure..................... $\sigma_{all} = 1.0$ ton/ft^2
Coefficient of base friction............. $\tan \phi_1 = 0.500$

Required factor of safety against sliding and overturning: $\eta = 1.5$. Calculate contact pressures on soil and stability against sliding and overturning. Plot contact pressure-distribution diagram.

chapter **11**

Calculation of Rectangular Plain Concrete Footings

11-1. PROCEDURE OF DESIGN

The procedure of footing design may be described briefly as follows:
1. Determine the size of the base contact area of the footing.
2. Select by trial and adjustment the effective thickness of the footing, or calculate it analytically, whichever the case may be.
3. Establish critical sections in the concrete footing for moment and shear as a measure for diagonal tension.
4. Determine and check stresses in footing materials.
5. Consult frequently the building code concerned.

The size of the necessary base contact area A_{nec} of the footing is calculated by dividing the maximum column load N (total dead and live loads plus the self-weight of the footing) by the safe, allowable soil (rock) bearing capacity σ_{all}:

$$A_{nec} = \frac{N}{\sigma_{all}}$$

The side length B of the square footing is calculated as

$$B = \sqrt{A_{nec}} = \sqrt{\frac{N}{\sigma_{all}}} = \sqrt{\frac{P+W}{\sigma_{all}}}$$

where all symbols are the same as before.

If the weight of the footing is not known, the necessary dimension B is obtained by approximation, first finding B_1 for the column load P. In such a case repeated calculations for B are unavoidable. After W and B have been found, one computes the contact pressure σ_o on soil from a normal load $N = P + W$:

$$\sigma_o = \frac{N}{B^2} \leq \sigma_{all}$$

Calculation of Rectangular Plain Concrete Footings

i.e., the contact pressure σ_o on soil (rock) should be less than the allowable bearing capacity σ_{all} on this soil or rock.

Problem

Given a massive, rigid square footing. It is loaded with a concentrated load of $P = 1.5$ ton. The unit weight of concrete is given as $\gamma_c = 150$ lb/ft³ $= 0.075$ ton/ft³. Assume weight of soil above widening of footing to have the same weight as the concrete.

Calculate the necessary width B for allowable soil pressures

$$\sigma_{1_{all}} = 0.50 \text{ ton/ft}^2$$

$$\sigma_{2_{all}} = 0.75 \text{ ton/ft}^2$$

$$\sigma_{3_{all}} = 1.00 \text{ ton/ft}^2$$

Solution

Weight of footing W:

$$W = \gamma_c B^2 z$$

where B = length of side of the square footing, and z = depth of base of footing below ground surface.

Reaction R:

$$R = P + W = P + \gamma_c B^2 z = \sigma_{all} B^2$$

Necessary length B of side of footing:

$$B = \sqrt{\frac{P}{\sigma_{all} - \gamma_c z}}$$

$$B_1 = \sqrt{\frac{1.5}{0.5 - (0.075)(5)}} = \sqrt{12.0} = 3.5 \text{ (ft)}$$

$$B_2 = \sqrt{\frac{1.5}{0.75 - (0.075)(5)}} = \sqrt{4.0} = 2.0 \text{ (ft)}$$

$$B_3 = \sqrt{\frac{1.5}{1.0 - (0.075)(5)}} = \sqrt{2.4} = 1.5 \text{ (ft)}$$

Ratio:

$$B_1 : B_2 : B_3 = 3.5 : 2.0 : 1.5 = 2.33 : 1.33 : 1$$

i.e., by increasing the allowable pressure on soil twice (from 0.5 ton/ft² to

1.0 ton/ft²), the length B of the side of the square footing decreases 2.33 times in this problem (from 3.5 ft to 1.5 ft).

Problem

The simplest, most widely used form of footing is that of a plain concrete wall footing, also called *strip footing*. The computation of a strip footing pertains to one unit of length of wall, $L = 1.0$.

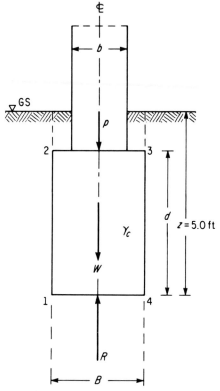

FIG. 11-1. Massive, rigid strip footing.

Given: a massive, rigid strip foundation loaded with a uniformly distributed linear centrical load $p = 1.5$ ton/ft and allowable pressures on soil:

$$\sigma_{1_{all}} = 0.50 \text{ ton/ft}^2$$
$$\sigma_{2_{all}} = 0.75 \text{ ton/ft}^2$$
$$\sigma_{3_{all}} = 1.00 \text{ ton/ft}^2$$

The unit weight of concrete is $\gamma_c = 150$ lb/ft³ $= 0.075$ ton/ft³. Calculate for these soil pressures the necessary width B of the strip foundation.

Calculation of Rectangular Plain Concrete Footings

Assume weight of soil above widening of footing to have same weight as the concrete (Fig. 11-1).

Solution
Weight W of foundation:
$$W = \gamma_c Bz(1)$$

Soil Reaction R:
$$R = \sigma_{\text{all}} B\,(1) = p + W = p + \gamma_c Bz\,(1)$$

Width B of footing:
$$B = \frac{p}{\sigma_{\text{all}} - \gamma_c z}$$

$$B_1 = \frac{1.5}{0.50 - (0.075)(5)} = 12.0 \quad \text{(ft)}$$

$$B_2 = \frac{1.5}{0.75 - (0.075)(5)} = 4.0 \quad \text{(ft)}$$

$$B_3 = \frac{1.5}{1.00 - (0.075)(5)} = 2.4 \quad \text{(ft)}$$

If $\gamma_c z = (150)(5) = 750$ lb/ft², and $\sigma_{1\text{all}} = 0.50$ ton/ft² $= 1{,}000$ lb/ft² then one notes that according to this method of calculation only 25 percent of $\sigma_{1\text{all}}$ can be apportioned for transmission of structural load p to the soil, and 75 percent for the weight W of the foundation block (1234).

The ratios of these widths are

$$B_1 : B_2 : B_3 = \frac{12.0}{2.4} : \frac{4.0}{2.4} : \frac{2.4}{2.4} = 5 : 1.55 : 1.$$

Thus the double increase in allowable soil pressure (from 0.50 ton/ft² to 1.00 ton/ft²) results in a fivefold decrease in footing width.

11-2. CRITICAL SECTIONS

Factors influencing the structural strength of a footing are bending moment, bond and shear at certain critical sections where failure because of excessive strains in flexure (bending) or diagonal tension (shear) may occur.

The critical sections can be decided upon and hence the moments and shears can be computed in several ways. However, for the sake of uniformity in computation and design, footings are designed empirically based on certain conventional rules.

The rules for selecting the critical sections are recommended by the American Concrete Institute, and are honored by many building codes. According to the ACI Building Code[1] recommended footing design:

The external moment on any section shall be determined by passing through the section a vertical plane which extends completely across the footing, and computing the moment of the forces acting over the entire area of the footing on one side of said plane ...

The greatest bending moment to be used in the design of an isolated footing shall be the moment computed in the manner prescribed above at sections located as follows:

1. At the face of the column, pedestal or wall, for footings supporting a concrete column, pedestal or wall (Fig. 11-2a).
2. Halfway between the middle and the edge of the wall, for footings under masonry walls (see Fig. 11-2b).
3. Halfway between the face of the column or pedestal and the edge of the metallic base, for footings under metallic bases [Fig. 11-2c]...

In computing the stresses in footings which support a round or octagonal concrete column or pedestal, the face of the column or pedestal may be taken as the side of a square having an area equal to the area enclosed within the perimeter of the column or pedestal,

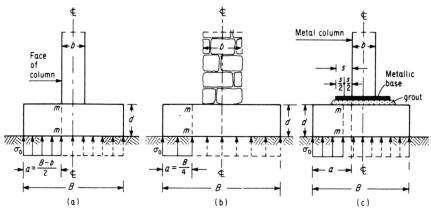

FIG. 11-2. Critical sections for bending moments for footings supporting (a) concrete column, pedestal or wall; (b) masonry wall, and (c) footings under metallic bases.

Strip Footing

If a strip footing (Fig. 11-3) is loaded with a uniformly distributed linear load p, then the contact pressure σ_o on soil from p and from weight W of footing per unit length of run of footing (perpendicular to drawing plane) is

$$\sigma_o = \frac{p + \gamma_c \left(bd + 2s\dfrac{d+d_1}{2}\right)}{(b+2a)(1)}$$

Calculation of Rectangular Plain Concrete Footings

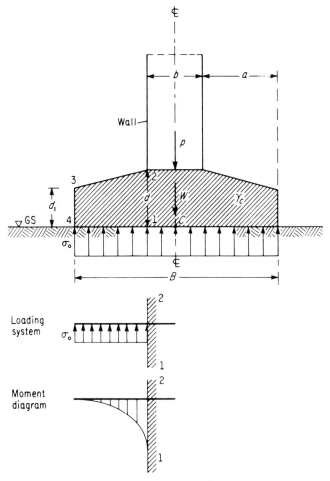

FIG. 11-3. Strip footing.

$$= \frac{p + \gamma_c d(b+a) + \gamma_c a d_1}{(b+2a)} \tag{11-1}$$

and the bending moment M in the critical section 1-2 is

$$M = \frac{\sigma_o a^2}{2} = \frac{a^2}{2} \frac{p + \gamma_c d(b+a) + \gamma_c a d_1}{d + 2a} \tag{11-2}$$

where γ_c = unit weight of concrete of footing.

In calculating the bending moment the weight (1234) of the widening (= cantilever) of the footing may be disregarded.

The widening or cantilever a of the footing is so proportioned that the contact pressure σ_o is just below the allowable soil bearing capacity.

If executed in plain concrete, the footing results in a big mass of concrete requiring a lot of material. Using a reinforced-concrete footing, the thickness of the footing can be reduced considerably. However, the reinforced-concrete footing should be sufficiently stiff to avoid significant bending of the footing. In the extreme case the still allowable length of the cantilever a for a given thickness d of the footing is that at which lifting (curling) of the edge of the end of the footing, because of the bending, would just impend. Such an uplift for a footing of a rectangular cross section, according to Hayashi,[2] sets in when the length of the cantilever is

$$a = \frac{\pi}{2} \sqrt[4]{\frac{4EI}{k}} \tag{11-3}$$

where E = modulus of elasticity of concrete
$\quad\quad\quad I$ = moment of inertia of footing with respect to length of footing equal to a unit of length
$\quad\quad\quad k$ = modulus of subgrade reaction, kg/cm^3 or $lb/in.^3$, whichever system of units of measurement is used

However, in proportioning the footing to obtain a more uniform pressure distribution which would correspond better to the calculations, it should be sufficiently thick that the ends of the cantilevers do not lift up.

11-3. CALCULATION OF MOMENT

The effective thickness d of the footing from a column (wall) load $P(p)$ depends on the strength of the column (wall) material as well as on the strength of the foundation material. The thickness d of the concentrically loaded footing is determined based on the following considerations (Fig. 11-4).

Imagine cantilever beams projecting in all directions from the single column. In the case of a strip foundation cantilever beams project on either side of the foundation wall. These cantilever beams resting on the soil may be assumed to be loaded upward by the uniformly distributed soil reaction load.

In strength calculations of the footing its weight W is usually neglected and so is the pressure (γz) of the backfill material on the footing (this is on the safe side). Also, the cleanliness layer is omitted from strength calculations of footings.

The soil reaction would tend to deflect and bend the cantilever part of the footing and to break the cantilever off from the column at the critical

Calculation of Rectangular Plain Concrete Footings

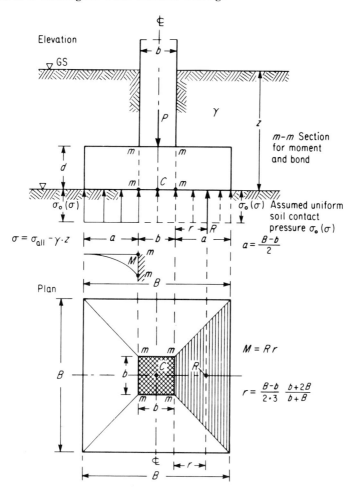

FIG. 11-4. Sketches pertaining to calculation of moment.

section m-m. This tendency is the greater the longer is the cantilever a. Upon bending, there are induced tensile stresses across the critical section in the footing material.

The maximum bending moment M from soil reaction to be used in the design of an isolated footing should be computed on a vertical section m-m at the face of the column or pedestal. The bending stresses σ_b in plain concrete footings are then computed in the same manner as for homogeneous cantilever beams.

The bending moment M from soil reaction may also be calculated about section m-m as

$$M = Rr$$

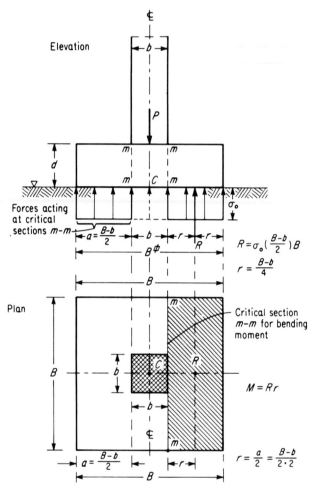

FIG. 11-5. Forces acting at critical section *m-m*.

where R = total upward soil reaction (Fig. 11-5), on one fourth of base area $\left(\dfrac{B^2 - b^2}{4}\right)$ of the footing:

$$R = \sigma_o \left(\frac{B^2 - b^2}{4}\right) \qquad (11\text{-}4)$$

r = moment arm of R from centroid C of the shaded trapezoid to critical section *m-m* at face of column:

$$r = \left(\frac{B-b}{2\cdot 3}\right) \cdot \left(\frac{2B+b}{B+b}\right) \qquad (11\text{-}5)$$

Calculation of Rectangular Plain Concrete Footings

With these values of R and r, the externally applied bending moment M is written as

$$M = Rr = \frac{1}{24}\sigma_o(B-b)^2(2B+b) \tag{11-6}$$

If the allowable bending stress of concrete is known, then the effective thickness d of the plain concrete footing to resist bending can now be calculated from the section modulus Z of the critical cross-section m-m:

$$Z = \frac{Bd^2}{6} = \frac{M}{\sigma_{\text{all}}} \tag{11-7}$$

Equation 11-7 may be used for fixing offsets $a = (B-b)/2$ of heavily, centrically loaded, stepped square plain concrete footings of given size B and b and for various stress ratios of $\sigma_o/\sigma_{b_{\text{all}}}$:

$$\frac{Bd^2}{6} = \frac{M}{\sigma_b} = \frac{1}{24}\frac{\sigma_o}{\sigma_{b_{\text{all}}}}(2a)^2(2B+b)$$

$$\left(\frac{d}{a}\right)^2 = \left(\frac{2B+b}{B}\right)\left(\frac{\sigma_o}{\sigma_{b_{\text{all}}}}\right)$$

$$\frac{d}{a} = \sqrt{\left(\frac{2B+b}{B}\right)\cdot\left(\frac{\sigma_o}{\sigma_{b_{\text{all}}}}\right)}$$

Effective thickness d:

$$d = a\sqrt{\left(\frac{2B+b}{B}\right)\left(\frac{\sigma_o}{\sigma_{b_{\text{all}}}}\right)} \tag{11-8}$$

According to the ACI Building Code[1] the bending moment should be computed on a vertical section m-m at the face of the column and extending completely across the footing, and all the forces acting on one side of the critical section should be included. The sought bending moment then calculates as

$$M_{m-m} = Rr = \frac{1}{8}\sigma_o B(B-b)^2 \tag{11-9}$$

(the footing not yet backfilled).

The most dangerous situation for a foundation is when there is a structure carrying its full load. For a ready structure carrying its full load, the most dangerous situation for a foundation is when there is a need during its service time to excavate the footing free in order to make repairs, or to install utilities along the footing or through the wall, or for underpinning purposes, for example.

The thickness d of the plain concrete footing to resist bending is calculated by equating the externally applied moment M (Eq. 11-9) to the moment of the internal resistance in critical section of the slab in bending, $\sigma_{b_{\text{all}}} \cdot Z$:

$$M = \sigma_{b_{\text{all}}} \cdot Z \qquad (11\text{-}10)$$

or

$$(1/8)\sigma_o B(B-b)^2 = (1/8)\sigma_o B(2a)^2 = \sigma_{b_{\text{all}}} \frac{Bd^2}{6}$$

$$\left(\frac{d}{a}\right)^2 = 3 \frac{\sigma_o}{\sigma_{b_{\text{all}}}}$$

$$d = (1.73)\, a \sqrt{\frac{\sigma_o}{\sigma_{b_{\text{all}}}}} \qquad (11\text{-}11)$$

11-4. CRITICAL SECTION FOR SHEAR

Another consideration for calculating the dimensions of a concrete footing is the piercing of the column through the footing slab by the concentrated column load P. The piercing fracture is a truncated cone or pyramid (Fig. 11-6).

Before the footing can be pierced through, bending stresses in the material must occur, and in their train, the complementary stresses of vertical and horizontal shear stresses, of which the resultant is the diagonal tensile stress. According to the Building Code of the American Concrete Institute, this diagonal tensile stress is assumed to have its maximum along a plane at 45° to the horizontal of the footing base extending from the column base (Fig. 11-7), so that the tendency will be a failure in the concrete at right angles to the plane of stress, which is also 45° to the horizontal.

The ACI Building Code of 1963[3] recommends that the critical section for shear to be used as a measure of diagonal tension shall be perpendicular to the plane of the slab and located at a distance $d/2$ from the periphery of the concentrated load or reaction area (or from the face of the column). Thus in Fig. 11-7b the critical section for shear is section s-s, and for bending and bond the critical section is m-m.

11-5. CALCULATION OF SHEAR STRESS

1. *For method of working stress design of footings.* In shear stress computation one distinguishes between two kinds of footing design, (a) the

Calculation of Rectangular Plain Concrete Footings

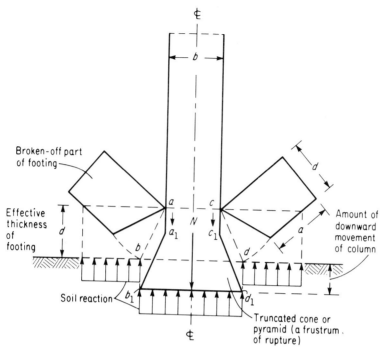

FIG. 11-6. Deformation of column footing.

working stress design, and (b) the ultimate strength design. In the working stress design method the nominal shear stress v should be computed according to the ACI Building Code of 1963[4] by

$$v = \frac{V}{b_o d} \leq 2\sqrt{\sigma_c} \qquad (11\text{-}12)$$

where V = shear force = (net soil bearing pressure) (area A_T)
b_o = length of periphery of critical section for shear
d = effective thickness of footing (or distance from extreme compression fiber to centroid of tension reinforcement in the case of a reinforced-concrete footing)
σ_c = allowable compressive stress of concrete

Here V and b_o are to be taken at the critical section for shear.

Note that for a homogeneous, rectangular section the maximum shear value is $v = (3/2)\dfrac{V}{b_o d}$. The ACI Building Code works with an average shear stress as in Eq. 11-12.

If the allowable shear stress v_{all} in concrete is known, then with V and B or $(d+b)$ known, the effective thickness d of the footing can be calculated.

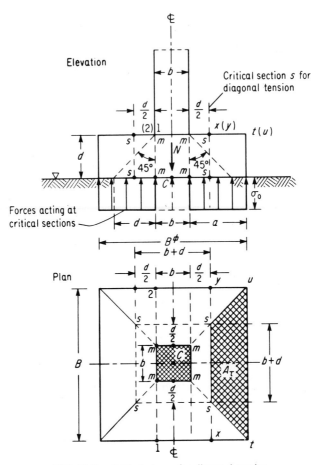

FIG. 11-7. Critical planes for diagonal tension.

If the shear stress v thus computed exceeds $2\sqrt{\sigma_c}$, shear reinforcement must be provided; in each case v should not exceed $3\sqrt{\sigma_c}$.

2. *For ultimate strength design of footings*, according to the ACI Building Code of 1963,[5] pp. 318-375, the nominal ultimate shear stress v_u as a measure of diagonal tension shall be computed by

$$v_u = \frac{V_u}{b_o d} \leqslant 4\phi\sqrt{\sigma_c} \tag{11-13}$$

where V_u = ultimate shear force
b_o = length of periphery of critical section for footings
d = effective thickness of footing (or distance from extreme compression fiber to centroid of tension reinforcement)

Calculation of Rectangular Plain Concrete Footings

$v_c = 4\phi\sqrt{\sigma_c}$ = shear stress carried by concrete
$\phi = $ a capacity reduction factor = 0.85 for diagonal tension
$\sigma_c = $ compressive strength of concrete

Here V_u and b_o are taken at the critical sections. If the computed ultimate shear strength v_u exceeds $v_c = 4\phi\sqrt{\sigma_c}$, shear reinforcement must be provided, in which case v_u should not exceed $6\phi\sqrt{\sigma_c}$.

Of course, in footing design other requirements, as set by the ACI Building Code must also be complied with.

For a square footing the magnitude of the total vertical shear force V at the critical section from soil reaction acting upwards on the hollow square (see hatching in Fig. 11-8) surrounding the critical section is calculated as

$$V = \sigma_n[B^2 - (b+d)^2] \tag{11-14}$$

where all symbols are the same as before, and $\sigma_n = $ net soil pressure.

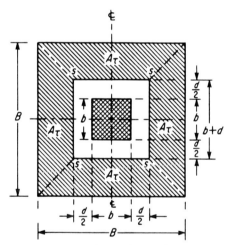

FIG. 11-8. Soil reaction acting on a hollow square.

The critical vertical shear stress v is calculated by using the four vertical sections formed upon the square which lies at a distance $d/2$ from the face of the column:

$$v = \frac{V}{A_{cr}} = \frac{\sigma_n(4A)}{b_o d} = \frac{\sigma_n[B^2 - (b+d)^2]}{4(b+d)d} \quad \text{(lb/ft}^2\text{)} \tag{11-15}$$

for a plain concrete footing, where

$A = $ hatched area upon which the net soil reaction σ_n for the total shear force V acts

$A_{cr} = 4(b+d)d$ = critical area for shear
$4(b+d) = b_o$ = length of periphery of critical sections for shear

Here σ_o, B, b, and d must have consistent units of measurement.

Knowing the allowable shear stress, the effective thickness d of the footing can be calculated.

The outside dimensions of the designed footing should be set in feet and whole inches. In practice, beginning with 1 ft = 12 in., the thickness of a footing is usually proportioned to the nearest 3-in. increment, for example, 15 in., or 18 in., until a satisfactory scale is reached.

The *effective thickness* to use in footing construction is the larger value of the two, computed either by means of the bending moment or by shear. The calculated thickness d of the footing is the effective thickness which does not include the thickness of the cleanliness course. The latter, because it is in contact with soil, is of inferior quality.

Problem

Given a square, plain concrete footing as shown in Fig. 11-9. The column transmits on the footing a centrical load of $P = 300{,}000$ lb. The safe allowable soil bearing capacity is given as $\sigma_{all} = 2.0$ ton/ft^2 = 4,000 lb/ft^2. The allowable compressive stress in concrete is given as $f_c = 1{,}125$ psi, the allowable tensile stress $f_t = 80$ psi, and shear stress as 100 psi. Assume weight of footing $W = 40{,}000$ lb.

Solution

The critical sections are
(a) for moment (flexure): *m-m*
(b) for diagonal tension (shear): 1-2.

The total shear as a measure of diagonal tension is the net soil reaction on the hatched area 1234.

Size of footing base contact area:

$$A = \frac{P+W}{\sigma_{all}} = \frac{300{,}000 + 40{,}000}{4{,}000} = 85 \text{ (ft}^2\text{)}$$

Length B of side of square footing is

$$B = \sqrt{A} = \sqrt{85} = 9.2 \text{ (ft)} = 9 \text{ ft } 2\tfrac{1}{2} \text{ in.}$$

say $B = 9$ ft 6 in. This gives an average, uniform net contact pressure on soil of

$$\sigma_n = \frac{P}{A} = \frac{300{,}000}{9.5^2} = 3{,}240 \text{ (lb/ft}^2\text{)}$$

Calculation of Rectangular Plain Concrete Footings

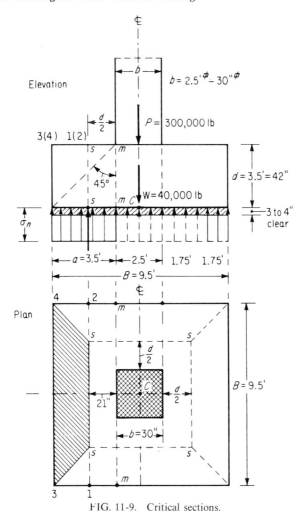

FIG. 11-9. Critical sections.

Bending moment at critical section m-m:

$$\text{Area } A_M = Ba = B\left(\frac{B-b}{2}\right) = (9.5)\left(\frac{9.5-2.5}{2}\right) = 33.25 \text{ (ft}^2\text{)}$$

$$M = \sigma_n A_M (a/2) = (3240)(33.25)(3.5/2) = 188,535.6 \text{ (ft-lb)}$$

Section modulus Z at m-m:

$$Z = \frac{Bd^2}{6} = \frac{(9.5)d^2}{6} = \frac{M}{f_t}$$

where f_t = bending stress in concrete in tension.

Shallow Foundations

Effective thickness d of footing:

$$d^2 = \frac{M}{f_t}\frac{6}{9.5} = \left[\sigma_n B\left(\frac{B-b}{2}\right)^2 \frac{1}{2}6\right] \Big/ (f_t)(9.5) = 10.3 \ (\text{ft}^2)$$

$$d = \sqrt{10.3} \approx 3.2 \ (\text{ft})$$

∴ Use 3 ft 6 in. = 3.5 ft

Critical section s-s for shear (Fig. 11-10):

$$s = a - \frac{d}{2} = 3.50 - \frac{3.50}{2} = 1.75 \ (\text{ft})$$

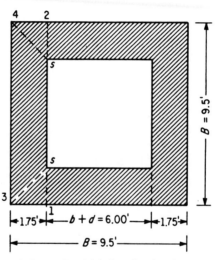

FIG. 11-10. Critical section for shear.

Shear stress v:

$$v = \frac{V}{A_v} = \frac{\sigma_n[B^2 - (b+d)^2]}{4(b+d)d}$$

$$= \frac{(3240)(9.50)^2 - (2.5+3.5)^2}{(4)(2.5+3.5)(3.5)}$$

$$= 2092.5 \ (\text{lb/ft}^2)$$

$$\approx \frac{2092.5}{144} \approx 14.5 \ (\text{lb/in.}^2) < 100 \ \text{lb/in.}^2$$

i.e., safe against shear.

11-6. RECTANGULAR PLAIN CONCRETE FOOTINGS

The critical sections in a rectangular plain concrete footing are (a) for moment (flexure): m-m and n-n; (b) for diagonal tension: s-s, and are indicated on Fig. 11-11. The total shear as a measure of diagonal tension is the net soil reaction on the hatched area 123456. The shear stress may be calculated as

$$v = \frac{V}{A_v} = \frac{\sigma_n[BL-(b+d)^2]}{4(b+d)d} \qquad (11\text{-}16)$$

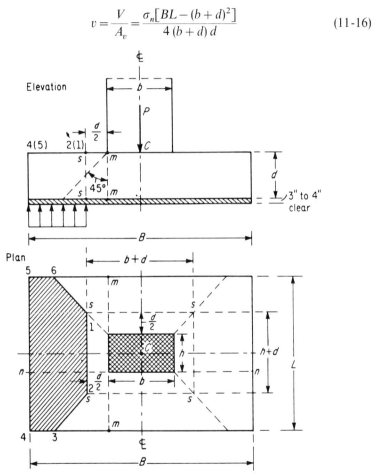

FIG. 11-11. Critical sections in a rectangular footing.

11-7. PLAIN CONCRETE WALL FOOTING

A plain concrete wall footing, also called a *strip footing*, is the simplest form of footing. A plain concrete wall footing is made up of a continuous, widened concrete slab. It projects as a cantilever on both sides of the wall

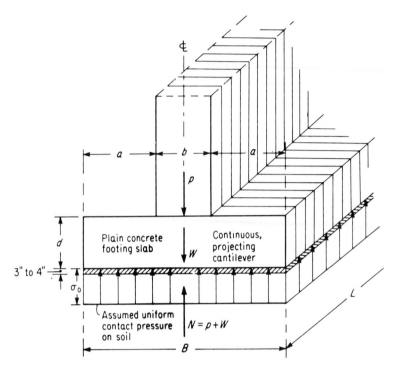

FIG. 11-12. Plain concrete strip footing.

(Fig. 11-12). For plain concrete footings the projected widenings are relatively small.

The thickness d of the footing slab is a function of the safe, allowable bearing capacity of the soil (rock), the length of the cantilevered projection, $(B-b)/2$, and the allowable tensile stress in the concrete.

The critical sections for moments and shear, per one unit length of run of wall footing, and the mode of calculating them are similar to those explained for square, plain concrete isolated footings. The contact pressure σ_o on soil is computed as $\sigma_o = N/A$, where $N = p + W$ = load from column and weight of footing, respectively, and A = base contact area of footing per unit length of run of wall.

The maximum externally applied bending moment M with respect to the critical section m-m in the footing slab from soil contact pressure σ_o is

$$M = \sigma_o a(1.0)\frac{a}{2} = (1/2)\sigma_o a^2 \quad \text{(ft-lb)} \qquad (11\text{-}17)$$

where $a = (B-b)/2$ = offset or projection of footing slab beyond the wall. Equating the externally applied moment M with the moment of the internal

Calculation of Rectangular Plain Concrete Footings

resistance of slab $(1/6)(\sigma_b d^2)$, one calculates the thickness d of the slab to resist bending as set forth:

$$M = \sigma_b Z = \sigma_b(1.0)\frac{d^2}{6}$$

or

$$(1/2)\sigma_o a^2 = (1/6)\sigma_b d^2$$

and the effective thickness d is

$$d = a\sqrt{3\sigma_o/\sigma_b} \qquad (11\text{-}18)$$

where σ_b = safe bending strength in tension of concrete
Z = section modulus of critical cross section m-m of footing
Equation 11-18 may be used for fixing offsets of stepped strip footings.
Instead of σ_o, the net pressure on soil σ_n may be used in these calculations:

$$\sigma_n = \sigma_{all} - \sigma_W$$

where σ_{all} = safe, allowable soil bearing capacity
$\sigma_W = W/A_f$ = contact pressure on soil from self-weight W of footing
A_f = contact area of footing

Plain concrete footings, under certain conditions, may result in a bulky, heavy design requiring considerable space, large amounts of forms for casting, and requiring a great volume of concrete, thus adding to the cost of the foundations. To avoid these disadvantages of bulky, plain concrete design engineers resort to lighter footing designs in reinforced concrete. Also, in practice, if the width of the plain concrete exceeds 3 ft, many engineers prefer to use reinforced-concrete footings.

Problem

Given a concrete wall 20 in. wide centered on a plain concrete strip footing, the width of which is 8 ft and which has an effective thickness of 3 ft (Fig. 11-13). The wall transmits to the footings a uniformly distributed linear load of $p = 25,000$ lb/ft = 12.5 ton/ft. The unit weight of concrete is $\gamma_c = 150$ lb/ft^3. The safe allowable bearing capacity of the soil is given as $\sigma_{all} = 2.0$ ton/ft^2. The allowable bending stress in a 2,500 lb/in^2 concrete in tension is given as $\sigma_{b_{all}} = 80$ lb/in^2 and that of shear stress in concrete is $\tau_s = 100$ lb/in^2.

Calculate
1. Contact pressure σ_o on soil.
2. Bending moment in footing slab.

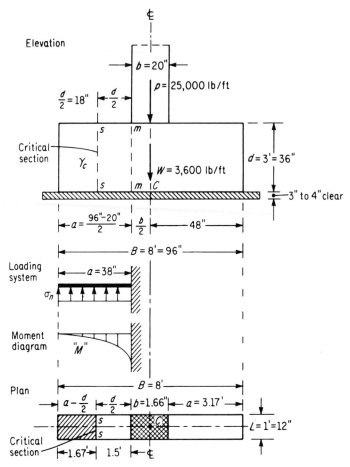

FIG. 11-13. Sketch pertaining to calculation of moment.

3. Bending, viz., tensile, stress in concrete.
4. Average and maximum shear stresses.

Compare the corresponding calculated values with the safe, allowable stresses.

For computing strength of footing, neglect weight of footing.

Solution

Normal load:
Weight of wall:

$$p = 25,000 \text{ lb/ft}$$

Calculation of Rectangular Plain Concrete Footings

Weight of concrete footing:

$$W = \gamma_c Bd(1.0) = (150)(8)(3)(1) = 3{,}600 \text{ lb/ft}$$

Total normal, vertical load $N = 25{,}000 + 3{,}600 = 28{,}600$ lb/ft

Contact pressure on soil:

$$\sigma_o = \frac{N}{A} = \frac{28{,}600}{(8)(1)} = 3{,}575 \text{ (lb/ft}^2) < 4{,}000 \text{ lb/ft}^2 \, (= \sigma_{\text{all}})$$

Net upward-bearing pressure on soil:

$$\sigma_n = \sigma_o - \sigma_W = 3{,}575 - (150)(1)(3) = 3{,}125 \text{ (lb/ft}^2)$$

where σ_n = net bearing pressure on soil
$\sigma_W = (\gamma_c)(1.0)(d) = (150)(1.0)(3.0) = 450$ (lb/ft^2) = pressure on soil from weight of footing
γ_c = unit weight of concrete

Here, in computing σ_n, the effect of the weight of the footing, σ_W, is neglected. The net soil pressure is the difference between the safe allowable soil contact pressure σ_{all} and the pressure on soil σ_W exerted by the self-weight of the footing proper.

Bending moment M with respect to critical section m-m:

$$M = \sigma_n(1.0)(a)(a/2) = (3{,}125)(3.17^2/2) = \underline{15{,}200.6 \text{ (ft-lb)}}$$

Bending stress σ_b in concrete:

$$\sigma_b = \frac{M}{\dfrac{Ld^2}{6}} = \frac{(15{,}200)(6)(12)}{(1 \times 12)(3^2)(144)} = \underline{70.4 \text{ (lb/in.}^2)} < 80 \text{ lb/in.}^2 (= \sigma_{b\,\text{all}})$$

Average shear stress v:

$$v = \frac{V}{b_o d} = \frac{\sigma_n L\left(a - \dfrac{d}{2}\right)}{Ld} = \frac{(3{,}125)(1.67)}{3.0} = 1{,}739.5 \text{ (lb/ft}^2)$$

$$\approx \underline{12.1 \text{ lb/in.}^2} < 100 \text{ lb/in.}^2 (= \tau_s)$$

Hence the footing is safe in vertical shear.
Maximum shear stress $v_{\max}$:

$$v_{\max} = (3/2)v = (1.5)(12.1) \approx 18.2 \text{ (lb/in.}^2) < 100 \text{ lb/in.}^2$$

Hence the footing is safe against maximum shear.

Problem

Given a plain concrete strip footing as shown in Fig. 11-14. The wall carries a linear load of 30,000 lb/ft. The safe, allowable soil bearing capacity is given as $\sigma_{\text{all}} = 2.5$ ton/ft². The allowable compressive strength of the concrete is $f'_c = 2500$ lb/in². Hence, the allowable tensile stress f_t in concrete is $f_t = (1.6)\sqrt{f'_c} = (1.6)\sqrt{2500} = 80$ (lb/in²) = 11,520 lb/ft², and the allowable compressive stress f_c in concrete is: $f_c = (0.45)f'_c = 1,125$ lb/in². Unit weight of concrete: $\gamma_c = 150$ lb/ft³. Calculate the effective thickness of the footing.

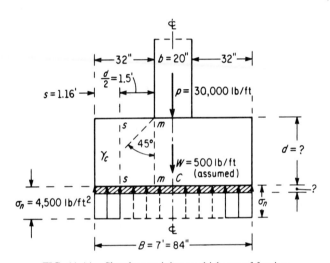

FIG. 11-14. Sketch pertaining to thickness of footing.

Solution

Assume that the pressure on soil from self-weight of the footing is $\sigma_W = 500$ lb/ft².

Net contact pressure σ_n on soil:

$$\sigma_n = \sigma_{\text{all}} - \sigma_W = 5{,}000 - 500 = 4{,}500 \text{ (lb/ft}^2\text{)}$$

Width B of footing:

$$B = \frac{p + W}{\sigma_n} = \frac{30{,}000 + 500}{4{,}500} = 6.77 \text{ (ft)}$$

or 6 ft $9\tfrac{1}{5}$ in. Use 7 ft 00 in. = 84 in.

Bending moment M at critical section m-m:

$$M = (1/2)\sigma_n(1.0)a^2 = (1/2)(4{,}500)(2.66^2) = 15{,}930 \text{ (ft-lb)}$$

Calculation of Rectangular Plain Concrete Footings

Effective thickness d of footing:

$$Z = \frac{Ld^2}{6} = \frac{M}{f_t}$$

$$d^2 = \frac{6M}{f_t L} = \frac{(6)(15{,}930)}{(80)(144)(1.0)} = 8.29 \ (\text{ft}^2)$$

$$d = \sqrt{8.29} = 2.88 \ (\text{ft}) = 34.6 \ \text{in.}$$

Use $d = 36$ in.

Checking self-weight W of footing:

$$W = \gamma_c LBd = (150)(1.0)(84/12)(3.0) = 3{,}150 \ (\text{lb})$$

$$\sigma_W = W/A = 3{,}150/(1.0)(7.0) = 450 \ (\text{lb/ft}^2) < 500 \ \text{lb/ft}^2$$

as was assumed at the outset of this problem.

Critical section s-s for shear:

$$s = (32/12) - (d/2) = 2.66 - 1.50 = 1.16 \ (\text{ft})$$

from edge of footing.

Total shear force V:

$$V = \sigma_n sL = (4.500)(1.16)(1.0) = \underline{5{,}220 \ (\text{lb})}$$

Average shear stress v (as a measure of diagonal tension):

$$v = \frac{V}{b_o d} = \frac{5{,}220}{(1.0)(3.0)} = \underline{1{,}740 \ (\text{lb/ft}^2)} < 2\sqrt{f_c'}(144) = 14{,}400 \ (\text{lb/ft}^2)$$

Maximum shear stress $v_{\max}$:

$$v_{\max} = (3/2)v = (1.5)(1{,}740) = \underline{2{,}610 \ (\text{lb/ft}^2)}$$

Problem

Given a plain concrete strip footing. It carries a 40 k/ft centrical linear load. The safe allowable soil bearing capacity is $\sigma_{\text{all}} = 6$ k/ft². Use a 2,500-psi ultimate strength concrete. The bending stress $f_t = 75$ psi, average vertical shear $v = (0.02)f_c' = 50$ psi; $\gamma_c = 145$ lb/ft³. The width of the wall is $b = 16$ in., and the width of the footing is $B = 8$ ft. The effective thickness d is given as $d = 34$ in.

Required. Determine the safety of the footing, and the maximum shear stress in the plain concrete footing.

Answers
1. Contact pressure on soil: $\sigma_o = 5.44$ k/ft^2 < 6.0 k/ft^2.
2. $M = (40/8)(3.33)(20) = 333$ k-in.
3. $f_t = M/Z = 143$ psi > 75 psi, hence unsafe.
4. $v = V/(b_o d) = 2{,}500/(12)(34) = 6.1$ psi < 50 psi (safe in vertical shear).
5. $\tau_{max} = (3/2)v = (1/5)(6.1) = 9.15$ psi < 50 psi (O.K.)
6. Redesign this footing.

$B_{nec} = (1.08)(40)/6 = 7.2$ (ft) (Try 7 ft 2 in.)

$d_{nec} = 43.5$ in. (Try 3 ft 10 in.), use 44.0 in.

Check shear v.

11-8. CALCULATION OF STRESSES IN A MASSIVE, PLAIN CONCRETE STEPPED OR PYRAMIDAL STRIP FOOTING

Stepped footings are proper for massive foundations.

The ACI Building Code provides that in sloped or stepped footings the angle of slope or depth and location of steps shall be such that the allowable stresses are not exceeded at any section. Sloped and stepped footings should be cast as a unit.

In calculating the thickness d of a plain concrete stepped strip footing and the stresses involved it is interesting to note that the vertical section v-v in the footing at the face of the column is not the most disadvantageous one but, as is shown by the following, it is an inclined one.

Referring to Fig. 11-15, pass through point D an inclined section x-x at an angle $\beta = \dfrac{(90° - \alpha)}{2}$ with the vertical v-v. The stresses in section x-x are now calculated under the following assumptions.

1. The stresses are calculated for a strip one unit of length per run of wall.
2. The part of the footing, DAD', is considered to be weightless.
3. The steps outside the line A-B are disregarded in the stress calculations.
4. The foundation is not yet backfilled.
5. The soil contact pressure σ_o is uniformly distributed under the base of the footing (the self-weight of the footing is disregarded in calculating σ_o).
6. No frictional force F in the contact plane between base of footing and soil is considered.

Calculation of Rectangular Plain Concrete Footings 375

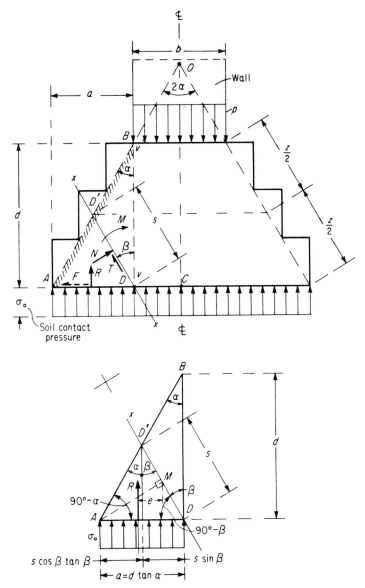

FIG. 11-15. Plain concrete stepped footing.

With these assumptions fixed, there acts to the left of section x-x the upward resultant R of the soil reaction, say, in the metric system of units of measurement:

$$R = \sigma_o d \tan \alpha \quad \text{(kg)} \tag{11-19}$$

Other geometric relationships in Fig. 11-15 are:

$$\tan \alpha = \frac{a}{d} \tag{11-20}$$

$$\beta = \frac{90° - \alpha}{2} \tag{11-21}$$

$$d \tan \alpha = s(\tan \alpha \cdot \cos \beta + \sin \beta) \tag{11-22}$$

$$s = \frac{d \tan \alpha}{\cos \beta (\tan \alpha + \tan \beta)} \tag{11-23}$$

Further, on the inclined section there act a normal force N, a tangential or shear force T, and a bending moment M_M (with respect to point M in the inclined section x-x).

The force components of R parallel and perpendicular to section x-x, and the bending moment M_M are:

1. Bending moment: $M_M = Re = (\sigma_o d \tan \alpha)(e)$ (kg-cm) (11-24)

$$= (1/2)\sigma_o d^2 \tan^2 \alpha \left(1 - \frac{\tan \beta}{\tan \alpha + \tan \beta}\right) \tag{11-25}$$

2. Shear force: $T = R \cos \beta = \sigma_o d \tan \alpha \cos \beta$ (kg) (11-26)
3. Normal force: $N = R \sin \beta = \sigma_o d \tan \alpha \sin \beta$ (kg) (11-27)

where e = moment arm of R with respect to point M:

$$e = \frac{d \tan \alpha}{2} - \frac{s}{2} \sin \beta$$

$$= \frac{d}{2} \tan \alpha - \frac{1}{2} \frac{d \tan \alpha}{\cos \beta (\tan \alpha + \tan \beta)} \sin \beta$$

$$= \frac{d}{2} \tan \alpha \left(1 - \frac{\tan \beta}{\tan \alpha + \tan \beta}\right) \text{ (cm)} \tag{11-28}$$

1. Bending stress

The bending moment M_M renders compressive and tensile bending stresses $\pm \sigma_b$ in section x-x:

$$\sigma_b = \pm \frac{M}{Z} = \pm \frac{\sigma_o d e \tan \alpha}{\frac{(1.0)s^2}{6}} \text{ (kg/cm}^2\text{)} \tag{11-29}$$

Calculation of Rectangular Plain Concrete Footings

where Z = section modulus in cm^3 for a rectangular area (width) (height)
$= (1.0)(s)$

Substitution of Eqs. 11-23 and 11-28 into Eq. 11-29 renders the general bending-stress σ_b equation as

$$\sigma_b = \pm 3\sigma_0 \tan \alpha \cos^2 \beta (\tan \alpha + \tan \beta) \qquad (11\text{-}30)$$

The maximum bending stress $\sigma_{b_{max}}$ occurs on that plane at an angle β for which the first derivative of the general bending-stress function with respect to angle β is zero:

$$\frac{d\sigma_b}{d\beta} = 0 = 3\sigma_0(-\sin 2\beta \tan^2 \alpha + \tan \alpha \cos 2\beta) \qquad (11\text{-}31)$$

$$\therefore \quad \tan \alpha = \cot 2\beta = \cot\left(\frac{\pi}{2} - \alpha\right)$$

$$\therefore \quad 2\beta = \frac{\pi}{2} - \alpha$$

and

$$\beta = \frac{\pi}{4} - \frac{\alpha}{2} \qquad (11\text{-}32)$$

When $\alpha = 0$, $\beta = \pi/4$. When $\beta = 0$, $\alpha = \pi/2$.

Because $\dfrac{d^2\sigma_b}{d\beta^2} < 0$, the σ_b-bending stress reaches a maximum when $\beta = \dfrac{\pi}{4} - \dfrac{\alpha}{2}$.

Substitution of Eq. 11-32 into Eq. 11-30 results in the following value of the maximum bending stress:

$$\sigma_{b_{max}} = 3\sigma_0 \tan \alpha \cos\left(\frac{\pi}{4} - \frac{\alpha}{2}\right)\left[\tan \alpha + \tan\left(\frac{\pi}{4} - \frac{\alpha}{2}\right)\right] \qquad (11\text{-}33)$$

2. Shear stress

The shear stress τ in section x-x is calculated as

$$\tau = \frac{T}{s(1)} = \sigma_0 \cos^2 \beta \,(\tan \alpha + \tan \beta) \qquad (11\text{-}34)$$

Because the shear-stress function, Eq. 11-34, differs from the bending-stress function, Eq. 11-30, only by the factor 3 tanα, it follows that when one sets $d\tau/d\beta = 0$, here the maximum shear stress also occurs when $\tan \alpha = \cot 2\beta$, i.e., when $\beta = \dfrac{\pi}{4} - \dfrac{\alpha}{2}$, then both the bending stress and the shear stress have their maximum values:

$$\tau_{max} = \sigma_0 \cos^2\left(\frac{\pi}{4} - \frac{\alpha}{2}\right)\left[\tan \alpha + \tan\left(\frac{\pi}{4} - \frac{\alpha}{2}\right)\right] \quad (\text{kg/cm}^2) \qquad (11\text{-}35)$$

3. Normal stress σ_n

The normal stress σ_n on section x-x is

$$\sigma_n = \sigma_o \sin \beta \cos \beta (\tan \alpha + \tan \beta) \quad (C) \quad (\text{kg/cm}^2). \tag{11-36}$$

The normal stresses, which are compressive stresses (C) acting perpendicularly to the section x-x, reduce the bending tensile stresses, i.e., $\sigma_b - \sigma_n$.

Summary of general stress functions:
Bending stress: $\quad \sigma_b = \pm 3 \sigma_o \tan \alpha \cos^2 \beta (\tan \alpha + \tan \beta) \quad (11\text{-}30)$
Shear stress: $\quad \tau = \quad \sigma_o \cos^2 \beta (\tan \alpha + \tan \beta) \quad (11\text{-}34)$
Normal stress: $\quad \sigma_n = \quad \sigma_o \sin \beta \cos \beta (\tan \alpha + \tan \beta) \quad (C) \quad (11\text{-}36)$

or

$$\sigma_b = \pm 3 \sigma_n \tan \alpha \cot \beta \tag{11-30a}$$
$$\tau = \sigma_n \cot \beta \tag{11-34a}$$
$$\sigma_n = \sigma_o \sin \beta \cos \beta (\tan \alpha + \tan \beta) \quad (C) \tag{11-36a}$$

Analysis of stresses
When $\beta = 0$ (a vertical section), then

$$\sigma_b = \pm 3 \sigma_o \tan^2 \alpha \tag{11-30b}$$
$$\tau = \sigma_o \tan \alpha \tag{11-34b}$$
$$\sigma_n = 0 \tag{11-36b}$$

The normal stress $\sigma_n = 0$ because $N = R \sin \beta = 0$ when $\beta = 0$.

This study of stresses on an inclined section x-x reveals that
1. For the magnitude of the stress only the inclination (angle β) of the section x-x is of importance, but not the position of the section x-x.
2. The vertical section v-v along the face of the wall, generally, does not render the greatest stresses in stepped footings.
3. The greatest stresses occur when $\beta = \dfrac{\pi}{4} - \dfrac{\alpha}{2}$, or when $\cot 2\beta = \tan \alpha$, i.e., on an inclined section x-x, where they are larger than on a vertical section (see Eqs. 11-30, 11-34, 11-36, 11-30a, 11-34a, 11-36a, 11-30b, 11-34b, 11-36b).
4. The outcrop of the triangular parts of a stepped plain concrete footing to the outside beyond the slope line A-B are of no importance for the maximum stresses under the assumptions made at the outset of this study.
5. The most unfavorable resultant principal tensile stress is

$$\sigma_{\max} = \sqrt{(\sigma_b - \sigma_n)^2 + \tau^2} \tag{11-37}$$

Calculation of Rectangular Plain Concrete Footings

Problem

Refer to Fig. 11-15 and compute stresses in the vertical section v-v per 1 m = 100 cm run of wall.

Given: $e = 15$ cm; $d = 40$ cm; $\gamma_c = 0.0022$ kg/cm^3 = unit weight of concrete; $\sigma_{c_{\text{all}}} = 0.50$ kg/cm^2; soil pressure = $\sigma_o = 1.0$ kg/cm$^2 \sim 1$ ton/ft^2.

Auxiliary values:

$$\cot \alpha = \frac{d}{e} = \frac{40}{15} = 2.67; \quad \alpha = 20°30'$$

Inclination of the most unfavorable section from the vertical:

$$\beta = \frac{90° - \alpha}{2} = \frac{90° - 20°30'}{2} = 34°45'$$

$\sin \beta = \sin 34°45' = 0.570$
$\cos \beta = \cos 34°45' = 0.821$
$\cos^2 \beta = \cos^2 34°45' = 0.674$
$\tan \beta = \tan 34°45' = 0.693$
$\tan \alpha = \tan 20°30' = 0.374$
$\tan \alpha + \tan \beta = 0.374 + 0.693 = 1.067$

Moment:

$$M_{v-v} = \frac{(100)\sigma_o e^2}{2} - \frac{(100)\gamma_c d e^2}{2}$$

$$= \frac{(100)(1.0)(15^2)}{2} - \frac{(100)(0.0022)(40)(15^2)}{2} = 11,250 - 990$$

$$= 10,260 \text{ (kg-cm)}$$

Section modulus Z:

$$Z = \frac{(100)d^2}{6} = \frac{(100)(40^2)}{6} = 26,670 \text{ (cm}^3\text{)}$$

Bending stresses:

$$\sigma_{\substack{\max \\ \min}} = \pm \frac{M}{Z} = \pm \frac{10,260}{26,670} = \pm 0.38 \text{ kg/cm}^2 \approx \pm 0.38 \text{ ton/ft}^2 < \sigma_{c_{\text{all}}}$$

$$= 0.50 \text{ kg-cm}^2$$

The stresses for $\sigma_o = 1.0$ kg/cm^2 are
Bending stresses:

$$\sigma_b = \pm (3)(1.0)(0.374)(0.674)(1.067) = \pm 0.81 \text{ (kg/cm}^2\text{)}$$

The bending tensile stress

$$\sigma_{bt} = -0.81 \ \text{kg/cm}^2$$

is reduced by the acting normal compressive stress σ_n.

Normal stress:

$$\sigma_n = (1.0)(0.570)(0.821)(1.067) = 0.48 \ (\text{kg/cm}^2)$$

Thus there remains a tensile stress of the magnitude of $-0.33 \ \text{kg/cm}^2$ acting perpendicularly to section x-x.

Shear stress:

$$\tau = (1.0)(0.674)(1.067) = 0.72 \ (\text{kg/cm}^2)$$

The most disadvantageous principal tensile stress in this inclined section x-x is

$$\sigma_{t_{max}} = \sqrt{(0.81 - 0.48)^2 + 0.72^2} = 0.85 \ (\text{kg/cm}^2)$$

Problem

If in Fig. 11-15 a section x-x is taken at an angle of $\beta = 45°$ with the vertical v-v, what are the magnitudes of σ_b, σ_n, τ, and $\sigma_{t_{max}}$ as functions of soil contact pressure σ_o for maximum bending stress conditions? Explain. (Ans.: (0); $(0.50)\sigma_o$; $(0.50)\sigma_o$; $(0.71)\sigma_o$)

Problem

If in Fig. 11-15 $\alpha = 90°$, what are the magnitudes of σ_b, σ_n, τ, and $\sigma_{t_{max}}$ as functions of soil contact pressure σ_o for maximum bending stress conditions? Explain.

Problem

What are the maximum principal compressive stresses in the foregoing illustrative problem?

REFERENCES

1. American Concrete Institute, *Standard Building Code Requirements for Reinforced Concrete*. Detroit, Mich.: 1963, pp. 318-104.
2. H. Hayashi, *Theorie des Trägers auf elastischer Unterlage und ihre Anwendung auf den Tiefbau*. Berlin: Springer, 1921.
3. American Concrete Institute, *Standard Building Code Requirements for Reinforced Concrete*. Detroit, Mich.: 1963, pp. 318-355.
4. *Ibid*, pp. 318-353.
5. *Ibid*, pp. 318-375.

Calculation of Rectangular Plain Concrete Footings

PROBLEMS

11-1. Explain and illustrate each step in designing a plain concrete strip footing.

11-2. What is understood by "critical sections" in plain concrete strip and squared footings, and how are the critical sections used in design?

11-3. Given a wall 12 in. thick. The centric load to be transferred from the wall to the footing is 8 kip/ft. The bearing capacity of the sand soil is given as 1 ton/ft^2. Design a plain concrete footing to support the given wall and to transmit safely the load to the soil. Assuming the unit weight of the concrete as $\gamma_c = 150$ lb/ft^3, check the contact pressure on soil and report.

11-4. Given all conditions as in Prob. 11-3. However, the wall load is applied at an eccentricity of 2 in. Design a concrete footing, and check contact pressure of soil. Owing to the eccentricity, is reinforcement of footing needed or is it not? How is the footing designed for no eccentricity? In the latter case, what are the contact pressures on soil under the edges of the footing?

11-5. A rectangular footing 12 ft long and 6 ft wide transmits a single, vertical, concentrated load of $N = 100$ tons to the soil. The load is applied on the short axis of symmetry (in plan) 2 ft from the edge of the long dimension. Determine the maximum and minimum edge pressures underneath the base of the footing. Would you design this footing in plain or reinforced concrete?

11-6. Design a rectangular, plain concrete footing for a central load of $P = 200$ kips. Given: $\sigma_{all} = 1.5$ tons/ft^2; $f_c = 1125$ psi; $f_t = 80$ psi; and the shear stress of concrete as 100 psi.

chapter **12**

Calculations of Rectangular Reinforced-Concrete Footings and Mats

Relative to calculations of reinforced-concrete footings, it is pertinent to say here that the necessary effective thickness d of the footing and the steel reinforcement may be computed by means of either the method of working stress design or the ultimate strength design method. In both methods the procedure of calculating the bending moments and the shear as a measure of diagonal tension are the same.

12-1. CENTRICALLY LOADED SQUARE REINFORCED-CONCRETE FOOTINGS

In the case of a reinforced-concrete footing the critical section s-s for shear as a measure of diagonal tension is to be located at a distance $d/2$ from the base of the column (see Fig. 12-1 and consult the Building Code of the ACI). According to the ACI Code, if the slab or footing acts essentially as a wide beam, then the critical section for shear is at the section a distance d from the face of the support.

The critical section m-m for bending moment (flexure) for a reinforced-concrete footing is the same as for the plain concrete footing.

Bond stresses are to be calculated at the same critical section m-m as for moments.

The effective thickness d of a footing slab shall be taken as the distance from the centroid of its tensile reinforcement to the extreme compression fiber (Fig. 12-1). Thus to design a concentrically loaded, square, reinforced-concrete footing, one calculates first the necessary thickness d of the footing in order safely to resist diagonal tension (viz., shear). Then the thickness of the footing necessary to resist bending is calculated and the amount of reinforcing steel computed to resist bending moments. The size of the reinforcing steel bars is chosen to satisfy the bond requirements.

Calculations of Rectangular Reinforced-Concrete Footings and Mats

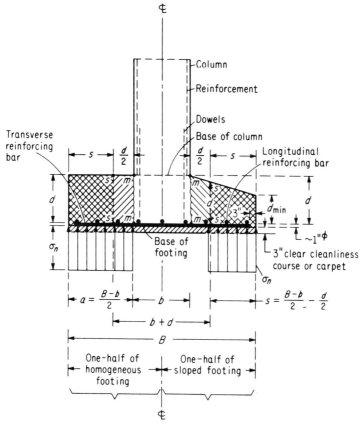

FIG. 12-1. Critical sections in square reinforced concrete footings.

One-way and two-way reinforced-concrete footings are recommended to be designed in accordance with the ACI Building Code. A reinforced square footing is usually reinforced two ways, i.e., the steel bars are placed in two layers, each at right angles to the other one.

The shear stress v is calculated as

$$v = \frac{V}{b_o d} \quad (\text{lb/ft}^2) \quad (12\text{-}1)$$

where V = total shear force
d = effective thickness (depth) of the reinforced-concrete footing
b_o = length of periphery of critical section s-s for shear

If the allowable shear stress of the concrete is given, the thickness d of the footing for shear can be calculated by Eq. 12-1.

The thickness d for bending is calculated by the internal couple method for the working stress design:

$$Bd^2 = \frac{M}{(1/2)(f_c kj)} = \frac{M}{R} \qquad (12\text{-}2)$$

and thickness d is

$$d = \sqrt{\frac{M}{RB}} \qquad (12\text{-}3)$$

where M = externally applied bending moment of the whole cantilever
$R = (1/2)(f_c kj)$ = coefficient of internal resistance in a reinforced-concrete section
f_c = allowable compressive stress of concrete, lb/ft^2
k = a decimal fraction of thickness d
j = a coefficient to d in concrete theory

In rectangular two-way reinforced-concrete footings, the steel must be computed separately in each direction.

The computation of reinforcement steel can then be performed as outlined in texts on reinforced-concrete design.

Thus, in general, to design a centrically loaded reinforced-concrete footing one proceeds as follows:

1. By means of the safe allowable soil bearing capacity calculate the necessary base contact area of the footing.
2. Determine the necessary thickness of the footing to resist diagonal shear.
3. Calculate the necessary thickness needed to resist bending.
4. Select reinforcement steel bar size to suit the bond requirement.

Equations for calculating thickness d and the steel reinforcement in the working-stress design method may be modified for use with the ultimate strength design method by replacing the allowable uniform soil pressure σ_{all} with the ultimate soil bearing capacity σ_u and R with ϕR_u, where ϕ is the capacity reduction factor, and index u to sigma and R means "ultimate." Thus $M_u = \phi R_u bd^2$ is the ultimate resisting moment, and R_u = ultimate internal resisting coefficient.

Here $\phi R_u = \phi[p f_y(1 - 0.59\, p f_y/f_c)]$ = a coefficient for concrete design, where $p = A_s/bd$ = steel ratio
A_s = area of tension reinforcement
b = width of compression face of flexural member
d = distance from extreme compression fiber to centroid of tension reinforcement

Calculations of Rectangular Reinforced-Concrete Footings and Mats

f_y = yield stress for steel, lb/in².
f'_c = ultimate compressive strength of concrete, lb/in².

By code, the coefficient ϕ is 0.90 for flexure. For diagonal tension, bond and anchorage $\phi = 0.85$.

For structures in such locations and of such proportions that the effects of wind and earthquake may be neglected the required ultimate design load capacity U is calculated as

$$U = (1.5)(D) + (1.8)(L) \tag{12-4}$$

where D = dead load
L = live load, specified in the general ACI Building Code

For structures in the design of which wind loading must be included, the design capacity U (loads, moments) is calculated as

$$U = [1.25(D + L + W)] \tag{12-5}$$

or

$$U = (0.9)(D) + (1.1)(W) \tag{12-6}$$

whichever is greater, provided that no member shall have a capacity less than required by Eq. 12-4. Here W = wind load.

For those structures in which earthquake loading must be considered, E shall be substituted for W in Eq. 12-5.

Problem

Given a 22 in. × 22 in. reinforced-concrete column carrying a vertical, centric load of $P = 400$ kips. The safe allowable soil bearing capacity is $\sigma_{all} = 3$ ton/ft² $= 6$ kip/ft². Refer to Fig. 12-2. Use a $f'_c = 2500$-psi ultimate compressive strength concrete, and consult the ACI Building Code. The reinforced-concrete specifications are given as $20{,}000 - 1{,}125 - 12$ $(= f_s - f'_c - n)$.

Required. Determine size of footing and calculate bending moment and shear necessary for the design of a square two-way reinforced unstepped concrete footing.

Solution (Working stress design method)
Size of footing:

Column load.	$P = 400.0$ k
Self-weight of footing W, estimated at 6 percent of column load $[W = (0.06)(400) = 24.0$ k$]$	$W = 24.0$ k
Total	$N = 424.0$ k

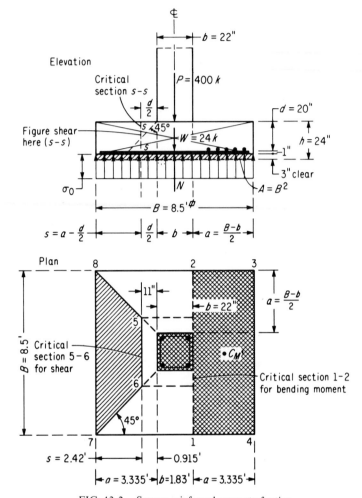

FIG. 12-2. Square reinforced concrete footing.

Necessary area A_{nec}:

$$A_{nec} = N/\sigma_{all} = 424.0/6.0 = 70.67 \text{ (ft}^2\text{)}$$

$$\text{Try } A = 8.5 \text{ ft} \times 8.5 \text{ ft} = 72.25 \text{ ft}^2$$

$$\therefore \quad B = 8.5 \text{ ft} = 102 \text{ in.}$$

Net upward soil pressure σ_{net}:

For calculating strength of footing, neglect its self-weight.

$$\sigma_{net} = P/A = 400.0/72.25 = 5.543 \text{ (k/ft}^2\text{)} = 5543 \text{ lb/ft}^2 = 38.5 \text{ (lb/in.}^2\text{)}$$

Calculations of Rectangular Reinforced-Concrete Footings and Mats

Shear force, shear stress and thickness d of footing:

The thickness d of the footing is determined by shear stress v_c in slabs and footings as a measure of diagonal tension:

$$v_c = \frac{V}{A_v} = \frac{\sigma_{net}[B^2 - (b+d)^2]}{4(b+d)d} \tag{12-7}$$

say in lb/in²., where A_v = critical area for shear, and all other symbols are as before.

If for a 2,500-psi ultimate strength concrete the shear stress $v_c = 2\sqrt{2500} = 100$ psi, then

$$v_c(4bd + 4d^2) = (38.5)[(102)^2 - (22+d)^2] \tag{12-8}$$

With $v_c = 100$ psi, $b = 22$ in., and $B = 102$ in., the necessary thickness d of the footing calculates as $\approx \underline{20 \text{ in.}}$

Bending moment M

$$M = \sigma_{net} B \left(\frac{B-b}{2}\right)\left(\frac{1}{2}\frac{B-b}{2}\right) = \frac{1}{8}\sigma_{net} B(B-b)^2 \tag{12-9}$$

Note again that for computing the strength of the footing the weight of the footing is neglected, hence the net upward soil contact pressure σ_{net} (pressure on soil from weight of footing neglected).

With $\sigma_{net} = 38.5$ lb/in²., $B = 102$ in., and $b = 22$ in., the magnitude of the whole cantilever bending moment M is

$$M = (0.125)(38.5)(102)(102 - 22)^2 = \underline{3{,}146{,}000 \text{ (in.-lb)}}$$

Necessary thickness d

$$d_{nec} = \sqrt{\frac{M}{RB}} = \sqrt{\frac{3{,}146{,}600}{(195)(102)}} = \underline{12.5 \text{ (in.)}}$$

Here $R = (1/2)f_c kj = f_s pj$ = coefficient of resistance, calculated or obtained from beam chart if f_s, f_c, n, and p are known from given specifications, usually written as f_s-f_c-n, or in numerical values as in this example, 20,000-1,125-12
$f_s = 20{,}000$ psi = allowable tensile stress in steel
$f_c = 1{,}125$ psi = allowable compressive stress in concrete
$k = \dfrac{1}{1 + \dfrac{f_s}{nf_c}} = \dfrac{1}{1 + \dfrac{20{,}000}{(12)(1{,}125)}} = 0.403$ = a decimal fraction of effective thickness d

kd = height of compressed area of concrete above neutral axis

$j = 1 - \dfrac{k}{3} = 1 - \dfrac{0.403}{3} = 0.866$ = ratio of distance between centroid of compression and centroid of tension to the depth = effective thickness d of slab (beam); in this problem, $j = 0.866$

jd = the lever arm of the internal couple (in this example $jd = 0.866d$)

$n = \dfrac{E_s}{E_c} = \dfrac{(3)(10^7)}{(2.5)(10^6)} = 12$ = ratio of modulus of elasticity of steel to that of concrete

$p = \dfrac{A_s}{Bd} = \dfrac{f_c}{2f_s}\, k = \dfrac{1}{2}\, \dfrac{1}{\dfrac{f_s}{f_c}\left(\dfrac{f_s}{f_c}+1\right)}$ = ratio of area of tension reinforcement to effective area of concrete, or simply the steel ratio

Effective area of concrete is the area of a section which lies between the centroid of the tension reinforcement and the compression face of the flexural member.

The k, j, and R values as functions of f_s, f_c, n, and p can be conveniently scaled off from specially prepared reinforced-concrete beam charts.

The total thickness h of the footing is governed by diagonal tension. Hence,

$$h = d_{nec} + \phi_{steel} + \text{cover}$$
$$= 20.0 \text{ in.} + 1.00 \text{ in.} + 3.00 \text{ in.} = 24.0 \text{ in.}$$

Make footing 24.0 in. thick.

12-2. CENTRICALLY LOADED RECTANGULAR REINFORCED-CONCRETE FOOTINGS

The moment with respect to the critical section m-m governs the thickness of the rectangular footing and the reinforcement steel parallel to the long sides. The amount of steel reinforcement bars parallel to the short side of the rectangle is determined from the bending moment on the critical section n-n (Fig. 12-3).

The critical section for shear in a rectangular reinforced-concrete footing is indicated in Fig. 12-3 by the solid line 1-2. It is located at a distance $d/2$ from the face of the column just as was described before in discussing plain concrete footings. Here d means the effective thickness of the footing. This critical section 1-2 should resist the shear force acting upward from the net soil reaction on the hatched-in trapezoidal area 123456. If the

Calculations of Rectangular Reinforced-Concrete Footings and Mats

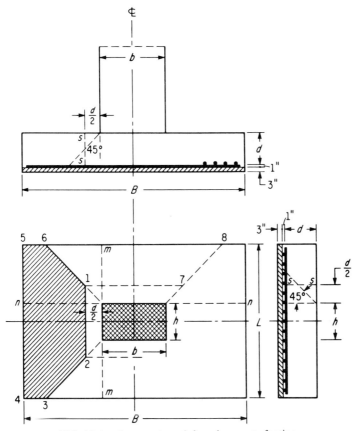

FIG. 12-3. Rectangular reinforced concrete footing.

cross section of the column is a rectangle (Fig. 12-3), then the shear stress v is calculated as

$$v = \frac{\sigma_{net}[BL - (b+d)(h+d)]}{2[(b+d) + (h+d)]d} \tag{12-10}$$

One would usually refrain from designing long, stretched-out rectangular footings because of the large percentage of steel required to make the footing stiff enough against lifting up.

Problem

Given: A reinforced-concrete column 28 in. square is required to transmit centrically a load of $P = 700$ kips to a rectangular reinforced-concrete footing. Its short dimension is 9 ft (Fig. 12-4). The allowable soil bearing capacity is 6 kip/ft². Use a 3,000 lb/ft² ultimate compressive strength

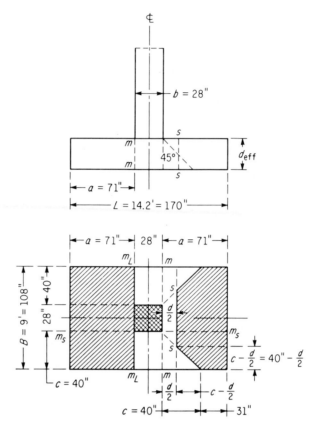

FIG. 12-4. Critical sections in a rectangular reinforced concrete footing.

concrete. Use net soil contact pressure from column load $\sigma_{net} = 5.5$ kip/ft^2. The allowable shear strength v_c of concrete is $v_c = 110$ lb/in^2. Consult and use the latest edition of the ACI Building Code. Use resistance $R = 261$ for calculating thickness of reinforced-concrete footing.

Required. Compute moments and shear forces for both directions of footing.

Solution
Length of footing:

$$L = \frac{P}{\sigma_{net} B} = \frac{700}{(5.5)(9.0)} = 14.14 \ (\text{ft})$$

Try $L = 14.2$ ft $\approx \underline{170 \text{ in.}}$

Calculations of Rectangular Reinforced-Concrete Footings and Mats

Bending moment M_L in the longitudinal direction with respect to section $m_L - m_L$:

$$M_L = \sigma_{net} \frac{a^2}{2} B = \left(\frac{5500}{144}\right) \left(\frac{71^2}{2}\right)(108) = \underline{10,398,589 \text{ (in.-lb)}}$$

Bending moment M_S in the lateral (short) direction with respect to section m_s-m_s:

$$M_S = \sigma_{net} \frac{c^2}{2} L = (38.2)\frac{(40^2)}{2}(170) = \underline{5,195,200 \text{ (in.-lb)}}$$

Effective thickness d of footing for moment M_L:

$$d = \sqrt{\frac{M_L}{RB}} = \sqrt{\frac{10,398,589}{(261)(108)}} = \underline{19.2 \text{ (in.)}}$$

Thickness of footing d for shear force V with respect to critical section s-s (here s-$s = b_o$):

$$v_c = \frac{V}{b_o d} = \sigma_{net} \frac{\left[\left(a - \frac{d}{2}\right)B - 2\frac{\left(c - \frac{d}{2}\right)^2}{2}\right]}{B - 2\left(c - \frac{d}{2}\right)d}$$

rendering the following quadratic equation:

$$\left(v_c + \frac{\sigma_{net}}{4}\right)d^2 + (v_c B - 2v_c c + \sigma_{net}\frac{B}{2} - \sigma_{net}c)d - \sigma_{net}(aB - c^2) = 0$$

$$\left(110 + \frac{38.2}{4}\right)d^2 + \left[(110)(108) - (2)(110)(40) + \left(\frac{38.2}{2}\right)(108) - (38.2)(40)\right]d$$

$$- (38.2)[(71)(108) - 40^2] = 0$$

$$d^2 + (30.2)d - 1940.5 = 0$$

$$d = 31.47 \text{ (in.)} \approx \underline{32.0 \text{ in.}}$$

Hence the diagonal tension governs the design of this rectangular footing. The steel reinforcement can now be computed.

Problem

Given a centrically loaded rectangular footing 18 ft × 11 ft in plan. Its total thickness is given as $d = 32$ in. The centrical load $P = 500k$ is transmitted to the footing by means of a square, reinforced-concrete column

24 in. × 24 in. in cross section. Use $f'_c = 3000$ lb/in²; allowable soil pressure $\sigma_{all} = 3.0$ k/ft². Analyze this footing.

Hint. Net soil pressure:

$$\sigma_{net} = \frac{500}{(11.0)(18.0)} = 2.525 \text{ (k/ft}^2) = \underline{2525 \text{ lb/ft}^2}$$

Weight W of footing:

$$W = \frac{d}{12}\gamma_c = (32/12)(150) = 400 \text{ (lb/ft}^2)$$

Total soil contact pressure:

$$\sigma_o = 2525 + 400 = 2925 \text{ (lb/ft}^2) < 3000 \text{ (lb/ft}^2)$$

Here $\gamma_c = 150$ lb/ft³ = unit weight of concrete.

For the short projection:

$$d = 32.0 \text{ in.} - 3.0 \text{ in. (clear)} - 1.0 \text{ in. (steel)} - 0.44 = \underline{27.56 \text{ (in.)}}$$

Required. Calculate V and the required d for shear; calculate M_L and V_L, and M_S.

12-3. CENTRICALLY LOADED REINFORCED-CONCRETE STRIP FOOTINGS

If by necessity the cantilever projections happen to become long, one rather resorts to a reinforced-concrete strip footing as being more economical than a bulky and heavy plain concrete strip footing.

For heavy walls, the design is quite similar to that of column footings, but simpler because the bending is in only one direction. For light walls, the design is frequently based on rather arbitrary minimum sizes. In such cases a plain concrete footing is often the appropriate design.

The critical sections for moments and shear per one unit of length of run of the strip foundation are the same as was explained for the square, reinforced-concrete isolated footing. The effective thickness d of reinforced-concrete strip footings, too, is determined according to the same rules described for plain concrete strip footings. The strip footing should be designed thick enough to avoid the need for diagonal tension reinforcement steel. The main or transverse reinforcement steel bars are placed perpendicular to the run of wall and spaced uniformly. Besides, in order to bridge the strip foundation over random spots of weak soil along the run of the wall, as well as to control temperature cracking, all strip footings, besides trans-

Calculations of Rectangular Reinforced-Concrete Footings and Mats

verse cross bars, should be provided with longitudinal reinforcing steel parallel to the wall. This steel is placed at right angles to the cross bars.

Problem

Given a reinforced-concrete strip footing (Fig. 12-5) carrying a centrical, uniformly distributed linear load of $p = 50$ kip/ft. The safe, allowable soil bearing capacity is $\sigma_{\text{all}} = 6$ k/ft^2 = 3 ton/ft^2.

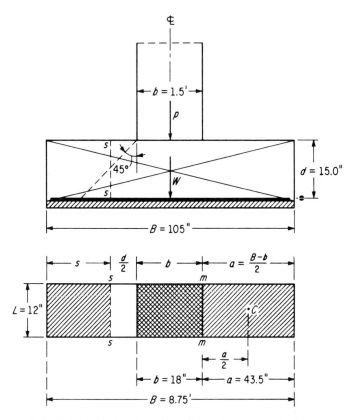

FIG. 12-5. Critical sections in a reinforced-concrete strip footing.

Use a 2,500-psi ultimate compressive strength concrete and the following specifications: $f_s - f_c - n = 20{,}000 - 1{,}125 - 12$, and $v = 100$ lb/in^2.

Required: Proportion this footing.

Solution

Width B of footing:

With estimated weight W of the footing at 4 percent of the superimposed load p, the required width B is:

$$B = \frac{(1.04)(p)}{\sigma_{all}} = \frac{(1.04)(50)}{6} = \underline{8.66 \text{ (ft)}}$$

Try $B = 8.75$ ft $= 8$ ft 9 in.
Area $= BL = (8.75)(1.0) = \underline{8.75 \text{ ft}^2}$

Net upward soil pressure σ_{net}:

$$\sigma_{net} = p/A = 50/8.75 = 5.714 \text{ (k/ft}^2) \approx 5{,}715 \text{ lb/ft}^2 = \underline{39.7 \text{ lb/in}^2}$$

Required thickness d:

$$d = \frac{V}{vL} = \frac{\sigma_{net}\left(a - \frac{d}{2}\right)L}{(100)(12)} = \frac{(39.7)\left(43.5 - \frac{d}{2}\right)(12)}{(100)(12)}$$

$$(239.7)d = 3{,}453.9$$

$$\therefore \quad d = \frac{3{,}453.9}{239.7} = 14.4 \text{ (in.)} \approx \underline{14.5 \text{ in.}}$$

Bending moment M:

$$d = \sqrt{\frac{M}{RL}}$$

Internal resisting moment M:

$$M = (1/2) f_c L(kd)(jd) = (1/2)(1{,}125)(12)(0.403)(d)(0.866)(d)$$
$$= (2{,}351) d^2 \text{ (in.-lb)}$$

External bending moment M_{m-m}:

$$M_{m-m} = \sigma_{net} aL(a/2) = (39.7)(43.5)^2(1/2)(12) = \underline{450{,}734 \text{ (in.-lb)}}$$

$$M = M_{m-m}$$

$$(2{,}351)d^2 = 450{,}734$$

$$d = \sqrt{\frac{M_{m-m}}{RL}} = \sqrt{\frac{450{,}734}{2{,}351}} = 13.84 \text{ (in.)} \approx 14.5 \text{ in.}$$

Hence the diagonal tension governs the thickness d of the footing. Make $\underline{d = 15.0 \text{ in.}}$ Recheck weight W of footing, net soil pressure σ_{net}, width B and thickness d, and design the footing for steel.

Calculations of Rectangular Reinforced-Concrete Footings and Mats

12-4. ECCENTRICALLY LOADED REINFORCED-CONCRETE FOOTINGS

Frequently footings are subjected to moments from a column or wall.

In cases where the footings are eccentrically loaded and/or the members being supported transmit a moment to the footing, the ACI Building Code says that proper allowance should be made for any variation that may exist in the intensities of reaction and applied load with the magnitude of the applied load and the amount of its actual or virtual eccentricity.

Whenever possible and feasible, eccentrically loaded footings should be designed so as to achieve a uniformly distributed soil contact pressure. This may be attained by offsetting the footing with respect to the column and/or wall. One draws the position of the total resultant vertical eccentric load (with respect to the column) and then designs the footing around it.

An example of an eccentrically loaded footing is that of an earth retaining wall. Here one usually adjusts its base so as to make the resultant load fall between the center and the third point of the base of the footing. This results in a trapezoidal soil pressure-distribution diagram. This modifies the magnitude of the moment and shear force, but does not affect the routine procedure in designing the footing.

If a footing is subjected to a vertical load P, a horizontal load H and a static moment M_1 (Fig. 12-6), then the equivalent eccentricity e of the load is

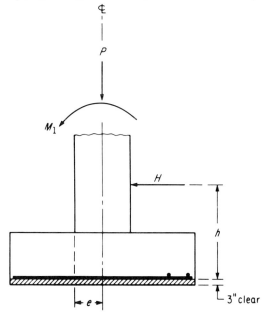

FIG. 12-6. Equivalent eccentricity e of load.

$$e = \frac{M_1 + Hh}{P} \qquad (12\text{-}11)$$

Figure 12-7 shows a reinforced-concrete footing eccentrically loaded by a reinforced concrete column load P and a clockwise moment $+M$. The weight of the footing is W. In this case the soil contact pressure is not uniform but varies linearly from a maximum at the side nearer to the center of gravity of the load to a minimum or zero at the opposite side. However, it is preferable to design the footing so that there is a compressive contact stress throughout the base of the footing under ordinary working conditions.

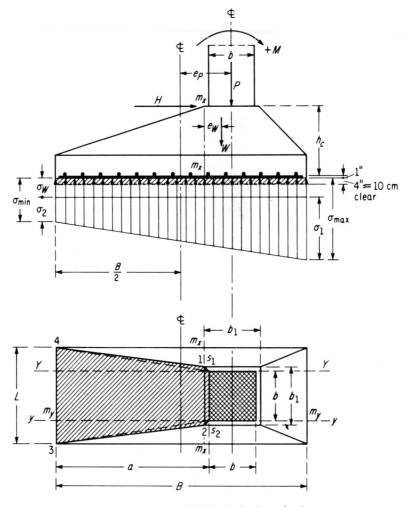

FIG. 12-7. Eccentrically loaded column footing.

Calculations of Rectangular Reinforced-Concrete Footings and Mats

The resultant eccentricity e brought about by the column load, weight of footing, the externally applied moment $+M$ and the moment $+M = Hh_c$ caused by the horizontal load H with its moment arm h_c (where h_c is the thickness of the footing at the column) is calculated as

$$e = \frac{M + Pe_p + We_W + Hh_c}{P + W} = \frac{\Sigma M}{\Sigma N} \quad (12\text{-}12)$$

where e_p = eccentricity of column load
e_W = eccentricity of weight of footing
Other symbols are shown in Fig. 12-7.
For a footing of uniform thickness $We_W = 0$.

Maximum and minimum edge pressures on soil:

$$\sigma_{max} = \left(\frac{P+W}{A}\right)\left(1 + \frac{6e}{B}\right) \quad (12\text{-}13)$$

$$\sigma_{min} = \left(\frac{P+W}{A}\right)\left(1 - \frac{6e}{B}\right) \quad (12\text{-}14)$$

where $A = BL$ = contact area of base of footing
B = length of long side of footing
L = length of short side of footing

The σ_{min} stress should be greater than or equal to zero, i.e., σ_{min} should be a compressive stress ($+$) but not a tensile stress ($-$).

Bending moments with respect to critical section x-x:
Because the moments are used for strength calculations of the footing, they are calculated from the column loads only, i.e., based on net soil pressure (no weight of footing W considered):

$$M_{x-x} = \sigma_2 \frac{a^2 L}{2} + \frac{\sigma_1 - \sigma_2}{B} aLa\frac{a}{3} = \frac{a^2 L}{2}\left[\sigma_2 + \frac{a}{2B}(\sigma_1 - \sigma_2)\right]$$

$$= \frac{a^2 L}{2}\left[(\sigma_{min} - \sigma_W) + \frac{a}{3B}(\sigma_{max} - \sigma_{min})\right] \quad (12\text{-}15)$$

where σ_W = stress from weight of footing.

Bending moments with respect to critical section y-y:

$$M_{y-y} = \frac{P}{LB}\left(\frac{L-b}{2}\right)B\left(\frac{L-b}{2\cdot 2}\right) = \frac{P}{8L}(L-b)^2 \quad (12\text{-}16)$$

Vertical shear force V:

For this asymmetrical footing the vertical shear force from soil reaction may be calculated approximately by the largest (hatched) trapezoidal area of the form (area 1234).

Referring to Fig. 12-8, the stereometric volume of wedge 12341'1 is:

$$V_s = (1/6)(2b + L)\sigma_1 a$$

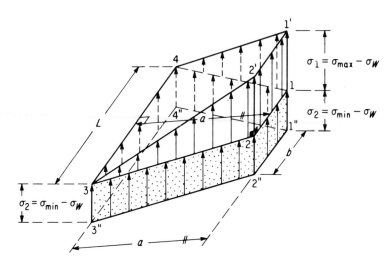

Stereometric volume of wedge 12341'1: $V = \frac{1}{6}(2b + L)\sigma_1 a$

FIG. 12-8. Volume of an obtuse stress wedge.

The magnitude of the total shear force V may be imagined as consisting of two stress volumes, namely:

$V_{1\ \text{stressed}} = (\sigma_2) \times$ (trapezoidal area 1234) and a sharp stressed wedge 122'341'1:

$$V_{2\ \text{stressed}} = (1/6)(2b + L)\sigma_1 a \quad \text{(Fig. 12-8)}$$

Thus

$$V = V_{1\ \text{stressed}} + V_{2\ \text{stressed}} = \sigma_2\left(\frac{L+b}{2}\right)a + (1/6)\sigma_1(2b+L)a$$

$$= (1/2)a[(1/3)\sigma_1(2b+L) + \sigma_2(L+b)] \quad (12\text{-}17)$$

where $\sigma_2 = \sigma_{\min} - \sigma_W$

$\sigma_1 = \sigma_{\max} - \sigma_W$

$\sigma_W = \dfrac{W}{BL} =$ average contact pressure on soil from self-weight of the footing

Calculations of Rectangular Reinforced-Concrete Footings and Mats

With the bending moments and shear force known the effective thickness h_c of the footing can now be calculated.

If the thickness d of the footing is constant, then the shear force V must be calculated according to the ACI Building Code specifications at a critical section at $d/2$ from the face of the column (or whatever the Code requires).

12-5. ECCENTRICALLY LOADED REINFORCED-CONCRETE WALL FOOTINGS

The bending moments and the shear force for an eccentrically loaded wall footing of reinforced concrete may be calculated as set forth:

Resultant bending moment M_c relative to center line of base of footing:

$$M_c = M + Hh_c - Pe$$

See Fig. 12-9.

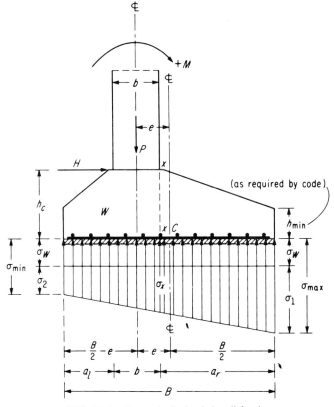

FIG. 12-9. Eccentrically loaded wall footing.

Edge pressures:

$$\sigma_{max} = \frac{P}{B(1)} + \frac{M}{Z} = \frac{1}{B}\left(P + \frac{6M_c}{B}\right) \tag{12-18}$$

$$\sigma_{min} = \frac{1}{B}\left(P - \frac{6M_c}{B}\right) \tag{12-19}$$

Moment M_x at critical section x-x:

$$M_x = \sigma_x a_r \frac{a_r}{2} + \frac{\sigma_{max} - \sigma_x}{2} a_r \frac{2a_r}{3} - \sigma_w \frac{a_r^2}{2}$$

$$= \frac{(a_r)^2}{6}(\sigma_x + 2\sigma_{max} - 3\sigma_w) \tag{12-20}$$

Shear force V at section x-x:

$$V = \frac{\sigma_x + \sigma_{max}}{2} a_r (1.0) - \sigma_w a_r = \frac{a_r}{2}(\sigma_x + \sigma_{max} - 2\sigma_w) \tag{12-21}$$

If the thickness d of the footing is constant, then the critical section for shear force must comply with the code requirements for such footing design.

12-6. FOOTINGS ON PILES

If the soil is weak and unable to support spread footings and mats, one resorts to *piled footings*. A piled footing is one whose load is transferred to firm soil or rock of sufficient bearing capacity at some distance below ground surface by means of piles.

The design of pile-supported footings does not differ much from that of footings resting directly on soil, except for some special features such as heavy shear and short grip lengths. These factors must be coped with. Reinforced-concrete footings are used with timber or reinforced concrete or steel piles.

Pile footings are designed to transfer large loads through the pile cap to individual piles. The piles, in their turn, transfer their loads to the soil (a) by means of mantle friction (skin friction) between the soil and the mantle surface of the piles, and (b) by point bearing (end bearing) of the piles.

Otherwise, the design of piled footings proceeds along lines similar to those outlined above for square and rectangular footings, except that calculations for moments and shear forces may be based on the assumption that the reaction from any pile is concentrated at the center of the pile.

Calculations of Rectangular Reinforced-Concrete Footings and Mats

Further, the load from the column is assumed to be uniformly distributed to all piles in the pile group or cluster. Therefore piles must be arranged symmetrically about the axis of the column.

Thus the net load per pile (column load divided by the number of piles) is assumed to act as an upward load on the footing, concentrated at the center of the pile (Fig. 12-10). In the case of eccentricity of the load,

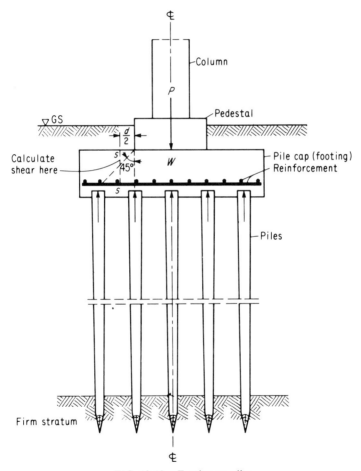

FIG. 12-10. Footing on piles.

piles should be spaced so that the load is distributed evenly among all piles in the cluster. This is not always possible to attain, but the variation in pile loads should be kept to a minimum. The design would be very unsatisfactory if any of the piles were subjected to tensile forces by the eccentricity-inducing moment.

The size of the pile cap or footing depends upon
1. The magnitude of the load to be transferred
2. The soil bearing capacity
3. The bearing capacity of the piles, singly and in group
4. The required number of piles and their spacing. The number of piles depends, of course, upon the external loads (vertical and horizontal) applied and the bearing capacity of the pile.

Here the critical section for bending moment and bond are the same as for footings directly on soil or rock. The critical section for shear is to be taken at a distance of $d/2$ from the face of the column of the pedestal, as the case may be.

The spacing of piles depends also on the soil conditions, common practice and sound engineering judgment. Friction piles are spaced not less than 3 ft apart for concrete piles, and not less than 2.5 ft for timber and steel H-piles. End-bearing piles are sometimes driven or installed at closer distances than friction piles.

For spacing of eccentrically loaded piled footings the reader is referred to the section on pile groups in this book.

The reinforcement steel is placed in two layers and in mutually perpendicular directions as for square and rectangular footings.

Minimum Clearances

The upper ends of the piles must be securely embedded and anchored in the footing. Many building codes require, and it is commonly practiced,

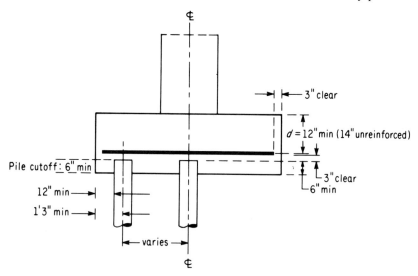

FIG. 12-11. Minimum clearances.

Calculations of Rectangular Reinforced-Concrete Footings and Mats

that the bottom of the footing should be located not less than 6 in. (15 cm) below the top of the piles. It is said that the pile cutoff is 6 in. (Fig. 12-11). This 6-in. concrete is to be neglected in calculating the strength of the footing.

The distance from the center of the outside pile to the edge of the footing is designed to be not less than 1.5 ft.

The steel reinforcement is located at a distance of 3 in. clear above the tops of the piles.

Anchoring of Piles into Pile Caps

To prevent piles from loosening because of shrinkage of concrete, as well as from lifting up of the cap footing by possible upward-acting hydrostatic

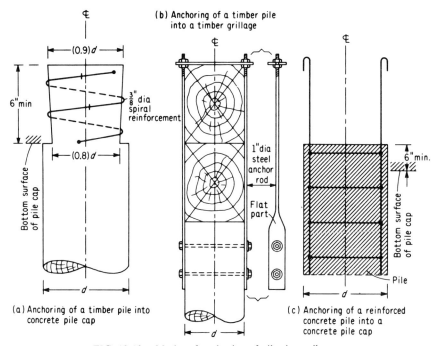

FIG. 12-12. Modes of anchoring of piles into pile caps.

pressure, the top of the piles must be anchored into the concrete pile cap (Fig. 12-12a). The anchorage also facilitates the transmittal of tensile forces in the pile if it can be subjected to tensile forces.

If the piles are connected with their tops to a timber grillage, the connecting elements may be steel anchor rods or flat straps, for example, such as shown in Fig. 12-12b.

Reinforced-concrete piles are anchored into the reinforced-concrete pile cap by means of hook bars, as shown in Fig. 12-12c. To get the necessary length for anchorage the top of the concrete pile must be stripped off, leaving the reinforcement for bending the hooks.

Two or Multiple Column Footings on Piles

If a piled footing supports two or multiple columns, the footing is calculated as a beam on two (column) supports loaded by concentrated pile

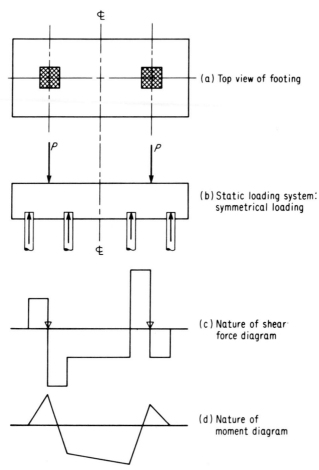

FIG. 12-13. Combined piled footing.

reactions. By means of the shear force and moment diagrams the effective thickness d of the footing can now be calculated (refer to Fig. 12-13).

Problem

Given a square, 20 in. × 20 in. reinforced-concrete column transmitting to its reinforced-concrete footing (or pile cap) on wooden piles the following

Calculations of Rectangular Reinforced-Concrete Footings and Mats 405

centrical loads and externally applied moments about the y-axis, as shown in Fig. 12-14.

Axial loads:

$$\text{Dead load} \dots \dots \dots \dots \dots \dots \dots \quad D = 140 \text{ kips}$$
$$\text{Live load} \dots \dots \dots \dots \dots \dots \dots \quad L = 60 \text{ kips}$$

$$\text{Total working load} \dots \dots \quad P = D + L = \underline{200 \text{ kips}}$$

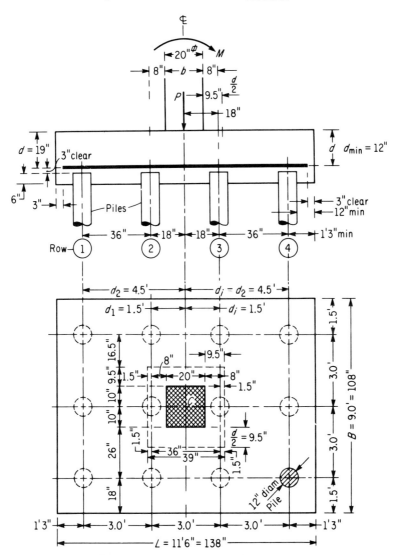

FIG. 12-14. Eccentrically loaded footing on piles.

External moments:

Dead load moment .... $M_D = 80$ ft-kips
Live load moment .... $M_L = 30$ ft-kips

Total working moment ... $M = M_D + M_L = 110$ ft-kips

The average diameters of the pile butts are given as 12 in. The ultimate bearing capacity of the piles is $P_u = 50.0$ kips for each pile. To design the pile cap,

1. Use an $f_c' = 3{,}000$ lb/in.² ultimate compressive strength concrete
2. Use a steel whose yield stress is $f_y = 40{,}000$ lb/in.²
3. Use a working load of 25.0 kips per pile.

Required. Design the pile cap.

Solution

Pile arrangement

After several sets of trial and adjustment, find that 12 piles are needed to carry the given loads and moments. The group arrangement of the 12-pile pattern (spacing of piles) in three longitudinal rows and four transverse rows is sketched in Fig. 12-14.

Load P_i on any pile:

$$P_i = \frac{P}{n} \pm \frac{Md_i}{\Sigma(d_y^2)} \tag{12-22}$$

where $n =$ number of piles in group
$d_i =$ distance of centroid of the farthest-spaced (i_{th}) pile from y-axis (here $d_i = 4.5$ ft)
$\Sigma(d_y^2) =$ the sum of distances squared of centroids of piles from y-axis

From Fig. 12-14, compute $\Sigma(d_y^2)$:

$$\Sigma(d_y^2) = (2)(n_1)(d_1^2) + (2)(n_2)(d_2^2)$$
$$= (2)(3)(1.5^2) + (2)(3)(4.5^2) = \underline{135.0 \ (ft^2)}$$

where $n_1 = 3 =$ number of piles in rows 2 and 3
$n_2 = 3 =$ number of piles in rows 1 and 4

Maximum load on a pile (in rows 1 and 4):

$$P_{i_{max}} = \frac{200}{12} + \frac{(110)(4.5)}{135.0} = 16.67 + 3.67 = 20.34 \ (kips)$$

Weight W of footing:

If the working load per pile is given as $P_{all} = 25.0$ kips, then the apportioned weight W_1 of footing per pile is

Calculations of Rectangular Reinforced-Concrete Footings and Mats

$$W_1 = P_{\text{all}} - P_{i_{\text{max}}} = 25.0 - 20.34 = 4.66 \text{ (kips per pile)}$$

The total weight W of the footing is

$$W = nW_1 = (12)(4.66) = 55.92 \text{ (kips)} \approx \underline{56.0 \text{ kips}}$$

Ultimate apportioned weight W_{1u} of footing per pile:

$$W_{1u} = \eta_D W_1 = (1.5)(4.66) = 6.99 \approx \underline{7.0 \text{ (kips/pile)}}$$

where $\eta_D = 1.5 = $ a load factor for dead weight in ultimate strength design as per the ACI Building Code.

Ultimate load P_u:

$$P_u = \eta_D D + \eta_L L = (1.5)(140) + (1.8)(60) = \underline{318.0 \text{ (kips)}}$$

where $\eta_D = 1.5 = $ a dead-load factor
$\eta_L = 1.8 = $ a live-load factor

Ultimate moment M_u:

$$M_u = \eta_D M_D + \eta_L M_L = (1.5)(80) + (1.8)(30) = \underline{174.0 \text{ (ft-kips)}}$$

Ultimate loads P_{ui} on piles without weight of caps:

By Eq. 12-22, the load on each pile in a row is:

$$P_{ui} = \frac{P_u}{n} \pm \frac{(M_u)(d_i)}{\Sigma(d_y^2)}$$

$$= \frac{318.0}{12} \pm \frac{(174.0)(d_i)}{135.0} = 26.5 \pm (1.288)(d_i)$$

In row 1: $P_{u1} = 26.5 - (1.288)(4.5) = 26.5 - 5.8 = 20.7$ (kips/pile)

In row 2: $P_{u2} = 26.5 - (1.288)(1.5) = 26.5 - 1.9 = 24.6$ (kips/pile)

In row 3: $P_{u3} = 26.5 + (1.288)(1.5) = 26.5 + 1.9 = 28.4$ (kips/pile)

In row 4: $P_{u4} = 26.5 + (1.288)(4.5) = 26.5 + 5.8 = 32.3$ (kips/pile)

Thus the most heavily loaded piles are those located in row 4.

Maximum ultimate load $P_{u_{\text{max}}}$ on pile, including the ultimate weight W_{1u} of the cap per pile:

$$P_{u_{\text{max}}} = P_{u4} + W_n = 32.3 + 7.0 = 39.3 \text{ (kips/pile)} < P_u = 50.0 \text{ kips/pile}$$

Hence, under the given loading conditions, in this problem failure of piles embedded in soil *should not be anticipated.*

Checking Pile Cap for Shear by Pile Punch

The next step in the design of this foundation system is to check whether the thickness d of the pile cap or footing is adequate against punching shear around the most heavily loaded piles, say a pile in row 4.

Assuming that after several trials the approximate thickness d of the footing was found to be $d = 19$ in., then the maximum ultimate shear stress v_u around the heavily loaded piles (Fig. 12-15) is calculated as

$$v_u = \frac{V_u}{b_o d} = \frac{P_{u4}}{\pi(\delta + d)d} \qquad (12\text{-}23)$$

$$= \frac{32{,}300}{(3.14)(12.0 + 19.0)(19.0)} = 17.5 \ (\text{lb/in.}^2)$$

where $b_o = \pi(\delta + d)$ = circular shear perimeter
  $\delta = 12$ in. = diameter of top of pile
  $d/2 = 19/2 = 9.5$ in. = distance of critical shear area ($b_o d$) from perimeter of pile

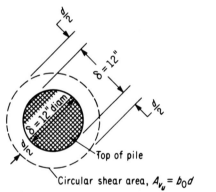

FIG. 12-15. Circular shear perimeter.

The allowable shear strength of concrete is given by the ACI Building Code as

$$v_{u_{\text{all}}} = 4\phi\sqrt{f'_c} = (4)(0.85)\sqrt{3000} = 186 \ (\text{lb/in.}^2) > v_u = 17.5 \ (\text{lb/in.}^2)$$

Thus the thickness of the pile cap is adequate against shear by a punch of the pile through the concrete cap.

Checking Pile Cap for Slab Shear

The critical section for shear may be assumed here to be spaced (Fig. 12-16) at a distance $d/2$ from the face of the column: $d/2 = 19.0/2 = 9.5$ (in.). This means that the middle piles in rows 2 and 3 are 1.5 in. inside the critical section. Therefore these two piles contribute only a fraction of their loads

Calculations of Rectangular Reinforced-Concrete Footings and Mats

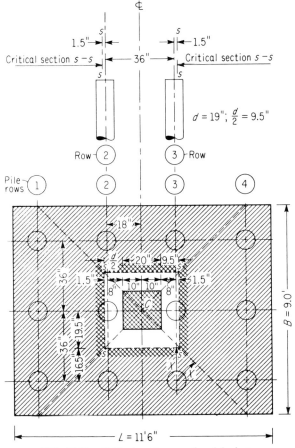

FIG. 12-16. Critical sections s-s for slab shear.

to the shear force V_u. The rest of the 10 piles contribute a full share to the shear force at the critical shear section.

The fractional amount of pile reactions contributing to the shear at the critical section in this problem may be assessed empirically as follows.

If the center of the pile is situated 6 in. and more outside the critical section s-s (Fig. 12-17), the pile contributes its full share of reaction to the shear force. If the center of the pile is situated 6 in. inside the critical section, it contributes no reactive pile load to the critical section s-s.

The contribution V_X a pile reaction makes toward the shear may be calculated by the following empirical equation:

$$V_X = \frac{P_{ui}}{2}\left(1 + \frac{X}{6}\right) \qquad (12\text{-}24)$$

say in pounds.

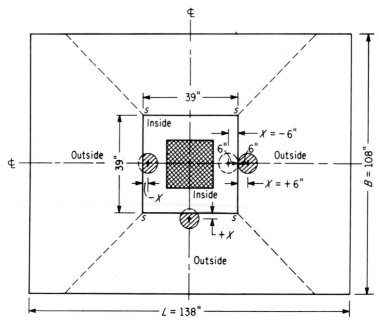

'FIG. 12-17. Contribution of piles to shear force.

When the pile is outside the critical section s-s, use $+X$; when inside, use $-X$. For example,

When $X = +6$ in., $V_X = P_{ui}$ (full reaction)
When $X = -6$ in., $V_X = 0$
When $X = 0$, $V_X = P_{ui}/2$
When $X > +6$, $V_X = P_{ui}$ (by logical contents of the matter)
When $X < -6$, $V_X = 0$ (by logical contents of the matter)

Thus the empirical equation 12-24 is valid in the interval between $X = -6$ in. and $X = +6$ in. Hence the six inches here mean merely a pile of 12-in. diameter whose radius is 6 in. With $X = -1.5$ in., $(1 - X/6) = 1 - 1.5/6 = 1 - 0.25 = 0.75$.

Ultimate shear force V_u:
The ultimate shear force V_u can now be calculated:

$$V_u = 3P_{u1} + 2P_{u2} + (1)\frac{P_{u2}}{2}\left(1 - \frac{1.5}{6}\right) + (1)\frac{P_{u3}}{2}\left(1 - \frac{1.5}{6}\right)$$
$$+ 2P_{u3} + 3P_{u4}$$
$$= (3)(20.7) + (2)(24.6) + (24.6/2)(0.75) + (28.4/2)(0.75)$$
$$+ (2)(28.4) + (3)(32.4) = \underline{285.18 \text{ (kips)}}$$

The shear perimeter b_o:

$$b_o = (4)[36 \text{ in.} + (2)(1.5 \text{ in.})] = (4)(39) = \underline{156 \text{ in.}}$$

The shear stress v_{cu} in concrete:

$$v_{cu} = \frac{V_u}{b_o d} = \frac{285{,}180}{(156)(19.0)} = \underline{96.2 \text{ (lb/in.}^2)} < v_{u_{\text{all}}} = 186.0 \text{ lb/in.}^2$$

Check for beam shear (short side):
Refer to Fig. 12-18.

$$V_u = 3P_{u4} = (3)(32.3) = \underline{96.9 \text{ (kips)}}$$

$$b_o d = Bd = (108)(19) = \underline{2{,}052 \text{ (in.)}}$$

$$v_u = \frac{V_u}{b_o d} = \frac{96{,}900}{2{,}052} = \underline{47.2 \text{ (lb/in.}^2)} < 93.0 \text{ lb/in.}^2 = v_{u_{\text{all}}}$$

where 93.0 lb/in.² is the allowable stress for beam action:

$$v_{u_{\text{all}}} = 2\phi\sqrt{f'_c} = (2)(0.85)\sqrt{3{,}000} = \underline{93.0 \text{ (lb/in.}^2)}$$

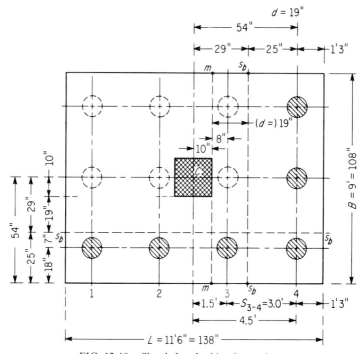

FIG. 12-18. Sketch for checking beam shear.

Check for beam shear (long side):

$$V_u = 20.7 + 24.6 + 28.4 + 32.3 = \underline{106.0 \text{ (kips)}}$$
$$b_o d = (L)(d) = (144)(19) = \underline{2{,}736 \text{ (in.)}}$$

Average shear stress:

$$v_u = \frac{V_u}{b_o d} = \frac{106{,}000}{2{,}736} = \underline{38.7 \text{ (lb/in}^2)} < v_{u\text{all}} = 93.0 \text{ lb/in}^2$$

Moment on critical section m-m:
    The critical section is at the face of the column across the footing slab.

Reactive moments M_u from piles on the heavy side relative to critical section m-m:

$$\begin{aligned}M_u = 3P_{u3}a_3 + 3P_{u4}a_4 &= (3)(28.4)(0.75) + (3)(32.3)(3.75) \\ &= \underline{4{,}273 \text{ (ft-kips)}} = 427{,}300 \text{ ft-lb} \\ &= \underline{5{,}127{,}600 \text{ in.-lb.}}\end{aligned}$$

Here $a_3 = 8$ in. $= 0.75$ ft $=$ moment arm of piles in the third row with respect to the critical section m-m
    $a_4 = a_3 + S_{3-4} = 0.75 + 3.00 = 3.75$ ft $=$ moment arm of piles in the fourth row
    $S_{3-4} = 36$ in. $= 3.00$ ft $=$ spacing of piles between third and fourth rows

$$\phi R_u = \frac{M_u}{Bd^2} = \frac{5{,}127{,}600}{(108)(19^2)} = \underline{131.5 \text{ (lb-in.)}}$$

Here $\phi R_u = \phi p f_y [1 - 0.59 p(f_y/f_c')] =$ a coefficient for concrete design
    $\phi =$ capacity-reduction factor
    $R_u = p f_y [1 - 0.59 p(f_y/f_c')]$
    $p =$ steel ratio
    $f_y =$ yield stress for steel, lb/in^2
    $f_c' =$ ultimate compressive strength for concrete, lb/in^2

The necessary steel reinforcement can now be calculated analytically or determined from specially prepared tables, and the cap redesigned if needed.

12-7. FOUNDATIONS RESTRICTED BY PROPERTY LINE

When because of various reasons the point of application of a force on a footing does not coincide with the centroid of the plan area of the base of the footing, one usually resorts to a one-sided, eccentrically loaded footing

Calculations of Rectangular Reinforced-Concrete Footings and Mats

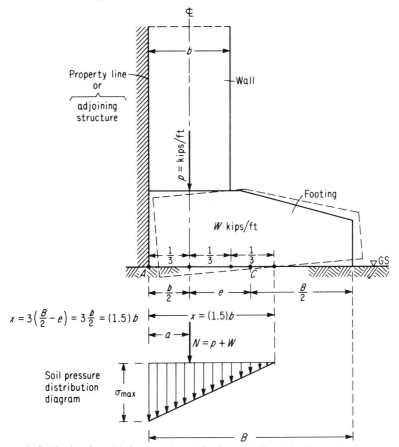

FIG. 12-19. One-sided, eccentric wall footing. Wall and footing designed as separate units.

such as shown in Fig. 12-19, for example. Such reasons may be a space limitation for a centrically loaded footing, say a strip footing; or perhaps a wall foundation of a new structure to be built must not cross the property line, or the adjacent structure may not permit constructing a symmetrical, centrically loaded foundation. Therefore, a one-sided widening or projection of footing slab under a wall or column is almost unavoidable, thus bringing about eccentricity of the load applied on the footing.

An isolated footing on the property line may have the following disadvantages:
1. Instability in a plane perpendicular to the property line
2. The soil pressure on soil is not uniform
3. It may bring about differential settlement as compared with other footings of the same structure

Figure 12-19 shows a wall and footing at the property line as separate units: the wall simply rests on the footing. In this case a one-sided widening of the foundation more than 1.5 times the width b of the wall is uneconomical because the excess width of the footing over $1.5b$ will not bring about a decrease in contact pressure on the soil at the edge heel point A of the footing. Such a footing probably could be laid only in a very good soil having a safe, admissible bearing capacity of, say, at least $\sigma_{all} = 2.5$ ton/ft^2 or more.

Problem

Refer to Fig. 12-19. Given wall and footing as separate units. $B = 30$ in. $= 2.5$ ft; $b = 18$ in. $= 1.5$ ft. Normal load $N = p + W = 5,750 + 450 = 6,200$ (lb/ft); $a = b/2 = 18$ in./2 $= 9$ in. $= 0.75$ ft; $x = (1.5)b = (1.5)(18$ in.$) = 27$ in. $= 2.25$ ft.
Here $p = 5.750$ lb/ft $=$ load from wall on footing
$W = 450$ lb/ft $=$ weight of footing
Calculate maximum contact pressure on soil at edge A (or heel) of footing.

Solution

A widening of footing beyond 2.25 ft to the right is not economical because the excess width does not transmit compressive stresses on soil.

The maximum edge pressure σ_{max} on soil at heel of footing is, by Eq. 10-12,

$$\sigma_{max} = \frac{2N}{3aL} = \frac{(2)(6200)}{(3)(0.75)(1)} = 5,520 \text{ (lb/ft}^2\text{)} = \underline{2.76 \text{ ton/ft}^2}$$

To continue the discussion, Fig. 12-20 shows a wall and footing at the property line designed to perform as a single, rigid, integral unit tied together firmly by reinforcement. The footing is eccentrically loaded and is constructed so as to transmit bending moment from soil reaction from the footing to the wall. Assume that the center of moments is at the center of the wall. For this case, when the section x-x can take a bending moment, it is assumed that the contact pressure σ_o on soil underneath the base of the footing distributes trapezoidally.

Should the footing for some reason tear off from the wall, say at section x-x, the load p of the torn-off wall from the footing will be transmitted to the soil one-sidedly or eccentrically and may bring about excessive edge pressure on the soil at edge point A.

The reinforcement of the footing-wall system (in the manner of a stub frame) has the effect of moving the resultant soil pressure reaction more toward the center of the width B of the footing, thus reducing the maximum edge pressure σ_{max} on the soil. Of course, the bending moment can be taken up by section x-x only when the wall can safely transfer this moment further

Calculations of Rectangular Reinforced-Concrete Footings and Mats

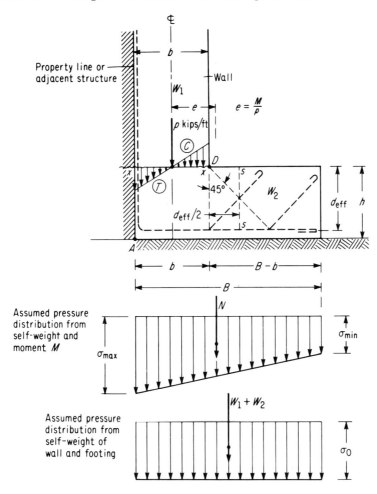

FIG. 12-20. One-sided, eccentrically loaded strip footing. Wall and footing designed as an integral unit.

to another part of the structure. In other words, for the existence of equilibrium of the system against rotation, and referring to Fig. 12-21, the bending moment $M = Pe$ must be resisted by an equal resisting (eccentric) moment $M = TH$, i.e.,

$$Pe = TH$$

Or the wall-footing system carrying an eccentric bending moment must be balanced by a similar (mirror) construction on the opposite side of the structure (Fig. 12-21).

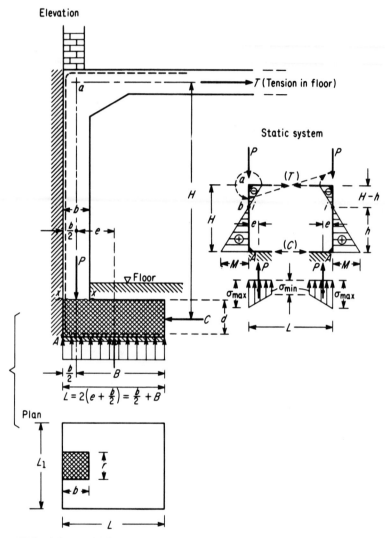

FIG. 12-21. Wall footing carrying a moment. The + moments are drawn on the tension side of the structural element.

If the rising wall is unable to transmit the moment, the pressure on soil may be reduced by extending the foundation slab to the next wall (to the right in this example). The structure to be designed would then rest upon a continuous slab. The moment is then taken up by the strength of the slab system buttressed by cross walls, or by the so-called *portal effect* of the columns.

If the wall footing is such that it induces a linear soil pressure distribution,

Calculations of Rectangular Reinforced-Concrete Footings and Mats

then the soil pressure-distribution diagram is a parallelogram (Fig. 12-20), and the restraining or fixing moment M_{x-x} may be approximately calculated as

$$M_{x-x} = \sigma_o B(1) \frac{B-b}{2}$$

$$= \sigma_o 2\left(e + \frac{b}{2}\right) \frac{2\left(e + \frac{b}{2}\right) - b}{2}$$

$$= \sigma_o(2e)\left(e + \frac{b}{2}\right) \tag{12-25}$$

where $B = 2\left(e + \frac{b}{2}\right)$

σ_o = average soil contact pressure

The shear as a measure of diagonal tension is checked at the critical section at $d/2$ from point D.

12-8. TRAPEZOIDAL FOOTING AT PROPERTY LINE

If instead of a strip footing at the property line a column footing trapezoidal in plan is to be used, such as shown in Fig. 12-22, then with the designations as on this figure, one calculates the *resultant moment* M_C on the footing system relative to the centroid C of the footing as

$$M_C - M + Hd \tag{12-26}$$

where $M = Pe$

H = a horizontal force on footing

d = effective thickness of footing

The footing should be centered so that the resultant of an assumed uniform soil contact pressure and the axial column load form a couple Pe equal and opposite to the moment M (Fig. 12-22).

The coordinate x_o (distance) of the centroid C (Fig. 12-22) of the trapezoid from the short side B_1 is

$$x_o = s + e = \frac{L}{3} \frac{B_1 + 2B_2}{B_1 + B_2} \tag{12-27}$$

where B_1 is the short side and B_2 the long side of the trapezoid parallel to B_1, respectively.

The necessary contact area A of the trapezoidal footing is calculated as $A = P/\sigma_{net}$, where σ_{net} is the upward soil pressure.

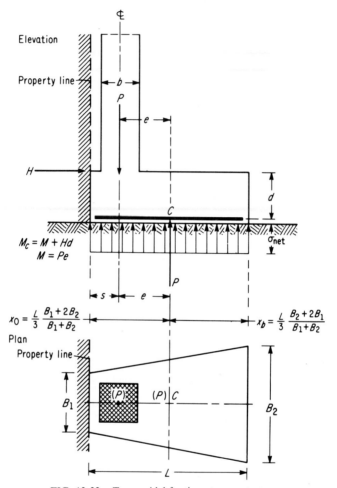

FIG. 12-22. Trapezoidal footing at property line.

If the position x_o of the centroid C and the length of the short side B_1 of a trapezoid are known, the length B_2 of the long side of the trapezoid and the length of the footing L, viz., height of the trapezoid, can be computed (Fig. 12-22) by solving simultaneously the two following equations:

$$x_o = \frac{L}{3} \frac{2B_2 + B_1}{B_1 + B_2} \tag{12-28}$$

$$P = \sigma_{all} A = \sigma_{all} L \frac{B_1 + B_2}{2} \tag{12-29}$$

Equation 12-28 gives the distance x_o for the location of the centroid C of the footing from the short side of the trapezoid, and Eq. 12-29 expresses

Calculations of Rectangular Reinforced-Concrete Footings and Mats

the applied load P balanced over the trapezoidal area by the total soil reaction (allowable soil bearing capacity σ_{all} times trapezoidal area A).

Substitution of L from Eq. 12-29 into Eq. 12-28 renders the following quadratic equation:

$$B_2^2 + 2\left(\frac{2A}{3x_o} - B_1\right)B_2 - \left(\frac{2A}{3x_o} - B_1\right)B_1 = 0 \qquad (12\text{-}30)$$

or, with

$$\frac{2A}{3x_o} - B_1 = \lambda \qquad (12\text{-}31)$$

$$B_2^2 + (2\lambda)B_2 - \lambda B_1 = 0 \qquad (12\text{-}32)$$

from which

$$B_2 = \lambda + \sqrt{\lambda^2 + \lambda B_1} \qquad \text{Q.E.D.} \qquad (12\text{-}33)$$

With B_2 now known, the length L of the footing (height of trapezoid) is calculated by Eq. 12-29 as

$$L = \frac{2A}{B_1 + B_2} \qquad (12\text{-}34)$$

Of course, if L and x_o are known (given by design), then B_1 and B_2 for the configuration of the given trapezoid (Fig. 12-22) can be calculated as

$$B_1 = \frac{2A}{L} - B_2 \qquad (12\text{-}35)$$

$$B_2 = \frac{2A}{L}\left(1 - \frac{3x_o}{L}\right) \qquad (12\text{-}36)$$

where the x_o-coordinate is measured from the short side B_1 of the trapezoid.

12-9. COMBINED FOOTINGS

A combined footing is one that usually carries an exterior column and one or more interior ones. In congested areas such as in cities the problems of design and laying of foundations may turn out to become complex in respect to property line restrictions and limitations, and adjacent buildings. If property rights prevent the use of footings projecting beyond the exterior walls of the designed building, combined footings, strap or connected footings, and mats or rafts are used so as not to project beyond the wall columns.

Combined footings are also used when soil bearing capacity is low, and also where single footings under heavily loaded columns would overlap or merge or nearly merge. In such instances combined footings under two or more columns are constructed, rendering greater stability of the foundations than single footings.

Combined footings are commonly designed and constructed in rectangular

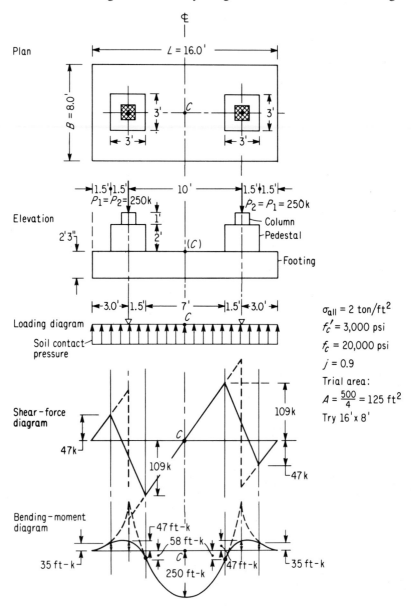

FIG. 12-23. Combined symmetrical rectangular footing.

Calculations of Rectangular Reinforced-Concrete Footings and Mats

[symmetrical and asymmetrical with respect to column (wall) loads], trapezoidal and other forms simple in plan.

Combined footings and mats supporting more than one column on a single footing or wall are to be designed according to the ACI Building Code as follows:

1. The soil pressures are assumed to act as uniformly distributed or varying linearly, except that other assumptions may be made consistent with the properties of the soil and the structure and with established principles of soil mechanics.
2. Shear as a measure of diagonal tension shall be computed in accordance with the ACI Building Code as described under plain and reinforced square concrete footings.

Figure 12-23 shows a combined footing rectangular in plan for equal and symmetrical load. Figure 12-24 shows a combined footing rectangular in plan for unequal and asymmetrical loads. Here the distance D fixes the center C of the asymmetrical slab ($L/2$) with respect to column loads (the resultant P of the two column loads coincides with the centroid of the

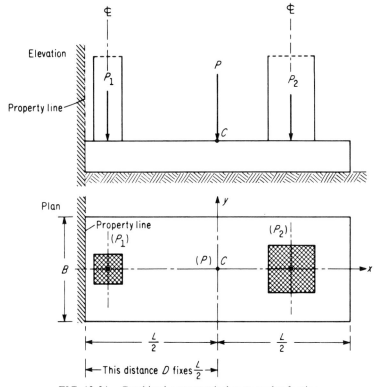

FIG. 12-24. Combined asymmetrical rectangular footing.

combined rectangular slab). The structural design of combined footings is based on the loads for which the columns (walls) on these footings are designed. In these footings bending moments and diagonal tension shears are determined at the face of the columns, especially at the external columns. The ACI Building Code specially states that it makes no recommendations for these footing designs.

The nature of shear and bending moment diagrams for a combined, symmetrical rectangular footing is shown in Fig. 12-23. The negative moments are designated here for the upward-acting, uniformly distributed, net soil contact pressures.

Figure 12-25 shows an enlarged combined footing at the property line. The pedestal stiffens the footing.

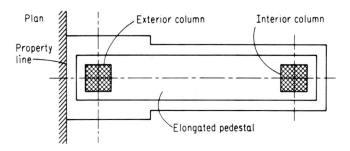

FIG. 12-25. Enlarged foooting.

12-10. COMBINED TRAPEZOIDAL FOOTING AT PROPERTY LINE

A combined trapezoidal footing at the property line is used when the exterior column transmits loads heavier than those of the interior column, permitting the resultant of the column loads to pass through the centroid C of the trapezoidal base contact area of the footing (Fig. 12-26).

Trapezoidal footings may also be constructed especially for industrial buildings accommodating heavy cranes. Again, the trapezoid is designed for uniform soil contact pressure distribution. The point of application of the resultant P of the two column loads P_1 and P_2 determines the position of the centroid C in plan of the trapezoidal footing.

12-11. CONNECTED FOOTINGS

Connected footings are also termed *strap footings*. To absorb the eccentricity moment, the separate exterior and interior footings may be connected into one by a tie or strap (Fig. 12-27), hence the term strap footings.

Calculations of Rectangular Reinforced-Concrete Footings and Mats

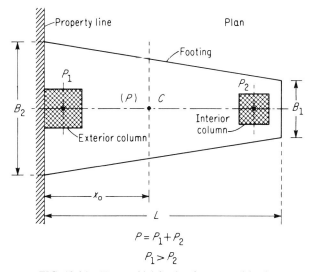

$$P = P_1 + P_2$$
$$P_1 > P_2$$

FIG. 12-26. Trapezoidal footing for unequal loads.

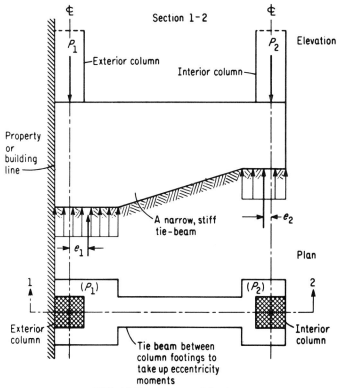

FIG. 12-27. Connected footing.

A *strap* is a narrow, stiff connecting beam designed to take bending. A strap footing is usually designed as an inverted continuous beam with a uniform or varying load, whichever is the case. The beam is attached to the footing, and performs as a freely supported beam on two supports.

Note in Fig. 12-28 that the footing A_1 under the exterior column is eccentric by the amount e from the column at the property line. The connecting tie beam is attached to the footings to balance the overturning moment $(P_1 e)$. This moment is carried over to the centric footing A_2 for its necessary balancing reaction. In other words, the tie beam counterbalances the eccentricity and maintains approximately uniform soil pressure distribution under the external footing.

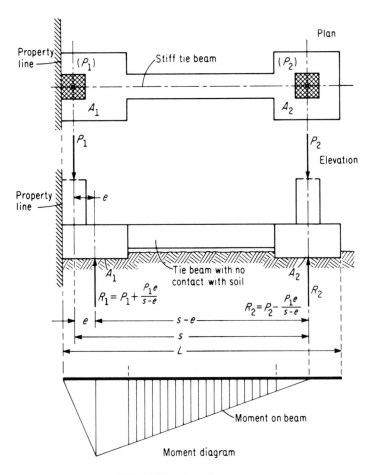

FIG. 12-28. Strap footing.

Calculations of Rectangular Reinforced-Concrete Footings and Mats

The connecting tie beam is designed based on two rough assumptions:
1. The only pressure on the soil is caused by the self-weight of the tie beam. No other loads bring about pressure on the soil under the tie beam. This means that the tie beam may be placed either directly on the soil (loose soil, which upon settlement leaves the beam free from soil) or it may be constructed free with no contact on the soil in the first place. In fact, the stiff tie beam functions most simply and effectively if it is relieved of soil pressure from below.
2. Further, it is assumed that the soil contact pressure distributions underneath of the bases of both column footings are uniform.

The magnitudes of soil reactions $R_1 = \sigma_{all} A_1$ and $R_2 = \sigma_{all} A_2$ are respectively:

$$R_1 = \frac{s}{s-e} P_1 = P_1 + \frac{P_1 e}{s-e} \qquad (12\text{-}37)$$

$$R_2 = P_2 - \frac{P_1 e}{s-e} \qquad (12\text{-}38)$$

Physically, the term $(P_1 e)/(s-e)$ means an uplift on footing A_2. Equations 12-37 and 12-38 also show that the footing area A_1 must be somewhat larger than needed for P_1 alone, and that footing area A_2 can be reduced somewhat below that needed for P_2 alone.

If the safe, allowable soil bearing capacity σ_{all} is known, the necessary size of the footing areas A_1 and A_2 is calculated as

$$A_1 = \frac{R_1}{\sigma_{all}} = \frac{P_1 + \frac{P_1 e}{s-e}}{\sigma_{all}} = \frac{P_1 s}{\sigma_{all}(s-e)} \qquad (12\text{-}39)$$

and

$$A_2 = \frac{R_2}{\sigma_{all}} = \frac{P_2 s - (P_1 + P_2) e}{\sigma_{all}(s-e)} \qquad (12\text{-}40)$$

For calculating footing base contact areas, R_1 and R_2 should include the weight of the footing. For strength calculations the weight of the footing may be neglected. The proportioning of the connecting tie beam is facilitated by means of its shear and moment diagrams.

The tie beam must be designed for a constant shear force $V = (P_1 e)/(s-e)$ and for a bending moment.

MAT FOUNDATIONS

12-12. CONTINUOUS FOUNDATIONS; MATS

If a foundation supports more than three independent loads whether or not in a straight line, it is termed a *continuous foundation*. An example of a continuous foundation is the mat, also known by the term "raft." It is nothing else but a continuous slab.

If the soil is weak and the column loads are large, single, independent, spread footings may necessitate large base contact areas, thus resulting in large, bulky footings of uneconomical proportions. In such a case one may resort either to a deep foundation or to a mat foundation.

According to *Suggested Design Procedures for Combined Footings and Mats*[1] a mat foundation is "a continuous footing supporting an array of columns in several rows in each direction, having a slablike shape with or without depressions or openings, covering an area of at least 75 percent of the total area within the outer limits of the assembly."

Thus, basically, a mat constitutes a number of closely spaced spread footings joined or fused to form one integral structural unit like a large concrete slab of adequate thickness working statically together rather than as a single, independent footing. The mat spreads the column loads to allowable soil bearing capacity, therefore acting as a combined footing.

Also, to reduce expenses in coping with groundwater, it is preferable to have a continuous foundation at a site rather than many small, isolated footings. Generally it is more advantageous to pump from one continuous trench (as for a strap or strip foundation, for example) or from a broad, shallow excavation than from a number of closely spaced, isolated foundation pits. This also pertains to economy in excavation earthworks as well as to formwork.

12-13. USES OF MAT FOUNDATIONS

Mat foundations are used
1. Where the sum of the individual footing base areas exceeds about one-half the total plan area of the building,
2. To distribute loads from the periphery of the building over the entire building area,
3. When the soil bearing capacity is too low to support the loads by other kinds of foundations,
4. To reduce concentrations of high soil pressure in hard soil areas,
5. In recognition of nonuniform soil conditions, to bridge over isolated "soft spots" in soil,

6. Where subsurface strata contain cavities or compressible lenses of soil difficult to define,
7. Where shallow shear strain settlements predominate and the mat would equalize differential settlements,
8. To resist a hydrostatic pressure of water (uplift),
9. Along property lines of adjacent sites and/or buildings.

12-14. DESIGN OF MAT FOUNDATIONS

Relative to static and strength calculations and design, mat foundations may be classed into two broad groups, namely: (1) mats designed by the so-called conventional method of analysis (rigid mats), and (2) flexible mats (mats on elastic support) designed on the basis of the theory of elasticity.

As to mats of conventional method of analysis, one may distinguish between the following principal types of mats, namely: (a) uniform mats, (b) flat slabs, and (c) ribbed slabs.

(a) A *uniform mat* is one of a uniform thickness supporting individual columns. The mat is made of reinforced concrete of uniform thickness (Fig. 12-29), and has two-way reinforcement at the top and at the bottom. The slab may be strengthened by additional reinforcement around the bases of the columns. It is assumed that the mat is so stiff and the load so constant that the plastic soil will compress and adjust itself so that each column load will spread almost uniformly under the mat. The mat is designed to be rigid, without excessive deflection or overstress. However, with nonuniform spacing of columns, and diverse loads on the columns, the mat may not be rigid enough to distribute the loads uniformly, viz., trapezoidally. Cost limitations, too, may limit the design to practical rather than full rigidity.

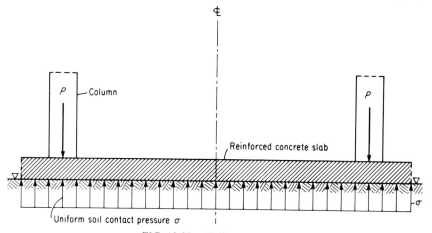

FIG. 12-29. Uniform slab mat.

(b) An example of a *flat-slab type of mat* foundation is a slab spread over the entire plan area of the structure. The slab is surmounted by truncated pyramidal slabs under all columns and trapezoidally shaped slabs under the walls (Fig. 12-30).

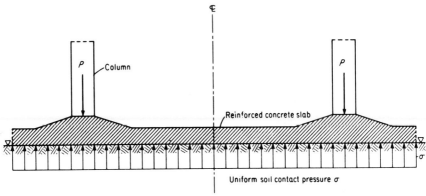

FIG. 12-30. Flat-slab type mat foundation.

(c) On long spans, or to reduce mat thickness, mats may be stiffened by ribs (beams and girders). A *ribbed mat* foundation consists of a reinforced concrete slab of adequate thickness spread over the entire plan area of the structure. The mat is stiffened monolithically by heavy, interconnected ribs—beams and girders—which run longitudinally and transversely (Fig. 12-31). The entire ribbed mat resembles an inverted floor-slab-beam system which carries a uniformly distributed load exerted by the soil contact pressure and which has known reactions at certain points—the column loads.

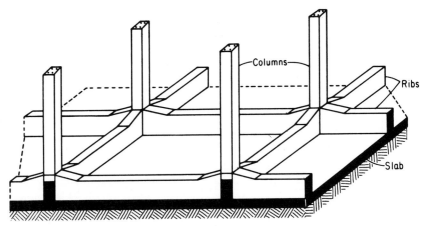

FIG. 12-31. Ribbed mat foundation.

Calculations of Rectangular Reinforced-Concrete Footings and Mats

The slab may be stiffened by ribs either above or below the slab. To avoid obstructions in the floor, the required thickness of the mat at the columns is realized by increasing the depth of excavation. It is a good practice to design ribs above the mat. Ribs below the mat are undesirable because excavation, the placing of forms and reinforcement, and the casting of the ribs usually cause disturbance in the soil adjacent to the ribs, thus bringing about nonuniform soil support for the mat.

Stiffness of mat may also be attained either by a rigid frame system or by a cellular construction. The latter consists of a bottom and top slab with intermediate ribs (Fig. 12-32).

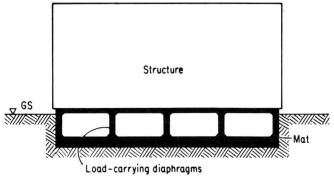

FIG. 12-32. Cellular mat.

The ribbed-mat type of foundation is probably the most reliable one and, therefore is most frequently resorted to where a spread slab foundation is called for.

The ribs of the mat are designed to resist the bending moments of the column loads to one side of any point on the beam, plus the moment of the uniformly distributed soil pressure contributing to the beam taken about the same point. Sometimes ribs are designed as T-beams.

The interior panels of the slab are usually designed (1) as continuous slabs supported on four sides, (2) as slabs supported by the transverse beams of the mat, or (3) as two-way slabs with fixed edges, the choice being that of the designer.

For very heavy structures, when tolerable settlement must be kept down to an absolute minimum, rigid frame foundations are designed and constructed. The merits of such a foundation must be, of course, compared with those offered by deep foundation, such as piles or caissons.

Great precision and refinements of preliminary calculations of mat foundations are not always justified or practicable because of the uncertainties of soil performance under load and water underneath the mat, and because of the soil pressure distribution underneath an extensive loading area such as the mat. In the final analysis refinements in calculations may be introduced.

All hitherto known and currently used design methods are merely approximate. A satisfactory, plausible method for mat foundation design has yet to be developed.

To avoid nonuniform soil contact pressure distribution, mats are sometimes designed to be flexible.

12-15. SOIL PRESSURE DISTRIBUTION UNDER MATS

The principle of static calculations of design of mat foundations is the assumption that soil bearing pressure is approximately uniform, or as varying uniformly under the entire area of the mat.

In order to attain the assumed soil contact pressure distribution, one should strive to achieve:
1. That the mat should be of adequate rigidity.
2. That the center of gravity of the column loads should coincide with the centroid of the area of the mat (uniform soil pressure distribution, viz., the resultant of the column loads should coincide with the point of application of the resultant of the upward-acting soil reaction).

If this is impossible, eccentricity e should be at a minimum:

To attain uniform soil pressure distribution, and the soil conditions at the site permitting, the slab and ribs may be extended as a cantilever beyond and outside of the column layout, so that the center of gravity of the column loads would coincide with the centroid of the mat area. Such a concentric loading (or nearly so) with no eccentricity ($e = 0$) is most appropriate for uniform mats and for flat-slab mats.

When the resultant normal load $R = \Sigma V$ coincides with the centroid of the mat area, the average soil contact pressure σ is computed as

$$\sigma = \frac{\Sigma V}{A} \tag{12-41}$$

If there is a two-way eccentricity present in the mat system, this may require a thick slab with a heavy reinforcement to provide for adequate rigidity of the mat. The maximum and minimum pressure on soil under a two-way eccentricity condition is calculated as

$$\sigma_{\substack{max \\ min}} = \frac{R}{A} \pm \frac{Re_y}{I_x} y \pm \frac{Re_x}{I_y} x \tag{12-42}$$

Calculations of Rectangular Reinforced-Concrete Footings and Mats

where $R = \Sigma V$ = resultant vertical (normal) load on the mat
 A = contact area of the mat with soil
 e_x, e_y = eccentricities of the resultant load R
 x, y = coordinates of any point on the mat relative to the axes through the centroid of the mat area
 I_x, I_y = centrifugal moment of inertia of the area of the mat with respect to the x and y axes, respectively

In the case of one-way eccentricity (say $e_y = 0$; $e_x = e_x$), the soil pressure is calculated as

$$\sigma_{\substack{max \\ min}} = \frac{R}{A} \pm \frac{Re_x}{I_y} x \tag{12-43}$$

Pressures under the edges or elsewhere beneath the mat cannot be computed by Eqs. 12-42 and 12-43 because the mat is not stiff enough to act as an integral unit. Hence, in the case of eccentricity and in order to attain a reasonably uniform or uniformly varying pressure distribution on soil, the mat is usually stiffened so that the center of gravity of the column loads coincides with the centroid of the mat; if not, the eccentricity should be at a minimum.

The method of assuming a uniform soil pressure distribution usually will give a safe design, but the mat may turn out to be very expensive in reinforcement.

Where soil pressure varies because of eccentricity, the mat thickness may be varied over the area of the mat.

Mats constructed below the groundwater table must be checked against hydrostatic uplift. This check will probably not affect the design as an integral unit, but will require special details for watertightness and strength of the side-wall connections.

12-16. STEPS FOR CONVENTIONAL DESIGN OF RIGID MAT FOUNDATIONS

The steps for mat foundation design may be summarized as follows.
1. Based on the predominance of the soil type in the depth zone below the base of the mat, adopt a type of mat to use (rigid or flexible) which would result in a minimum, uniform, tolerable settlement of the structure.
2. Determine the necessary depth below ground surface for laying the mat.
3. Determine the allowable bearing capacity of the soil.

4. Assume a uniform or uniformly varying soil contact pressure distribution.
5. Develop a mat design by providing for a general arrangement of the elements (ribs, viz., slab, beams, girders) in the mat system and also for clear height of the ribs which will distribute the applied loads as assumed under (4).
6. Determine the various combinations of dead and live loads on columns.
7. Determine the soil pressures σ

$$\sigma_{\substack{max \\ min}} = (R/A) \pm (M/Z)$$

where R = weight of structure
A = plan area of mat
M = wind moment
Z = section modulus for whichever part of mat is being checked
8. Determine forces for each of the mat beams for the worst loading condition.
9. Plot shear and moment diagrams for the worst possible condition for the individual beams and the slab.
10. By means of the bending moments and shears, calculate the depth of the mat beams and an adequate, economical thickness of the mat slab, and arrange for a suitable reinforcement.

Suggestions for reinforced-concrete mat design are given by the American Concrete Institute as set forth in Ref. 1.

12-17. DESIGN OF A RIGID, UNIFORM MAT BASED ON STATICS

The approximate design of a mat may be most readily elucidated by means of an illustrative example.

Example No. 1

Given a plan of arrangement of columns, beams and girders, and column loads as shown in Fig. 12-33. Assume that the mat is stiff, the column loads are constant, and that the loads are distributed almost uniformly on soil under the mat. The allowable soil bearing capacity is given as $\sigma_{all} = 0.5$ ton/ft² = 1.0 kip/ft². The ribs are to be spaced at one-third points of span.

Design a monolithic, ribbed, reinforced concrete mat foundation. The stiffening girders should be designed above the top surface of the slab.

Calculations of Rectangular Reinforced-Concrete Footings and Mats

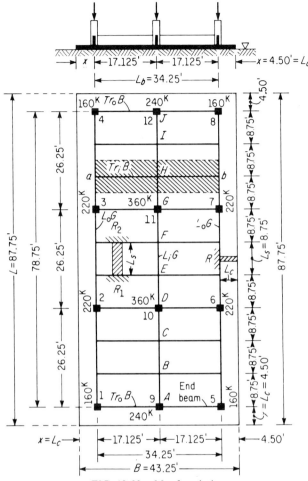

FIG. 12-33. Mat foundation.

Solution

The column loads are symmetrical. Hence their resultant $R = \Sigma P_c$ coincides with the centroid of the mat slab.

1. Cantilever projection x of slab beyond the center line of the outside row of columns

    (a) Column loads ΣP_c:

    $$\Sigma P_c = (4)(160) + (2)(240) + (4)(220) + (2)(360) = 2{,}720 \text{ (kips)}$$

    (b) Net upward soil pressure for supporting column loads (ΣP_c):
    The net upward soil pressure σ_{net} for supporting column loads

of $\Sigma P_c = 2{,}720$ kips is to be calculated by excluding the pressure from the weight of the mat slab (σ_{mat}) from the safe, allowable soil pressure (σ_{all}), i.e.,

$$\sigma_{net} = \sigma_{all} - \sigma_{mat}$$

Safe, allowable soil bearing capacity. . . . . $\sigma_{all} = 1.000$ kip/ft^2
Assume weight of slab 18 in. thick and fill (estimated) . $\sigma_{mat} = 0.285$ kip/ft^2

Net upward pressure for supporting column loads. $\sigma_{net} = 0.715$ kip/ft^2

The self-weight of the mat in this calculation is not reckoned with because it is assumed that the mat is supported directly in contact by the soil, and because it acts against the bending loads. Because the weight of the slab of the mat does not influence either the shear forces or the bending moments in the slab, only column loads need to be considered.

(c) Cantilever projection x:

$$(\sigma_{net})(34.25 + 2x)(78.75 + 2x) = 2{,}720 \text{ (kips)}$$
$$x^2 + 56.5x - 276.75 = 0$$
$$\therefore \quad x = 4.5 \text{ (ft)}$$

2. Size of slab

$$[34.25 + (2)(4.50)][78.75 + (2)(4.50)] = (43.25)(87.75)$$
$$= 3{,}795 \text{ (ft}^2\text{)}$$

3. Calculations for slab design

(a) Bending moment M_{isl} in the interior slab between ribs per foot width of slab (uniformly loaded beam fixed at each end).

The interior slab between the interior transverse beams may be designed as a slab supported on four sides, or as a slab fixed at each end in two adjacent transverse beams. The designer makes the choice. The mat slab is uniformly loaded by the net upward soil contact pressure, $\sigma_{net} = 0.715$ k/ft^2 per foot width of slab, and is supported by the columns as reactions (Fig. 12-34).

$$M_{isl} = M_1 = M_2 = \frac{\sigma_{net} L_s^2}{12} = \frac{(715)(8.75^2)}{12}$$

$$= 4{,}562 \text{ (ft-lb)} = (4{,}562)(12)$$
$$= 54{,}744 \text{ (in.-lb)} \approx \underline{54{,}800 \text{ in.-lb.}}$$

Calculations of Rectangular Reinforced-Concrete Footings and Mats

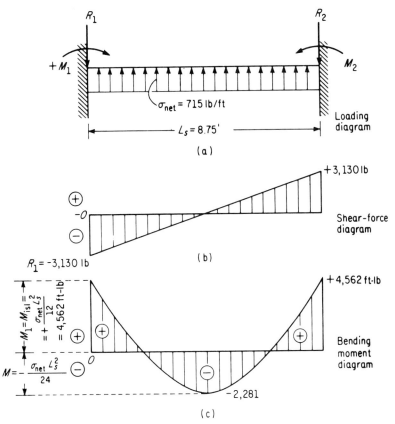

FIG. 12-34. Interior slab diagrams.

The effective depth of the slab, d_{eff}, the necessary reinforcement, and the total thickness of the slab can now be calculated.

(b) Bending moment M_{csl} in 4.5-ft cantilever slab projection (Fig. 12-35)

$$M_{\text{csl}} = \frac{\sigma_{\text{net}} L_c^2}{2} = \frac{(715)(4.5^2)}{2} = 7{,}240 \text{ (ft-lb)}$$

$$= (7{,}240)(12) = 86{,}880 \text{ (in.-lb)} \approx \underline{86{,}900 \text{ in.-lb.}}$$

4. Transverse beams
   (a) Load q on inner transverse beam, $\text{Tr}_i B$

   $$q = \sigma_{\text{net}} L_s = (715)(8.75) = 6{,}256 \text{ (lb/ft)} \approx \underline{6{,}260 \text{ lb/ft}}$$

   where $L_s = 8.75$ ft is the span of the slab.
   (b) Total upward load $P_{\text{Tr}_i B}$ acting on inner transverse beam $\text{Tr}_i B$

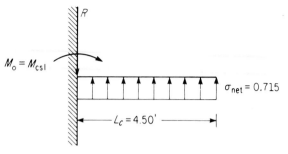

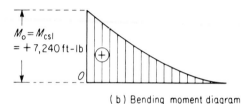

(b) Bending moment diagram

FIG. 12-35. Cantilever projection slab diagrams.

$$P_{Tr_iB} = qL_b = (6{,}260)(34.25) = 214{,}405 \text{ (lb)} \approx \underline{214{,}400 \text{ (lb)}}$$

where $L_b = 34.25$ ft is the length of the two-span transverse beam.

(c) Total upward load P_{Tr_oB} on outer transverse beam Tr_oB

$$P_{Tr_oB} = \sigma_{net}\left(\frac{L_s}{2} + x\right) L_b = (715)[(8.75/2) + 4.5](34.25)$$

$$= \underline{217{,}350 \text{ (lb)}}$$

Hence, the outer transverse beam Tr_oB is loaded by

$$\frac{217{,}350}{214{,}400} = 1.0375$$

or by 3.75 percent more than the inner transverse beam Tr_iB, i.e.,

$$Tr_oB = (1.037)\ Tr_iB$$

To continue, it is now assumed that the sum of the upward center reactions from eight inner transverse beams (Tr_iB) and from two outer (end) transverse beams (Tr_oB) equals the sum of the loads of the four central columns (9, 10, 11, 12).

Calculations of Rectangular Reinforced-Concrete Footings and Mats

(d) Reactions

$$(8)(Tr_iB) + (2)(Tr_oB) = (8)(Tr_iB) + (2)(1.037)(Tr_iB)$$
$$= (10.074)(Tr_iB)$$

where Tr_iB and Tr_oB mean here center reactions from transverse beams.

(e) Total load of columns 9, 10, 11, 12

(9, 12).........	(2)(240) = 480 (kips) =	480,000 lb
(10, 11).........	(2)(360) = 720 (kips) =	720,000 lb
Total.......	1,200 (kips) =	1,200,000 lb

Equating d and e, obtain:

$$1{,}200{,}000 = (10.074)(\text{reactions from } Tr_iB \text{ and } Tr_oB) = (10.074)(R_{ci}),$$

where R_{ci} = center reactions from eight inner transverse beams.

$$R_{ci} = (1{,}200{,}000/10.074) = 119{,}120 \text{ (lb)} = \underline{119.12 \text{ kips}}$$

Center reaction R_{co} from two outer (end) transverse beams

$$R_{co} = (1.037)(119{,}120) = 123{,}520 \text{ (lb)} = \underline{123.52 \text{ kips}}$$

(f) End reactions R_{ei} for inner transverse beam on outer longitudinal girders L_oG

$$R_{ei} = \frac{P_{Tr_iB} - R_{ci}}{2} = \frac{214{,}400 - 119{,}120}{2} = \frac{95{,}290}{2}$$

$$= 47{,}644 \text{ (lb)} = \underline{47.64 \text{ k}}$$

End reactions R_{eo} for outer (end) transverse beam on outer longitudinal girders L_oG

$$R_{eo} = \frac{P_{Tr_oB} - R_{co}}{2} = \frac{217{,}350 - 123{,}520}{2} = \frac{93{,}830}{2}$$

$$= 46{,}915 \text{ (lb)} = \underline{46.92 \text{ k}}$$

With the reactions now known, the shear force and bending moment diagrams for the inner and end transverse beams can be drawn. The loading, shear force and bending moment diagrams are shown in Fig. 12-36.

(g) Inner transverse beam, Tr_iB

This is a uniformly loaded beam ($q = 6{,}260$ lb/ft, ↑) between the two outer longitudinal girders L_oG. The beam is supported at its midspan ($L_b/2 = 34.25$ ft/2) by the inner longitudinal

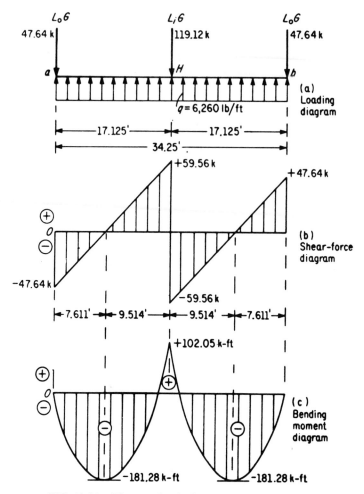

FIG. 12-36. Diagrams for the inner transverse beam Tr_iB.

girder L_iG, the load of which is 119,12 kips (↓). The end loads are 47.64 kips (↓). The diagrams for the design of this girder are shown in Fig. 12-36.

5. Inner longitudinal girder L_iG

In general, the loading of the longitudinal girders is composed of the column loads (9, 10, 11 and 12, for example) and the reactions from R_{ci} and R_{co} from the inner transverse beams and the outer (end) transverse beams, respectively. The shear force and moment diagrams for this girder are calculated as for continuous beams (Fig. 12-37).

Calculations of Rectangular Reinforced-Concrete Footings and Mats

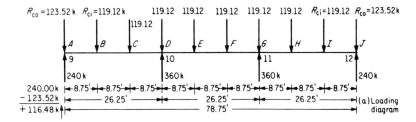

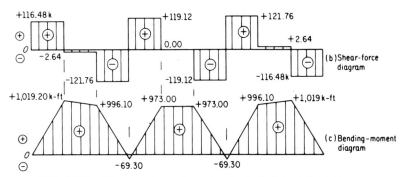

FIG. 12-37. Diagrams for the inner longitudinal girder L_iG (A-J).

6. Outer longitudinal girder, L_oG

The outer longitudinal girder is loaded with the column loads (5, 6, 7, 8, for example), reactions from transverse beams and girders and also by the upthrust of soil pressure on the cantilevered part of the slab.

Example No. 2

In this example, indications are given for the preliminary, approximate design of an eccentrically loaded mat foundation. The safe, allowable soil bearing capacity is given as $\sigma_{all} = 2.0 \text{ ton/ft}^2 = 4.0 \text{ kip/ft}^2$.

The practice of mat design is as follows: If the magnitudes of the adjacent *column* loads do not differ among themselves by more than approximately 20 percent, the mat slab may be designed as a slab strip 1 ft wide.

In this example, it is given that the mat slab is loaded with *continuous wall* loads as shown in Fig. 12-38. Because of the given wall loads, the mat may be analyzed for a strip of slab one foot wide.

Because of the unequal column load distribution, one first calculates the magnitude of the eccentricity e in this loading system:

$$Rx = \Sigma M_H$$

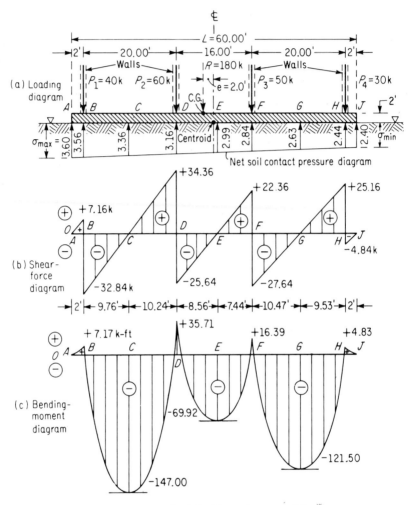

FIG. 12-38. Mat loaded with continuous wall loads.

or

$$Rx = (40)(56) + (60)(36) + (50)(20) = 5{,}400 \text{ (k-ft)}$$

where x = distance of resultant vertical load from a predetermined point, in this case, say, from point H
$R = 180$ kips so given

$$x = \frac{\Sigma M_H}{R} = \frac{5{,}400}{180} = \underline{30} \text{ (ft)}$$

Calculations of Rectangular Reinforced-Concrete Footings and Mats

Hence the resultant load $R = 180$ kips acts 30 feet to the left of point H, and the eccentricity e is

$$e = 30 - 28 = 2 \text{ (ft)}$$

to the left of the centerline (viz., centroid) of the mat slab.

The average soil contact pressure σ is

$$\sigma = R/A = 180/60 = 3.00 \text{ (k/ft}^2\text{)}$$

where A = contact area of slab on soil.

Pressure on soil from average and eccentric loading σ_L (exclusive of the weight of the slab):

$$\sigma_{L\max \atop L\min} = \frac{R}{A} \pm \frac{M}{Z} = \frac{180}{60} \pm \frac{(180)(2.0)}{(1.0)(60^2)/6} = 3.00 \pm 0.60$$

$$\sigma_{L\max} = 3.60 \text{ (k/ft}^2\text{) (C)}$$

$$\sigma_{L\min} = 2.40 \text{ (k/ft}^2\text{) (C)}$$

Weight of the mat, 2 ft thick:

$$0.30 \text{ (k/ft}^2\text{) (C)}$$

Total pressure on soil:

$$\sigma_{\max} = 3.60 + 0.30 = 3.90 \text{ (k/ft}^2\text{)} < \sigma_{\text{all}}$$

$$\sigma_{\min} = 2.40 + 0.30 = 2.70 \text{ (k/ft}^2\text{)} < \sigma_{\text{all}}.$$

The loading, shear and bending moment diagrams can now be constructed and the mat designed. These diagrams are shown in Fig. 12-38.

The eccentricity e brings about a linear variation in soil pressure resulting in a trapezoidal soil pressure distribution diagram. For calculating shear forces and bending moments only the net soil pressure diagram (exclusive of weight of slab) is used.

For convenience of calculation, intermediate soil pressure ordinates are shown in the net soil pressure distribution diagram.

For preliminary calculations, as a *first approximation*, the shear forces, their course of distribution, and bending moments may be calculated *approximately* and most readily from the *loading* diagram (net soil pressure and concentrated wall loads).

After preliminary calculations, refine, if necessary, the calculations of shear forces, bending moments and the sizes of the structural elements.

Flexible strip footings, flexible slabs, and flexible mat foundations on elastic support must be designed on the basis of the theory of elasticity by the method of subgrade reaction, or by the method of modulus of elasticity of soil, or any other pertinent analytical method.[2-9]

12-18. MATS RESISTING HYDROSTATIC UPLIFT PRESSURE

The question as to how soil pressure reacts statically on the structural load, viz., the slab, is one of the most difficult to answer in foundation engineering.

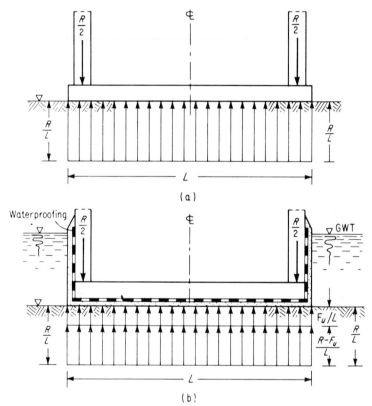

FIG. 12-39. Rectangular soil pressure distribution diagram under slab (a) on dry soil, (b) in groundwater.

However, to simplify calculations, assume, as done before, that the total reactive soil pressure is equal to the applied resultant load R from the structure. Assuming uniform soil pressure distribution on the soil, the total physical area of the soil pressure distribution on the soil, the total physical area of the soil pressure distribution diagram is R, according to Fig. 12-39a:

Calculations of Rectangular Reinforced-Concrete Footings and Mats

$$\frac{R}{(1)(L)} L = R \tag{12-44}$$

where L = width of slab.

Upon immersing the structure in water, the sum of the acting forces on the base of the slab does not change, because the loss of weight of the structure is exactly balanced by hydrostatic uplift. The size of the pressure distribution area is as above (Fig. 12-39b).

$$\left(\frac{R - F_u}{L} + \frac{F_u}{L}\right) L = R \tag{12-45}$$

where F_u = hydrostatic uplift force.

If one elects to calculate with a triangular pressure distribution diagram (Fig. 12-40a), then, upon immersion of the structure in water, the form of the pressure distribution diagram changes (Fig. 12-40b), while the magnitude of the total pressure does not change. Thus for the structure founded on dry soil (Fig. 12-40a),

$$(1/2)(2) \frac{2R}{L} \frac{L}{2} = R \tag{12-46}$$

For a basement mat immersed in water (Fig. 12-39b),

$$(1/2)(2) \frac{2(R - F_u)}{L} \frac{L}{2} + \frac{F_u}{L} L = R - F_u + F_u = R \tag{12-47}$$

Thus the magnitude of the upward pressure to use in the static analysis of the mat slab depends only upon the structural dead and live loads R, but not, as frequently is assumed, upon uplift. Uplift merely exerts its influence upon the pressure distribution.

One of the remedies in coping with hydrostatic uplift and buoyancy of mats is increasing thickness and reinforcement of the mat to resist shear and bending moments in the mat. Or the size of the slab may be increased by cantilevered projections like a footing, such as described in the preceding sections. The backfilled weight of soil on the cantilevered projections resists the uplift.

Also, the side walls of the basement may be thickened to increase the strength of the mat against uplift and hydrostatic pressure. Uplift pressure on a slab may be relieved by properly functioning underdrainage.

In poor soil and for high loads, basements are constructed as rigid frame boxes with stiffeners to increase the resistance to water pressure, lateral earth pressure and reaction from soil contact pressure.

If the pressure on soil cannot be increased by thickening and/or widening of the mat slab, one resorts to the so-called *floating foundation* (hollow,

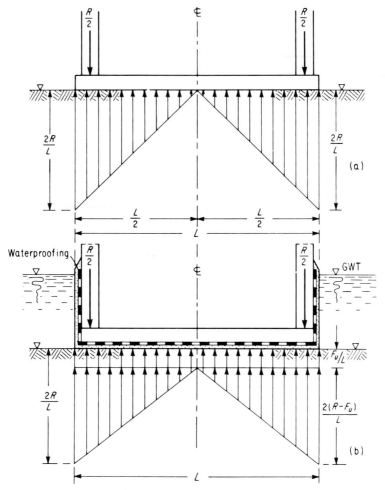

FIG. 12-40. Triangular soil pressure distribution diagram under slab (a) on dry soil, (b) in groundwater.

cellular, reinforced-concrete box). The principle involved in designing a floating foundation is that the weight of the excavated soil for the mat foundation should be equal to the equivalent weight of the new structure. The structure is said to *float*. The intent of such a design is to reduce excessive, intolerable settlement of the soil, viz., structure.

12-19. WATERPROOFING OF FOUNDATIONS

The submerged parts of foundations and basements must be protected against the influx of groundwater and soil moisture by means of suitable membranes, waterstops, and moisture barriers. The foundations should also be protected against the action of aggressive groundwater and harmful matter encountered in the soil.

One may distinguish among several methods for protecting foundations against water, namely: (1) drainage, (2) dampproofing against upward and laterally migrating soil moisture, and (3) waterproofing against groundwater and water under hydrostatic pressure.

1. *Drainage.* Groundwater is coped with by means of drainage and waterproofing membranes. To relieve lateral and upward hydrostatic pressure of groundwater on walls and basement floor slabs, respectively, peripheral drainage systems containing footing drains are provided around the structure. If clogging of footing drains by fine particles of soil can be anticipated, underfloor drain pipes are installed (Fig. 12-41). The underfloor drains are also used to cope with unavoidable water pockets under the slab.

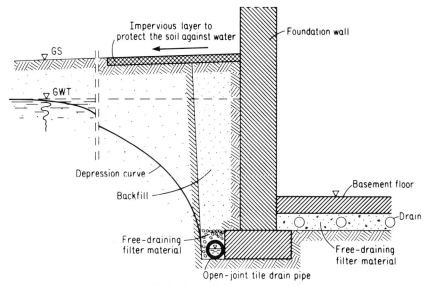

FIG. 12-41. Footing drain.

The minimum thickness of an underfloor-drainage course may be taken as about 10 in. The underfloor drainage also relieves hydraulic uplift pressure on the slab from below.

Around drain pipes there is placed free-draining filter material such as sound, clean gravel or crushed rock, 3/4 in. to 2 in. in size.

To avoid decreasing the bearing capacity of the soil because of water in the soil upon which the footings rest, intercepting drains may be used (Fig. 12-42).

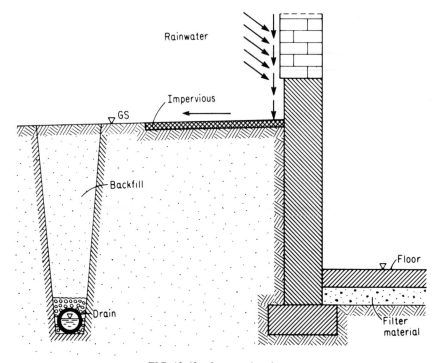

FIG. 12-42. Intercepting drain.

To protect the soil from precipitation water and that flowing over the gutters of the roof and down to the soil, the ground surface adjacent to the foundation must be provided with an impervious layer of soil. This layer must slope away from the foundation.

Needless to say, the drainage waters must be carried away over the shortest possible route either to an outfall or to a pump sump.

Also, construction, expansion and contraction joints, and any other joints in the foundation below groundwater table must be well waterproofed by suitable waterstops.

Waterstops. Waterstops are devices which are placed in the joints to provide for an obstacle against flow of water through the joints. Waterstops are made of corrosion-resistant soft metal such as copper, for example, rubber and plastic. Soft metal waterstops are used in joints where little movement of adjoining slabs is anticipated.

Some commonly used forms of waterstops are illustrated in Fig. 12-43. These are straight metal strips, V-shaped metal strips, flexible rubber dumbbells, and various other plastic materials.

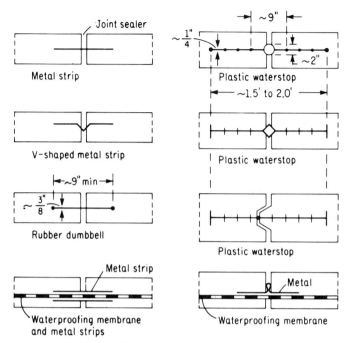

FIG. 12-43. Various kinds of waterstops.

The flexible waterstops made of rubber are inert to normal movement of the joint, whereas those made of plastic tolerate some joint movement, of the order of magnitude of, say, 0.5 in., for example. The tolerance in movement of these waterstops depends upon the design and kind of plastic used, of course.

Usually 20-oz annealed soft copper plates are used for metal waterstops. The minimum width of these waterstops is taken as 12 in. To be effective, all copper plate waterstop joints must be soldered.

The minimum width of a rubber waterstop is $\approx$ 9 in., the thickness is about 3/8 in., and, to be effective, all rubber joints must be vulcanized.

Plastic waterstops are made a minimum of 9 in. wide, 3/16 in. thick, and, to be effective, all plastic joints must be heat-fused.

Of course, depending upon the design of the structure (retaining walls, locks, docks, for example) any other kind and size of waterstops different from those mentioned above may be made and used.

2. *Dampproofing.* Foundations and basements are protected against moisture and dampness by so-called dampproofing, or insulation. For this purpose a course or two of tarred roofing paper is placed under the basement floor and on top of the foundation wall. This insulation is to be so placed

(viz., the surface of the working slab and that of the top of the foundation wall must be thus prepared) that the tarred roofing paper will not be pierced through by protrusion of uneven parts of the concrete or any other sharp objects.

To prevent soil moisture migrating laterally from the outside into the foundation masonry, viz., basement, the pores of the outside foundation wall are closed by applying on the outside foundation face several coatings of hot bituminous or tar compounds in the form of solutions and emulsions. For best results, these coatings must be applied on an air-dry surface.

A dampproofing barrier may also be applied on the interior face of the foundation wall of the basement (hot coal tar). On rough, uneven exterior surfaces, hot bituminous coatings (minimum thickness 1/8 in.) to protect the moisture barrier from damage must be applied. Usually these are built up in successive coats. The soil backfill material should be free of stones and/or rock fragments.

Such dampproofing is satisfactory when the highest groundwater table is at or below the footing and below the basement floor (Fig. 12-44). This method is effective for a head of water pressure of < 2 ft.

3. *Waterproofing.* Under hydrostatic pressure water may leak into the basement through joints and cracks. In the case when for physical and

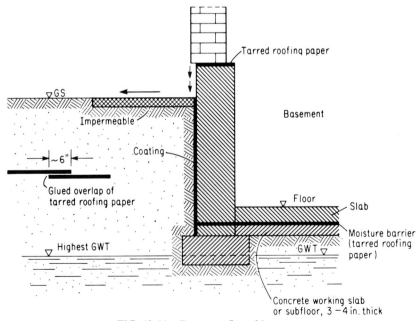

FIG. 12-44. Dampproofing of basement.

Calculations of Rectangular Reinforced-Concrete Footings and Mats

technical reasons a drainage system may prove to be ineffective or economically infeasible, or if the groundwater table rises above the lower insulation course of the basement floor, then waterproofing of foundation and basement should be provided by means of waterproofing membranes according to the principle illustrated in Fig. 12-45. Note here that the side wall and waterproofing membrane and the basement floor insulation are subjected to full hydrostatic pressure. Hence they must be designed to resist this force. The foundation wall must also be designed for active earth pressure.

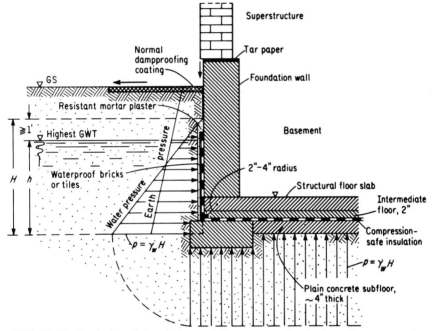

FIG. 12-45. Insulation of foundation and basement against hydrostatic pressure of groundwater.

Waterproofing materials are
  Asphalt mixtures and emulsions
  Bituminous and pitch-type tar compounds
  Waterproof cement compounds
  Coal tar
  Mixtures for decreasing permeability to water of the concrete
  Impregnated construction felt
  Tarred roofing papers
  Thermoplastic sheets (plastic)
  Waterproofing mortars

Water-repellent compounds, mixed with the concrete

The waterproofing membrane may consist of several layers of tarpaper, or bituminous fabrics (usually two or three layers) which are applied and hot-glued together. First, however, the foundation receives a hot base coating on an air-dry surface upon which the first layer of the waterproofing membrane is glued and well pressed for good bonding.

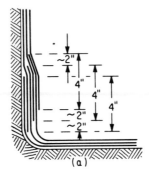

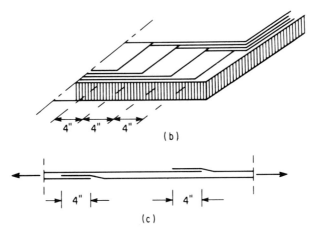

FIG. 12-46. Overlapping of waterproofing membrane layers: (a) overlapping at corner, (b) overlapping on floor, (c) overlapping for tension.

It is recommended that waterproofing work should not be done at temperatures below $+6°C$. Normally the waterproof coatings are relatively flexible, hence perform their function upon tolerable expansions of the wall and floor, and even upon a reasonably small amount of cracking.

Where waterproofing membranes are subjected to tensile force, the membranes are glued continuously tight by overlapping the insulating course as shown in Fig. 12-46c, for example.

Calculations of Rectangular Reinforced-Concrete Footings and Mats 451

The outside waterproofing membrane on the wall is protected by a water-resistant mortar plaster. Over the latter, a protective cover course of waterproofing bricks or tiles is affixed, laid in rich cement or bituminous mortar. Above the highest normal groundwater table, the outside wall, up to the ground surface, is provided with a normal dampproofing coating.

The structural basement floor is placed upon a compression-safe insulation, such as bituminized construction felt. The felt, in turn, is usually placed upon plain concrete. The felt is protected against damage by an intermediate floor 2 in. thick. Upon this course, the structural floor slab is cast.

From the foregoing description of waterproofing one should become aware that waterproofing of foundations and basements by a water-resistant subfloor (cleanliness layer, usually 4 in. thick) requires rigid design of particular details, and, to be effective, great patience in execution of the waterproofing.

Detailed design of waterproofing similar to that for walls must also be made for interior columns resting on a mat or foundation slab.

Some methods of overlapping of membrane layers are shown in Fig. 12-46.

12-20. GRILLAGES

Grillages may be constructed of timber, steel and reinforced concrete. The assumptions made in grillage footing design are

1. The pressure from the footing is uniformly distributed over the soil or rock, or plain concrete block.
2. The pressure of one tier of beams is uniformly distributed on the tier below.
3. Each tier acts independently of the other tiers.
4. The load from one tier of beams is uniformly distributed on the tier below.
5. The concrete encasing the grillage beams does not carry any load but serves merely as a protection for the steel grillage beams against corrosion.
6. The beams in a grillage are computed as cantilever beams loaded with a uniformly distributed soil contact pressure.

12-21. TIMBER GRILLAGE

A timber grillage consists of two or more tiers of heavy timber impregnated with creosote. Each tier is placed at right angles to the one above and below (Fig. 12-47). The several tiers are well tied together with bolts. The timber grillage is suitable for supporting temporary buildings.

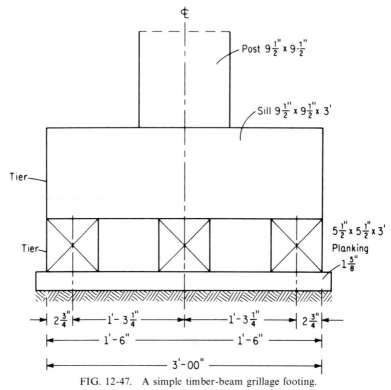

FIG. 12-47. A simple timber-beam grillage footing.

12-22. STEEL GRILLAGES

Steel grillages are constructed for the distribution on soil, rock or a concrete block of heavy loads transferred from multistory buildings such as skyscrapers, for example, by means of isolated columns. The grillages may be constructed in several tiers and encased in concrete (Fig. 12-48).

12-23. GRADE BEAMS

The general purpose of grade beams is to avoid deep foundation walls. Grade beams constitute a type of foundation which is usually used for industrial buildings in mild climates. These beams are supported on round or rectangular columns or piers. The latter are placed on a firm stratum that lies several feet below the ground surface. The piers are also used to transmit heavy localized loads through unreliable, weak layers of soil to a firm stratum. A masonry wall of a building is placed on the grade beam.

A grade beam should be stiff and reinforced as a simply supported structural member if it is to be unaffected by any differential settlement of

Calculations of Rectangular Reinforced-Concrete Footings and Mats 453

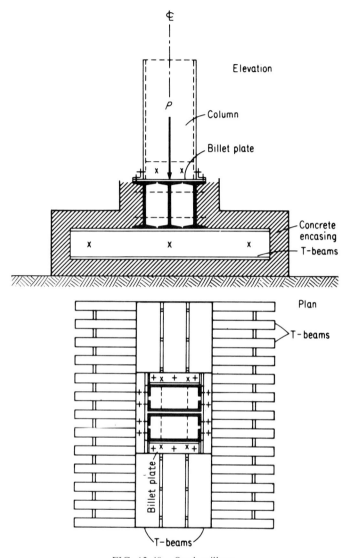

FIG. 12-48. Steel grillage.

the piers. On the other hand, the series of grade beams and piers may be tied together except at contraction joints. In such a case it is well to use a continuous beam with adequate stiffness so that it will support the masonry wall without having its deflection cause cracks in the masonry.

The grade beam should be designed as a self-supporting structural member, even though the soil may offer some bearing resistance underneath it.

To reduce the bearing resistance under the grade beam if appreciable settlement of the main footing can be anticipated, the bottom of the grade beam may be made triangular, to act as a cutting edge (Fig. 12-49). However, in cold climates the cutting edge of the grade beam is of little value. The grade beam should be designed deep enough below the frost penetration depth to prevent heaving.

Where the soil properties vary considerably from one part of the construction site to another, it may also be assumed that they vary quickly with depth, hence a deep foundation will render a better solution than a shallow one.

To minimize damage to the grade beam and the wall it supports (because of excessive deflection of the grade beam, viz., possible settlement of the soil and foundation), the grade beam is usually designed as a sufficiently stiff self-supporting member stiff enough, namely, as a simple, one-span beam

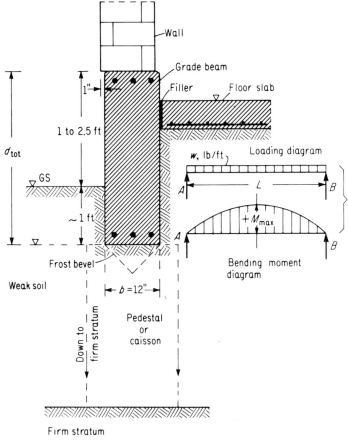

FIG. 12-49. Grade beam.

Calculations of Rectangular Reinforced-Concrete Footings and Mats

between the beam-supporting columns, or pedestals, or caissons (Fig. 12-49). The load on the grade beam is the dead load only. For example, assume a grade beam $b \times d_{tot} = (1.0) \times (3.0)$ ft^2 in cross section. If the dead load $w = w_w + w_g = 400 + (1.0)(3.0)(150) = 850$ lb per linear foot of grade beam, then the maximum bending moment M_{max} is

$$M_{max} = \frac{wL^2}{8} = \frac{(850)(18^2)}{8} \approx 34{,}430 \text{ (ft-lb)}$$

where w_w = weight of wall on grade beam, per linear foot of run of wall
w_g = weight of grade beam
3 = full depth d_{tot} of grade beam (in ft)
(1.0)(3.0) = cross section of grade beam
L = 18 ft span of grade beam

With the maximum bending moment known, the minimum required effective depth d_{eff} for the assumed 1 ft × 3 ft grade beam as well as its reinforcement can now be calculated and the stresses in the materials checked as discussed in the previous sections of footing design. If these calculations reveal inappropriate results, the grade beam must be redesigned.

REFERENCES

1. S. V. De Simone (chairman), American Concrete Institute's Committee Report No. 436, "Suggested Design Procedures for Combined Footings and Mats." *A.C.I.J.* (October 1966), pp. 1041-1057.
2. E. Winkler, *Die Lehre der Elastizität und Festigkeit*, Prague: Dominicus, 1867.
3. H. Zimmermann, *Die Berechnung des Eisenbahnoberbaues*. 2nd ed. Berlin: Ernst, 1930.
4. K. Hayashi, *Theorie des Trägers auf elastischer Unterlage und ihre Anwendung auf den Tiefbau*, Berlin; Springer, 1921.
5. F. Schleicher, "Zur Theorie des Baugrundes," *Der Bauingenieur* (1926), p. 931.
6. M. A. Biot, "Bending of an Infinite Beam on an Elastic Foundation," *J. App. Mech. Trans. ASME*, Vol. 59 (1937), pp. A1-A7.
7. J. Ohde, "Die Berechnung der Sohldruckverteilung unter Gründungskörpern," *Der Bauingenieur*, Vol. 23 (1942), pp. 99-107, 122-127.
8. M. Hetényi, *Beams on Elastic Foundation*. Ann Arbor: U. of Michigan Press, 1946.
9. K. Terzaghi, "Evaluation of Coefficient of Subgrade Reaction," *Géotechnique* (London), Vol. 5, No. 4 (1955), pp. 297-326.
10. A. B. Vesić, "Beams on Elastic Subgrade and Winkler's Hypothesis," Proceedings of the Fifth International Conference on Soil Mechanics and Foundation Engineering, Paris (1961), Paper 3A/48, Vol. 1. Paris: Dunod, 1961, pp. 845-850.

QUESTIONS

12-1. What are the stability requirements of a foundation?
12-2. What is the ultimate bearing capacity of soil?

12-3. What is safe or allowable bearing capacity of soil?
12-4. Discuss why the foundation designer must know the safe bearing capacity of a soil.
12-5. Formulate and illustrate the following terms: (a) foundation, (b) footing, and (c) pedestal.
12-6. Why are foundations constructed?
12-7. What are the objectives of foundation footing design?
12-8. Why is the proper design of the foundation important for the stability of the entire structure?
12-9. What are the consequences if a foundation fails?
12-10. What are the advantages of reinforced concrete footings? (Economy: savings in amount of excavation, materials and weight of footing.)
12-11. What kind of stresses govern the design of a plain concrete footing?
12-12. What kind of stresses govern the design of a reinforced-concrete footing?
12-13. What is understood by "critical sections" in squared, reinforced-concrete footing design and how are the "critical sections" used?
12-14. Explain and illustrate each step in designing a plain concrete wall footing.
12-15. Explain and illustrate each step in designing a plain concrete column footing.
12-16. What are the advantages of reinforced-concrete footings over plain concrete, timber and steel footings?
12-17. When would one use a reinforced-concrete mat?
12-18. What factors govern the outside dimensions of footings?
(Allowable soil bearing capacity; thickness to resist bending and the diagonal tension resulting from vertical shear; the total applied load.)
12-19. When do plain concrete footings serve well? (When loads are light and soil conditions excellent.)
12-20. State clearly how to calculate the factors of safety against sliding on its base of a foundation, against overturning of the foundation, and against failure of soil bearing capacity (groundbreak) of a soil-foundation system.

PROBLEMS

12-1. Design a centrally loaded plain concrete strip footing for a concrete wall 24 in. thick which carries a uniformly distributed load of 20 kips per linear foot of wall, including the weight of the wall. The footing is to rest on fine sand.
   1. Determine the necessary loads and allowable material stresses to be used in this design problem.
   2. Determine the safe bearing capacity of the soil.
   3. Determine the dimensions of the footing.
   4. Check the contact pressure on the soil.
   5. Check the stresses in the concrete footing.
   6. In this design, what governs the dimensions of the footing—the pressure on the soil, the shearing force or the bending stresses? (*Note*: consult the National Building Code and the specifications of the ACI.)

Calculations of Rectangular Reinforced-Concrete Footings and Mats 457

12-2. A concentrically loaded reinforced-concrete column footing supports a tied reinforced-concrete column whose horizontal cross section is 22 in. by 22 in. The column carries 400 kips of concentrated load. Design a square two-way reinforced-concrete column footing. The safe bearing capacity of the soil is given as 6 kip/ft^2. Consult the ACI Building Code. Use a 2,500-psi concrete. Other specifications are: 20,000-1,125-12.
    1. Determine footing area.
    2. Design for diagonal tension.
    3. Design for bending.
    4. Design for bond.
    5. Perform recalculations with corrected dead load.
    6. Check diagonal tension.
    7. Determine dowels for anchorage of column.
    8. What is the maximum aggregate size to use?
    9. Which factor determines the minimum thickness of the footing to avoid diagonal tension reinforcement in footing?

12-3. A wall footing as shown in the accompanying figure carries centrally 35 kip/ft of wall load, and is subjected to a clockwise 35 kip-ft moment. The dead load of the footing is 2.1 kip/ft. Determine analytically and show graphically the contact pressure distribution on soil caused by the loading conditions and report whether or not there is an uplift of the footing from the contact surface of the soil. (*Note*: 1 kip = 1,000 lb, a convenient unit of measurement used in structural engineering.)

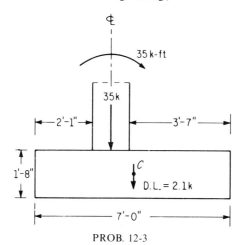

PROB. 12-3

12-4. A footing, the size of which is 5 ft × 9 ft in plan, carries a vertical load of 135 kips, and is to be subjected to a moment. To how large an axial moment in the direction of the long axis can the footing be subjected before the contact pressure on the soil exceeds (a) 5 kip·ft^2 and (b) 8 kip·ft^2.

Ans. a) $e = 1$ ft. $\sigma_{min} = 1000$ lb/ft². b) $e = 2.5$ ft. $M = Pe = 135 \times 2.5 = 337.5$ (k-ft)
$\sigma_{min} = -2$ (k-ft)²(tension; not good for $\sigma_{max} = 8$ k/ft²)
To have $\sigma_{min} = 0$, P must be on the third point ($e = 1.5$ ft)
Allowable moment:

$$M = Pe = (135)(1.5) = 202.5 \text{ (k-ft)}$$

$$\sigma_{max} = \frac{P}{A}\left(1 + \frac{6e}{L}\right) = (3)\left(1 + \frac{6 \times 1.5}{9}\right) = \underline{6.0 \text{ (k/ft}^2)}$$

For $\sigma_{max} = 5$ k/ft², the sought axial moment is

$$M = Pe = (135)(1) = \underline{135 \text{ k-ft}}$$

12-5. An 18 in. × 18 in. exterior column with a total load of 200 kips and an interior column 22 in. × 22 in. with a total load of 400 kips are to be supported on a combined rectangular footing whose exterior abuts a property line and cannot protrude beyond the outer face of the exterior column. The two columns are centered 15 ft apart. The safe soil bearing capacity is given as 6 kip/ft². Use 2,500 psi concrete. Consult ACI specifications. Design the necessary combined footing for uniform soil pressure.
  1. Determine footing area.
  2. Determine the size of the footing.
  3. Draw a principal sketch of the footing.
  4. Draw statical diagram for the footing.
  5. Draw shear diagram.
  6. Draw moment diagram of the footing.
  7. Determine and place reinforcement.

Ans.
C.G. 10 ft from center of 18 in. × 18 in. column (200 kip load)
Contact area $A = 110$ ft²
Net upward soil pressure: $600/21.5 = 27.9$ k/ft²
Maximum shears: -158 k; $+214$ k
Zero shear: 7.17 from left end
Maximum moment: $-6,800,000$ in.-lb

$$d = \sqrt{\frac{M}{Rb}} = \sqrt{\frac{68,000,000}{(196)(63)}} = 23.5 \text{ (in.)}$$

$$d = \frac{V}{vbj} = \frac{188,500}{(75)(63)(0.875)} = 45.6 \text{ (in.)}$$

Check transverse beam.

12-6. In the illustrative Example 1, on mats, calculate and plot shear force and bending moment diagrams for the outer (end) transverse beam.

12-7. In the illustrative Example 1, on mats, calculate and plot shear force and bending moment diagrams for the outer longitudinal girder 1_oG.

chapter 13

Calculations of Circular Footings

13-1. CIRCULAR FOOTINGS

Nowadays technological design frequently requires the use of circular substructures, viz., footings. Heavy, concentrated loads, circular bridge piers, gasholders, oil and other storage tanks for liquids, silos, television towers, columns, industrial chimneys, low-lift pumping plants, and (lately) round apartment houses are examples requiring circular foundations. Because of the structural advantage of an arch, round basement walls can withstand large active earth pressures as compared with straight walls of the same height (requiring unreasonably thick walls).

Circular footings are economical in the quantity of materials needed. However, this economy may be partly offset by the expense of the formwork and labor for preparing complicated types of steel reinforcement.

The calculation of a circular footing is more complex than that of a square or rectangular footing, and does not belong to the discipline of simple strength of materials but must be derived from the mathematical theory of elasticity. In the theory of circular footing design there exist a number of involved theories which are difficult to simplify, mainly because of lack of precise knowledge concerning the various factors encountered in them.

Because of lack of available, plausible knowledge on how to design circular footing slabs and rings, in practice they are designed with the assumption that the base contact pressure of the centrically loaded circular footing distributes approximately uniformly on the soil. In the case of eccentric loading one assumes linear soil stress distribution (trapezoidal stress distribution diagram). Such calculations, however, should enable the designer to satisfy himself that a given design is sound.

13-2. STATIC SYSTEM

In circular footing design one distinguishes between centrically loaded and eccentrically loaded footings. The design of a circular footing itself is based on the theory of elasticity, assuming a thin plate and that the deflections of the thin plate (viz., settlement of soil) are small. Further, it is assumed that the material of the plate is ideally elastic, homogeneous and isotropic, and stress is directly proportional to strain. Also, the plate rests on elastic support (elastic soil) with a constant coefficient of proportionality. Thus the soil pressure p (reaction) is assumed to be directly proportional to the deflection w of the plate at the corresponding point: $p = kw$, where k is the coefficient of proportionality—the so-called coefficient of subgrade reaction.

For designations of the various forces and moments acting on a circular footing element, refer to Fig. 13-1.

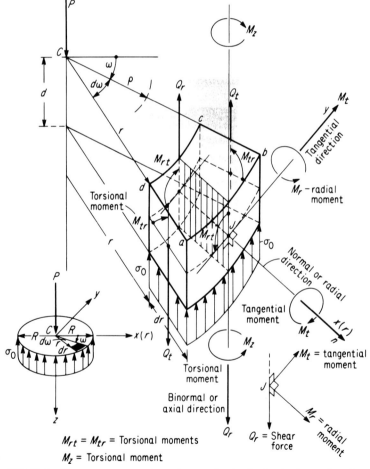

FIG. 13-1. Forces on moments in an element of a circular footing (general case).

Calculations of Circular Footings

Centrically loaded circular footings are subjected to
1. Radial bending moments M_r per unit of length which tend to produce rotation in the radial direction about tangential axes
2. Tangential moments M_t per unit of length which tend to produce rotation in the tangential direction about radial axes
3. Vertical shear forces Q_r acting on a circumferential section per unit of length

In the case of eccentric loading there also occur torsional moments M_{rt} acting perpendicular to the M_r and M_t moments (Fig. 13-1). The torsional moments are induced because of the large differential shear forces tending to deflect (vault) the slab.

Because of the symmetry of the circular slab in the case of centrical loading, the torsional moment is $M_{rt} = 0$.

In this theory the radial and tangential moments and the shear forces are functions of radius r and amplitude ω. From symmetry it can be concluded that the shear forces that may act on the element of the slab must vanish in diametric sections of the slab but that they are usually present in cylindrical sections (circumferential sections), such as sides cd and ab of the element.

The sign convention as used here is:
1. Deflections are positive when downward,
2. Moments are positive when they produce tensile stresses below the middle plane of the plate,
3. Shear forces are positive when the inner part of the plate tends to move up with respect to the outer part.

13-3. EQUATIONS FOR CALCULATING A CIRCULAR FOOTING OF UNIFORM THICKNESS, CENTRICALLY AND SYMMETRICALLY LOADED

With reference to Fig. 13-2, the equations for calculating a circular column footing loaded centrically and symmetrically according to Beyer[1] are

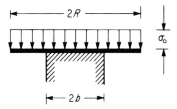

FIG. 13-2. Symmetrically-loaded circular slab system.

Radial moment

$$M_r = \frac{\sigma_o b^2}{16}[2(1-2\beta^2+\psi)-(3+\mu)C_1+(1-\mu)\psi C_4-4(1+\mu)\beta^2 C_3]$$

(13-1)

Tangential moment

$$M_t = \frac{\sigma_o b^2}{16}[2\mu(1-2\beta^2+\psi)-(1+3\mu)C_1-(1-\mu)\psi C_4-4(1+\mu)\beta^2 C_3]$$

(13-2)

There are no torsional moments present in this footing system.

Shear force

$$Q_r = -\frac{\sigma_o b}{2}\left(\rho - \frac{\beta^2}{\rho}\right)$$

(13-3)

where σ_o = average uniformly distributed soil contact pressure
$2b$ = length of side of column
μ = Poisson's ratio; for reinforced concrete, choose $\mu = 1/6$
$\beta = R/b$, a ratio of radius R of circular footing to length of half side b of column

$$\psi = \frac{k_1 + \psi_1}{k_2}\beta^2 = \text{a coefficient}$$

$$k_1 = (1+\mu)+(1-\mu)\beta^2 = \frac{7+5\beta^2}{6}$$

(13-4)

$$k_2 = (1-\mu)+(1+\mu)\beta^2 = \frac{5+7\beta^2}{6}$$

(13-5)

$$\psi_1 = 4(1+\mu)\beta^2 \ln \beta = \frac{14}{3}\beta^2 \ln \beta = \frac{14}{3}\frac{R^2}{b^2}\ln\frac{R}{b}$$

(13-6)

$$\psi = \frac{\frac{7+5\beta^2}{6}+\frac{14}{3}\beta^2 \ln \beta}{\frac{5+7\beta^2}{6}}\beta^2 = \frac{7+5\beta^2+28\beta^2 \ln \beta}{5+7\beta^2}\beta^2$$

(13-7)

$$B = 1-2\beta^2+\psi = \text{a coefficient}$$

(13-8)

$3+\mu = 19/6$ when $\mu = 1/6$

$$C_1 = 1-\rho^2$$

(13-9)

$$C_3 = \ln \rho$$

(13-10)

Calculations of Circular Footings

$$C_4 = \frac{1}{\rho^2} - 1 \tag{13-11}$$

$$\rho = r/b \tag{13-12}$$

$r =$ variable radius from center of footing

$$1 - \mu = 5/6 \text{ when } \mu = 1/6$$
$$4(1 + \mu) = 14/3 \text{ when } \mu = 1/6$$
$$2\mu B = (1/3)B$$
$$1 + 3\mu = 3/2$$

With the above designations and for $\mu = 1/6$, the moment and shear force equations may be rewritten as

$$M_r = \frac{\sigma_o b^2}{16}\left(2B - \frac{19}{6}C_1 + \frac{5}{6}\psi C_4 - \frac{14}{3}\beta^2 C_3\right) \tag{13-13}$$

$$M_t = \frac{\sigma_o b^2}{16}\left(\frac{1}{3}B - \frac{3}{2}C_1 - \frac{5}{6}\psi C_4 - \frac{14}{3}\beta^2 C_3\right) \tag{13-14}$$

$$Q_r = -\frac{\sigma_o b}{2}\left(\rho - \frac{\beta^2}{\rho}\right) \tag{13-15}$$

These equations are the results of solving the general differential equation of the bending of circular plates of constant thickness on an elastic support by the conventional method used in the theory of elasticity, as shown by A. Föppl,[2,3] Michell,[4] Melan.[5] Schleicher,[6] Flügge,[7] Reissner,[8] Timoshenko and Voinowsky-Krieger,[9] Girkmann[10] and other authors. A compilation of the various solutions for a circular slab of various systems is given by Beyer.[1]

The application of these equations may be learned most readily by an exercise in solving a circular footing design problem.

Problem

Given a column 20 in. × 20 in. to transmit a load of $P = 120$ short tons. The column rests on a circular reinforced concrete footing. The footing, in its turn, is to be laid on fine sand whose safe, allowable bearing capacity is $\sigma_{\text{allow}} = 2 \text{ ton/ft}^2$. Calculate the necessary size and thickness of the footing and plot the moment and shear force diagrams for points at the face (b) of the column, midway (at $a/2$) of the cantilevered part (a) of the footing, and at the edge (R) of the circular footing (points C, M, and E, respectively) (see Fig. 13-3). Assume uniform soil contact pressure distribution σ_o. Poisson's ratio for concrete is here given as $\mu = 1/6$. Estimate weight W of footing at 4 percent of column load, and check calculated thicknesses of footing.

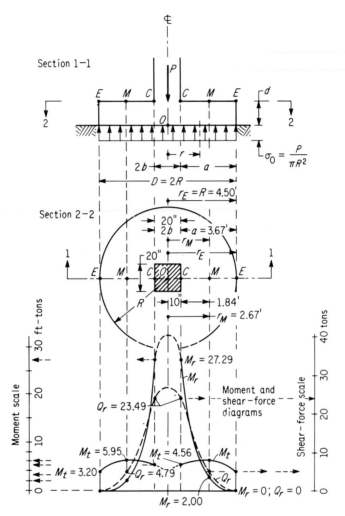

FIG. 13-3. Symmetrically-loaded circular footing.

Solution

Necessary contact area A_{nec}:

$$A_{\text{nec}} = \frac{P+W}{\sigma_{\text{allow}}} = \frac{120 + (120)(0.04)}{2.0} = \underline{62.5 \ (\text{ft}^2)}$$

Diameter of footing:

$$D = 2R = \sqrt{\frac{4A_{\text{nec}}}{\pi}} = \sqrt{\frac{(4)(62.5)}{3.14}} = 8.92 \ (\text{ft}) \approx \underline{9.0 \ \text{ft}}$$

Calculations of Circular Footings

$R = 4.50$ ft

Auxiliary quantities:

With $b = 10.0$ in. $= 10/12 = 0.833$ ft, and $R = 4.50$ ft

$$\beta = R/b = \frac{(4.50)(12)}{10.0} = 5.40; \quad \beta^2 = 29.16; \quad \ln \beta = \ln 5.40 = 1.68640 \approx 1.69$$

$$k_1 = \frac{7 + 5\beta^2}{6} = \frac{7 + (5)(5.40)^2}{6} = 25.46 \approx 25.5$$

$$k_2 = \frac{5 + 7\beta^2}{6} = \frac{5 + (7)(5.40)^2}{6} = 34.86 \approx 34.9$$

$$\psi_1 = \frac{14}{3}\beta^2 \ln \beta = (14/3)(29.2)(\ln 5.40) = 231.319$$

$$\psi = \frac{k_1 + \psi_1}{k_2}\beta^2 = \frac{25.5 + 231.3}{34.9}(29.2) = 214.62$$

$$B = 1 - 2\beta^2 + \psi = 1 - 58.4 + 214.6 = 157.2$$

C_1, C_4, and C_3 values:

1. At Face of Column	2. At Middle of Cantilever	3. At Edge of Footing
$\rho = r/b = 0.833/0.833 = 1$	$\rho = r_M/b = 2.67/0.83 = 3.22$	$\rho = r_E/b = R/b = 4.50/0.833$ $= 5.40$
$\rho^2 = 1.00$	$\rho^2 = 10.35$	$\rho^2 = 29.2$
$1 - \rho^2 = 0$	$1 - \rho^2 = 1 - 10.35 = -9.35$	$1 - \rho^2 = 1 - 29.2 = -28.2$
$1/\rho^2 = 1/1^2 = 1$	$1/\rho^2 = 1/10.35 = 0.0965$	$1/\rho^2 = 1/29.2 = 0.0343$
$\ln \rho = \ln (1) = 0$	$\ln \rho = \ln (3.22) = 1.16938$	$\ln \rho = \ln (5.40) = 1.68640$
$C_1 = 1 - \rho^2 = 1 - 1 = 0$	$C_1 = 1 - \rho^2 = 1 - 10.35$ $= -9.35$	$C_1 = 1 - \rho^2$ $= 1 - 29.2 = -28.2$
$C_4 = \frac{1}{\rho^2} - 1 = 1 - 1 = 0$	$C_4 = \frac{1}{\rho^2} - 1 = 0.0965 - 1$ $= -9.9035$	$C_4 = \frac{1}{\rho^2} - 1$ $= 0.0343 - 1 = -0.966$
$C_3 = \ln \rho = 0$	$C_3 = \ln \rho = \ln (3.22)$ $= 1.16938$	$C_3 = \ln \rho = \ln (5.40)$ $= 1.68640 \approx 1.69$

Calculation of moment and shear-force values

The values of moments and shear forces will now be calculated at the following places in the footing. With the auxiliary values as above, and introducing soil pressure $\sigma_o = 2$ ton/ft^2, proceed as set forth.

$$\frac{\sigma_o b^2}{16} = \frac{(2.0)(0.833^2)}{16} = 0.0868$$

$$\frac{\sigma_o b}{2} = \frac{(2.0)(0.833)}{2} = 0.833$$

Radial moment M_r, by Eq. 13-13:

$$M_r = 0.0868[2B - (3.16)C_1 + (178.85)C_4 - (136.30)C_3] \quad (13\text{-}16)$$

At face of column (C):

$$M_r = (0.0868)(2)(157.2) = \underline{27.29} \text{ (ft-tons)}$$

At midpoint M:

$$M_r = 0.0868[(2)(157.2) - (3.16)(-9.35) + (178.85)(-0.9035) - (136.30)(1.17)]$$
$$= \underline{2.00} \text{ (ft-tons)}$$

At edge E of footing:

$$M_r = (0.0868)[(2)(157.2) - (3.16)(-28.16) + (178.85)(-0.966) - (136.30)(1.69)]$$
$$= 0.0868(403.38 - 403.37) \approx 0 \quad \text{(checks out O.K!)}$$

Tangential moment M_t, by Eq. 13-14:

$$M_t = (0.0868)[52.40 - (1.5)C_1 - (178.85)C_4 - (136.30)C_3] \quad (13\text{-}17)$$

At face of column (C):

$$M_t = (0.0868)(52.40) = \underline{4.56} \text{ (ft-tons)}$$

At midpoint M:

$$M_t = 0.0868[52.40 - (1.5)(-9.35) - (178.85)(-0.9035) - (136.30)(1.17)]$$
$$= \underline{5.95} \text{ (ft-tons)}$$

At edge E of footing:

$$M_t = 0.0868[52.40 - (1.5)(-28.16) - (178.85)(-0.966) - (136.30)(1.69)]$$
$$= \underline{3.20} \text{ (ft-tons)}$$

Shear force Q_r, by Eq. 13-15:

$$Q_r = -0.833 \left(\rho - \frac{29.2}{\rho}\right) \quad (13\text{-}18)$$

At face of column (C):

$$Q_r = -(0.833)\left(1 - \frac{29.2}{1}\right) = \underline{+23.49} \text{ (tons)}$$

Calculations of Circular Footings

At midpoint M:

$$Q_r = -(0.833)\left(3.22 - \frac{29.2}{3.22}\right) = \underline{+4.79 \text{ (tons)}}$$

At edge E of footing:

$$Q_r = -(0.833)\left(5.40 - \frac{29.2}{5.40}\right) = -(0.833)(5.40 - 5.40) = 0 \quad \text{(checks out O.K.!)}$$

Compilation of moments and shear forces

M_r, M_t and Q_r	At Face of Column C	At Midpoint M	At Edge E
Radial moments, M_r	27.29	2.00	0.00
Tangential moments, M_t	4.56	5.95	3.20
Shear forces, Q_r	23.49	4.79	0.00

The moment and shear force diagrams are shown in Fig. 13-3. The necessary thickness of the footing is calculated according to the principles of reinforced-concrete footing design. According to the ACI Building Code, in reinforced-concrete footings the minimum thickness above the reinforcement at the edge should be not less than 6 in. for footings on soil.

Suppose that these calculations resulted in a footing whose weight is 7 tons. Then $P + W = 120 + 7.0 = 127.0$ tons. The soil contact pressure σ_o is checked as

$$\sigma_o = \frac{P + W}{A} = \frac{127.0}{\pi R^2} = \frac{127.0}{(3.14)(4.5^2)} = 1.99 \text{ (ton/ft}^2) \approx \underline{2.0 \text{ ton/ft}} \ (= \sigma_{all})$$

13-4. ECCENTRICALLY LOADED CIRCULAR FOOTINGS OF UNIFORM THICKNESS

The system of the footing under consideration is shown in Fig. 13-4. The loading consists of a centrical load P and a moment M.

The soil contact pressure σ_e at the edge of the circular footing caused by the moment M, is assumed to be linear and antimetric as shown in Fig. 13-4. It is computed from the moment equilibrium equation relative to center O of the circular footing:

$$V_{\sigma_e}(2x_o) = M \qquad (13\text{-}19)$$

where $V_{\sigma_e} = (2/3)(D^2/4)(\sigma_e) = $ "stress volume" $\qquad (13\text{-}20)$

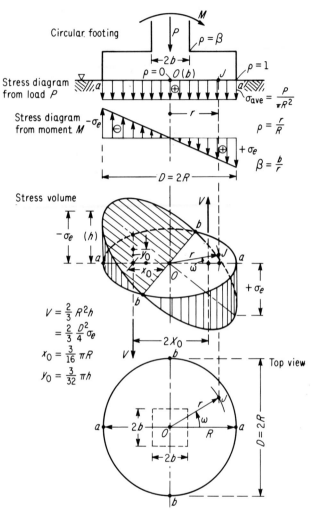

FIG. 13-4. Assumed soil-contact pressure distribution under an eccentrically loaded circular footing caused by moment alone.

$$2x_o = (2)(3/16)(\pi)(D/2) = (3/16)\pi D = \text{moment arm} \quad (13\text{-}21)$$

σ_e = edge pressure on soil from moment M

With these values the moment equation 13-19 is rewritten as

$$M = \sigma_e(2/3)(D^2/4)(2)(3/16)(\pi)(D/2) \quad (13\text{-}22)$$

and the edge pressure σ_e from moment M is

Calculations of Circular Footings

$$\sigma_e = \frac{32}{\pi} \frac{M}{D^3} \qquad (13\text{-}23)$$

This computation, of course, presupposes that because of the stiffness of the footing the soil contact pressure from the moment M varies linearly from $-\sigma_e$ to $+\sigma_e$ as indicated in Fig. 13-4.

Because of the moment (= eccentric loading), besides radial and tangential moments M_r and M_t, respectively, the circular footing is also subjected to a torsional moment M_{rt}. The torsional moment M_{rt} acts perpendicularly to the radial and tangential moments (Fig. 13-1).

The moment and shear force equations from moment M alone, after Beyer[1] for the design of eccentrically loaded circular footings, are

Radial moment M_r:

$$M_r = \frac{\sigma_e R^2}{48 K_1} [(5+\mu)K_1\rho^3 - (3+\mu)K_2\rho + 3(1+\mu)K_1\rho^{-1} - (1-\mu)K_3\rho^{-3}]\cos\omega$$

$$(13\text{-}24)$$

Tangential moment M_t:

$$M_t = \frac{\sigma_e R^2}{48 K_1} [(1+5\mu)K_1\rho^3 - (1+3\mu)K_2\rho + 3(1+\mu)K_1\rho^{-1} + (1-\mu)K_3\rho^{-3}]\cos\omega$$

$$(13\text{-}25)$$

Torsional moment M_{rt}:

$$M_{rt} = -\frac{\sigma_e R^2}{48 K_1}(1-\mu)[K_1\rho^3 - K_3\rho + 3K_1\rho^{-1} + K_3\rho^{-3}]\sin\omega \qquad (13\text{-}26)$$

In these equations, μ is Poisson's ratio, and ω is the angle of amplitude (Fig. 13-1).

Shear force in radial direction:

$$Q_r = \frac{\sigma_e R}{24}\left(9\rho^2 - 2\frac{K_3}{K_1} - 3\rho^{-2}\right)\cos\omega \qquad (13\text{-}27)$$

Shear force in tangential direction:

$$Q_t = -\frac{\sigma_e R}{24}\left(3\rho^3 - 2\frac{K_2}{K_1}\rho + 3\rho^{-1}\right)\sin\omega \qquad (13\text{-}28)$$

Auxiliary quantities

In Eqs. 13-24 to 13-28, the auxiliary quantities are

$$\beta = b/R; \quad \rho = r/R$$

$$K_1 = (3+\mu) + (1-\mu)\beta^4 \qquad (13\text{-}29)$$

$$K_2 = 4(2+\mu) + (1-\mu)(3+\beta^4)\beta^2 \qquad (13\text{-}30)$$

$$K_3 = 4(2+\mu)\beta^4 - (3+\mu)(3+\beta^4)\beta^2 \qquad (13\text{-}31)$$

In these auxiliary quantities, μ is Poisson's ratio.

Problem

Given a column 20 in. × 20 in. to transmit a centrical load of $P_1 = 120$ short tons, and an eccentric vertical load of $P_2 = 12$ short tons from a crane at an eccentricity of $e = 1.50$ ft, as shown in Fig. 13-5. At the head of the crane rail 13 ft above the base of the column there also acts a horizontal load $H = 1.2$ short tons. The column rests on a circular reinforced concrete footing of a uniform thickness. The footing, in its turn, is to be laid on a medium sand whose safe, allowable bearing capacity is $\sigma_{all} = 3.0$ ton/ft².

Calculate the necessary size and thickness of the footing and plot the moment and shear diagrams for points at the face (b) of the column, midway (at $a/2$) of the cantilevered part (a) of the footing, and at the edge (R) of the circular footing. Assume linear soil contact pressure distribution.

Poisson's ratio for concrete is here given as $\mu = 1/6$. Estimate weight W of footing at 6 percent of total vertical load $P = P_1 + P_2$ and check calculated thickness d of footing.

Solution

Assume diameter D of footing to be $D = 2R = 9.00$ ft.

Average soil contact pressure σ_o from centrical column load $P_1 = 120$ tons:

$$\sigma_o = \frac{P_1 + (0.06)(P_1 + P_2)}{A} = \frac{127.9}{(3.14)(4.50^2)} = 2.01 \approx 2.00 \ (\text{ton/ft}^2)$$

Moment M from eccentric and horizontal loads, P_2 and H, respectively:

$$M = P_2 e + Hh = (12.0)(1.50) + (1.2)(13.00) = 18.0 + 15.6 = \underline{33.6} \ (\text{ft-tons})$$

Here $e = 1.5$ ft is the eccentricity of P_2.

Edge pressure σ_e on soil from moment M, by Eq. 13-23:

$$\sigma_e = \frac{4}{\pi}\frac{M}{R^3} = (1.275)\frac{33.6}{(4.5)^3} = \underline{0.47} \ (\text{ton/ft}^2)$$

Calculations of Circular Footings

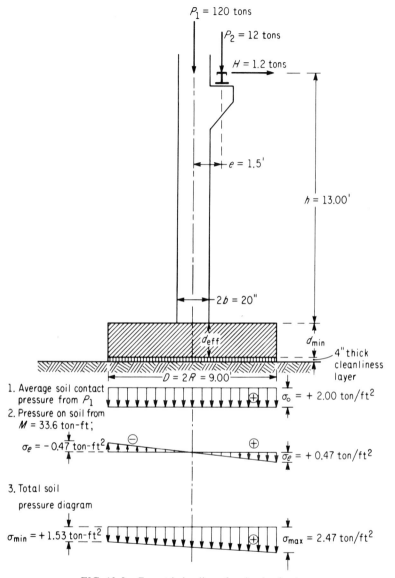

FIG. 13-5. Eccentric loading of a circular footing.

Maximum edge pressure σ_{max} on soil:

$$\sigma_{max} = \sigma_o + \sigma_e = 2.00 + 0.47 = \underline{2.47 \text{ (ton/ft}^2)} < \sigma_{all} = 3.0 \text{ ton/ft}^2$$

Auxiliary quantities:

With $R = 4.50$ ft and $b = 10$ in. $= 0.833$ ft, and with a Poisson's ratio for reinforced concrete of $\mu = 1/6$, the auxiliary quantities are

$$\beta = b/R = 0.833/4.50 = 0.185; \quad \beta^2 = 0.034235; \quad \beta^4 = 0.0011713$$

$$K_1 = (3 + \mu) + (1 - \mu)\beta^4 = \left(3 + \frac{1}{6}\right) + \left(1 - \frac{1}{6}\right)(0.185^4) = \underline{3.168} \quad (13\text{-}29)$$

$$K_2 = 4(2 + \mu) + (1 - \mu)(3 + \beta^4)\beta^2 = (4)\left(2 + \frac{1}{6}\right) + \left(1 - \frac{1}{6}\right)(3 + 0.00117)(0.03423)$$

$$= \underline{8.752} \quad (13\text{-}30)$$

$$K_3 = 4(2 + \mu)\beta^4 - (3 + \mu)(3 + \beta^4)\beta^2 = 4\left(2 + \frac{1}{6}\right)(0.00117)$$

$$- \left(3 + \frac{1}{6}\right)(3 + 0.00117)(0.03423)$$

$$= \underline{-0.3150} \quad (13\text{-}31)$$

With these auxiliary quantities on hand, the moment and shear force equations are now rewritten as

Radial moment M_r:

$$M_r = [(1.025)\rho^3 - (1.735)\rho + (0.694)\rho^{-1} + (0.0164)\rho^{-3}]\cos \omega \quad (13\text{-}32)$$

Tangential moment M_t:

$$M_t = [(0.364)\rho^3 - (0.822)\rho + (0.694)\rho^{-1} - (0.0164)\rho^{-3}]\cos \omega \quad (13\text{-}33)$$

Torsional moment M_{rt}:

$$M_{rt} = -[(0.165)\rho^3 + (0.0164)\rho + (0.495)\rho^{-1} - (0.0164)\rho^{-3}]\sin \omega \quad (13\text{-}34)$$

Shear forces:

$$Q_r = [(0.793)\rho^2 + (0.0175) - (0.2643)\rho^{-2}]\cos \omega \quad (13\text{-}35)$$

$$Q_t = -[(0.2643)\rho^3 - (0.487)\rho + (0.2643)\rho^{-1}]\sin \omega \quad (13\text{-}36)$$

(a) When $r = 0.833$ ft, $\rho = r/R = 0.833/4.500 = 0.185$ (at face of column)
(b) When $r = 4.5$ ft, $\rho = r/R = 4.5/4.5 = 1.00$ (at footing periphery = edge)

(a)

$$\rho^2 = (0.185)^2 = 0.03422$$

$$\rho^3 = (0.185)^3 = 0.00633$$

Calculations of Circular Footings

$$\rho^{-1} = 1/0.185 = 5.40$$
$$\rho^{-2} = 1/(0.185^2) = 29.2$$
$$\rho^{-3} = 1/(0.185^3) = 158.00$$

(b)
$$\rho = 1; \quad \rho^2 = 1.00; \quad \rho^3 = 1.00; \quad \rho^{-1} = 1.00; \quad \rho^{-2} = 1.00; \quad \rho^{-3} = 1.00$$

Further, for the diametral plane a-a, $\omega = 0$, $\cos \omega = \cos 0 = 1.00$, and $\sin \omega = \sin 0 = 0$. For $\omega = 90°$, $\cos 90° = 0$, and $\sin 90° = 1.00$.

(a) Moments at face of column ($\rho = 0.185$):

  (i) Radial moment M_r for section a-a ($\omega = 0°$). . . . . $M_r = $ 6.02 (ft-tons)
  (ii) Radial moment M_r for section b-b ($\omega = 90°$). . . . . $M_r = $ 0 (ft-tons)
  (iii) Tangential moment M_t for section a-a ($\omega = 0°$) . . . . $M_t = $ 1.01 (ft-tons)
  (iv) Tangential moment M_t for section b-b ($\omega = 90°$) . . . . $M_t = $ 0 (ft-tons)
  (v) Torsional moment M_{rt} for section a-a ($\omega = 0°$). . . . . $M_{rt} = $ 0 (ft-tons)
  (vi) Torsional moment M_{rt} for section b-b ($\omega = 90°$). . . . . $M_{rt} = $ -0.09 (ft-tons)

(a) Shear forces at face of column ($\rho = 0.185$):

  (i) Shear force Q_r in radial direction for section a-a ($\omega = 0°$). $Q_r = $ -7.67 (tons)
  (ii) Shear force Q_r in radial direction for section b-b ($\omega = 90°$) $Q_r = $ 0 (tons)
  (iii) Shear Q_t in tangential direction for section a-a ($\omega = 0°$) . . $Q_t = $ 0 (tons)
  (iv) Shear Q_t in section b-b ($\omega = 90°$). $Q_t = $ -1.34 (tons)

(b) Moments at edge of circular footing (point E, $\rho = 1.00$):

  (i) Radial moment M_r for section a-a ($\omega = 0°$). . . . . $M_r = $ 0 (ft-tons)
  (ii) Radial moment M_r for section b-b ($\omega = 90°$). . . . . $M_r = $ 0 (ft-tons)
  (iii) Tangential moment M_t for section a-a ($\omega = 0°$) . . . . $M_t = $ 0.22 (ft-tons)
  (iv) Tangential moment M_t for section b-b ($\omega = 90°$) . . . . $M_t = $ 0 (ft-tons)
  (v) Torsional moment M_{rt} for section a-a ($\omega = 0°$). . . . . $M_{rt} = $ 0 (ft-tons)
  (vi) Torsional moment M_{rt} for section b-b ($\omega = 90°$). . . . . $M_{rt} = $ -0.76 (ft-tons)

(b) Shear forces at edge of footing ($\rho = 1.00$):

  (i) Shear force Q_r in radial direction for section a-a ($\omega = 0°$). $Q_r = $ 0.55 (tons)
  (ii) Shear force Q_r in radial direction for section b-b ($\omega = 90°$) $Q_r = $ 0 (tons)
  (iii) Shear force Q_t in tangential direction for section a-a ($\omega = 0°$). . . . . $Q_t = $ 0 (tons)
  (iv) Shear force Q_t in tangential direction for section b-b ($\omega = 90°$). . . . . $Q_t = $ -0.04 (tons)

(c) Moments at midpoint M ($\rho = r/R = 2.67/4.50 = 0.593$):

  (i) For section a-a ($\omega = 0°$) . . . . $M_r = $ 0.43 (ft-tons)
  $M_t = $ 0.68 (ft-tons)
  $M_{rt} = $ 0 (ft-tons)

(ii) For section $b\text{-}b$ ($\omega = 90°$) .... $M_r = 0$ (ft-tons)
$M_t = 0$ (ft-tons)
$M_{rt} = -0.73$ (ft-tons)

(c) Shear forces at point M ($\rho = 0.593$):

(i) For section $a\text{-}a$ ($\omega = 0°$) .... $Q_r = -0.44$ (tons)
$Q_t = 0$ (tons)
(ii) For section $b\text{-}b$ ($\omega = 90°$) .... $Q_r = 0$ (tons)
$Q_t = -0.25$ (tons)

The moments and shear forces from centrical loading P and from resulting moment M alone are compiled in Table 13-1. The moment and shear force diagrams are shown in Fig. 13-6.

The maximum moments are

$$M_r = +33.31 \text{ ft-tons at face of the column } (C)$$
$$M_t = +6.62 \text{ ft-tons midway (point } M)$$
$$M_{rt} = -0.76 \text{ ft-tons at edge of footing (point } E)$$

TABLE 13-1
Compilation of Moments and Shear Forces

Moments and Shear Forces		In Section $a\text{-}a$ $\omega = 0°$			In Section $b\text{-}b$ $\omega = 90°$		
		At Face of Column C	Midway M	At Edge E	At face of Column C	Midway M	At Edge E
M_r	from P	+ 27.29	+ 2.00	0.00	+ 27.29	+ 2.00	0.00
	from M	+ 6.02	+ 0.43	0.00	0.00	0.00	0.00
		+ 33.31	+ 2.43	0.00	+ 27.29	+ 2.00	0.00
M_t	from P	+ 4.56	+ 5.95	+ 3.20	+ 4.56	+ 5.95	+ 3.20
	from M	+ 1.01	+ 0.68	+ 0.22	0.00	0.00	0.00
		+ 5.57	+ 6.63	+ 3.42	+ 4.56	+ 5.95	+ 3.20
M_{rt}	from P	0.00	0.00	0.00	0.00	0.00	0.00
	from M	0.00	0.00	0.00	− 0.09	− 0.73	− 0.76
		0.00	0.00	0.00	− 0.09	− 0.73	− 0.76
Q_r	from P	+ 23.49	+ 4.79	0.00	+ 23.49	+ 4.79	0.00
	from M	− 7.67	0.00	+ 0.55	0.00	− 0.25	0.00
		+ 15.82	+ 4.79	+ 0.55	+ 23.49	+ 4.54	0.00
Q_t	from P	0.00	0.00	0.00	0.00	0.00	0.00
	from M	0.00	0.00	0.00	− 1.34	− 0.25	− 0.04
		0.00	0.00	0.00	− 1.34	− 0.25	− 0.04

Note: The values for moments and shear forces from concentric load $P = 120$ tons are taken from compilation in the problem in Sec. 13-3.

Calculations of Circular Footings

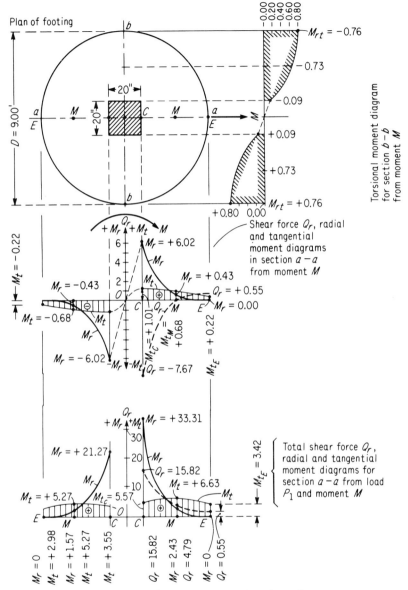

FIG. 13-6. Resultant moment and shear force diagrams.

The maximum shear forces are

$$Q_r = +23.49 \text{ tons at face of the column (point } C)$$
$$Q_t = -1.34 \text{ tons at face of the column (point } C)$$

With the magnitudes of the moments and shear forces known, the eccentrically loaded circular footing can be proportioned and the necessary steel reinforcement calculated.

Example: Circular Structure Founded on Soil

If the eccentrically loaded structure is to be founded on soil, then the diameter D of the circular footing can be calculated by means of the following equation:

$$\sigma_{max} = \frac{\Sigma V}{A} + \frac{\Sigma M}{Z} = \frac{\Sigma V}{\pi D^2/4} + \frac{\Sigma M}{\pi D^3/32} \qquad (13\text{-}37)$$

where σ_{max} = maximum soil contact pressure, kips/ft², not to exceed safe allowable soil bearing capacity σ_{all}
ΣV = sum of all vertical loads acting on bearing area A
A = bearing area ($A = \pi D^2/4$), ft²
ΣM = sum of all moments, kip-ft
$Z = \pi D^3/32$ = section modulus of bearing area A, ft³

For example, if $\Sigma V = 1{,}000$ kips, $\Sigma M = 1{,}400$ kip-ft, and $\sigma_{max} = \sigma_{all} = 4.0$ kip/ft², then Eq. 13-37 can be rewritten as

$$4.0 = \frac{1{,}000}{(\pi/4)D^2} + \frac{1{,}400}{(\pi/32)D^3}$$

resulting in the following cubic equation:

$$D^3 - (318)D - (3{,}566.9) = 0 \qquad (13\text{-}38)$$

$$\therefore \quad D \approx 21.9 \text{ (ft) or, say, } \underline{D = 22.0 \text{ ft}}$$

The area of the circular footing is thus

$$A = (\pi/4)(22.0)^2 = \underline{396 \text{ ft}^2}$$

The section modulus Z of this circular area is thus

$$Z = (\pi/32)D^3 = (0.09812)(10{,}648) = 1{,}044.78 \text{ (ft}^3) \approx \underline{1{,}045 \text{ ft}^3}$$

The thickness d of the circular footing depends upon the bending moment stresses brought about by the net upward pressure of the soil. The net upward pressure of soil σ_{net} is computed by subtracting the weight of the circular slab (mat) (here for the sake of illustration given as, say $W_{mat} = 75$ kips) from the total pressure ΣV, i.e.,

$$\sigma_{net} = \frac{(\Sigma V - W_{mat})}{A} \pm \frac{\Sigma M}{1{,}045} = \frac{1{,}000 - 75}{396} \pm \frac{1{,}400}{1{,}045}$$

$$= \frac{925}{396} \pm \frac{1,400}{1,045} \text{ (kip/ft}^2)$$

$$\sigma_{max} = 2.337 + 1.339 = \underline{3.68} \text{ (kip/ft}^2)$$

$$\sigma_{min} = 2.337 - 1.339 = \underline{1.00} \text{ (kip/ft}^2)$$

REFERENCES

1. K. Beyer. *Die Statik im Eisenbetonbau*, 2nd ed. Berlin: Springer, 1956, p. 649.
2. A. Föppl, "Die Biegung einer kreisförmigen Platte," *Sitzungsberichte der Bayerischen Akademie der Wissenschaften* (1912), p. 155.
3. A. Föppl and L. Föppl, *Drang und Zwang*, 3rd. ed. Munich and Berlin: Oldenbourg, 1941, Vol. 1.
4. J. H. Michell, *Proceedings of the London Mathematical Society*, Vol. 34 (1902), p. 223.
5. E. Melan, "Die Berechnung einer exzentrisch durch eine Einzellast belasteten Kreisplatte," *Eisenbau*, No. 17 (1926), p. 190.
6. F. Schleicher, *Kreisplatten auf elastischer Unterlage*. Berlin: Springer, 1926.
7. W. Flügge, *Die strenge Lösung von Kreisplatten unter Einzellasten*. Berlin: Springer, 1928.
8. H. Reissner, "Über die Biegung der Kreisplatte mit exzentrischer Einzellast," *Math. Annalen*, Vol. 111 (1935), p. 777.
9. S. Timoshenko and S. Voinowsky-Krieger, *Theory of Plates and Shells*, 2nd ed. New York: McGraw-Hill, 1959, pp. 51-78.
10. K. Girkmann, *Flächentragwerke*, 6th ed. Vienna: Springer, 1963.

part **IV**

DEEP FOUNDATIONS

chapter **14**

Open Caissons

14-1. DEFINITION AND DESCRIPTION

In practice, a foundation is considered, arbitrarily, to be deep if its depth d exceeds about twice the width B of the base of the footing, i.e., if $d > 2B$.

Deep foundations are usually used where soil properties are inferior for shallow foundations. Also, if a firm stratum of soil or competent rock is so deep that it cannot feasibly be reached economically by ordinary types of foundations (say, shallow foundations), then one resorts to deep foundations. Thus deep foundations are those laid at considerable depth below the ground surface to reach, rest on, and to transmit structural loads safely to a firm geological stratum of adequate bearing capacity.

The depth limit of deep foundations so far attained is about 260 ft = 79.3 meters (for the foundation of the Tagus River suspension bridge at Lisbon, Portugal, completed in 1966, where the depth of one of the several caissons is 260 ft).

A caisson for the San Francisco-Oakland bridge (completed in 1937) was sunk to a depth of 242 ft = 73.8 m. One of the open caissons for the bridge at Mackinac Straits, Michigan, is founded 200 ft = 61.0 m below water, and one of the Verrazano-Narrows Bridge pier foundations is placed at a depth of 172.65 ft = 52.6 m below water level.

For purpose of comparison, it is interesting to note that the thickness of the earth's crust is estimated to be a minimum of 30-40 km and a maximum of 120 km.

Deep foundations are used in the following applications:

1. For the transmission of structural loads through weak, nonuniform, compressible soils and/or deep water to a firm soil.
2. When a groundbreak or rupture of soil underneath a shallow foundations can be anticipated.
3. When the soil conditions are such that a washout and erosion of soil from underneath a shallow foundation may take place.

4. When a competent load-bearing soil lies at great but attainable depth consistent with economy and present-day technology.
5. When the design of a structure calls for unequal and nonuniform loads on soil.
6. When the plan of the structure is irregular relative to its outline and load distribution, presenting the possibility of nonuniform, intolerable settlements of the soil and thus of the structure.

The commonest types of deep foundations are caissons, deep piers, solid piles, and tubular or cylinder piles.

14-2. TYPES OF CAISSONS

By definition, a caisson (Fr. *caisson* = box) is a watertight box or chamber used for laying foundations under water, as in harbors, rivers, lakes, and even on land. A caisson is one of the principal types of deep foundations.

Caissons are utilized for laying foundations at greater depths than shallow foundations. The heavy loads of bridges and multistory structures, concentrated at certain points, can be transmitted to deep-lying, firm soil and/or rock by means of caissons. In open water and shallow depth, caissons may be sunk from an artificial sand island.

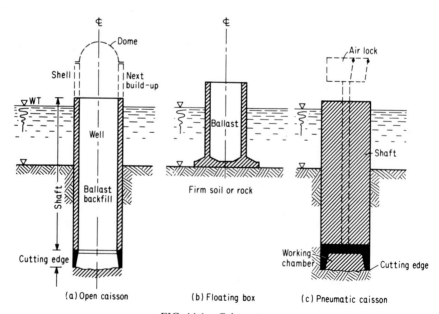

FIG. 14-1. Caisson types.

Open Caissons

There are two principal types of caissons:

1. Open caissons
   (a) Well type (top and bottom open to air).
   (b) Floating type (open top and closed bottom). During sinking operations, the chamber of an open caisson is exposed to atmospheric pressure.
2. Pneumatic caissons
   During sinking operations there is maintained in the chamber of the pneumatic caisson a pressure (compressed air) greater than atmospheric in order to keep water from entering the caisson.

Caissons may be cylindrical, elliptical, or rectangular in plan.

The fundamental difference among these three types of caissons is illustrated in Fig. 14-1.

Other types of caissons are: belled-out caissons (Fig. 14-2), step caissons,

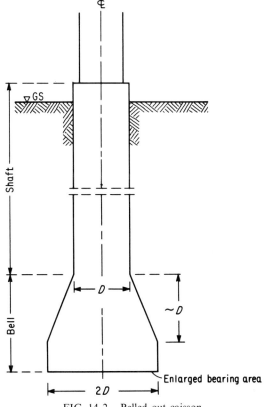

FIG. 14-2. Belled-out caisson.

inclined caissons, and special caissons. All caissons, however, have a common element, namely, a permanent shell, which is an integral part of the foundations of the bridge, building or hydraulic structure.

14-3. OPEN CAISSONS

Open caissons are hollow cylinders made of timber, brick, masonry, concrete, reinforced concrete, or steel. Today, reinforced concrete and sometimes steel are the most usual materials applied in constructing a caisson.

Open caissons are used where cofferdams are infeasible, or when a firm load-supporting soil is located deeper than approximately 15 ft (≈ 5 m) below a saturated soil. Here the 15-ft depth is meant merely for one's orientation, as a kind of yardstick in classification for the transition between shallow and deep foundations.

Also, open caissons find their application in groundwater when the firm, load-bearing soil is situated fairly deep below ground surface. Or they are used where other kinds of foundation would be uneconomical if the foundation required is small in plan area in relation to the depth of water.

Open caissons are not recommended when the soil to be penetrated contains obstructions such as tree trunks, logs, roots, boulders, structural remnants, and other obstacles. Also, they are not applicable at sites adjacent to existing structures which are not founded as deeply as the caissons is to be penetrated; upon excavating the soil from inside of the caisson, the soil outside of the caisson may yield easily and thus impair the adjacent structures.

14-4. ADVANTAGES AND DISADVANTAGES OF OPEN CAISSONS

Some advantages of open caissons as compared with pneumatic are
1. Speed
2. Penetration to great, practically unlimited depth
3. Economy

Some disadvantages of open caissons are
1. Ineffectiveness where an influx of soil into the excavation takes place faster than it can be dredged out. In coping with such conditions the pneumatic caisson (for depths corresponding to less than 4 atm pressure) is more effective than the open one.
2. Difficulty of cleaning at bottom of caissons through water.
3. Problems in placing a concrete seal in the caisson through water.

Open Caissons 485

4. Delay in sinking when obstacles to penetration are encountered in soil.
5. Open caissons cannot be as easily controlled as pneumatic caissons, nor as accurately located. Generally, the larger the caisson the easier it is to keep plumb and the more accurately it can be landed. To keep small open caissons in plumb, they must be shored.

Open caissons can be sunk to great depths. Therefore, as a kind of deep foundation, the open caisson,—as compared with all other types of deep foundations,—is superior not merely technically but also economically.

14-5. ELEMENTS OF OPEN CAISSONS

The basic elements of an open caisson are
1. Its *shell* (wall), termed also the *shaft*
2. A knife-edge or cutting edge provided at the lower end of the caisson
3. Its backfill (ballast)

Shaft

The shaft is needed for excavating the soil in order to enable the caisson to penetrate into the soil, for access of the working crew to and from the well of the caisson, and for the conveyance of construction material such as concrete for the seal and ballast for the caisson.

Cutting Edge

A cutting edge of a caisson, open or pneumatic, is the terminating rim of the caisson. It is in direct contact with the soil. The purpose of the cutting edge is to facilitate the sinking (penetration) of the caisson through the weak strata down to a firm, load-bearing soil or rock. Also, it strengthens the lower part of the well (shaft).

The work of sinking the caissons begins with placing the cutting edge at the bottom of the foundation pit (or on a river bed, or on an artificial island). Shell sections are added during the sinking operations of the caisson.

The soil is excavated through water from the inside of the caisson and removed by means of grab buckets, for example. On removing the soil from underneath the cutting edge, the caisson sinks down deeper into the soil, normally by its own weight. The sinking of the caisson should be done uniformly, with no jerks, and vertically to plumb.

Sinking of a caisson may be aided by a dead weight placed on top of the well. During its penetration, the caisson is built up by joining additional sections to the shaft. The weight of the added sections also assists in sinking the well.

Ballast

When the well has attained its designed depth, a concrete seal is poured at the bottom of the well by a method of underwater concreting. After setting the concrete seal, the water is pumped out of the well; the caisson is completed. The caisson is now filled with a ballast such as concrete, or sand, or rocks, for example. After the concrete ballast has been poured in, the top of the caisson is covered with a reinforced concrete slab on which the structural load is supported.

14-6. FORM OF OPEN CAISSONS

In cross section, the caisson cylinders may be circular, elliptical, quadratic, rectangular or of any other geometric form. The most advantageous cross-sectional form of an open caisson is the circular one because

1. Its perimeter is the minimum of all known geometric figures. This means the least frictional mantle resistance to soil upon sinking the caisson into the soil.
2. Upon excavation in a spiraled fashion (Fig. 14-3) the soil falls uniformly from all sides toward the center of the well. This affords a uniform, untilted penetration of the caisson.

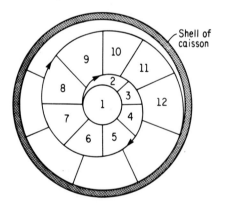

FIG. 14-3. Excavation in a spiraled fashion.

3. It takes the external lateral pressure of soil and water well, inducing compressive stresses only in the circular wall material of the well.

The main disadvantage of a circular-symmetrical open caisson is that upon penetration it has a tendency for a spiraled axial rotation which may result in its getting out of alignment in a row with other caissons. An

Open Caissons

eventual tilting of the open circular caisson is hardly to be conceived because of its complete symmetry, and if any tilting occurs, it is relatively easy to correct by righting it back into plumb.

The use of asymmetric caissons should be avoided if possible because of the many difficulties encountered in their sinking process and nonuniform stress and/or moment distributions in the caisson material, although elliptic, quadratic, and rectangular caissons have indeed been sunk successfully.

To reduce side friction between soil and shell upon sinking, caissons are sometimes tapered longitudinally with a slope of 1/15 to 1/7.5 of their height, flaring out at the bottom. However, the flaring-out may impede the downward guidance of the caisson. The sinking and penetration of open and pneumatic caissons may be facilitated by the use of thixotropic fluids.

The empty space between the soil and cutting edge (viz., outside wall of caisson) brought about in the penetration process of the caisson is filled with a thixotropic fluid (bentonite slurry, for example). During the sinking of the caisson, the thixotropic fluid reduces the friction on the external mantle surface of the caisson shell and stabilizes the steep bank of the soil through which the caisson is penetrating. Reduction of friction reduces weight or ballast needed for penetration of the caisson. After completion of the sinking operation, the slurry is removed and the space is backfilled with soil. One must be watchful, though, of slanting and righting problems of the caisson if such external circumferential spaces between soil and caisson occur.

14-7. SIZE OF OPEN CAISSON

The size of the base, viz., contact area, of an open caisson, is proportioned for the load to be transmitted to the soil, taking into consideration the allowable soil bearing capacity. In these calculations the wall friction between soil and shaft material is usually ignored.

The minimum inside diameter of the shaft or well must be such as to facilitate easy hoisting of soil and materials, and to provide a convenient working space.

In foundation work it is advantageous to have fewer large caissons than many little ones. This is because

1. Large caissons permit construction of thick, strong walls, hence they are easier to sink—sometimes even without extra surcharge—than small caissons.
2. Large caissons can be spaced farther apart than small ones, hence less side deviation from alignment occurs than would be the case with many small caissons.

Small caissons, however, have the advantage that their length of shaft may be easily extended with consecutive sections of rings.

The thickness δ in meters of the wall of an open masonry or concrete caisson is chosen empirically by experience as

$$\delta \approx 0.10 + (0.10)D_o \quad (m) \quad (14\text{-}1)$$

where D_o = external (outer) diameter, in meters, of a circular cylindrical caisson.

For reinforced-concrete shells,

$$\delta \approx 0.10 + (1/12)D_o \quad (m) \quad (14\text{-}2)$$

The length or depth of an open caisson is determined by its penetration depth through soil and/or water plus a freeboard. The penetration depth, in its turn, is determined from the geological and soil profile at the site. Thus, for example, the caisson 197 ft × 97 ft in cross section in the plan of the central anchor pier of the San Francisco-Oakland Bay Bridge is 220 ft below water level,[1-4] and one of the Mackinac Straits Bridge open circular steel caissons for a bridge pier 19, 116 ft in diameter was sunk to bedrock 206 ft deep below the surface of the straits.[5]

14-8. EXCAVATION; SINKING OF OPEN CAISSON

The sinking of an open caisson may have to be accomplished

1. By sinking the caisson well through open water and supporting it on soil or bedrock
2. By sinking the well through water and penetrating it into the soil
3. By penetrating the caisson in the dry, or from a dewatered construction site, or from an artificial sand island.

The sinking operation of a caisson itself is accomplished by its own weight plus that of its next build-up (this assists sinking as the excavation proceeds), or by placing on top of the caisson a ballast, and by excavating the soil from inside the caisson, say, by means of a clamshell bucket through the open shaft (Fig. 14-4).

Normally, excavation from an open caisson should always be done from the middle of the caisson. The caisson, by its self-weight, should cut off and push the soil from under the cutting edge toward the middle (already excavated part) of the well. This is possible in small wells. In large wells the excavation may be pursued from the middle of the well spirally outward toward the shell of the well. Elliptic wells may be excavated as indicated in Fig. 14-5.

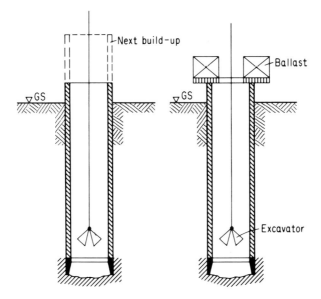

FIG. 14-4. Sinking of caissons.

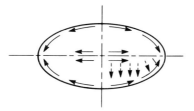

(a) Excavation of an open, elliptic caisson

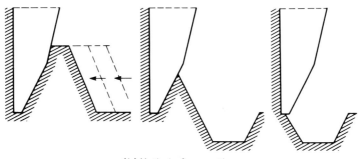

(b) Method of excavation

FIG. 14-5. Excavation of an open caisson.

490   Deep Foundations

Excavation of soil from a caisson manually is undesirable because the caisson may rest for too long a time and may "seat in" in soil—that is, bind into or adhere to the soil before further penetration can be brought about. Fast excavation is best accomplished by means of mechanical excavators.

14-9. FORMS OF CUTTING EDGES

In sinking a caisson its weight is at first carried on its cutting edge, which may have various shapes as shown in Fig. 14-6. As the caisson gradually penetrates the soil, more of the weight is resisted by friction on the mantle

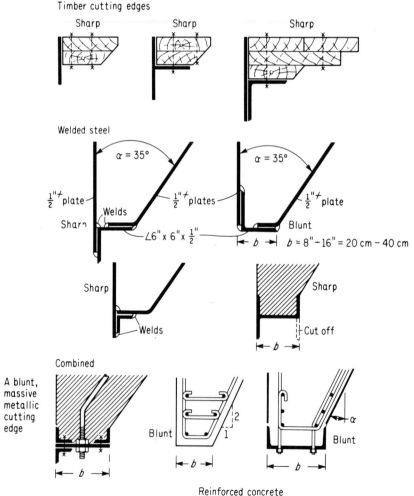

FIG. 14-6. Cutting edges.

Open Caissons

surface of the shaft and by bearing on the cutting edge. Thus the cutting edge is always in contact with the caisson-supporting soil.

The cutting edge not only supports the caisson but also facilitates its easy penetration in soil, and protects the shell of the well against impact and various obstacles encountered in its path of penetration (boulders, rock, tree trunks).

Depending upon the size and weight of open caissons, and the load they have to sustain, cutting edges are made of timber or angles and plates of structural steel, or reinforced concrete, or of a combination of timber and steel, or reinforced concrete and steel (Fig. 14-6).

Sharp cutting edges are easily damaged on hitting obstructions. The blunt cutting edge is in this respect less vulnerable, hence it is more widely used than the sharp forms.

The angle α of the inside slope of the cutting edge should not be larger than $35°$. Sometimes it is indicated as $h:v = 1:2$. However, a too-steep slope causes large stresses in the cutting edge by wedge action on penetrating the soil.

To avoid tearing off the cutting edge or any lower parts of the shell of the well because of a change in mantle friction, the shell masonry must be anchored (or tied together) to the cutting edge.

14-10. THICKNESS OF CONCRETE SEAL

The height of the concrete seal must withstand the hydrostatic uplift of the water. Referring to Fig. 14-7, the thickness t of the seal is calculated as

$$t = \frac{\gamma_w H}{\gamma_c} \tag{14-3}$$

where γ_w = unit weight of water
H = hydrostatic head
γ_c = unit weight of concrete

Before pouring the ballast into the caisson, the surface of the concrete seal must be cleaned of scum, cement milk, and other kinds of impurities and dirt.

If the concrete seal can be considered as a uniformly loaded circular slab supported along its periphery against the beveled cutting edge of the well, then, by the theory of elasticity, the thickness t of the seal is calculated as

$$t = \frac{1}{2}\sqrt{\frac{3}{2}\frac{W}{\pi\sigma_f}\left(\frac{3m+1}{m}\right)} \tag{14-4}$$

where W = total load on seal of caisson

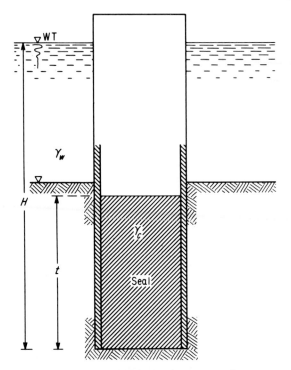

FIG. 14-7. Thickness of concrete seal.

$\sigma_f \approx 250$ lb/in.² = flexural strength of concrete seal (from 200 to 500 lb/in.²)

$m = 1/\mu = 1/0.15 = 6.66$ for concrete (Poisson's number)

R_s = radius of base of circular seal (Fig. 14-8)

The load W is the uplift by water, and is calculated as

$$W = \gamma_w H \frac{\pi}{4} 4 R_s^2 = \pi \gamma_w H R_s^2 \tag{14-5}$$

The radius to use in Eq. 14-5, whether R_s or R_o, depends upon the design and execution of the attained position of the caisson after its sinking is completed (Fig. 14-8). In the case of R_i, use Eq. 14-3.

For a rectangular caisson, the thickness t of its seal is calculated as

$$t = \frac{b}{2}\sqrt{\frac{3}{1+1.61\lambda}\frac{\sigma_o}{\sigma_f}} \tag{14-6}$$

where b = short side of rectangular caisson
λ = ratio of short side to long side of caisson

Open Caissons

σ_o = soil (rock) contact pressure at base of caisson and other symbols are the same as before

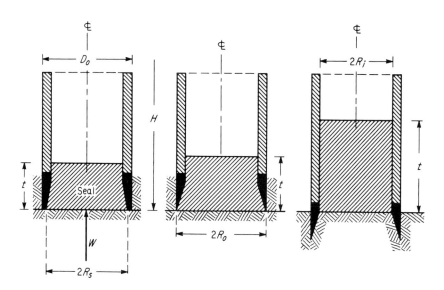

FIG. 14-8. Radii of concrete seals.

14-11. OPEN CAISSONS WITH DREDGING WELLS

Large rectangular open caissons are provided with circular dredging wells. For example, the caisson of the west anchorage pier of the first Delaware Memorial Bridge, 95 ft × 221 ft in size, contained five rows of 12 dredge wells each 15 ft in diameter. The cutting edge of the caisson was 13 ft high and the outside walls were 4 ft thick. The final depth of the caisson was −92.7 ft when the caisson was concreted in with a seal 15 ft thick above the cutting edge. The concrete was then carried upward in 5-ft lifts, and a reinforced-concrete load-distribution block 10 ft thick was placed.[6,7]

Figure 14-9 is a photograph showing the 20-ft-high cutting edge and four of 28 dredging wells 18 ft in diameter of an all-walled caisson 151 ft by 88 ft in size weighing 1,000 tons. The caisson was fabricated at Dravo Corporation's Neville Island plant, near Pittsburgh, and formed the base of Pier 2 of the Mississippi River Bridge at New Orleans, built by the company in the mid-1950's. The caisson rests on a stratum of hard clay 180 ft below water level, and was placed on a fascine mattress to prevent scour around the bridge pier.

The word *fascine* originates from Latin *fascina*, a bundle of sticks of wood; a bundle of twigs—shoots of branches of trees (usually willow).

Thus, as an engineering material, fascines are bundles of twigs approximately 5 to 30 ft long bound together, and are used in river regulation works, for making revetments for river banks, for earth dike construction; fortifications, and in other hydraulic engineering structures. Also, fascines

FIG. 14-9. All-welded steel cutting edge. (Courtesy of Dravo Corporation.)

are bound together to form wide mats or mattresses which are used in hydraulic structures and in foundation engineering to prevent erosion of soil and scour at bridge piers, for example.

The open-caisson suspension cable support pier No. 18 of the Mackinac Straits Bridge, 92 ft × 44 ft in cross section, has 21 dredging wells, each 9 ft in diameter (Fig. 14-10). Such a caisson was sunk through water and bottom sediments by dead weight of water, rocks, and concrete filled around the hollow walls. As the caisson was sunk, the bottom sediments were dredged out through the open wells. After the caisson attained its final depth of 130 ft and was sitting firmly on bedrock, a concrete seal was poured. Then the circular steel wells were cut off above the seal and salvaged. The inside of the caisson was then filled with stone and concrete. The lower 48 ft section of this pier was constructed in a shipyard and towed to the site.

Open Caissons

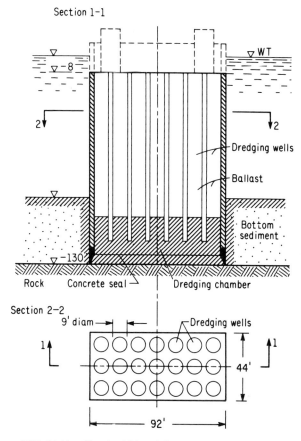

FIG. 14-10. Sketch of Pier of the Mackinac Straits Bridge.

Open, rectangular caissons with circular cells over square-cell dredging wells were used for the construction of the tower piers of the Verrazano-Narrows Bridge in New York. The caissons, 229 ft × 129 ft in plan,[8] were sunk to the required elevations by dredging the overburden with clamshell buckets through the dredging wells of the caisson, which were open top and bottom, and the bottom of the caisson was sealed with tremie concrete, 20 ft and 22 ft thick in the Staten Island and Brooklyn caissons respectively. The wells were then closed at the top with precast concrete covers. The wells were left filled with water. A distribution slab 4 ft thick was cast over the entire area of the top of the caisson. The caissons were sunk 107.65 ft deep below water surface at the Staten Island side, and 172.65 ft deep below water surface at the Brooklyn side. The walls of the caissons are 5 ft thick.

The sinking of the caissons was started out from artificial "sand islands." The sand-island fill was enclosed and kept by a series of sand-filled cellular sheet-pile cofferdams, and was kept dry by a wellpoint dewatering system.

A large circular open caisson for the North Main Tower pier of the Mackinac Straits Bridge is shown in Fig. 14-11. This caisson, 116 ft in

FIG. 14-11. Dredging of bottom sediment and sinking of the circular caisson for the North Main Tower of the Mackinac Straits Bridge to bedrock 200 feet below the water surface. (Courtesy and by permission of Steinman, Boyton Gronquist and London, Consulting Engineers, New York.)

outer diameter, is formed by two concentric shells. The space between these shells is subdivided by radial diaphragms into eight watertight compartments each 15 ft wide. The inner space (86 ft in diameter) is, thus, one huge dredging well. The two concentric shells taper down to a thin cutting edge at the toe. The sinking of the caisson was brought about by flooding the eight side compartments. As the sinking of the caisson progressed, its shaft was extended above the water by building it up with prefabricated panels, and the caisson continued to be sunk to bedrock, whereupon the dredging well as well as the eight side compartments between the two concentric shells were progressively concreted full to a certain

Open Caissons

design elevation. This pier was sunk 200 ft below the water surface to bedrock.

14-12. DOME-CAPPED CAISSONS

Sometimes the ordinary cylinders or dredging wells encountered in all open caissons are capped with removable domes to allow compressed air pressure to be put in the cylinders (Fig. 14-12). This aids the floating, sinking, and stabilization operations of the caisson by increasing or decreasing the pressure in selected groups of cylinders. Also, this method allows easy conversion into pneumatic-type caissons, allowing the floor to be inspected without having to dive down. Such air-domed cylinder caissons were used for the foundations of the San Francisco-Oakland Bay Bridge piers.[9,10] This type of caisson is credited to D. E. Moran.[9,10]

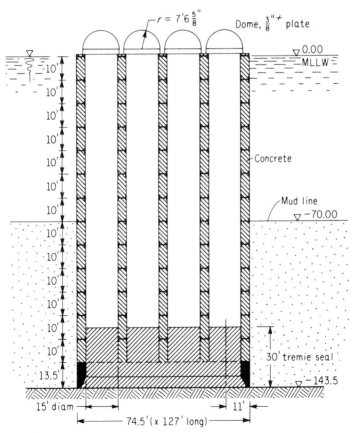

FIG. 14-12. Moran's dome-capped caisson. Sketched after Fig. 5, p. 11, Ref. 9.

The domed caisson of the west pier No. 4 of that bridge had a size of 197 ft × 92 ft in plan with 55 domed cylinders, and was sunk to a depth of 180 ft. Pier E-3 of the same bridge was sunk to a depth of 240 ft.

The sinking of the caissons was accomplished by introducing water into the cylinders, thus lowering the center of gravity of the caisson, and maintaining the air pressure.

After contact with the bay deposits, such as silt and mud, was made the air was released, and the caisson commenced to sink into the mud by its own weight. The dredging of the mud took place through the cylindrical dredging wells (domes off), whereas some of the air-contained domed cylinders were retained for stabilization and righting of the caisson during its sinking.

Another conspicuous example of using dome-capped caissons was the construction of the foundations and piers of the Tagus River Bridge at Lisbon, Portugal[11] (Fig. 14-13). There the sinking depth was 260 ft below the water surface to bedrock, or 18 ft deeper than the record, held by one of the main tower piers of the San Francisco-Oakland Bay Bridge.

As with the San Francisco-Oakland Bay Bridge, so in the case of the Tagus River Bridge the compressed air in the domed cylinders was also utilized

FIG. 14-13. Dome-capped caisson for the Tagus River Bridge at Lisbon, Portugal. (Courtesy and by permission of Steinman, Boynton, Gronquist and London, Consulting Engineers, New York.)

Open Caissons

for maintaining buoyancy and for stabilizing the caisson during its sinking. By varying the pressure in different well cylinders, sinking and righting the list of the caisson could be controlled.

Sometimes, to cope with an irregular riverbed configuration, the base of a caisson must be adapted to a similar irregularity at its lower end. Figure 14-14 shows the caisson of the tower pier No. 4 of the Tagus River Bridge built on a sloping rock line. The cutting edge of the caisson had to be made to conform to that slope.

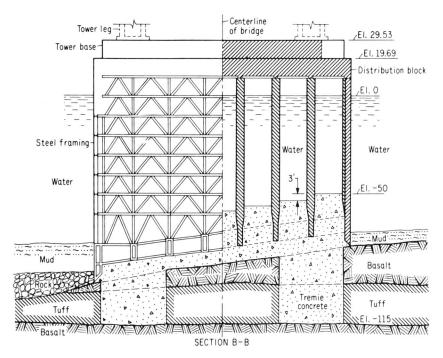

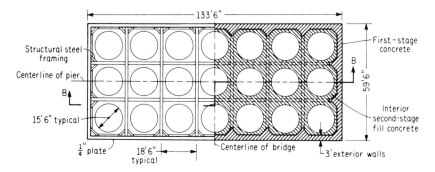

FIG. 14-14. The Tagus River Bridge pier built on a sloping rock. (Ref. 11.)

REFERENCES

1. C. H. Purcell, C. E. Andrew, and G. B. Woodruff, "Deep Open Caissons for Bay Bridge," *Engineering News-Record*, Vol. 113, No. 8 (Aug. 23, 1934), pp. 227-233.
2. C. S. Proctor, "Foundation Design for Trans-Bay Bridge," *Civil Engineering*, Vol. 4, No. 12 (December 1934), pp. 617-621.
3. C. S. Proctor, "Constructing Foundations of Trans-Bay Bridge," *Civil Engineering*, Vol. 5, No. 2 (February 1935), pp. 91-95.
4. C. H. Purcell, C. E. Andrew, and G. B. Woodruff, "Difficult Caisson Problems Overcome in Sinking Deep Piers," *Engineering News-Record*, Vol. 114, No. 7 (Feb. 14, 1935), pp. 239-242.
5. Anon., "Mackinac Bridge—Incredible but True," *Engineering News-Record*, Jan. 27, 1955, pp. 35-44.
6. R. C. De Simone, "Delaware Memorial Bridge Substructure," Boston Society of Civil Engineers, July 1952, pp. 271-283.
7. H. R. Seely, "Construction of the Delaware Memorial Bridge," *Proc. ASCE*, August 1952, pp. 40-43.
8. L. H. Just, I. G. Levy, and V. Obrcian, "Verrazano-Narrows Bridge: Design of Tower Foundations and Anchorages," Conference Preprint No. 121, presented at the ASCE Structural Engineering Conference and Annual Meeting, Oct. 19-23, 1964.
9. C. H. Purcell, C. E. Andrew, and G. B. Woodruff, "Bay Bridge Foundations Built with Unique Domed Caissons," *Engineering News-Record*, Vol. 112, No. 14 (Apr. 5, 1934), pp. 431-436; special edition, pp. 8-13.
10. C. H. Purcell, First Annual Progress Report, San Francisco-Oakland Bay Bridge, (July 1, 1934), pp. 1-75.
11. L. W. Riggs, "Tagus River Bridge-Tower Piers," *Civil Engineering*, Vol. 36, No. 2 (February 1966), pp. 41-45.

chapter **15**

Open Caisson Statics

15-1. FORCES ON OPEN CAISSON

The forces acting on an open caisson are
1. Self-weight W of caisson
2. Weight of ballast in caisson
3. Surcharge on caisson (equipment, materials, soil, rock, pig iron, water tanks, bricks)
4. Lateral earth pressure E_a on outside wall (shaft) of open caisson
5. Hydrostatic pressure of water, P_w
6. Soil reaction, R_s
7. Uplift at base of caisson
8. Buoyancy B (where it applies)
9. Suspension forces S (when a caisson is suspended from a barge or a scaffold)
10. Forces on caisson during various stages of its sinking (friction, R_f, for example)
11. Ice pressure, p_i
12. Seismic forces

The calculations of dimensions of open caissons in plan are based on equilibrium conditions of forces acting on the caisson (Fig. 15-1):

$$N + W - R_f - R_s = 0 \qquad (15\text{-}1)$$

where N = externally applied normal (vertical) structural load
W = self-weight of caisson, including backfilled ballast
R_f = total frictional resistance in outer mantle surface of shaft of caisson (= wall friction)
R_s = vertical component of soil reaction on caisson

The self-weight of the caisson is calculated as

$$W = \gamma_c A H \qquad (15\text{-}2)$$

where $\gamma_c = 125$ lb/ft³ ≈ 2 t/m³ = average unit weight of ready caisson, including ballast, weight of shell, and cutting edge

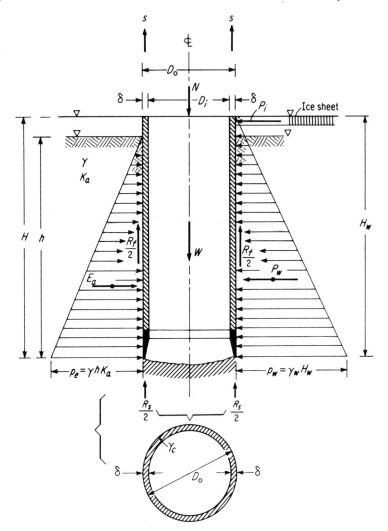

FIG. 15-1. Some forces acting on an open caisson.

$A = (\pi/4)D_o^2$ = cross-sectional area (viz., supporting area of caisson)
H = full height of caisson

In equilibrium calculation, for the submerged part of the weight of the caisson, buoyancy of water must be considered where it applies.

The soil reaction R_s is computed as

$$R_s = A\sigma_{\text{all}} \tag{15-3}$$

where $A = (\pi/4)D_o^2$ = contact or supporting area of caisson on soil or rock
σ_{all} = allowable soil bearing capacity at elevation of cutting edge

Open Caisson Statics

The mantle friction R_f is calculated as

$$R_f = U\mu_1 E_a \tag{15-4}$$

where U = outside perimeter of caisson
μ_1 = weighted average coefficient of wall friction against soil
E_a = active earth pressure on caisson

The wall-friction force R_f may be calculated by means of Coulomb's earth-pressure theory:[1-3]

$$E_a = \frac{1}{2}\gamma h^2 K_a \tag{15-5}$$

where E_a = active earth pressure on wall
γ = unit weight of soil
$K_a = \tan^2\left(\dfrac{\pi}{4} - \dfrac{\phi}{2}\right)$ = coefficient of active earth pressure (see *Active and Passive Earth Pressure Coefficient Tables*, Ref. 3)
ϕ = angle of internal friction of soil

If the soil penetrated consists of several layers each of which is $h_1, h_2, h_3, \ldots, h_n$ units thick, and each of the corresponding individual layers has a unit weight $\gamma_1, \gamma_2, \gamma_3, \ldots, \gamma_n$, an angle of internal friction $\phi_1, \phi_2, \phi_3, \ldots, \phi_n$, and a wall friction $\mu_1 = f_1, f_2, f_3, \ldots, f_n$, (Fig. 15-2), then the average weighted unit weight γ of soil may be calculated as

$$\gamma = \frac{\gamma_1 h_1 + \gamma_2 h_2 + \gamma_3 h_3 + \cdots + \gamma_n h_n}{h_1 + h_2 + h_3 + \cdots + h_n} = \frac{\sum_1^n (\gamma h)}{\sum_1^n h} \tag{15-6}$$

(to be used in Eq. 15-5).

The average weighted angle of internal friction ϕ is calculated approximately as

$$\phi = \frac{\phi_1 h_1 + \phi_2 h_2 + \phi_3 h_3 + \cdots + \phi_n h_n}{h_1 + h_2 + h_3 + \cdots + h_n} = \frac{\sum_1^n (\phi h)}{\sum_1^n h} \tag{15-7}$$

(to be used in the K_a-equation).

The average weighted mantle friction $\mu_1 = f$ is calculated as

$$f = \mu_1 = \frac{f_1 h_1 + f_2 h_2 + f_3 h_3 + \cdots + f_n h_n}{h_1 + h_2 + h_3 + \cdots + h_n} = \frac{\sum_1^n (fh)}{\sum_1^n h} \tag{15-8}$$

(to be used in Eq. 15-4).

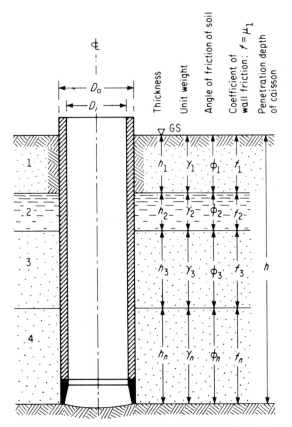

FIG. 15-2. Penetration of caisson through a layered system of soil.

In Eqs. 15-3 and 15-4 the cross-sectional area A and the perimeter U of the caisson are generally unknown, and must be found by trial and adjustment. However, for a circular caisson, the equilibrium equation 15-1 renders a quadratic equation which must be solved for the outer diameter D_o of the caisson. Substitution of Eqs. 15-3 and 15-4 and $A = (\pi/4)D_o^2$, $U = \pi D_o$, into Eq. 15-1 renders the following quadratic equation

$$N + \gamma_c(\pi/4)HD_o^2 - \pi FhD_o - (\pi/4)\sigma_{\text{all}}D_o^2 = 0 \tag{15-9}$$

or

$$(\pi/4)(\gamma_c H - \sigma_{\text{all}})D_o^2 - \pi FhD_o + N = 0 \tag{15-9a}$$

and

Open Caisson Statics

$$D_o = \frac{\pi Fh \pm \sqrt{(\pi Fh)^2 - (4)(\pi/4)(\gamma_c H - \sigma_{all})N}}{(2)(\pi/4)(\gamma_c H - \sigma_{all})} \quad (15\text{-}9b)$$

where $F = fE_a$, and $f = \mu =$ coefficient of internal friction of soil.

Now that the outer diameter D_o of the caisson is found, the thickness δ of the shell can be calculated by Eq. 14-1 or Eq. 14-2.

Example.

Derive a formula for calculating the length D_s of the outer side of a square open caisson.

Solution. In this problem, the contact area A_s is: $A_s = D_s^2$, and the perimeter of the caisson is $U = 4D_s$.

Equilibrium, by Eq. 15-1:

$$N + \gamma_c D_s^2 H - R_f - R_s = 0 \quad (15\text{-}1)$$

or

$$N + \gamma_c D_s^2 H - 4D_s hF - D_s^2 \sigma_{all} = 0 \quad (15\text{-}10)$$

or

$$(\gamma_c H - \sigma_{all})D_s^2 - 4FhD_s + N = 0 \quad (15\text{-}10a)$$

and

$$D_s = \frac{4Fh \pm \sqrt{(4Fh)^2 - (4)(\gamma_c H - \sigma_{all})N}}{2(\gamma_c H - \sigma_{all})} \quad \text{Q.E.D.} \quad (15\text{-}10b)$$

Example

Determine the size of a circular, open, reinforced-concrete caisson (requiring no surcharge) for sinking through a loose sand whose average wall friction is $F = \mu_1 E_a = 500$ lb/ft^2 = 0.25 ton/ft^2.

Solution. This problem requires that the minimum weight of the reinforced-concrete caisson should be equal to the total mantle friction:

$$(\pi/4)\gamma_c(D_o^2 - D_i^2)H = \pi FHD_o \quad (15\text{-}11)$$

and

$$D_i = \sqrt{D_o^2 - (4/\gamma_c)FD_o} \quad (15\text{-}12)$$

where $D_i = D_o - 2\delta =$ inner diameter
$\delta =$ thickness of shell

15-2. THICKNESS OF SHELL OF ROUND OPEN CAISSON

In calculating the thickness of the shell of an open caisson we distinguish between two cases of lateral pressure conditions acting on the shaft, namely:

Case 1. Assume that the caisson is full of water.

Case 2. Assume that the caisson has attained its designed depth, the concrete seal has been poured into the caisson, and the well of the caisson has been pumped out (Figs. 15-3a and 15-3b).

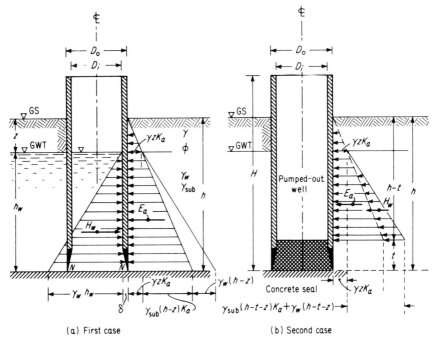

(a) First case (b) Second case

FIG. 15-3. Lateral pressure conditions on caisson.

In Case 1 the maximum lateral pressure p on the shell at depth h is

$$p = p_E - p_w \tag{15-13}$$

where $p_w = \gamma_w h$ = hydrostatic pressure

$p_E = [(\gamma - \gamma_w)K_a + \gamma_w]h$ = buoyant pressure from soil plus full hydrostatic pressure from water on shaft (15-14)

γ_w = unit weight of water

γ = unit weight of soil

$K_a = \tan^2\left(\dfrac{\pi}{4} - \dfrac{\phi}{2}\right)$ = coefficient of active earth pressure

ϕ = angle of internal friction of soil

Then

$$p = (\gamma - \gamma_w)K_a h \qquad (15\text{-}15)$$

In case 2 only the lateral earth pressure p_E acts on the shell of the caisson. By Eq. 15-14, and referring to Fig. 15-3,

$$p_E = [\gamma_w + (\gamma - \gamma_w)K_a](h - t) \qquad (15\text{-}16)$$

The larger pressure, p or p_E, must be used in calculating the necessary thickness δ of the shell of caisson.

If the open caisson is a round one, then the thickness δ of its shell may be computed by means of the well-known Lamé's formula (originally given for calculating tubes subjected to external pressure). The thickness of the shell is $\delta = r_o - r_i$, where

$$r_o = r_i \sqrt{\frac{\sigma_{all}}{\sigma_{all} - 2p}} \qquad (15\text{-}17)$$

in which r_o = outer radius of caisson
r_i = inner radius of caisson
σ_{all} = allowable compressive stress on external mantle surface of shaft
p = intensity of resultant lateral earth pressure of soil on shaft of caisson

Pressure p is calculated by Eq. 15-13 using for p_E Eq. 15-14 or Eq. 15-16, whichever case obtains.

15-3. THICKNESS OF WALL OF RECTANGULAR OPEN CAISSON

The thickness of wall of an open rectangular caisson may be calculated by means of the special box-frame theory. Referring to Fig. 15-4, the thickness of the wall may also be calculated approximately as set forth. Consider in the wall of the open caisson a slice of wall 1 unit of length high, δ units wide, and L units long (a small beam on two supports). This beam is loaded with a uniformly distributed load p from resultant lateral pressure.

The bending moment M for midspan of the beam for this loading is

$$M = \frac{pL^2}{8} \qquad (15\text{-}18)$$

The allowable bending moment is $M_{all} = \sigma_t Z$, where σ_t is the allowable bending stress in tension of the wall material, and $Z = (1)(\delta^2)/6$ is the section modulus of the small beam.

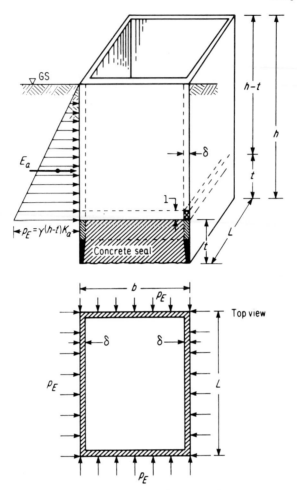

FIG. 15-4. A rectangular open caisson.

For equilibrium, set

$$M = M_{\text{all}}$$

or

$$\frac{pL^2}{8} = \sigma_t \frac{\delta^2}{6} \qquad (15\text{-}19)$$

From here,

$$\delta = \frac{L}{2}\sqrt{3\frac{p}{\sigma_t}} \qquad (15\text{-}20)$$

Open Caisson Statics 509

Of course, the rigidity of the corners of the caisson must also be secured.

For steel caissons, the outer and inner shells are made of steel plates $1/8$ in. to $1/2$ in. thick, braced together with steel frames. The interstices between outer and inner shells are filled with concrete for structural rigidity of the caisson.

Reinforced-concrete caissons with one or several partitioning walls (diaphragms) are calculated as closed frames by the methods outlined in structural analyses.

15-4. SOIL LATERAL EARTH PRESSURE ON VERTICAL, CIRCULAR, CYLINDRICAL CAISSON

For the spatial problem of earth pressure on curved, viz., cylindrical walls such as circular caissons, Steinfeld[4] derived the following earth-pressure E_a equation:

$$E_a = \gamma \pi h^2 \tan(\rho - \phi) \left(\frac{h}{3 \tan^2 \rho} + \frac{R}{\tan \rho} \right) \quad (15\text{-}21)$$

where γ = unit weight of soil
h = height of penetrated caisson mantle
ρ = angle of inclination of dangerous rupture surface with horizontal
ϕ = angle of internal friction of soil

The angle ρ of the dangerous rupture surface, in its turn, is calculated from the following cubic equation:

$$\tan^3 \rho + 2\left(\frac{1}{3}\frac{h}{r} - \tan \phi\right) \tan^2 \rho$$

$$- \left(1 - \frac{1}{3}\frac{h}{r}\frac{1}{\tan \phi} + \frac{h}{r} \tan \phi\right) \tan \rho - \frac{2}{3}\frac{h}{r} = 0 \quad (15\text{-}22)$$

where r = radius of caisson.

15-5. PRESSURE DISTRIBUTION AROUND CUTTING EDGE (SOIL REACTION)

In order to determine the distribution of soil pressures around the cutting edge (viz., soil reaction on cutting edge), it is necessary to know the penetration depth d of the cutting edge into the soil. If the cutting edge is not deep enough, lateral expulsion of soil from underneath the edge takes place. The depth d at equilibrium at which soil expulsion is impending may be determined grapho-analytically by means of Krey's circle method,[5] as shown in Fig. 15-5 and explained in Ref. 1. This method is by trial and

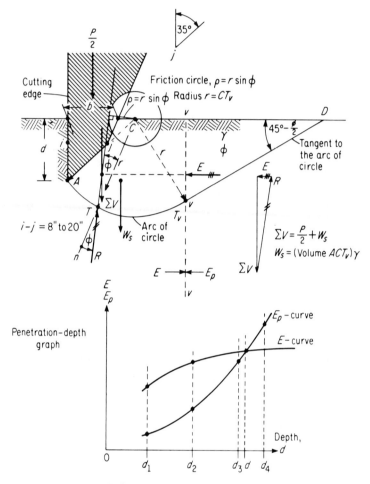

FIG. 15-5. Krey's circle method.

adjustment. We assume several arbitrary d-values, and, by drawing a circle with an arbitrary chosen point C as the center, we find on the vertical plane v-v that corresponding lateral earth pressure E necessary for maintaining equilibrium of the three forces ΣV, R, and E. The corresponding d and E values are then plotted on a graph to give the E-curve (Fig. 15-5). Then, for the same d-values, one calculates the passive earth pressures E_p and plots them on the same graph to obtain the E_p-curve. The intersection of the E-curve and E_p-curve renders the sought depth d of penetration into the soil of the cutting edge at equilibrium. For more definite analysis than that using the circle, a logarithmic spiral can be used. For complete logarithmic spiral tables see Ref. 6. The advantage of using the spiral as the rupture

Open Caisson Statics

curve lies in the fact that the spiral has a definite position and size for a given ϕ, whereas in the circle method the position of the center of the rupture surface must be assumed and varied.

This method of determining penetration depth d is applicable only when each of the rupture surfaces under each cutting edge can develop fully with no overlapping. For small caissons, however, overlapping of rupture surfaces would always take place, as may be understood by a simple experiment as shown in Fig. 15-6.

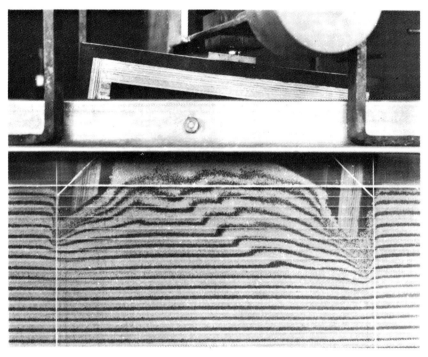

FIG. 15-6. Overlapping or rupture surfaces in sand under model caisson cutting edges. (Author's study.)

15-6. STATIC CONDITIONS OF OPEN CAISSONS

In static calculations of open caissons we distinguish between the following two static conditions:
1. Caisson during penetration through soil
2. Completed caisson in service (Fig. 15-7)

For succesful sinking, the weight $W + S$ (S = surcharge) should be greater than the allowable soil bearing capacity (to be attained by increasing surcharge S and by excavating away soil from the cutting edge = decrease of supporting

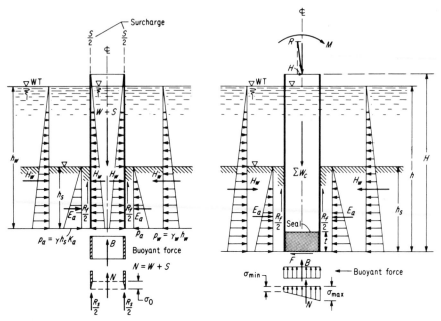

FIG. 15-7. Open caisson static conditions.

area). Here R_s = reaction from caisson and surcharge; R_f is taken undiminished by buoyancy and friction. Wave pressure, ice pressure, and flow pressures should be added to the system where applicable (they contribute overturning moments).

15-7. ARTIFICIAL SAND ISLAND

An artificial island is made of sand for the purpose of raising the ground surface above the water level, thus obtaining a dry area for sinking a caisson. The artificial sand island has been used in foundation engineering in India for many years, where open caissons were sunk from sand islands in 8 ft of water.[7,8]

The artificial sand island is poured at 45° slopes in open water of shallow depth; from there the caissons are then sunk as on land. At sites where great wave action prevails, the slopes of the island are protected against erosion with a course of fascine revetment work; otherwise, the island may be enclosed by a wall of sheet piling. Where scouring can be anticipated, a fascine mat, prior to sand filling, is placed and ballasted at the bottom of the body of water over an area larger than the size of the artificial island. The sheet piling is then driven, the mat inside the sheet-piling enclosure is cut and removed, sand is filled into the sheet-piling enclosure, and sinking

Open Caisson Statics

of the caisson commenced. After sinking the caisson in position, the sheet piling can be salvaged, and the island-forming sand may be dredged away.

The advantages of artificial sand islands are:
1. Working on dry ground, hence good access in sinking the caisson.
2. Possibility of locating the position of the caisson correctly.
3. Good control over tilting of the caisson.
4. Work can proceed without interruption by flooding of the river, because the shell is carried above the flood level.
5. The risk of a subsoil blow-in under the cutting edge during sinking is reduced.

The disadvantage of the artificial sand island is the increased expense for laying of foundations.

Examples of successful sinking of open-dredged caissons by the sand island method in the United States are those of the pier foundations for the Southern Pacific Railroad Martinez-Benicia Bridge across Suisun Bay, California, in 1930,[9] and for the Mississippi River Bridge at New Orleans[10] (Fig. 15-8) in 1935.

In the latter case, the shell enclosing the sand island, 122 ft in diameter, was assembled in rings, each made up of 30-ft curved steel plates 10 ft high and reinforced with angles. Each 30-ft section was supported on the falsework, and lowered into water onto willow mattresses 450 ft × 450 ft in size at the river bottom and extended to a safe height above water level. The willow mattress was then cut around the inside diameter of the cofferdam shell and removed by clamshell buckets. The river bottom was then dredged inside the shell to allow the shell to sink about 5 ft in water, and riprap was placed around the outside to prevent scour. The shell was then filled with sand to a safe elevation above the water level. On the artificial sand island thus formed, the steel cutting edge of the caisson was erected. The caisson shell was made of concrete.

Artificial sand islands were also used for the construction of pier foundations of the Verrazano-Narrows Bridge where the water is only 30 ft deep. The artificial island was formed by installing a series of sheet-pile cells 49 ft in diameter to form a cellular cofferdam encompassing the location of the caisson, and sand was poured into the enclosure to form the artificial island as a dry working site for the construction and sinking of the caisson.

The Haringvliet dike under construction southwest of Rotterdam in Holland—a part of the so-called Delta Works—was constructed on a man-made island. When the sluice gates were installed, the artificial island was washed away.

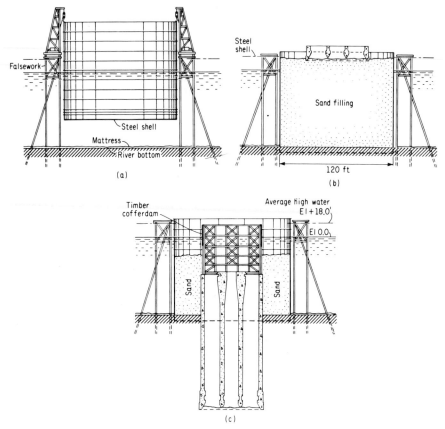

FIG. 15-8. Three steps in sinking a caisson using the sand-island method. (Ref. 10.) (a) Steel shell for sand island being lowered. (b) Shell completely filled with sand and first lift of caisson in place. (c) Timber cofferdam constructed-dredging operations nearly finished.

In deep waters the pouring of an artificial sand island may turn out to be impractical and uneconomical. In such cases a scaffold is constructed, the piles of which must be driven deeper than the base of the caisson. The caisson is floated in, suspended to the scaffold. and lowered into water and sunk into position. The scaffold must be provided with platforms for work and storing material.

15-8. SPECIAL CAISSONS

In Japan, in order to attain reasonable safety against earthquake effects, it has become the practice since about 1938 to build large structures of reinforced concrete. Such structures are frequently founded on large open

Open Caisson Statics 515

caissons. Some of the advantages of such caissons under Japanese conditions are:
1. Savings by avoiding piled foundations
2. Accelerated execution of construction work

Penetration of a caisson in inclined position was recommended in the United States in 1924 for the Walt Whitman Bridge across the Delaware River. However, because of lack of experience to draw on at that time, and because of the relatively great sinking depth, this recommended method was not adopted.[11]

15-9. PROBLEM

Calculate the outside diameter D_o of an open reinforced concrete caisson for a bridge pier to be sunk through $h = H = 120$ ft of sand and water to bedrock, the safe allowable bearing capacity of which is $\sigma_{all} = 15$ ton/ft^2 $= 30{,}000$ lb/ft^2. The caisson receives a load of 4,000 tons from the structure above. Also, compute the thickness of the concrete seal to be poured at the bottom of the caisson when resting on the bedrock. The mantle friction may be taken as $F = 450$ lb/ft^2. Note that the caisson is submerged in water; therefore it is subjected to hydrostatic uplift, $(\pi/4)\gamma_w H D_o^2$. The average unit weight of concrete for caisson and backfill concrete, including seal and cutting edge, may be taken as 145 lb/ft^3. Also, check the feasibility of sinking this open caisson relative to friction. $\sigma_f = 250$ lb/in^2 $= 36{,}000$ lb/ft^2, $m = 6.66$.

Solution

1. By Eq. 15-9, and with $h = H$, and including uplift, the outside diameter is:

$$\frac{\pi}{4}(\gamma_c H - \sigma_{all} - \gamma_w H)D_o^2 - \pi F H D_o + N = 0 \qquad (15\text{-}23)$$

and

$$D_o = \frac{\pi F H \pm \sqrt{(\pi F H)^2 - 4\frac{\pi}{4}(\gamma_c H - \sigma_{all} - \gamma_w H)N}}{2\frac{\pi}{4}(\gamma_c H - \sigma_{all} - \gamma_w H)} \qquad (15\text{-}23a)$$

$$= \{(3.14)(450)(120)$$
$$\pm \sqrt{[(3.14)(450)(120)]^2 - 3.14[(145)(120) - 3{,}000 - (62.4)(120)](4{,}000)(2{,}000)}\,\}$$
$$/\{(3.14/2)[(145)(120) - 30{,}000 - (62.4)(120)]\}$$

$$= \frac{19.956 \pm 72.900}{-3.14}$$

516   Deep Foundations

$$\therefore D_o = (-56.00)/(-3.14) = 17.85 \text{ (ft)} \approx \underline{18 \text{ ft}} \tag{15-23b}$$

2. By Eq. 15-11, and for no uplift, the inside diameter D_i of the caisson is

$$\frac{\pi}{4}\gamma_c(D_o^2 - D_i^2)H = \pi FHD_o \tag{15-11}$$

$$D_i = \sqrt{D_o^2 - \frac{4}{\gamma_c}FD_o}$$

$$= \sqrt{(18.0^2) - \frac{4}{145}(450)(18.0)} = 10.1 \text{ (ft)} \approx \underline{10.0 \text{ ft}}$$

3. The thickness δ of the wall of the caisson for sinking by its own weight with no surcharge, to overcome mantle friction only, is

$$\delta = \frac{D_o - D_i}{2} = \frac{18 - 10}{2} = \underline{4 \text{ (ft)}}$$

4. Thickness of seal:

   (i) by scheme (i) (Fig. 14-8) and Eq. 14-3,

$$t = \frac{\gamma_w H}{\gamma_c} = \frac{(62.4)(120)}{150} = \underline{50 \text{ (ft)}} \tag{14-3}$$

   (ii) By scheme (ii) and Eq. 14-4,

$$t = \frac{1}{2}\sqrt{\frac{3}{2}\frac{1}{\pi}\left(\frac{1+3m}{m}\right)\frac{W}{\sigma_f}} \tag{14-4}$$

$$= \frac{1}{2}\sqrt{\left(\frac{1.5}{3.14}\right)\left(\frac{20.98}{6.66}\right)\left(\frac{1,150,000}{36,000}\right)} = 3.47 \text{ (ft)} \approx \underline{4.00 \text{ ft}}$$

The total hydrostatic pressure W upon the circular seal, the radius of which is R_s (Fig. 14-8) is

$$W = \pi\gamma_w HR_s^2 = (3.14)(62.4)(120)(7^2) = 1,150,000 \text{ lb}$$
$$2R_s = D_i + \delta = 10.0 + 4.0 = 14.0 \text{ (ft)}$$

and

$$R_s = 14.0/2 = \underline{7.0 \text{ (ft)}}$$

5. Check of contact pressure σ_o on rock:
   (a) backfilled with lean concrete ($\gamma_c = 125 \text{ lb/ft}^3$), no uplift, no mantle friction:

Open Caisson Statics 517

$$\sigma_o = \gamma_c H = (125)(120) = \underline{15{,}000 \ (lb/ft^2)} < 30{,}000 \ lb/ft^2 (= \sigma_{all})$$

(b) same as under (a) but with uplift:

$$\sigma_o = (\gamma_c - \gamma_w)H = (125.0 - 62.4)(120) = \underline{7{,}500 \ (lb/ft^2)} < 30{,}000 \ lb/ft^2$$

6. Checking the possibility of sinking the caisson through soil: Weight of caisson, open bottom and open top:

$$G = (\pi/4)(D_o^2 - D_i^2)H\gamma_c$$

$$= (1.275)(18.0^2 - 10.0^2)(120)(125) = 3{,}284{,}000 \ (lb)$$

Mantle friction:

$$\Sigma F = \pi D_o H f = (3.14)(18.0)(120)(450) = 13{,}062{,}000 \ (lb)$$

Because the mantle friction ΣF is less than the weight of the caisson, the latter can be penetrated without a ballast B. The ballast B plus the weight of the empty caisson should be greater than mantle friction, i.e., it should be that

$$(G + B) > \Sigma F$$

REFERENCES

1. A. R. Jumikis, *Soil Mechanics*, Princeton, N.J.: Van Nostrand, 1962.
2. A. R. Jumikis, *Mechanics of Soils: Fundamentals for Advanced Study*, Princeton, N.J.: Van Nostrand, 1964.
3. A. R. Jumikis, *Active and Passive Earth Pressure Coefficient Tables*, New Brunswick, N.J.: Bureau of Engineering Research, College of Engineering, Rutgers University, Engineering Research Publication No. 43, 1962.
4. K. Steinfeld, "Über den räumlichen Erdwiderstand," Mitteilungen der Hannoverschen Versuchsanstalt für Grundbau und Wasserbau, Franzius Institut für Grundbau und Wasserbau der Technischen Hochschule Hannover, No. 3 (1953).
5. H. D. Krey, *Erddruck, Erdwiderstand und Tragfähigkeit des Baugrundes*. Berlin: Ernst, 1936, pp. 143-148, 197-201.
6. A. R. Jumikis, *Stability Analyses of Soil-Foundation Systems*. New Brunswick, N.J.: Bureau of Engineering Research, College of Engineering, Rutgers University, Engineering Research Publication No. 44, 1965.
7. R. R. Gales, "The Curzon Bridge at Allahabad," Paper No. 3626, *Proceedings of the Institution of Civil Engineers* (*London*), Vol. 174 (1908), pp. 1-40.
8. A. S. Napier, "The Netravati Bridge at Mangalore," *Proceedings of the Institution of Civil Engineers* (*London*), Vol. 174 (1908), pp. 41-52.
9. W. H. Kirkbridge, "The Martínez-Benecia Bridge," *Trans., ASCE*, Vol. 99 (1934), p. 154.

10. N. F. Helmers, "Pier Foundations for the New Orleans Bridge," *Civil Engineering,* Vol. 6, No. 7 (July 1936), pp. 442-444.
11. Anon., "Sinking of Pneumatic and Open Caisson Foundations for Philadelphia-Camden Bridge," *Engineering News-Record,* June 4, 1925.

PROBLEMS

15-1. Design an open caisson foundation for a load of $P = 1,500$ tons (see accompanying figure). Determine the diameter, thickness δ of wall (shell), and compute the various constructive elements of the caissons. What is the contact pressure on soil? *Given*:

    (a) for silt: $\gamma = 110$ lb/ft^3; $\phi = 30°$; $K_a = 0.333$; skin friction $f_1 = 200$ psf; $\phi_1 = 0°$

    (b) for fine-particled silty sand: $\gamma = 100$ lb/ft^3; $\phi = 25°$; $K_a = 0.406$; $\phi_1 = 0°$; $n = 30\%$ (porosity); skin friction $f_2 = 300$.

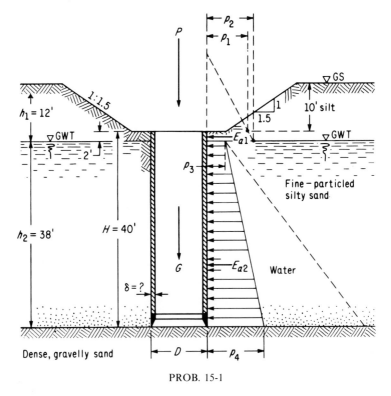

PROB. 15-1

15-2. Given a circular, open caisson whose outside diameter is $D_o = 20$ ft. The caisson is to be founded in a thick deposit of cohesive soil whose properties are

Open Caisson Statics

$$\gamma_{soil} = 120 \text{ lb/ft}^3$$
$$\phi = 10°$$
$$c = 1,500 \text{ lb/ft}^2$$
$$f = 250 \text{ lb/ft}^2 = \text{skin friction, supporting value}$$

The caisson transmits a load of $P = 2,200$ kips $= 1,100$ tons to the soil. The caisson extends 5 ft above the ground surface. *Required*: Calculate the sinking depth z of the caisson into soil.

Solution (*Hint*: make use of Eq. 10-4 of the increase in soil bearing capacity with increase in depth).

Ultimate contact pressure σ_u at depth z:

$$\sigma_u = \gamma \cdot z \cdot \tan^4\left(\frac{\pi}{4}+\frac{\phi}{2}\right) + 2c\tan\left(\frac{\pi}{4}+\frac{\phi}{2}\right)\left[1+\tan^2\left(\frac{\pi}{4}+\frac{\phi}{2}\right)\right] \quad (1)$$

or

$$\sigma_u = \gamma z K_p^2 + 2c\sqrt{K_p}\,(1+K_p) \quad (2)$$

where K_p = coefficient of passive earth resistance.

Auxiliary calculations:

$$K_p = \tan^2[45° + (10°/2)] = \tan^2 50° = (1.19175)^2 = 1.420$$
$$K_p^2 = (1.420)^2 = 2.016$$
$$\sqrt{K_p} = \sqrt{1.420} = 1.319$$
$$\gamma z K_p^2 = (120)(2.016)(z) = (241.9)(z) \approx (242)(z)$$
$$2c\sqrt{K_p}(1+K_p) = (2)(1500)(1.319)(2.420) = 9,576 \text{ (lb/ft}^2)$$

Substitute these auxiliary values into Eq. 2 to obtain:

$$\sigma_u = (242)(z) + (9,576) \quad (3)$$

Choosing a factor of safety $\eta = 2.0 = \sigma_u/\sigma_{all}$, obtain a general expression for the allowable soil bearing capacity σ_{all} at depth z:

$$\sigma_{all} = (141)(z) + 4,788 \quad (4)$$

The total soil bearing pressure S at the base of the $D = 20$ ft diameter caisson is

$$S = \sigma_{all}A = A[(141)(z) + 4,788] \quad (5)$$

where

$$A = \pi D^2/4 = (0.785)(20^2) = 314 \text{ (ft}^2) = \text{base area of caisson}$$

Total skin friction $F_s = \pi D z f = (3.14)(20.0)(250)(z)$

$$= (15,700)(z) \text{ (lb)} \quad (6)$$

Total supporting capacity R:

The total supporting capacity R of the caisson is the sum of the soil bearing S at its base and the total skin frictional resistance F_s between the soil and the outside mantle surface of the caisson over its embedment depth z:

$$R = S + F_s = (314)(141)(z) + (314)(4{,}788) + (15{,}700)(z)$$
$$= (59{,}974)(z) + 1{,}503{,}432 \text{ (lb)} \tag{7}$$

Weight G of concrete caisson:
$$G = (\pi D^2/4)(z+5)\gamma_c = (314)(150)(z) + (314)(5)(150)$$
$$= (47{,}100)(z) + 227{,}650 \text{ (lb)} \tag{8}$$

where $\gamma_c = 150$ lb/ft^3 is the unit weight of concrete.
Thus the total load acting downward is
$$W = P + G \tag{9}$$

Penetration depth of caisson:

Equate Eqs. 7 and 9, and solve for z to obtain the sought penetration or embedment depth z of the caisson: $R = W$, or
$$z = \frac{924{,}218}{12{,}874} = 71.78 \approx 72.0 \text{ (ft)}$$

By Eq. 8, $G = 1{,}810$ tons.
By Eq. 9, the total load W on soil is: given by $W = 1{,}100 + 1{,}810 = 2{,}910$ tons.
By Eq. 4, the allowable soil bearing capacity at depth $z = 72$ ft is:
$$\sigma_{\text{all}} = 7.64 \text{ ton/ft}^2$$
By Eq. 6, the amount of skin friction is $F_s = 565$ tons.
By Eq. 5, the amount of total soil bearing pressure is $S = 2{,}345$ tons.

15-3. In Prob. 15-2, determine the shell thickness of the open caisson and report whether or not the weight of the shell of the caisson is sufficient to overcome the skin friction.

15-4. If $\phi = 0$, and all other conditions are the same as given in Prob. 15-2, what is the necessary depth z for sinking the open caisson?

15-5. If $\phi = 30°$, and all other conditions are the same as given in Prob. 15-2, what is the necessary depth z for sinking the caisson?

chapter **16**

Floating Caissons

16-1. DESCRIPTION

Floating caissons, sometimes called box caissons, are hollow bodies used in foundation engineering to support superstructures. They are constituent parts of the foundation. Floating caissons differ from open and pneumatic ones in that the latter, in order to reach a firm stratum of soil or rock, must penetrate the soil above, from which the weak soil is then excavated. In contradistinction, the floating caisson does not penetrate soil.

The floating caisson has a closed bottom and an open top. Because the floating caisson does not penetrate the soil but rests upon it, it usually has a wide base area and is therefore suitable for floating. It is now usually made of reinforced concrete. After floating into position, a floating caisson is sunk by filling it with ballast (sand, concrete, or water).

The most important advantage of a floating caisson however, is its quick and convenient construction on land (on slips, or in docks, or on floating barges, or on sand islands), and its inexpensive transport by floating it to the site by water. Since their parts are prefabricated, floating caissons enjoy the indisputable advantage of quality construction.

Sometimes, for floating purposes, caissons are built with temporary false bottoms and sunk in position as open caissons after removing the false bottoms.

The floating caisson requires careful placing on a natural or prepared level ground surface for good pressure distribution on its supporting soil or rock.

Floating caissons are used where quick progress in construction is desired, or where, with great depth of water, great resistance to horizontal forces (wave action, hydrostatic pressure, earth pressure) is necessary, such as in the construction of harbor piers, moles (breakwaters), wharves, quays, or bridge piers. They are also used for closures in dike construction, and are suitable for laying foundations in deep-lying strata, on gravel strata, and where no danger of scour exists.

Floating caissons are used where there is construction of other types of caissons or where cofferdams are impractical or economically prohibitive.

If floating caissons are used to form a waterfront structure—a quay, for example—the caissons are sunk close together with a key between, which is sealed after the caissons are in place. Care is required to sink and seat the caissons plumb.

Floating caissons may be constructed of timber, concrete, reinforced concrete, steel, or a combination of these materials. The plan cross section of a floating caisson may be circular, elliptical, square, or rectangular. To reduce shrinkage cracks in concrete caissons, they may be designed by prestressing their various concrete elements. Also, these caisson are designed by means of the rules applied in shipbuilding for floating and against tipping over, against lateral translation when in position, and against rupture of soil (groundbreak).

It is of advantage to design and construct small, monocellular caissons. Large caissons are subdivided into multiple cells, 10 ft × 10 ft, or to any other convenient size. The thickness of the walls and diaphragms of the caissons is determined by strength considerations, as well as those of weathering, wave action, and (sometimes) warfare.

The main advantages of a floating caisson are
1. Quick and convenient construction
2. Prefabrication possibilities
3. Economy in relatively small construction cost
4. Inexpensive transport on water

The main disadvantages are
1. Costly for deep excavations
2. Necessity for preparation of level soil or rock surface for supporting the caisson

16-2. BUOYANCY OF CAISSON

For floating, not only must the caisson be subjected to calculation of its strength, but also to checking of its stability during flotation. During its transportation on water the caisson is subject to buoyancy and flotation.

The Archimedes principles of buoyancy and flotation are stated as follows:
1. A body immersed in water is buoyed up by a force equal to the weight of water displaced by the body.
2. A floating body displaces its own weight of the water in which it floats.

The vertical equilibrium of the floating body (Fig. 16-1) is expressed by
$$B - W = 0 \tag{16-1}$$
where B = buoyant force (up)
W = weight of caisson (down)

Floating Caissons

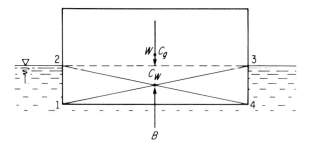

FIG. 16-1. Equilibrium of a prismatic floating body.

Also, $B = \gamma_w V_{1234}$. Thus $W = \gamma_w V_{1234}$; note that the body displaces its own weight of the water in which it floats. Here γ_w = unit weight of water, and V_{1234} is the volume of displaced water.

In general, the stability of submerged or floating bodies is dependent on the relative position of the buoyant force B and the weight W of the body. The buoyant force acts upward through the center of gravity of the displaced volume of water, C_W. The weight W of the floating body acts downward at the center of gravity C_g of the body. Upon an angular displacement (= tilting) of the floating body, its center of gravity remains in the same position relative to the floating body itself. The position of the center of buoyancy, however, changes because of the displaced volume of water upon tilt. A small angular displacement θ of the floating caissons brings about a static moment, or couple, $M_W = Wz$ (Fig. 16-2).

Stability or instability of a floating body will be determined by whether a righting or overturning moment is developed when the center of gravity and center of buoyancy move out of vertical alignment.

In surface vessels such as ships, scows, and floating caissons stability exists because of movement of the center of buoyancy C_B to a position outboard of the center of gravity C_g as the ship "heels over," thus producing a righting moment. An overturning moment, resulting in capsizing, occurs if the center of gravity moves outboard of the center of buoyancy.

16-3. STABILITY OF A FLOATING BODY

When the prismatic body (viz., a floating caisson) is tipped (Fig. 16-2) by an angular displacement θ, the center of buoyancy C_B is at the centroid B_t of the trapezoid $CDJN$. The buoyant force B acts upward through point C_B. The weight W acts downward through point C_g. The intersection of the vertical buoyant force B through C_B with the original center line $\mathcal{C}_L$ above C_g gives point M, called the *metacenter*. When, upon counterclockwise tilting, point M is above C_g, a restoring couple, $M_W = Wz = Wm \sin \theta$, is

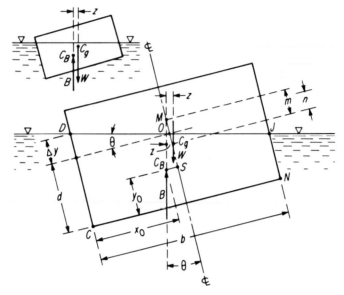

Fig. 16-2. Tilting of a prismatic floating body.

produced, and the floating box is in stable equilibrium. When the metacenter M is below C_g, the floating box is unstable. When the metacenter M coincides with the center of gravity C_g, the floating body is in neutral equilibrium. The distance $m = \overline{MC_g}$ is called the metacentric height. The metacentric height m is a direct measure of the stability of the floating body. When the caisson is unstable, it must be redesigned, or ballast added to bring the metacenter up above the center of gravity of the caisson.

If the weight of the caisson with a closed bottom is W lb, its length is L ft and width is b ft, and if the center of gravity C_g is n units above the water surface, then the metacentric height m and restoring couple M_W for a tilt of Δy is calculated as follows.

Depth of submergence d in water:

$$d = \frac{W}{bL\gamma_w} \text{ (ft)} \quad (16\text{-}2)$$

where γ_w = unit weight of water.

The coordinates of the centroid C_B in the tipped position are calculated by moments about DC and CN:

$$x_0 = \frac{(d - \Delta y) b \frac{b}{2} + \frac{1}{2} 2 \Delta y \, b \frac{b}{3}}{bd} \quad (16\text{-}3)$$

Floating Caissons

$$y_o = \frac{\left[(d - \Delta y) b \cdot \frac{d - \Delta y}{2}\right] + \frac{1}{2} 2 \Delta y \, b \left[(d - \Delta y) + \frac{2}{3} \Delta y\right]}{bd} \quad (16\text{-}4)$$

By geometry in Fig. 16-2,

$$\tan \theta = \frac{\Delta y}{b/2} = \frac{\overline{C_B S}}{\overline{MS}} \quad (16\text{-}5)$$

$$\overline{C_B S} = b/2 - x_o \quad (16\text{-}6)$$

$$\overline{MS} = \frac{\overline{C_B S}(b/2)}{\Delta y} = \frac{[(b/2) - x_o] b/2}{\Delta y} \quad (16\text{-}7)$$

$$\overline{C_g S} = (d - n) - y_o \quad (16\text{-}8)$$

$$m = \overline{MS} - \overline{C_g S} \quad (16\text{-}9)$$

The caisson is stable when m is positive.

Problem. Given a prismatic floating body: $b = 20$ ft; $L = 60$ ft; $W = 225$ short tons $= 450{,}000$ lb; $\Delta y = 1.0$ ft; height of center of gravity above the water surface is $n = 1$ ft.

Required: Calculate (1) the metacentric height m, and (2) the magnitude of restoring couple M_W when tip $\Delta y = 1.0$ ft.

Ans.:
1. Depth of submergence: $d = 6.0$ ft
2. Centroid in the tipped position: $x_o = 9.46$ ft; $y_o = 3.03$ ft
3. $\overline{C_B S} = 0.54$ ft; $\overline{MS} = 5.40$ ft
4. $\overline{C_g S} = 3.97$ ft.
5. $\overline{MC_g} = m = +1.43$ ft
6. $M_W = 64{,}000$ lb-ft
7. $\tan \theta = 0.1000$; $\theta = 5°43'$

For a floating nonprismatic object of irregular cross section (Fig. 16-3), the metacentric height m is determined for very small angles θ of rotation.

The submerged hatched wedge ① of water brings about an upward buoyant force ΔB on the left side of the floating body—a change in the buoyant force B. The right-hand wedge ② decreases the buoyant force B by an equal amount ΔB on the right.

The horizontal shift a of the buoyant force B from C_B to C_{B_1} is calculated by taking static moments about C_B:

$$(\Delta B) r = W a \quad (16\text{-}10)$$

since $Ba = Wa$.

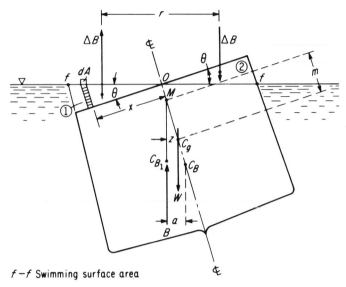

FIG. 16-3. Floating body of irregular cross section.

An elementary volume of the submerged wedge is expressed as

$$(dA)(\cos\theta)(y) = x(\sin\theta)dA \approx \theta x\, dA \tag{16-11}$$

Here dA is an elementary area on the horizontal section of the floating body at the swimming surface $f\text{-}f$ of the water, and x and y are the rectangular dimensions of the submerged elementary wedge (Fig. 16-3).

The buoyant force of this elementary wedge is

$$\Delta B = \gamma_w \theta x\, dA \tag{16-12}$$

and its static moment about point 0 is

$$M_{\Delta B_O} = \gamma_w \theta x^2\, dA \tag{16-12a}$$

The total moment of the change in buoyant force over the entire swimming surface area is

$$M_B = (\Delta B)r = \gamma_w \theta \int_O^A x^2\, dA = \gamma_w \theta I_y \tag{16-13}$$

where $I_y =$ moment of inertia of the floating surface area $f\text{-}f$ of the floating body about the horizontal axis $y\text{-}y$ through point O (perpendicular to the drawing plane).

Because $M = Ba = Wa$,

$$\gamma_w \theta I_y - Wa - \gamma_w V a \tag{16-14}$$

Floating Caissons

from which the shift a is calculated as

$$a = \frac{\theta I_y}{V} \qquad (16\text{-}15)$$

where W = weight of floating body
V = total volume of displaced water

For very small b the shift a may be calculated as

$$a = \overline{MC_B} \sin \theta \approx \overline{MC_B}\, \theta \qquad (16\text{-}16)$$

The metacentric height m is then calculated as

$$m = \overline{MC_g} = \overline{MC_B} \pm \overline{C_g C_B} \qquad (16\text{-}17)$$

With $\overline{MC_B} = \dfrac{a}{\theta} = \dfrac{I_y}{V}$ from Eq. 16-16, the metacentric height m is

$$m = \frac{I_y}{V} \pm \overline{C_g C_B} \qquad (16\text{-}18)$$

The $+$ sign is used when C_g is below C_B. The $-$ sign is used when C_g is above C_B.

16-4. LAUNCHING OF A FLOATING CAISSON

Floating caissons are usually built ashore, floated into position and sunk. Upon launching a prefabricated caisson from a launching slip into water the caisson is subjected to the following forces, which must be taken into account in design (Fig. 16-4):
1. B = buoyancy
2. W = weight of floating caisson
3. R = pressure on the launchway, viz., sleds

The equilibrium conditions are:

$$V = 0: \quad W - B - R = 0 \qquad (16\text{-}19)$$
$$M_A = 0: \quad Ba - Wb - Rx = 0 \qquad (16\text{-}20)$$

where R and x should be $\geqslant 0$.

When $W > B$, the caisson sinks. It sinks until $B = W$. As soon as $B = W$, the caisson begins to float. If, upon full immersion, $W > B$, then the caisson cannot float but goes under. The depth of water at the launching site must be such that upon hitting the bottom the caisson will not be subjected to distressing stresses. The depth of water is usually assessed at twice the draft of the caisson.

In order that the closed-bottom caisson may float, it should be that

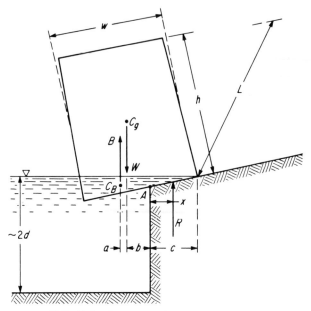

FIG. 16-4. Forces on caisson upon launching.

$W = B$. This condition corresponds to a certain depth of flotation called the *draft d* of the caisson. For such a draft, the buoyant force B is:

$$B = \gamma_w w L d \quad (16\text{-}21)$$

where γ_w = unit weight of water
w = average width of caisson
L = length of caisson
d = draft of caisson

With $W = B$, the draft is

$$d = \frac{W}{\gamma_w w L} \quad (16\text{-}22)$$

If the draft d is less than the height h of the caisson, i.e., when $d < h$, then the caisson is capable of floating.

In preparing the caisson for floating, its stability must be checked during its towing on water to the site of its sinking. One must also remember that upon towing, as well as because of flow pressure and wind pressure, the caisson rotates more or less. To reduce these effects, the towing must be slow. To prevent the floating ballast from shifting in the caisson during transportation, the floating caisson is built with cells, viz., compartments, as noncommunicating vessels.

Floating Caissons

In quiet waters the freeboard of the floating caissons is designed usually as 1 ft. In rough waters the freeboard must be greater to prevent splashing wave water from filling and overflooding the cells. This may cause capsizing of the caisson, or even its premature sinking.

16-5. COMPLETION OF CAISSONED FOUNDATION

After the caisson has been sunk into position, the cells are filled with sand and gravel, or lean concrete, and a course of concrete is poured on top of the caisson for receiving structural loads from above. To prevent scour underneath the caisson, protective riprap is placed along and/or around it (Fig. 16-5).

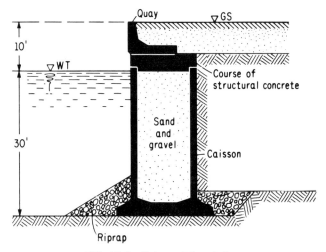

FIG. 16-5. Caissoned foundation.

For sinking to a shallow depth, open caissons are constructed without cross-walls and diaphragms. Large caissons, however, to provide for strength and stiffness, are designed and constructed with cross-walls and longitudinal walls, thus dividing the caisson into cells or compartments, approximately 10 to 20 ft square. Such steel or reinforced-concrete cellular cofferdams are designed by the method of moment distribution, for example. In this method all corners are treated as monolithic and reinforced for these moments.

If a floating caisson is to be constructed in a dry, open-cut excavation, it must be cut so deep that the water in it later will be deeper than the draft of the caisson, to afford successful flotation of the caisson when ready.

16-6. LITTLE BELT BRIDGE CAISSONS

Sometimes floating caissons are constructed on slipways upside down and launched into water; then the inverted caissons are turned over (careened) into proper floating position. Celebrated examples of inverted caissons are those built for the Little Belt Bridge in Denmark.[1,2]

Figure 16-6a shows an upside-down floating caisson for the Little Belt Bridge in three positions: after launching, floating, and inverting.

FIG. 16-6a. An upside-down floating caisson for the Little Belt Bridge, Dennmark. (Courtesy of Monberg and Thorsen, Consulting Engineers, Copenhagen.)

Floating Caissons

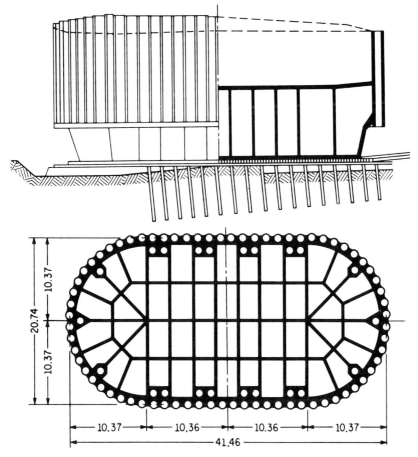

FIG. 16-6b. Little Belt Bridge floating caisson. (Ref. 1.)

A horizontal cross section of one of these floating caissons is shown in Fig. 16-6b. The caisson was made of reinforced concrete with a steel pocket-tube apron around its periphery. The size of these tubes was 1.05 m and they were used for tilting and turning the caisson in water. The caisson was 18 m high and weighed 6,000 metric tons. The ends of the steel pocket tubes were constructed so as to fit the uneven configuration of the bottom of the sea at the site.

The caisson was constructed in an inverted position because its upper, smooth, horizontal deck afforded easy launching, and also to reduce its draft when afloat. It would have been very difficult to float the caisson with its "horns" (pocket tubes) down into the water (Fig. 16-7). Another reason for constructing the caisson upside down on the slipway was because

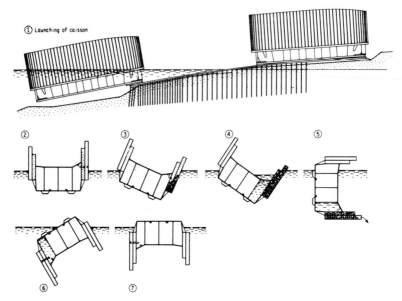

FIG. 16-7. Launching and turning of an inverted floating caisson for the Little Belt Bridge in Denmark. (Courtesy of Monberg and Thorsen, Consulting Engineers, Copenhagen.)

the steel pipe apron around the caisson had to be constructed to fit the configuration of the bottom of the sea. Hence these tubes around the caisson were of unequal length. Supporting such a caisson upright on a few of its pipes would have been very unsafe in respect to both stability and stress concentration.

The turning of the launched, inverted caisson was accomplished by means of water and sand ballast. Water ballast was introduced and regulated in the compartments of the caisson, whereas the sand ballast was poured into the apron tubes as needed (Fig. 16-7). With turning in progress the sand and gravel ballast poured out of the tubes and kept the turned caisson afloat.

Figure 16-7 illustrates the launching of the floating caisson from a slipway, as well as the sequence of turning of the floating caissons for the piers of the Little Belt Bridge, Denmark. The sequences are as follows.
1. Launching of an inverted caisson from a slipway.
2. The floating caisson is in an inverted condition. The sand ballast in the short tubes keeps the caisson in a level, balanced position.
3. The sand ballast is removed from the short tubes. The long tubes are filled with stone ballast, and the turning of the caisson begins.
4. Water (as a ballast) is admitted into the cells of the work-chamber by opening of valves.

5. The caisson is now turned through 90°. The stones begin to run out of the tubes.
6. The turning of the caisson is almost completed. All of the stones have run out.
7. The caisson is in a normal floating condition after completion of the turning operation. The water ballast has been blown out by means of air pressure.

After being turned upright, the caisson, still afloat, was extended to a height of 35 m (weight 11,000 tons). Then it was positioned and sunk to the bottom of the Little Belt. The caisson was sitting on a layer of uniform, stiff, fine-particled clay about 100 m thick. The clay was then drilled or augured out through the steel tubes, whereupon the caisson penetrated 6.5 m into the clay. No compressed air nor any other means for holding back water was needed. In the working chamber of the caisson the soil was excavated 3.5 m deep, a concrete seal was poured on the clay, and the chamber was pumped out.

The caisson was sunk 41.0 m below water level. Then the bridge pier shaft was built up.

The maximum pressure on the clay was 9.5 kg/cm^2. The magnitude of settlements at the time this work was reported [1,2] was one half of those predicted.

16-7. STATIC CALCULATIONS FOR FLOATING CAISSONS

The method of static calculations for floating caissons should include all possible situations of force play in the structural elements of the caisson which may set in upon its construction, during its floating, suspension, penetration and service conditions.

The forces acting on a floating caisson are
1. Self-weight of caisson
2. Its ballast
3. Hydrostatic pressure
4. Uplift on bottom of caisson
5. Ice pressure (in regions where this applies)

In sunken condition, uneven support of the caisson brings about a different force play upon it, for which the floating caisson must be properly designed.

The side walls are designed as closed, single- or multiple-span frames. The bottom of the floating box is calculated as for freely supported plates (slabs) on all sides with or without a cantilever slab.

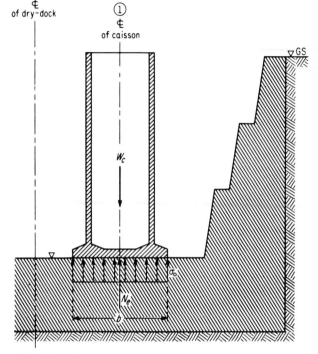

FIG. 16-8. Forces on caisson upon construction in a dock (or on land).

In static calculations for floating caissons four principal static conditions may be distinguished, namely:
1. Caisson in dock, or on a slipway, or on land
2. Caisson during floating
3. Caisson sunken in position
4. Caisson in service condition

The various loading conditions of a floating caisson are shown in Figs. 16-8 through 16-11.

1. Forces on caisson upon construction in a dock (or on land) (Fig. 16-8):
W_c = self-weight of caisson

$$N_e = A\sigma_o = W_c = \text{reaction}$$

where A = contact area
σ_o = uniformly distributed pressure from self-weight of caisson
b = width of base of caisson

2. Forces on caisson during floating (Fig. 16-9):
W_o = self-weight of caisson and its ballast (including buoyancy where it applies)

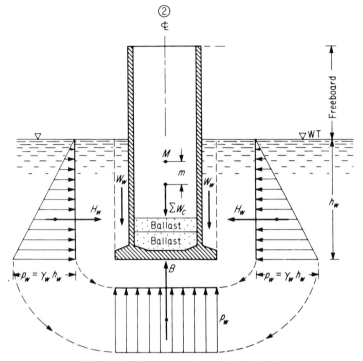

FIG. 16-9. Forces on caisson during floating.

B = uplift or buoyant force
$H_w = (1/2)\gamma_w h_w$ = hydrostatic pressure
W_w = weight of water on cantilever part of bottom slab
C_g = center of gravity of caisson
M_c = metacenter
m = metacentric height
I = moment of inertia of cross section of caisson in the water table

3. Caisson sunken in position (Fig. 16-10):
Uplift $\geqslant$ vertical loading
$B = \gamma_w bh$ = buoyant force
σ_0 = uniformly distributed contact pressure
E_s = silo pressure
H_w = hydrostatic pressure
All other forces are as shown under (2).

4. Service condition (Fig. 16-11):

$$\sigma_{\substack{max\\min}} = \frac{N}{A}\left(1 \pm \frac{6e}{b}\right) = \text{edge pressures on soil (rock)}$$

e = eccentricity of N

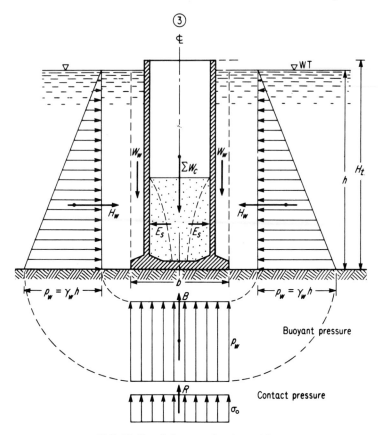

FIG. 16-10. Caisson sunken in position.

$F = \mu_{c/s} N$ = friction force at base of caisson
$\mu_{c/s}$ = coefficient of friction between caisson and soil (rock) material
N = normal component of soil reaction
M = externally applied moment
V = vertical component of externally applied inclined force R
H = horizontal component of externally applied inclined force R
H_t = total height of caisson
$H_w = (1/2)\gamma'_w h^2$ = hydrostatic pressure
E_s = silo pressure

The greatest stresses in floating caisson elements such as walls, corners, stiffening membranes and diaphragms and rods occur usually during its launching, subjecting the caisson to bending, tension, and warping.

For a floating condition of the floating caisson, besides its analysis for

Floating Caissons

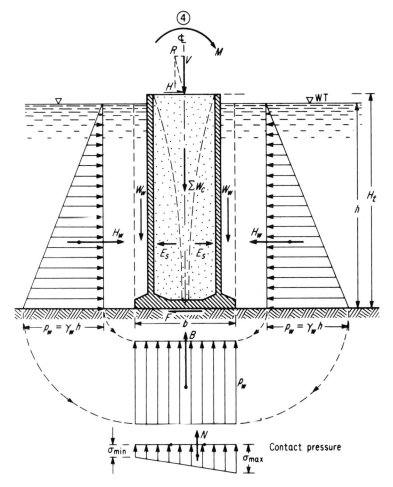

FIG. 16-11. Service condition.

strength, the caisson should also be subjected to stability analysis. The caisson floats in water stably only when the center of gravity C_g of selfweight of caisson and that of the expelled water are situated vertically one above the other, and when upon rotation out of this position the weight of the caisson and the weight of the expelled water form a moment righting the caisson back into its original stable position. The latter, stable position or floating of the caisson exists when the center of gravity C_g is situated below the center of gravity of the expelled mass of water.

Problem. Calculate stability of floating and the sinking of a floating caisson the size of which is shown in Fig. 16-12. The length L of the caisson is given as $L = 21.0$ m. The weight W of the caisson is given as $W = 860.5$

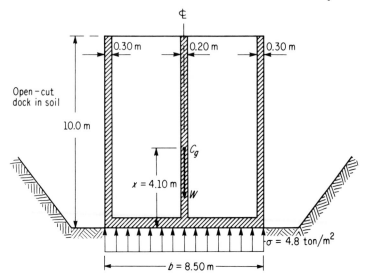

FIG. 16-12. Caisson in open-cut dock.

tons. It is composed of the following:

Weight of the bottom	192.5 tons
Weight of the outside walls	398.0
Weight of inside walls	270.0
Total	$W = 860.5$ tons

The unit weight of sea water is $\gamma_w = 1.025$ t/m³. What are the pressures

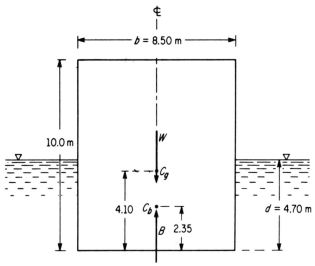

FIG. 16-13. Stable flotation.

Floating Caissons

on the soil for the various conditions of the caisson? The caisson must be sunk in position for an immersion of 8.50 m below the water table.

The solution to this problem is based on the ideas of O. Streck, Ref. 3.

Answers
1. In open-cut dock:
   (a) Center of gravity: $x_g = 4.10$ m
   (b) Pressure on soil: $\sigma_o = W/A = \underline{0.48 \text{ kg/cm}^2}$
2. Flotation
   (a) For stable flotation, the buoyant force B is: $B = W = \underline{860.5 \text{ tons}}$
   (b) Draft $d = \underline{4.70 \text{ m}}$
   (c) Center of buoyancy: $x_b = d/2 = \underline{2.35 \text{ m}}$ (Fig. 16-13)
3. Tilting
   (a) Couple: $M_W = Wa$ (Fig. 16-14)
   (b) Distance $\overline{MC_B}$ of metacenter M from center of buoyancy C_B:
   $$\overline{MC_B} = \frac{I}{V} = \frac{\frac{Lb^3}{12}}{bLd} = \frac{b^2}{12d} = \underline{1.28 \text{ m}}$$
   (c) Distance between C_a and C_B:
   $$\overline{C_g C_B} = x_g - x_b = 4.10 - 2.35 = \underline{1.75 \text{ m}}$$

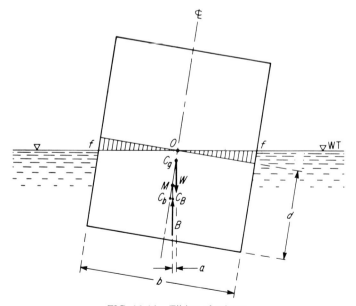

FIG. 16-14. Tilting of caisson.

(d) Because $(\overline{MC_B}) < (\overline{C_g C_B})$, or 1.28 m < 1.75 m, the metacenter M is situated below the center of gravity C_g of the caisson. Hence the caisson is unsafe with respect to tilting. To attain stability against tilting, the caisson must be redesigned wider for greater $\overline{MC_B} = m = b^2/(12)d$, or the center of gravity C_g must be lowered by adding ballast in the caisson. Because widening of the caisson may be uneconomical, and for its function the structure does not need to be wider, add ballast to the caisson for improving its stability against tilting when afloat. Because at the site the caisson will be filled with sand, sand ($\gamma_s = 2.2$ t/m^3) will now be utilized for stabilizing the caisson against tilting. Setting up the requirement that the metacentric height $\overline{MC_g} = m = \overline{MC_B} - \overline{C_g C_B}$ should be approximately 0.45 m above the center of gravity C_g, calculate the weight of sand [try a layer of sand 0.60 m thick, the weight of which is

$$W_s = (8.50 - 0.80)(21.0 - 1.60)(0.60)(2.20) = 197.0 \text{(tons)}$$

(e) Position of center of gravity: $x_g = 3.48$ (m)   (see Fig. 16-15)
(f) Draft: $d = 5.78$ (m)
(g) Distance of metacenter M from center of buoyancy: $\overline{MC_b} = 1.04$ (m)
(h) Metacentric height: $m = \overline{MC_g} = \overline{MC_B} - (x_g - \frac{d}{2})$
$= 1.04 - [3.48 - (5.78/2)] = \underline{0.45}$ (m)

This metacentric height of $m = 0.45$ meter is satisfactory for

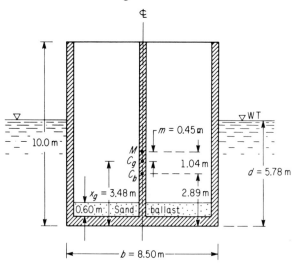

FIG. 16-15. Stabilizing the caisson against tilting.

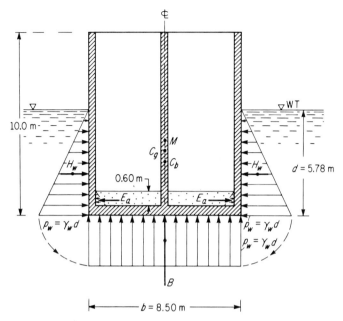

FIG. 16-16. Hydrostatic pressure on caisson.

slow towing in calm weather and calm waters. The bottom of the open-cut dock should be at least 0.60 m > d below water level in the dock to permit flotation.

(i) Hydrostatic pressure. This pressure is measured per 1 linear meter of length of caisson:
$$H_w = (1/2)\, \gamma_w\, d^2 = \underline{17.1\ (t/m)}$$
(see Fig. 16-16).

(j) Buoyant force (uplift), B:
$$B = \gamma_w\, bd(1) = \underline{49.1\ (t/m)}$$

(k) Sand pressure σ_s on bottom of caisson:
$$\sigma_s = W_s/A = 197.0/21.0 = \underline{9.38\ (t/m^2)}$$
for each strip of caisson 1 linear meter wide and 21.0 m long.

(l) Active sand pressure on outside walls of caisson. With $\gamma_s = 2.2\ t/m^3$, $\phi = 35°$, $\delta = 0$ and $\alpha = 0$, $K_a = 0.271$, the earth pressure per 1 m of length of caisson is
$$E_a = (1/2)\, \gamma_s\, d^2 K_a = \underline{0.11\ (t/m)}$$

3. Sinking of caisson
   (a) Ballast
   For sinking the caisson more sand ballast is poured into it

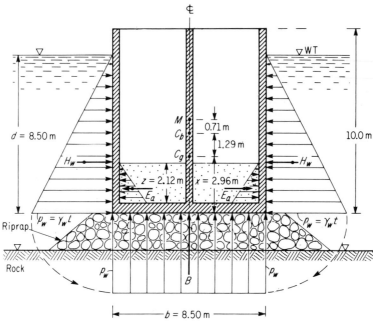

FIG. 16-17. Sinking of caisson.

until it "sits" on a prepared base of, say, crushed rock, and is submerged at the design depth of $d = 8.50$ m. For this condition of flotation $W = B$. Here W consists of self-weight of caisson and the weight W_{sz} of the sand ballast to be poured into the caisson in a course z meters thick (Fig. 16-17):

$$W_{sz} = W_s \frac{z}{0.60} = 197.0 \left(\frac{z}{0.60}\right)$$

(b) Uplift B for $d = 8.50$ m: $B = \gamma_w bL\, d = \underline{1,550}$ (tons)
(c) Total weight W of caisson plus ballast W_{sz}:

$$860.5 + (197.0)(z/0.60) = \underline{1,550} \text{ (tons)}$$

(d) Thickness z of course of ballast: $\underline{z = 2.12}$ (m)
(e) Center of gravity C_g of caisson and ballast:

$$(860.5)(4.10) + (197.0)(2.12/0.60)\left(0.45 + \frac{2.12}{2}\right)$$

$$= \left[(860.5) + \left(197.0\, \frac{2.12}{0.60}\right)\right] x$$

$$\underline{x = 2.96} \text{ (m)}$$

Floating Caissons

(f) Metacentric distance $\overline{MC_B}$ from center of buoyancy C_B:

$$\overline{MC_B} = \frac{b^2}{(12)d} = \frac{(8.50)^2}{(12)(8.50)} = \underline{0.71 \text{ (m)}}$$

Observe now that the center of gravity C_g lies $\frac{d}{2} - x = \frac{8.50}{2} - 2.96$
$= 1.29$ (m) *below* the center of buoyancy

(g) Loads on caisson:

$$H_w = (1/2)\gamma_w d^2 = \underline{37.0 \text{ (t/m)}}$$

$$B = 1{,}550/21.0 = \underline{74.0 \text{ (t/m)}}$$

$$E_a = (1/2)\gamma_s z^2 K_a = \underline{1.35 \text{ (t/m)}}$$

4. *Completion of structure.* After sinking into position the caisson is filled with sand, and the quay is placed upon the caisson. Hence the weight $W = P$ and earth pressure E_a increases, whereas the uplift B and hydrostatic pressure H_w remain constant. After completion, the total weight of this waterfront structure is given as $P = 4{,}557.0$ tons.

(a) Position of application of load is given as $x = 4.31$ m, and eccentricity as $e = x - \frac{b}{2} = 4.31 - (8.50/2) = 0.06$ (m) (Fig. 16-18).

(b) Edge pressures:

$$\sigma_{\substack{max \\ min}} = \frac{P}{A}\left(1 \pm \frac{6e}{b}\right) = \frac{4{,}557}{(21.0)(8.50)}\left[1 \pm \frac{(6)(0.06)}{8.50}\right]$$

$$= 25.5(1 \pm 0.042)$$

$$\sigma_{max} = (25.5)(1.042) = \underline{26.6 \text{ (t/m}^2\text{)}}$$

$$\sigma_{min} = (25.5)(0.958) = \underline{24.4 \text{ (t/m}^2\text{)}}$$

(c) Uplift: $\sigma_b = \gamma_w d = (1.025)(8.50) = \underline{8.7 \text{ (t/m}^2\text{)}}$

(d) Soil reaction:

$$\sigma_{s\,max} = \sigma_{max} - \sigma_b = 26.6 - 8.7 = \underline{17.9 \text{ (t/m}^2\text{)}}$$

$$\sigma_{s\,min} = \sigma_{min} - \sigma_b = 24.4 - 8.7 = \underline{15.7 \text{ (t/m}^2\text{)}}$$

(e) Hydrostatic pressure at depth 8.5 m:

$$\sigma_w = \sigma_b = \underline{8.7 \text{ (t/m}^2\text{)}}$$

$$H_w = (1/2)\gamma_w d^2 = (1/2)(1.025)(8.50)^2 = \underline{37 \text{ (t/m)}}$$

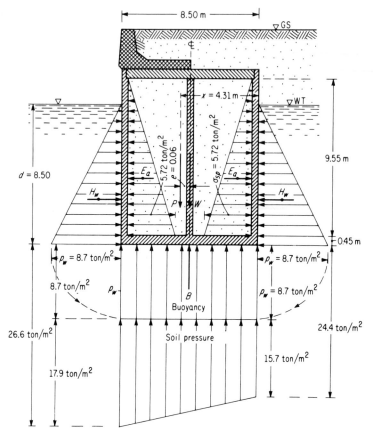

FIG. 16-18. Stresses at base of completed caisson.

(f) Silo pressure σ_{sp} from sand:

$$\sigma_{sp} = \gamma_s H K_a = (2.2)(9.55)(0.271) = \underline{5.72} \ (t/m^2)$$

$$E_a = (1/2)\sigma_{sp}H = (1/2)(5.72)(9.55) = \underline{27.3} \ (t/m)$$

REFERENCES

1. G. Schaper, "Vom Bau der Eisenbahn- und Strassenbrücke über den Kleinen Belt," *Die Bautechnik*, Vol. 9 (1931), pp. 72, 683.
2. A. Engelund, "Some Examples of the Design and Construction of Bridges in Denmark," *Structural Engineer*, February 1939, p. 134.
3. O. Streck, *Grund-und Wasserbau in praktischen Beispielen*, Berlin: Springer, 1956, Vol. 1, p. 234.

chapter **17**

Pneumatic Caissons

17-1. DESCRIPTION

Where for any reason the soil enclosed in an open caisson cannot be excavated through its shaft satisfactorily under water, or when it can be anticipated that upon sinking an open caisson frequent obstacles in the soil may be encountered, the use of a pneumatic caisson is resorted to. Pneumatic caissons are also used in a soft, running soil which cannot be excavated in the dry. In places with groundwater, the pressure of the compressed air will keep water out of the working chamber of the pneumatic caisson.

A pneumatic caisson may be popularly described as a rigid, inverted box with the bottom omitted. The box is kept free of water and entering mud by compressed air during the sinking. In this pneumatic box soil excavation work and the pouring of concrete is carried out in dry under compressed air. The space where workmen enter and work under pressure is called the *working chamber*. The pneumatic caisson is a permanent part of the monolith of a deep foundation.

The principle of a pneumatic caisson is based on the following well-known physics experiment. If in a vessel partly filled with sand in water an inverted glass cylinder is placed as shown in Fig. 17-1 and air is forced into the inverted cylinder, the air expels all of the trapped water from the cylinder (pneumatic caisson) and keeps the surface of the sand dry. The pressure p_a necessary to keep the water out of the inverted glass cylinder (viz., working chamber) is calculated as

$$p_a = \gamma_w H \qquad (17\text{-}1)$$

where γ_w = unit weight of water
H = pressure head

The pressure p_a must be sufficient to balance the full hydrostatic pressure from the surface of the water to the cutting edge (rim of the inverted glass cylinder).

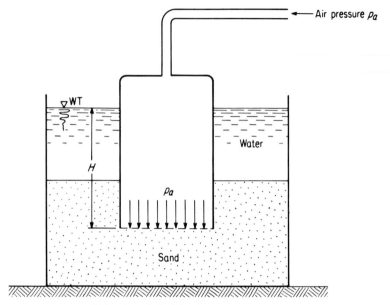

FIG. 17-1. Principle of a pneumatic caisson.

The idea behind this physics experiment was first made use of by the French mining engineer Triger in 1841 for the expulsion of water from a bottomless box by compressed air in penetrating a mining shaft in water-bearing sand. Triger also built air locks for locking of working crews and materials in and out of caissons. Since that time the use of pneumatic caissons for the construction of deep foundations spread rapidly throughout all parts of the world.

Sometimes pneumatic caissons are started out as open caissons and so continued until the occurrence of water and/or other obstacles requires the application of compressed air.

Pneumatic caissons are used for building deep foundations for bridge piers and abutments, lighthouses, wharves, quays and other waterfront facilities, building foundations, hydraulic structures, and underwater tunnels.

Pneumatic caissons are used in open water, or when there is a great influx of water, or when difficult obstructions in soil can be anticipated (tree trunks, boulders), or when other types of caissons and foundations are infeasible because of danger of scour and erosion. In these respects, the pneumatic caisson is the *ultima ratio* for the execution of a deep foundation.

In general, pneumatic caissons are more certain in performance against unknown obstacles such as boulders in clays of doubtful origin, more effective

Pneumatic Caissons

in penetrating harder soil or very coarse ballast, and the actual soil formation is visible for inspection.

The maximum depth of a pneumatic caisson is dictated by the endurance limit of man in working under compressed air pressure. Depending upon the individual's constitution and health, this limit is 3 to 4 atm, or about 100 ft to 120 ft (35 to 40 m) under water. At this endurance depth, the air pressure in the working chamber is 4×14.7 psi $= 58,8$ psi, or about 4 kg/cm^2 ≈ 4 ton/ft$^2 \approx 8,000$ lb/ft^2.

Working under a pressure > 4 atm, caisson sickness increases to such an extent that, for health, the 4 atm pressure in a caisson is considered to be the maximum allowable limit of human endurance in caisson work.

Pressures of 35 to 40 psi are about the limit of normal working, and pressures of 20 psi to 25 psi (46 ft to 57 ft head of water) are common.

The maximum depth attained for pneumatic work in the United States is 125 ft below water surface for pier 2 of the bridge over the Kennebec River at Bath, Maine (1941). Two pneumatic caissons were sunk through 180 ft of alternating water-bearing and impervious materials at the Merriman Dam of the Delaware water supply for New York City.

17-2. CONSTRUCTION OF PNEUMATIC CAISSON

Pneumatic caissons may be made of timber, masonry, concrete, reinforced concrete, steel, and a combination of the above materials.

A metal caisson may be riveted or welded, and is of advantage for floating purposes because of its relatively small weight.

Today the commonest caissons are those made of reinforced concrete. Its advantage lies mostly in its monolithic construction.

Today timber caissons are not used as extensively as formerly. The main disadvantage of a timber caisson is the fire hazard (intensive burning, as in concentrated oxygen). Their squeaking upon sinking creates a bad psychological effect (fear) upon workmen.

Pneumatic caissons may be constructed at the site, or floating, or lowered from overhead frames supported on falsework, or on barges, or on artificial sand islands. A sketch of a pneumatic caisson is shown in Fig. 17-2.

After the cutting edge is positioned and provided with a shaft, air locks and compressors, compressed air is introduced into the working chamber, expelling water from the caisson. The working chamber begins to become dewatered when the air pressure in the working chamber has attained the weight of the external column of water. After dewatering, workmen are admitted through the air lock into the working chamber for excavation work in dry, and the gradual sinking of the caisson.

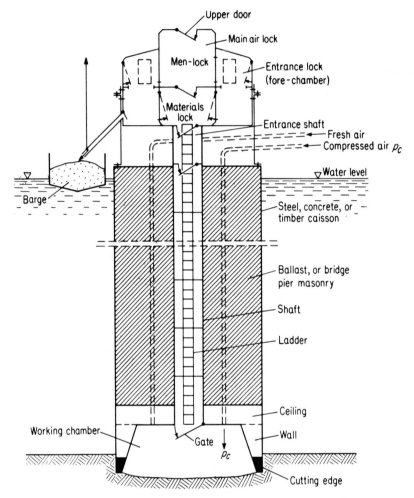

FIG. 17-2. Sketch of a pneumatic caisson.

Pressure in the working chamber should be sufficient to counteract that of water and silt at the cutting edge. As the caisson sinks, the compressed air pressure in the caisson must be increased because of the increase in head of water.

The excavated material is removed from the caisson through the shaft and the materials' lock. Upon excavation of the soil the caisson sinks deeper into the soil by its own weight against friction, uplift and air pressure. The caisson is usually sunk vertically, sometimes inclined. As rapidly as the caisson is sunk concreting of the caisson takes place simultaneously above the ceiling of the working chamber (ballast for penetrating the caisson is added) keeping the air lock at the top above water. Thus the caisson is sunk by adding more weight. After the caisson has attained its design depth, it is filled with concrete. It is of particular importance that the working chamber is filled with concrete in tight contact up to the ceiling, with no empty space left. Empty space causes settlement of the superimposed structure upon transferring its full load to the foundation.

After concreting the working chamber, the lower part of the shaft is filled with a cement mortar and the air pressure is increased in the shaft. Thus the cement grout fills up all the empty spaces between concrete and ceiling. This operation is repeated until it is impossible to force more cement grout from the lower part of the shaft.

After completion of concreting and grouting of the working chamber, the shaft tubes are dismantled and the shaft itself is filled up with a lean concrete. The caisson work is completed.

17-3. CONSTITUENT PARTS OF A PNEUMATIC CAISSON

The principal parts of a pneumatic caisson are (Fig. 17-3):
1. The working chamber
2. The shaft
3. The air and material locks
4. Compressors and appropriate service facilities

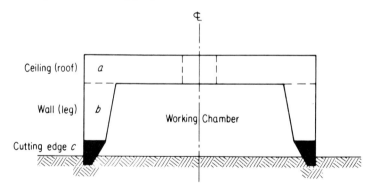

FIG. 17-3. Elements of a working chamber.

1. Working Chamber

The working chamber consists of (a) the ceiling or roof, (b) walls or legs, and (c) the cutting edge.

The working chamber is constructed so as to be airtight. Communication to and from this chamber with the outside takes place via the shaft with an air lock above the level of water. The lock separates the shaft from the atmosphere.

In the working chamber excavation of soil is performed under air pressure for sinking the caisson and keeping out water. Therefore, the working chamber must be leakproof to air. The minimum height of the chamber should be 8 ft for easy work.

(a) *The ceiling.* The ceiling supports superimposed loads of the caisson masonry, ballast, structural load, water pressure, earth pressure, soil reaction, and the maximum uplift force from pressure of the compressed air used for work. Therefore the ceiling must be very strong. It is built as a grillage of beams supported on the side walls (legs). A reinforced concrete slab is placed upon the beams. The ceiling and walls may also be constructed as a frame.

(b) *The walls.* Because of the large superimposed loads on the working chamber the walls must be thick and strong, and also leakproof to air. The walls carry the ceiling of the working chamber. The walls may also be designed as legs of a frame.

Sometimes the outside of the wall is given a batter of $\geqslant$ 1 : 20 with the idea of reducing frictional resistance against the outside surface of the caisson during its sinking through the soil. To keep the frictional resistance down, walls should be made smooth.

(c) *Cutting edge.* The purpose of the cutting edge is to facilitate the penetration of the caisson, and to deflect obstacles encountered in the soil during the course of sinking the caisson. The steel-reinforced cutting edge is particularly suitable for this purpose. The various types of cutting edges are shown in Fig. 14-6.

The cutting edge has usually a blunt end. This distributes soil reactive pressure uniformly under the cutting edge and protects the caisson against tilting during its penetration.

2. Shaft

The shaft tube is usually built vertically and provides access to and exit from the working chamber for workmen and materials, and serves for the transport of the excavated soil upward. The upper end of the shaft is closed by the air lock. The shaft and lock are removable. As the caisson sinks into the soil, the shaft must be extended to keep the lock above water. During this operation the working chamber is closed by a gate plate at the lower end of the shaft.

In a small caisson one shaft suffices. In large caissons two or more shafts are provided to rationalize locking and material transportation work,

Pneumatic Caissons

say one for people, one for soil and one for concrete, for example. The shafts are made of steel, round or elliptical in cross section. The sections of the shaft tubes are 5 to 10 ft long, and the joints of the shaft tubes are provided with rubber gaskets for airtightness. Each shaft is provided with its own air lock.

3. Air Lock

Everything entering or leaving the shaft must pass through an air lock to prevent loss of pressure. The air lock, a steel chamber, is mounted at the upper end of the shaft above water. At the lock there are one, two, or more forechambers or conveyence locks through which the crew is locked into the compressed air. The doors and lids of the lock and its chambers open to the inside in order that the compressed air may press the doors tightly against the seats of their frames. The doors are airtight when closed.

Because of the uplift force from air pressure, the lock must be anchored securely to the mass of the caisson.

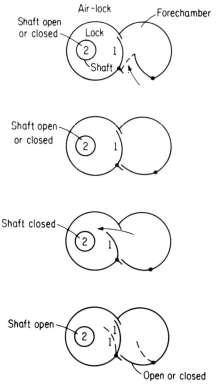

FIG. 17-4. A one-chamber lock.

The air lock is the most responsible part of the caisson equipment. The purpose of the air lock is (1) to prevent loss of pressure in the working chamber and shaft, and (2) for equalizing pressure gradually and so slowly that nobody feels the change in pressure (compression and decompression of people).

The single-shaft air lock has two doors: one to the outside atmosphere (or to the forechamber), and one to the shaft (Fig. 17-4).

The working chamber and the shaft are provided with air pressure slightly above that of the water pressure at the cutting edge.

4. Compressors and Service Facilities

The motors, compressors and pressure pumps are usually located on the shore. Pressure to the working chamber is supplied by pressure lines. In order to cope with eventual equipment failure and power interruption, two independent sets of pressure-producing equipment are a must in pneumatic caisson work—one for normal work, the other as a spare.

The discussion of the actual compressor equipment, pumps, locks, and other caisson service facilities is beyond the scope of this book, as they belong to the subject of construction machinery. Hence this equipment will not be discussed here.

17-4. WORK IN A PNEUMATIC CAISSON

The work in a pneumatic caisson under compressed-air pressure is regulated by special labor laws, and requires the supervision of the health of the workers by a special physician. Some labor laws permit only healthy workers, 20 to 50 years of age, to be employed for work in pneumatic caissons. For each worker a certain amount of fresh air should be supplied in the working chamber. The temperature in the caisson should be between 10°C and 25°C. The maximum pressure normally in the working chamber is approximately 3.5 atm. The length of working hours in a working chamber under air pressure is 8 to 4 hours per day, decreasing with increase in pressure. Only incandescent electric light is permitted in the caisson. Smoking is forbidden.

Caisson disease is marked by neuralgic pains and paralysis, believed to be induced by too-rapid decompression after a stay in a compressed atmosphere, as in a caisson, for example. The "bends" and "aeroembolism" are some of the designations of caisson disease.

Upon leaving the caisson too fast, or with rapid decompression, air is liberated in the human body. While oxygen is readily absorbed by the blood, the nitrogen is not. Nitrogen must be expelled or it forms bubbles, thus impeding circulation of blood, muscular movements and the senses.

Pneumatic Caissons

The bubbles may lodge in the heart, brain, or spinal column. Or enlarged bubbles expand and may rupture blood vessels, resulting in paralysis and death. Therefore, during a slow decompression joints should be subjected to massage. If the human body is decompressed slowly, the nitrogen gas does not form bubbles but escapes gradually and harmlessly from the surface of the lungs. These are the main (health) reasons why length of work under air pressure is reduced considerably. If work must be done in excess of 2 atm pressure, then crews work in two-hour or one-hour shifts. For high pressures the decompression time may be so long that a workman can work only one hour out of eight![1]

17-5. CALCULATIONS OF PNEUMATIC CAISSONS

The various kinds of forces acting on a pneumatic caisson and shown in Fig. 17-5 are:

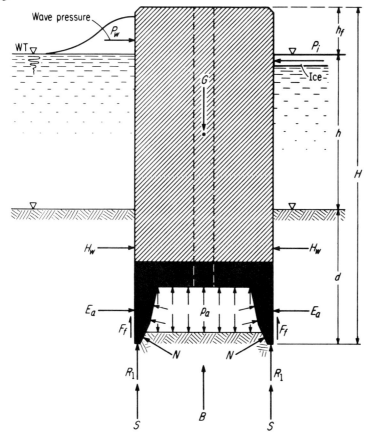

FIG. 17-5. Forces acting on a pneumatic caisson.

1. Self-weight of the caisson, W
2. Weight of ballast, $W_b = G - W$, where G = total weight of pier
3. Hydrostatic pressure, H_w
4. Buoyant force, B
5. Active earth pressure, E_a
6. Soil reaction, $R = 2R_1$
7. Mantle friction, $R_f = 2F_f$
8. Compressed air pressure, p_a
9. Wave pressure, P_w
10. Ice pressure, P_i
11. Dynamic forces, F_d
12. Seismic forces, F_s
13. Suspension forces (in trusses or anchors for suspending, lifting, and righting the caisson), S

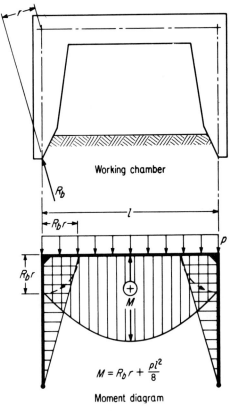

$$M = R_b r + \frac{pl^2}{8}$$

Moment diagram

FIG. 17-6. Bending moments in walls and ceiling of working chamber.

Pneumatic Caissons

In calculating a pneumatic caisson one must consider the various loading conditions during the various stages of sinking, as well as after the sinking of the caisson is accomplished. In the case of a caisson with a long-stretched plan, its rotation too must be considered. The calculations pertain to one unit of length of caisson perpendicular to the drawing plane.

The various stages of calculation may be listed as follows:

Stage I. The caisson is in the forms, and sets. No stresses. The forms are supported on the soil.

Stage II. The forms are removed, and the self-weight of the caisson rests now with its cutting edge on the soil. The entire weight of the caisson is taken up by soil reaction $R = W$. The maximum positive bending moment is in the middle of the span of the ceiling. The soil reactive forces on the cutting edge cause an extra moment in the ceiling (Fig. 17-6).

Stage III. The caisson and part of its final shaft masonry (massive concrete) has penetrated the soil to some depth d_c (Fig. 17-7);

$$W = R + R_f + p_a A \qquad (17\text{-}2)$$

i.e., the caisson sinks upon digging small ditches along the cutting edge inside the caisson. Here $R_f = U d_c H_n \tan \phi_1$, where all symbols are as before.

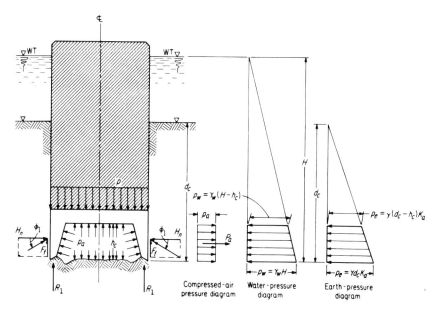

FIG. 17-7. Loads on walls of working chamber.

Stage IV. The caisson has attained its design depth, and rests complete with its full load from self-weight and superstructure on the soil. In this stage the pressure on the soil is checked and compared with its bearing capacity as usual.

Ceiling

The ceiling may be calculated as a continuous slab supported by the walls. The ribs (if any) of the slab may be calculated as freely supported beams on two supports.

The bending moments of the ceiling are calculated in two basic sections: for the central vertical section 1-1, and for the horizontal section 2-2 (Fig. 17-8), where the wall is fixed to the ceiling beam.

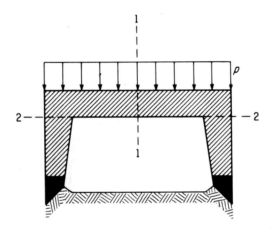

FIG. 17-8. Basic sections for calculating bending moments.

The magnitude of the vertical load from the brick or stone shaft of the pier (ballast) of the caisson may be ascertained from a parabolic or triangular loading on the ceiling as shown in Fig. 17-9. If z/l is the ratio of the vertex of the parabola to its chord l, then with $\tan \alpha = \frac{8}{3}\frac{z}{l}$ the triangular weight of the load on the ceiling will be equal to the parabolic weight (Fig. 17-9):

$$\text{Area of parabola:} \quad a_p = \frac{4}{3} z \frac{l}{2}$$

$$\text{Area of triangle:} \quad a_t = \frac{1}{2} l x.$$

With $a_p = a_t$, $x = \frac{4}{3} z$, and $\tan \alpha = \frac{8}{3}\frac{z}{l}$ Q.E.D.

Pneumatic Caissons

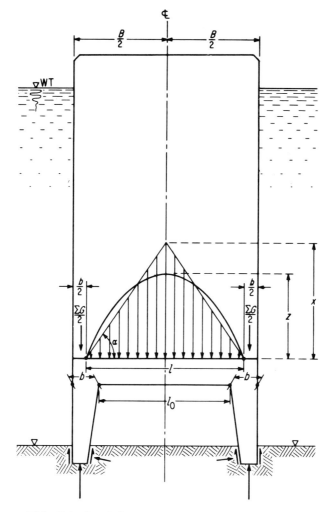

FIG. 17-9. Parabolic and triangular loading from brick on ceiling of caisson.

For a bulk concrete pier, however, its full superimposed load must be reckoned with.

Forces on Cutting Edge

The forces (soil reactions) on the various forms of cutting edges are shown in Fig. 17-10. This figure shows the determination of the horizontal and vertical components, H and R_2, respectively, of soil reaction for a blunt cutting edge penetrated d units deep into the soil. From σ_{max}, the penetration depth d can be calculated. The point of application of R_{ϕ_1} is at $d/3$.

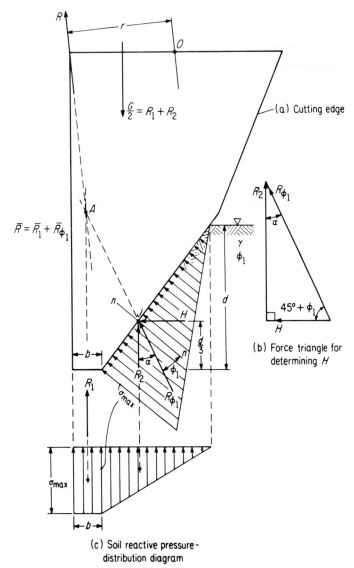

(a) Cutting edge

(b) Force triangle for determining H

(c) Soil reactive pressure-distribution diagram

FIG. 17-10. Forces on cutting edge.

$R_1 = \sigma_{max} b$. The necessary width b of the cutting edge is $b = R/\sigma$, where $\bar{R} = \bar{R}_1 + \bar{R}_{\phi_1}$, and σ may be chosen close to ultimate compressive stress of soil. The force R forms a bending moment $M = Rr$ about point O.

Earth Pressure Distribution on Cutting Edge

The distribution of earth pressure on the cutting edge may be assumed according to Schoklitsch[2] to be increasingly proportional with depth (Figs. 17-11, 17-12, and 17-13).

For the cutting edge in Fig. 17-11 (one side open),

$$\frac{\Sigma G}{2} = N \sin \alpha_1 + \mu N \cos \alpha_1 \qquad (17\text{-}3)$$

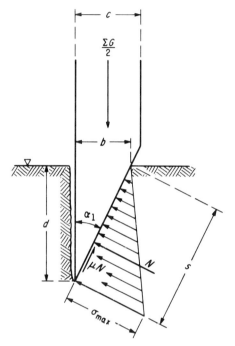

FIG. 17-11. Acute cutting edge (one side open).

But

$$N = \sigma_{max} \frac{s}{2} = \frac{1}{2} \sigma_{max} \frac{d}{\cos \alpha_1} \qquad (17\text{-}4)$$

$$\therefore \frac{\Sigma G}{2} = \frac{1}{2} \sigma_{max} d (\tan \alpha_1 + \mu) \qquad (17\text{-}5)$$

and

$$\sigma_{max} = \frac{\Sigma G}{d (\tan \alpha_1 + \mu)} \qquad (17\text{-}6)$$

where σ_{max} = maximum stress from earth pressure on cutting edge

G = total load on caisson per one unit length of cutting edge, in tons, kips, pounds, or in metric tons or kilograms
N = normal earth resistance
$\mu = \tan \phi$ = wall friction coefficient between soil and cutting edge material. For average conditions, μ may be taken from 0.6 to 0.7

For the same type of cutting edge, but both sides embedded (Fig. 17-12):

$$\frac{\Sigma G}{2} - \frac{F_f}{2} = N \sin \alpha_1 + \mu N \cos \alpha_1 \tag{17-7}$$

$$\frac{\Sigma G}{2} - \frac{F_f}{2} = \frac{1}{2}\sigma_{max} d (\tan \alpha_1 + \mu) \tag{17-8}$$

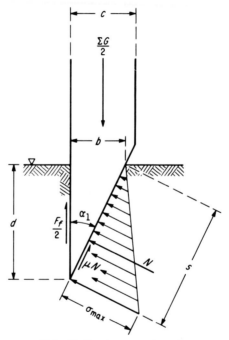

FIG. 17-12. Acute cutting edge (both sides embedded).

and

$$\sigma_{max} = \frac{(\Sigma G) - F_f}{d (\tan \alpha_1 + \mu)} \tag{17-9}$$

where F_f = frictional force on outside of the embedded part of the cutting edge.

Pneumatic Caissons

For a blunt cutting edge (Fig. 17-13), and by the same way of reasoning as before,

$$\sigma_{max} = \frac{(\Sigma G) - F_f}{[2b + d(\tan \alpha_1 + \mu)]} \quad (17\text{-}10)$$

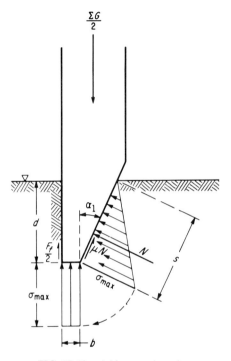

FIG. 17-13. A blunt cutting edge.

Soil reaction distribution for a blunt, obtuse-acute cutting edge: Refer to Fig. 17-14.

Vertical equilibrium:

$$\left(\Sigma \frac{G}{2}\right) - \frac{F_f}{2} = R_B + N_1 (\sin \alpha_1 + \mu \cos \alpha_1) + N_2 (\sin \alpha_2 + \mu \cos \alpha_2) \quad (17\text{-}11)$$

Set

$$N_1 = \frac{1}{2} \sigma_1 s_1 \quad \text{and} \quad N_2 = \frac{\sigma_1 + \sigma_{max}}{2} s_2$$

and observe that

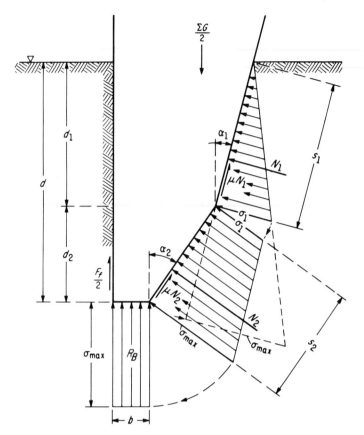

FIG. 17-14. A blunt, obtuse-acute cutting edge.

1. $$\frac{\sigma_1}{s_1} = \frac{\sigma_{max}}{s_1 + s_2}$$

$$\sigma_1 = \sigma_{max}\frac{s_1}{s_1 + s_2}$$

2. $$s_1 = \frac{d - d_2}{\cos \alpha_1}$$

3. $$s_2 = \frac{d_2}{\cos \alpha_2}$$

$$(\Sigma G) - F_f = 2b\sigma_{max} + \sigma_1 s_1 (\sin \alpha_1 + \mu \cos \alpha_1) \\ + (\sigma_1 + \sigma_{max})s_2(\sin \alpha_2 + \mu \cos \alpha_2) \tag{17-12}$$

and

$$\sigma_{max} = \frac{(\Sigma G) - F_f}{2b + \dfrac{1}{s_1 + s_2}[s_1 d_1 (\tan \alpha_1 + \mu) + (2s_1 + s_2)d_2 (\tan \alpha_2 + \mu)]} \quad (17\text{-}13)$$

The horizontal component of N is

$$H = H_1 + H_2$$
$$= N_1 \cos \alpha_1 + N_2 \cos \alpha_2$$

or

$$H = \frac{1}{2}\sigma_{max}\left(\frac{s_1^2}{s_1 + s_2}\right)\cos \alpha_1 + \frac{1}{2}\sigma_{max}\frac{(2s_1 + s_2)}{s_1 + s_2}s_2 \cos \alpha_2 \quad (17\text{-}14)$$

17-6. ADVANTAGES AND DISADVANTAGES OF PNEUMATIC CAISSONS

Some of the advantages and disadvantages of pneumatic caissons may be listed as follows.

Advantages
1. Easy access to the bottom of the caisson.
2. Obstructions under cutting edge encountered in sinking a pneumatic caisson are relatively easy and quick to remove.
3. Excavation and pouring of concrete can be readily done in the dry.
4. At the design elevation the soil in the working chamber can be conveniently inspected.
5. Soil samples can be taken.
6. Soil bearing capacity in the working chamber can be determined by the method of loading plates and jacks.
7. The working chamber can be concreted in the dry.
8. Sinking the caisson brings about no vibrations.
9. The position of the groundwater table remains unaltered, hence no settlement of the caissons or adjacent structures occurs because of the groundwater.
10. The construction site requires less steel than a sheet-piled foundation pit, and less equipment (cranes, excavators) than in the case of open caissons.

Disadvantages
1. Inconvenience of working in the caisson under compressed-air pressure.

2. The penetration depth of the pneumatic caisson below water table is limited to man's endurance limit of working under pressure of 3 to 4 atm, or ~ 100-120 ft below water level.
3. High construction costs for wages for work under air pressure as compared with open caissons.
4. Large amount of manual work involved.
5. Caisson disease.
6. Special labor laws must be observed.
7. Dangerous during seismic activities.

Because of the high cost of labor and the large amount of manual work involved, today the pneumatic caisson is no longer used as much as formerly.

REFERENCES

1. E. E. Seelye, *Foundations: Design and Practice*. New York: Wiley, 1956, pp. 11-16.
2. A. Schoklitsch, *Der Grundbau*, 2nd ed. Vienna: Springer, 1952, p. 382.

OTHER USEFUL REFERENCES

1. G. At Hool and W. S. Kinne (eds.), *Foundations, Abutments and Footings*, New York: McGraw-Hill, 1923.
2. *Engineering News-Record*, Feb. 12, 1931, pp. 275-281.
3. H. S. Jacoby and R. P. Davis, *Foundations of Bridges and Buildings*, 3rd ed., New York: McGraw-Hill, 1941, pp. 318-381.
4. C. W. Dunham, *Foundations of Structures*, 2nd ed., New York: McGraw-Hill, 1962, pp. 511-543.
5. E. Paproth, "Druckluftgründungen," in *Grundbau Taschenbuch*. Berlin: Ernst, 1955, pp. 603-636.
6. W. Mack Angas and C. S. Proctor, "Caissons," in R. W. Abbett(ed.), *American Civil Engineering Practice*. New York: Wiley, 1956, Vol. 1, Sec. 8, pp. 8-28.
7. R. E. White, "Caissons and Cofferdams," in G. A. Leonards (ed.), *Foundation Engineering*. New York: McGraw-Hill, 1962, Chap. 10, pp. 894-964.

PROBLEMS

17-1. A pneumatic caisson of a bridge pier is to be sunk into a riverbed through a 36-ft depth of water and through an 84-ft depth of silt (Prob. 17-1). Calculate, in atmospheres, how great a minimum air pressure p_a is needed inside the caisson to keep the caisson free of water thus permitting working in the dry. (52 psi.)

17-2. What is the magnitude (algebraically) of the total vertical pressure on the roof and on the walls (legs) of the working chamber? $(\gamma_w HA)$

Pneumatic Caissons

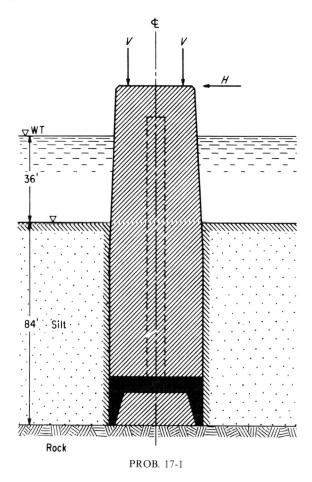

PROB. 17-1

When this force is introduced into calculations, the weight of the masonry should be taken without subtracting the uplift.

17-3. What is the algebraic magnitude (per 1 ft of run of caisson) of the horizontal force on the chamber walls inside of the caisson? $\quad(\gamma_w H h_{ch})$

17-4. Given a system of a pneumatic caisson of steel and a concrete bridge pier shaft as shown in the figure. Plot effective and total lateral pressure distribution diagrams and calculate the magnitudes of lateral and frictional forces per one unit of length of run of caisson (a) if the outside wall is vertical, (b) if the outside wall is slanted 10 : 1. $\quad(F_f = \mu_1 E_a)$

17-5. In Prob. 17-4, what is the magnitude of the lateral pressures W_{ch} from water and earth $E_{a_{ch}}$ on the leg of the working chamber? Also, calculate the bending moment M_{d-d} for bending of leg with respect to section d-d.

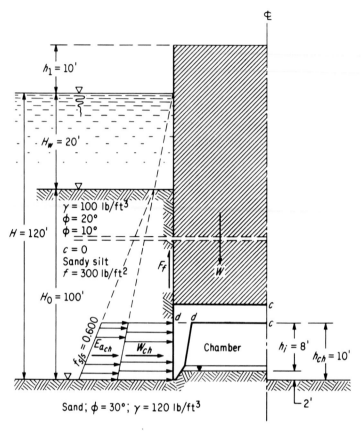

PROB. 17-4

17-6. Refer to Fig. 17-11. *Given*: $(\Sigma G)/2 = 30$ kips/ft of run of cutting edge; $\phi = 30°$; $\alpha_1 = 30°$; $\mu_1 = \tan \phi_1 = 0.7$; $\gamma_{\text{soil}} = 120$ lb/ft^3. The width of the cutting edge is $b = 2$ ft. Determine the penetration depth d of the cutting edge.

17-7. Refer to Fig. 17-12. All conditions are the same as in Prob. 17-6. Determine the penetration depth of the cutting edge.

17-8. Refer to the accompanying figure. Plot the soil pressure distribution diagram for the cutting edge. Determine the magnitude of the horizontal force H, and the position x of the resultant R_v from V_1 and V_2.

$$x = \frac{V_1 \dfrac{a}{2} + V_2 \left(a + \dfrac{b}{3}\right)}{R_v}$$

$$R_v = V_1 + V_2$$

Pneumatic Caissons

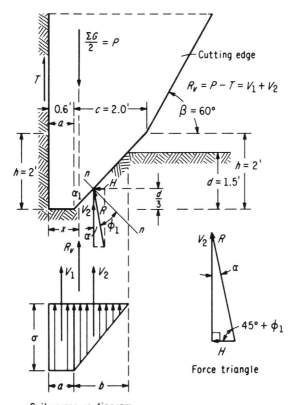

Soil-pressure diagram

PROB. 17-8

$$\frac{V_2}{V_1} = \frac{b\sigma}{2a\sigma} = \frac{V_2}{R_v - V_2}$$

$$\therefore \quad V_2 = \frac{bR_v}{2a + b}$$

chapter **18**

Piles

18-1. FUNCTION OF PILES

A pile is a relatively long columnar construction element made of wood, or reinforced concrete, or metal, or a combination thereof. It is embedded into soil to receive and transmit vertical and inclined loads into the soil or to rock below the ground surface at an economically feasible depth in such a way that these soil layers (or rock) can sustain the loads without causing intolerable settlements to the soil (rock) and structure.

Some of the main functions of piles are:
1. To support superimposed structural loads.
2. To transfer structural loads to layers of soil or rock of good bearing capacity, and below the extent of scour and erosion where this applies.
3. To resist and transfer lateral pressures of loads such as those from structures, soil, water, wind, ice, lateral impact, as well as against uplift by water.
4. To densify the soil to increase its bearing capacity.
5. To form a sheet-pile wall against the influx of water, dry pouring (running) sand and/or plastic soil.
6. To reduce excessive settlement of the structure.

18-2. CLASSIFICATION OF PILES

Piles may be classed by their mode of embedment in soil (free-standing piles and group piles); by the way they derive their resistance to loads (end- or point-bearing piles, and friction piles); by their function (compression piles and tension piles); by the material the piles are made of (timber, concrete, reinforced concrete, cast iron, steel, composite); by the geometrical form of their longitudinal profile (cylindrical; tapered; see Fig. 18-1); solid bearing piles and sheet piles; by the method of their manufacture: ready made (cast in plant) and cast in place, and by the method of embedment (or introduction)

Piles

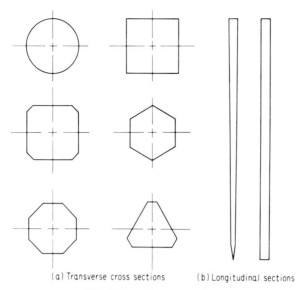

(a) Transverse cross sections   (b) Longitudinal sections

FIG. 18-1. Pile cross sections.

into the soil (piles driven, jetted, jacked-in, vibrated, or prepared in a casing in situ).

The upper end of the pile (when in its final position) is the *head*. The lower end is the *foot*. The *butt* is the larger end of the pile, and the *tip* its smaller end. The latter two terms apply to a tapered pile. When the entire length of the pile is embedded in the soil, such a pile is called a *ground pile*. A pile, when part of its lower end is in soil and the upper end is free, is called a *free-standing pile*. Inclined piles transmitting lateral loads are termed *batter piles*. The inclinations of batter piles usually are from 1 horizontal to 6 vertical (1 : 6) to 5 horizontal to 12 vertical (5 : 12).

Bearing Piles

Any pile sustaining and transmitting a superimposed load is termed a *bearing pile*. Depending upon the manner in which piles transfer loads to the soil, bearing piles are usually classed into *end-bearing piles* and *friction piles*. End-bearing piles derive their support from the underlying firm layer of soil (gravel, sand, stiff clay, soft clay, rock). These piles transmit their loads through their bottom tips. Hence they are termed end-bearing piles or *point-resistance piles*. This is the resistance which the soil (rock) exerts against further penetration of the pile into the soil. In very dense soil the end resistance is large. In very soft soil this resistance may be negligibly small. When piles derive their support from mantle friction, they are termed *friction piles*.

In frictional soils tapered piles provide for a better bearing capacity than do straight, cylindrical piles. Taper utilizes well the mantle friction of the pile, as well as the vertical component of the soil reaction.

To avoid uplift by ice, frost action, water, and the action of tensile forces, wood piles may be driven butt end down. In such cases the advantage of the taper lies in its beneficial mobilization of the soil resistance against pulling the piles out of the soil.

The various types of piles are schematically illustrated in Figs. 18-2, 18-3, and 18-4.

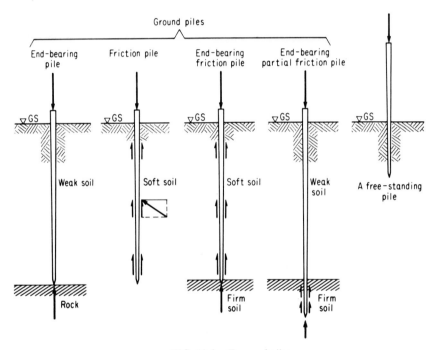

FIG. 18-2. Types of piles.

End-bearing piles resting on rock can support heavier loads than those resting on a firm, geologically unconsolidated soil bearing material. The soil below the elevation of the tips of the piles must be able to support the loads transmitted by piles into these layers.

In gravelly and sandy soils pile penetration may be aided by means of jetting. The jet water may loosen up the pile-surrounding soil to such a degree that the pile may penetrate by its own weight. About 3 to 5 ft before the final penetration the water jet is turned off and the rest of the penetration distance down to the design elevation of the pile is achieved by driving it

Piles 571

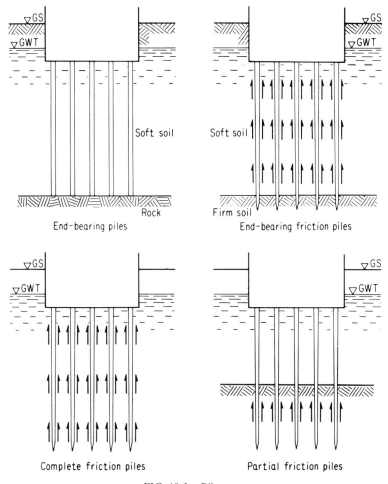

FIG. 18-3. Pile types.

till refusal. Jetting of piles should not be practiced near foundations of existing adjacent structures, subsurface structures, or utilities if jetting would bring about settlement.

Sheet Piles

Sheet piles are used to install a shell or a wall. Sheet piles are interlocking members made of wood, steel, or reinforced concrete. They are embedded individually by driving, or vibrated into sandy soils[1] to form a shell or wall for the purpose of obstructing the influx of water or running sand into the foundation excavation. Sheet piles are used for construction of cofferdams, and are utilized in waterfront structures.

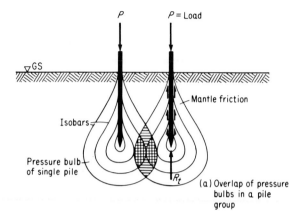

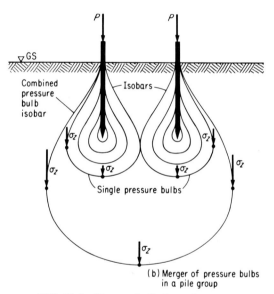

FIG. 18-4. Pressure bulbs in a pile group.

18-3. NEGATIVE MANTLE FRICTION

If a pile is driven through a soil which is subject to a continuous process of consolidation due to its own self-weight, the settlement of the soil (motion down) brings about an oppositely acting *skin friction* relative to the pile. This skin friction is also called the *negative mantle friction*. It tends to pull the pile upward out of the soil. Thus the phenomenon of the negative mantle friction is undesirable in laying foundations by means of friction piles.

18-4. PILE MATERIAL

Piles may be of timber, of round or square cross section, concrete and reinforced concrete—mainly of two kinds: precast, or ready-made, and cast in situ; metal, tubular cylindrical and tapered; smooth mantle surface and corrugated mantle surface; screw piles, made of cast iron and of steel; open-end steel pipe piles; steel H-piles; sheet piles; composite piles, and sand piles.

18-5. TIMBER PILES

Chronologically, plain timber piles are the oldest material used for construction of foundations in and above water in antiquity (lake dwellers on piles, piled structures, trestles through swamps, bridges). The scriptures read that King Solomon built his temple of cedar piles from Lebanon.

Timber piles are trunks of trees from forest regions. Their shape naturally suggests itself for the purpose.

A good pile must be straight, free from sharp bends, large knots, splits, and decay, and must have a uniform taper from butt to tip. Wood piles are usually prepared from the following species of trees:

Soft: willow, poplar: cedars are soft, but resistant to decay

Semi-hard: cypress (straight, well-bodied); longleaf yellow pine

Hardwood piles: of oak, elm, beech, birch, ash, chestnut; mahogany is of high strength and durability

Oregon fir piles; Baltic fir or pine is elastic, strong and durable.

Yellow pine from the southern states (straight, well-bodied): yellow pines are relatively free from large branches and are available in long lengths.

Douglas fir piles originate from the Western forests.

Because of the reduction of our forests, well-bodied wood piles are becoming a scarce commodity.

For standard specifications for timber piles the reader should consult Ref. 2.

In chemically inert soil, and when under water, plain (untreated or unimpregnated) wood piles normally never deteriorate, nor does their strength become impaired. One of the basic requirements for the use of untreated wood piles is that piles must be designed and driven and the pile cutoff must be below the lowest fluctuating water and/or groundwater table. It is within the zone of the fluctuating groundwater table that untreated and/or unprotected wood piles decay.

Where marked fluctuation of water level and/or groundwater table takes place, and where marine borers are known to be at work, wood piles must be protected by an impregnation process such as creosoting, for example—hence the term "treated" piles. Creosote-treated Douglas fir piles have an average service life of 36 years in coastal waters infested with teredo and limnoria (marine borers), where untreated wood piles are often completely destroyed by these marine borers within a year.[3]

Figure 18-5 shows the decay of tops of timber piles caused by a fluctuating groundwater table.

FIG. 18-5. Decayed tops of timber piles caused by fluctuating groundwater table. (Courtesy of the Trustees of the Boston Public Library.)

The decay of wood piles is brought about by fungi—a form of plant life which tears down the cellular structure of the pile.

By driving piles below the lowest groundwater table the fungus is deprived of air, hence it cannot contribute to the decay of the wood pile. These fungi cannot thrive in a dry environment either. Their pile-deteriorating work is effective only in an air-temperature environment of fluctuating moisture content.

For insect attack on wood piles (termites, wood-borer beetles, and marine borers) see Ref. 4.

In areas where future local and regional lowering of the groundwater table by drainage is anticipated, or where the installation of sewer systems is planned, the use of timber is not advised, especially if the lowest position of the lowered groundwater table in that region is not known.

Length of Timber Piles

The length of timber piles—and indeed any material such as concrete and steel—must be determined based on adequate soil investigation and tests made prior to pile-driving operations in order to transmit structural loads to soil safely. Needless to say, successful pile-driving and pile-foundation design requires carefully prepared soil-profile diagrams.

The character of the soil must be known along the driven pile as well as below the pile. Adequate soil borings and soil-profile diagrams are invaluable.

It has happened a good many times that because of lack of adequate soil information piles have been ordered in extravagant lengths the excess of which had later to be cut off, thus contributing merely to the waste of material. The ordinary market length of wood piles is roughly 20-90 ft, although Douglas fir piles 125 ft long have been used. The length of wood piles is limited by the length of railroad platforms and trucks and thus by the transportation cost. These factors govern price and time of delivery of piles.

To increase the length of piles they may be spliced. Although splicing of timber piles is not a good practice, it is sometimes done.

Two piles may be spliced together end to end, bolted together on four sides of the piles (Fig. 18-6a). Another method of splicing is by means of a metal sleeve in the form of a heavy pipe to provide for lateral strength and stiffness (Fig. 18-6b). Sometimes in soft, swampy ground one pile is driven on top of another with only a steel dowel pin connecting the two (Fig. 18-6c). The splicing by the method of a half-lap joint (Fig. 18-6d) is not too satisfactory because of lack of strength.

Cross Section of Wood Piles

Wood piles may be round, square, and sometimes rectangular in cross section. Square wood piles are usually used in countries which are deficient in forests and hence must import wood piles. The central part in cross section of a trunk of a tree is harder than the periphery. Thus only the best of a trunk can sustain somehow the long transportation and its cost. The

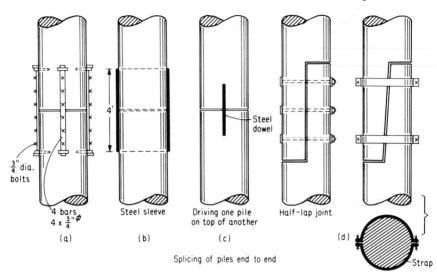

FIG. 18-6. Some methods of splicing of timber piles.

soft parts remain in the sawmill where the square piles were cut. Square piles can be easily kept in line while driving, and provide a good bearing for end-bearing piles.

Diameter of Wood Piles

Timber pile diameters are measured 3 ft from butt and at the tip.

Length of Piles, ft	Diameter of Piles, in.	
	Butt	Tip
40	12-20	8-10
40-90		6- 9
90		5- 6

Tips of Timber Piles

In soft soil where driving is easy it is not necessary to sharpen the tips of the timber piles. In driving a pile with a blunt end, a cone of densified soil forms under the tip, acting as a wedge, just as if the tip were pointed.

Successful driving in dense soil materials requires sharpening of the tip of the pile, as shown in Fig. 18-7. The length of the sharpened point is usually made about twice the diameter of the tip.

Piles

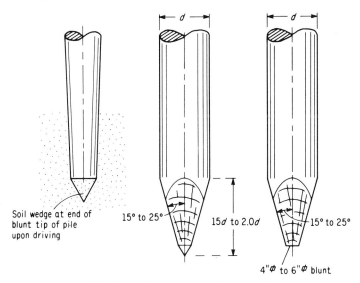

FIG. 18-7. Sharpened tips of timber piles.

Pile-Driving Shoes

To protect the tip of the timber pile from damage upon driving through hard soil, debris, and old grillages, and to aid in penetration of the pile into dense formations of soil, the tips are provided with steel shoes (Fig. 18-8). Various designs of these shoes are on the market. Some shapes of pile-driving shoes for concrete and steel pipe piles are also shown in Fig. 18-8.

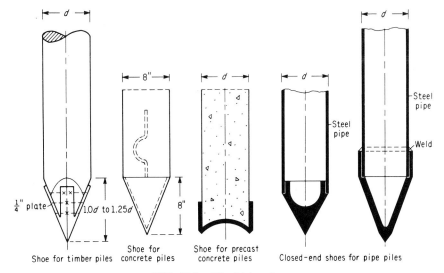

FIG. 18-8. Pile-driving shoes.

Storage of Timber Piles

Stored wood piles should be protected against decay and marine borers (where applicable). While stored, piles should be free from bending. Piles should be loaded, unloaded, and otherwise handled with care so as not to break and split them, or inflict on them any other damage that might cause rejection.

Economy of Wood Piles

Wood piles are economical in constructing pile foundations in regions rich in forests. In regions where there are natural water courses suitable for floating timber, the transport of the latter (because of its relatively small weight) to the site of pile projects located downstream is simple and inexpensive. Also, the careful handling of timber at the construction site by rolling prevents the danger of damage. The transport of timber, viz., wood piles, to distant places may, however, turn out to be an expensive enterprise.

18-6. CONCRETE PILES

Concrete piles are used
1. Where groundwater conditions make the use of timber piles undesirable (if the groundwater table is deep below the ground surface, for example).
2. Under heavy structures where timber piles cannot be used because of their insufficient bearing capacity.
3. In regions where no timber piles are available.

Reinforced-concrete piles are independent of the position of the groundwater table and its fluctuation.

A comparison difference in timber and reinforced-concrete pile foundations for the same soil profile and the same position of the groundwater table is illustrated in Fig. 18-9. Note that in both instances piles are driven into the firm soil.

One distinguishes between two basic types of concrete piles, namely, (a) precast or ready-made piles, and (b) cast-in-place concrete piles.

Precast Concrete Piles

Precast concrete piles are designed and provided with steel reinforcement which is needed for strength only while handling (slinging, lifting, and hoisting) and driving the piles. Otherwise (theoretically), for piles working under axial static compressive load in a structural foundation reinforcement is not needed. The reinforcement is needed, however, when the piles work as columns, or when subjected to bending.

Piles

Precast concrete piles are manufactured at a central pile plant, are delivered at the construction site and then driven into position. Otherwise they may be made at the construction site.

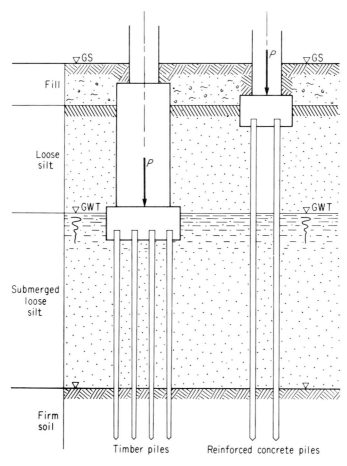

FIG. 18-9. Timber and reinforced concrete piles for identical load, soil and groundwater conditions.

Precast concrete piles are used to support vertical loads; to carry combined vertical and lateral loads. There is no deterioration of concrete piles at the water table if concrete is properly made, poured, and cured. Concrete piles are not damaged by termites, nor by marine borers.

A precast reinforced-concrete pile is shown in Fig. 18-10.

Precast concrete piles are square, octagonal, solid-circular, or hollow-circular in cross section. Concrete piles are usually made to order.

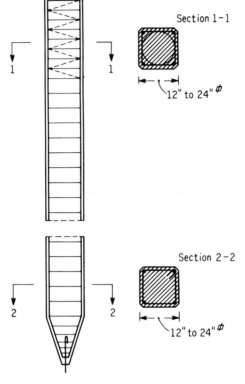

FIG. 18-10. A precast reinforced concrete pile.

Precast reinforced-concrete pile lengths are made up to 100 ft in length, the optimum length attaining 40-50 ft. The maximum length of prestressed concrete piles is about 200 ft.

Reinforced-concrete piles have large bearing capacities as compared with timber piles.

The recommended stresses for precast concrete piles are 15 percent of 28-day strength of concrete, but not more than 700 lb/in.^{2} For prestressed concrete piles the recommended stresses are 15 percent of 28-day strength of concrete but not more than 800 lb/in.^{2} in excess of prestress.

The maximum pile load for usual conditions may be taken as 100 tons for precast piles, the optimum load range being from 40 to 60 tons, and 200 tons for prestressed concrete piles.

The dimensions of reinforced-concrete piles are from 12 to 24 in. (side or diameter), and the hollow-circular, tubular piles are made from 12 to 54 in. in diameter.

The tip of the concrete pile may be adapted to the soil and the method of driving. For plastic soil a blunt tip is suitable. For noncohesive soils (gravel and sand), a long tapered point is desirable. Sometimes it is desirable to provide a metal shoe for the tip. Extra lateral reinforcement should be provided near the tip.

The Handling of Precast Concrete Piles

It has happened on construction jobs that many precast reinforced-concrete piles have cracked or broken and hence been rejected before even being driven, mainly because of improper handling, or by negligence in reading and/or understanding instructions for handling reinforced-concrete piles before commencement of pile-driving operations.

The bending stresses caused by picking up and handling precast concrete piles are greatly influenced by the number of points of attachment of the slings and the positions of the points. To save precast reinforced-concrete piles from cracking and breaking during their handling, piles are designed for longitudinal reinforcement in order to take up bending moments induced by handling of piles—stacking, loading, pitching, lifting and hoisting, transportation and while being stored—as well as for coping with stresses induced by buckling upon driving of piles. The lateral reinforcement serves the purpose of increasing the axial compressive strength of the pile.

To insure minimum bending moments in the pile commensurate with convenient handling, precast reinforced-concrete piles are provided with fixed suspension points such as lugs, or hooks, or rings in place which are cut off after the pile is in driving position, or notches for pitching and hoisting. In the absence of lugs, the piles are usually marked with dots showing where to attach hoisting slings to produce the minimum bending moments for which that pile is designed. Hence, it is forbidden to lift piles in a manner different from that called for in the instructions in order to avoid cracking as a result of induced unequal bending moments from improper support of piles (resulting in bending moments larger than allowable).

Figure 18-11 illustrates single and double pickup-point locations. If designed for these pickup points, the pickup of the piles will induce equal positive and negative bending moments.[5] Note in Fig. 18-11 that when the pile is designed for pickup at 0.33 L from the upper end of the pile, its equal + and − moments are only 1/3 of those when picked up at its end.

For lifting long piles, double or multiple points of suspension are used. Piles must also be stored with due regard for possible cracking.

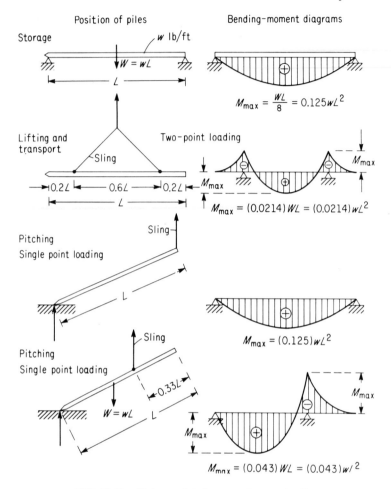

FIG. 18-11. Pickup points for precast concrete piles.

18-7. REINFORCED-CONCRETE PIPE PILES

These are piles which are cast in a special centrifugal machine at 300 to 1,000 rpm. The reinforcement consists of high-grade steel, and may have single or double spirals. The tips are conical to facilitate driving. These tubular piles are usually filled in with concrete.

An interesting example of the use of open-end reinforced-concrete pipe piles is that for the piers of the Lidingö Bridge at Stockholm, Sweden, across the little Värtan (bay).[6] At the site of the bridge, the bay is 750 m wide, and 18 to 20 m deep. Rock is 35 to 40 m, and in some places even 60 m, below the average water level. The rock is covered with a course of gravel

Piles 583

several meters thick. The gravel in turn is overlain by a soft, fine-particled layer of clay.

The bridge piers were placed on a cluster of inclined and vertical reinforced concrete pipe piles (Figs. 18-12 and 18-13). The diameter of these piles was 93 cm, consisting of a spiraled reinforced-concrete mantle filled with

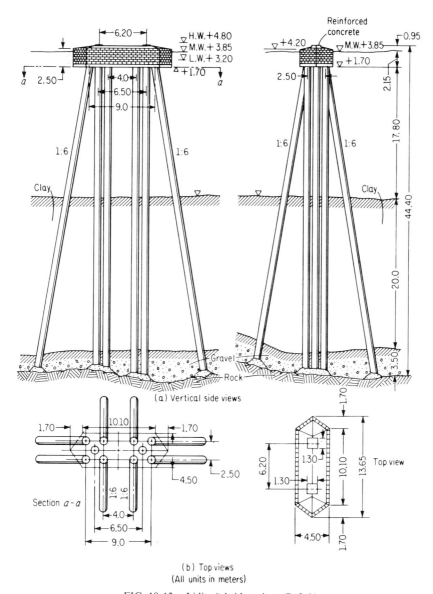

FIG. 18-12. Lidingö bridge piers (Ref. 6).

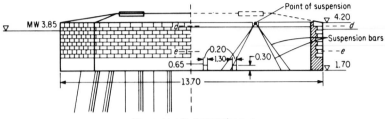

View and longitudinal section a–a

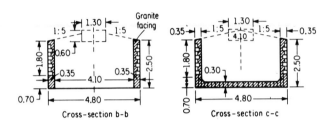

Cross-section b–b   Cross-section c–c

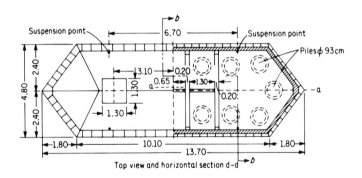

Top view and horizontal section d–d

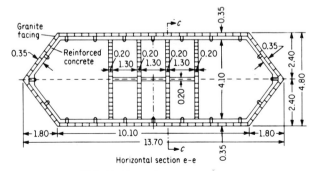

Horizontal section e–e

FIG. 18-13a. Pier cap (Ref. 6).

Piles

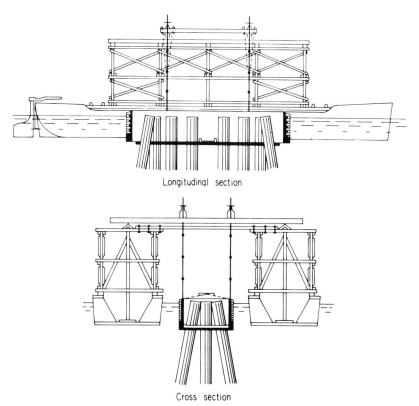

FIG. 18-13b. Placing of pier cap (Ref. 6). Barges for placing pier cap on piles (top); pier cap lowered and placed on piles (bottom).

concrete. The piles were driven to rock, the soil was removed from the inside of the pipe piles by means of an air-jet pump, and the resulting hollow core concreted. These pipes were manufactured on land and then floated with both ends closed to the site. To sink the tip of the pile, some water had to be introduced into the pile. After fixing the pile to the guides of the pile-driving rig the lids were removed from the ends of the pile, and driven into position. The washing out of the pipes had the advantage of ascertaining that the piles were resting on the rock.

The bridge piers were placed on eight inclined and two vertical pipe piles 44 m ≈ 144 ft long. A pier cap of reinforced concrete with granite facing was placed on the pier cluster.

18-8. CAST-IN-PLACE CONCRETE PILES

As the term suggests, cast-in-place piles are cast at the site (in situ) where they are designed to be.

Cast-in-place concrete piles may be classed as follows:
1. Thin shell driven with mandrel
2. Shell driven without mandrel
3. Shells withdrawn
4. Concrete-filled steel pipe piles.

Cast-in-place concrete piles are used in foundations to transmit vertical loads.

A preferred method of making cast-in-place piles is by driving a tapered metal tube into the soil. Then the tube is withdrawn and the space thus left in the soil filled with concrete for the pile "in place" (in situ).

Other cast-in-place concrete pile systems utilize a cylindrical pipe, provided at its lower end with a solid metal driving shoe (Fig. 18-8). This pipe is driven into the soil. Then the pipe is gradually pulled out of the soil (the shoe being left behind), concrete is poured into the vacant space below the lower end of the withdrawing pipe, and steel reinforcement is introduced into the concrete if needed.

Some cast-in-place concrete-pile systems drive an open-end metal pipe into the soil. The soil is removed from the inside of the metal cylinder, and the pipe is then filled with concrete. The pipe is left in the soil as a shell of the concrete pile.

The Pentagon at Arlington, Virginia, is founded on 41,492 cast-in-place concrete piles.

Some cast-in-place concrete-filled steel pipe piles have a high cost, but have the best control during their installation, have a high bearing capacity, and are relatively easy to splice.

A method and process for making cast-in-place concrete piles originated by Intrusion-Prepact Corporation, in Cleveland, Ohio, has gained popularity in the Midwest. According to this method, a hollow-stemmed helical auger is bored to the pile depth and slowly withdrawn as grout is pumped through the stem into the bore hole. Aluminum powder is used in the grout to cause expansion. The advantages of this method are: (1) there are no driving vibrations, and (2) there is no need for casing of the bore holes.

A disadvantage in this method is that pumping pressures and volumes should be monitored to prevent gaps in the pile.*

Another method is to auger a hole, case if needed, and fill through a tremie. These bore holes are frequently called "caissons," particularly if belled by use of a special auger.

---

*This information is by courtesy of Dr. R. Handy, Professor of Civil Engineering and Director of the Soil Mechanics Laboratory of the Iowa State University of Science and Technology at Ames, Iowa.

18-9. METAL PILES

Metal piles may be divided into cast iron piles and steel piles.

Cast-Iron Piles

Although cast iron is a brittle material, and continuous blows from heavy driving hammers may crack the metal, cast-iron piles are still occasionally used. Cast iron has some favorable properties—it is practically immune to oxidation, does not corrode from salt water, and is safe against attack by marine borers. Tench[7] writes that a pier for the British Admiralty was supported on circular, hollow cast-iron piles, screw-driven into the river bed. The cast iron piles were cast in sections of convenient lengths

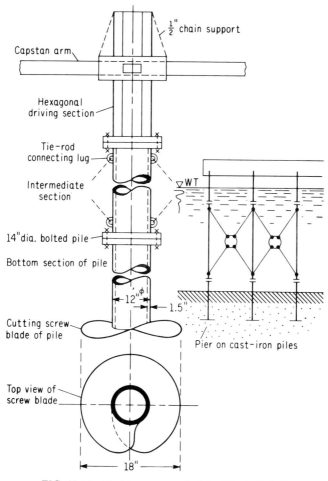

FIG. 18-14. Cast-iron screw pile (after Tench, Ref. 7).

with flanges at the ends for connecting additional units (Fig. 18-14). At the bottom of the lower section of each pile two cutting blades (the so-called *screw*) were cast as an integral part of the pile. The pile was penetrated into the soil by rotating it so that the cutting blades pulled it into the soil. Rotation was accomplished by applying torque to the pile by means of a capstan. The capstan was connected to a hoist drum by a manila rope wound on the drum. Power for screwing the pile into the soil was obtained from a hoisting device.

Screw Piles

Screw piles have been used successfully for building piers of bridges in alluvial beds. Screw piles were also used for founding lighthouses and harbor piers in the eighteenth century in England. They were made of cast iron. A screw pile with its threads and discs affords a large bearing area. When rotated, the screw causes the pile to descend. A screw pile is shown in Fig. 18-15.

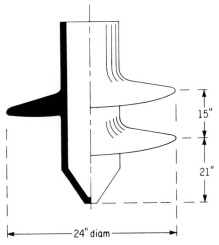

FIG. 18-15. Pile screw.

Today steel screw piles are occasionally still used for wharf piers and as a means of anchorage, particularly in temporary structures.

From Ref. 8, the maximum vertical pressure which the soil can resist at the base of a screw pile at the base of a cylindrical pile z units of length deep below the ground surface without a rupture of the soil may be calculated as an end support pile (the screw-pile blades are pressed against the soil = passive case, or major principal stress). For example, with $\gamma = 100$ lb/ft^3, $c = 0.1$ ton/ft^2, $\phi = 30°$, and $z = 6.0$ ft, the ultimate pressure σ_u the soil can sustain is

Piles

$$\sigma_u = \sigma_{III} \tan^4\left(\frac{\pi}{4} + \frac{\phi}{2}\right) + 2c \tan^3\left(\frac{\pi}{4} + \frac{\phi}{2}\right) + 2c \tan\left(\frac{\pi}{4} + \frac{\phi}{2}\right) \quad (18\text{-}1)$$

where $\sigma_u = \sigma_1$ = ultimate contact pressure on the soil
$\sigma_{III} = \gamma z$ = minor principal stress in soil
γ = unit weight of soil
ϕ = angle of internal friction of soil
c = cohesion of soil

or

$$\sigma_u = (0.05)(6.0)\,(\tan^4 60°) + (0.2)\,(\tan^3 60°) + (0.2)\,(\tan 60°)$$
$$= 4.085 \approx 4.1 \ (\text{ton/ft}^2)$$

If the area of the horizontal projection of the screw pile is A ft^2, then the ultimate bearing capacity of the screw pile due to the soil is

$$P_u = \sigma_u A \ (\text{tons})$$

The safe bearing capacity P of the screw pile is obtained by dividing the ultimate bearing capacity P_u by a factor of safety η:

$$P = P_u/\eta$$

The effect of friction along the sides or mantle area of the pile shaft is usually neglected in these calculations. It is very difficult to ascertain the distribution of the friction along the sides of a pile. Besides, in penetrating the pile in a moist soil hydrostatic pore water (neutral) pressure may temporarily be built up, a phenomenon which may at least temporarily reduce the cohesive bond between the soil and the metal plate.

Steel Piles

These are piles that depend mainly upon steel for their resistance to axial and lateral loads. Steel piles are strong as columns. When well seated, steel piles give the highest load values of any form of pile foundation. Note again that buckling reduces allowable stresses in piles. Steel pipe piles are used with open ends and closed ends.

The maximum length of steel piles is practically unlimited. Their optimum length varies from 40 to 100 ft. The recommended maximum stress for noncorrosive locations is 9,000 lb/in.^{2}, and the maximum load on the steel pile is the maximum allowable stress multiplied by the cross section. For corrosion, compensate 1/16 in. of cross section of steel. The optimum load range of steel piles is 40 to 120 ton/pile. Steel piles are best suited as end-bearing piles on rock. For cased and uncased steel piles, refer to the Building Laws of the City of New York,[9] or to other pertinent codes.

A steel-pipe driving shoe and the splicing of a steel-pipe pile by a welded inside sleeve are shown in Fig. 18-16.

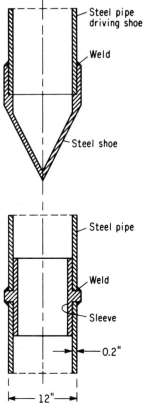

FIG. 18-16. Steel-pipe driving shoe and splicing a steel-pipe pile.

Steel H-Piles

In practice, the most common type of structural steel piles has an H profile. They are heavy rolled H sections. Steel H-piles penetrate the soil easily. They are used for buildings, bridges and other engineering structures. Fluctuating groundwater conditions are insignificant in the use of steel H-piles. The H-beams are more often used as end-bearing than as friction piles. The most commonly used H-beams for use as bearing piles are those between 10 and 14 in. H-beam piles should normally be considered only when bearing on rock or gravel hardpan. Railroad rails may also be used as steel piles.

The United States Steel Corporation recommends limiting the fiber stress in H-pile design to a maximum of 12,000 lb/in.^{2} under working load instead of the 15,000 lb/in.^{2} used in structural design, to allow for corrosion,

Piles

whereas the building laws of the City of New York limit the fiber stress in H-piles to 9,000 lb/in.² H-beams drive easily, and can be spliced readily by welding, to make piles of great length.

Steel H-piles are sometimes enlarged at their lower ends by welding to the main pile various short steel profiles (lags) to give a greater bearing capacity than that provided by a single, slender pile. The process of covering piles with lags is termed lagging of piles. This idea is shown in Fig. 18-17.

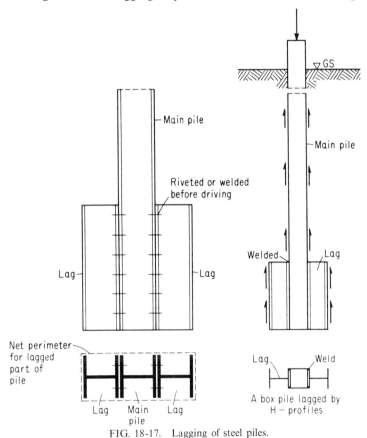

FIG. 18-17. Lagging of steel piles.

For transmitting large structural loads to the soil, sometimes heavy-duty concrete-filled steel pipes with a heavy H-pile core are used (Fig. 18-18a). Figure 18-18b shows a heavy-duty concrete and steel pipe composite pile whose design bearing capacity is more than 2,000 tons.

Corrosion

The rate of corrosion of steel piles varies greatly with the texture and composition of the soil, the depth of embedment, the soil moisture content,

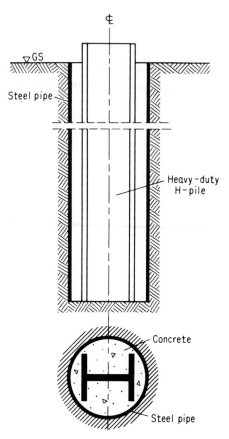

FIG. 18-18a. Heavy-duty concreted steel pipe with a heavy-duty section H-pile.

and the number of years of service. Swamps, peat bogs, and alkaline spots (industrial mine wastes, new cinder fill, acids from coal storage) are considered corrosive soils. In coarse-particled soils, because of air circulation, the intensity of corrosion may approach that occurring in atmosphere. Deficiency of oxygen in heavy clay results in a slow rate of corrosion. Steel piles projecting into air from soil may corrode near the line of the ground surface and for a short distance below (atmospheric corrosion). Steel piles between mud line and low-tide level in seawater corrode most in the upper part where more oxygen is present in the air.

Painting and/or concrete encasements are used to cope with atmospheric corrosion.

Catalogues from steel manufacturing companies carry the information that[10]

Piles 593

FIG. 18-18b. Heavy-duty H-core. (Courtesy of the Thomas Crimmins Contracting Company, New York.)

1. Steel H-piles embedded in any neutral soil suffer little or no loss by corrosion even where the soil is alternately wet and dry.
2. Electrolysis rarely attacks a protected steel pile.

18-10. COMPOSITE PILES

Composite piles are usually made of timber and reinforced concrete. They are used when they are to project above the groundwater table. The lower part of the pile is made of timber, the upper part of concrete because of its driving strength and resistance to deterioration. The joint (splice) between the timber pile and the reinforced-concrete pile must be a good

TABLE 18-1
Advantages and Disadvantages of Various Kinds of Piles

Kind of Pile	Advantages	Disadvantages
Timber	Usually available; relatively inexpensive; light weight; easy to transport, store, handle and to drive. Satisfactory strength. Relatively simple pile-driving machinery needed. Piles can be extended in length during driving operations. Can be sawed below water. Rolling of piles is safe. Long service below groundwater table and in waterlogged soils.	Subject to decay and attack by fungi above water table and in zones of varying moisture content, and to attack by marine borers where this applies. Bearing capacity per pile is low, about 20 ton/pile. Driving end of pile has tendency to broom, requiring driving caps. Can be overdriven.
Concrete (cast-in-place)	Stronger than wood. Relatively large bearing capacity. Less subject to deterioration as compared with wood. Unaffected by position of groundwater table, hence permanent. Initial economy.	Concrete is sensitive to attack by chemicals such as alkali in soils or to action of aggressive groundwater; great weight and cost; difficult to handle and to store. May break with unskilled handling; may crush upon driving. Difficult to splice after concreting.
Reinforced Concrete (ready-made)	Same as under concrete. Desired shape and length possible. Large bearing capacity. Unaffected by position of groundwater table. Corrosion resistance can be attained. Take hard driving.	Same as under concrete. Difficult to handle unless prestressed. Difficult to splice because of steel and prestressing. Large consumption of steel reinforcement. Heavy driving equipment needed. Sensitive in aggressive groundwater. High initial cost.
Steel	Can take very large load (about 100 ton/pile admissible). Can be made very long because easy to splice. Can stand hard driving. Able to penetrate through light obstructions. Can be cut to length under water.	Great weight; high cost. Vulnerable to corrosion. Storage requires protection against rusting. May be damaged or deflected by great obstructions.
Composite	Possible to obtain great length. Timber part performs well under water table.	Joints between timber and concrete pose problems. The strength of the pile is limited by the strength of the timber, viz., the weakest compound part of the composite pile. Handling is more complex than with piles made of timber, or steel, or concrete.

one, and designed to be below the lowest or permanent water table (whichever the case is) to prevent the joint from decaying. The bearing capacity of the composite pile is limited by the strength of the timber part of the pile.

Advantages and disadvantages of timber, concrete, reinforced-concrete, steel, and composite piles are given in Table 18-1.

REFERENCES

1. A. R. Jumikis, *Theoretical Soil Mechanics*. Princeton, N.J.: Van Nostrand, 1969.
2. American Society of Civil Engineers, *Manual of Engineering Practice No 17*. Specification for Timber Piles, 1939.
3. H. R. Peterson, in Shu-t'ien Li, "Criteria of Relative Merits of Construction Materials as Used in Waterfront Facilities on the Basis of Inspection Tests and Service Records," American Railway Engineering Association's *Report on Relative Merits and Economics of Construction Materials Used in Waterfront Facilities*, Bulletin 539, Vol. 58 (November 1957).
4. R. D. Chellis, *Pile Foundations*. New York: McGraw-Hill, 1951, pp. 298-340.
5. Portland Cement Association, *Concrete Piles*. Chicago: Portland Cement Association, 1951, pp. 44-50.
6. G. Schaper, "Bau der Lidingöbrücke bei Stockholm," *Die Bautechnik*, 1924, p. 405.
7. R. Tench, "Cast Iron Piles Screw-Driven to Rock," *Engineering News-Record*, Vol. 133, No. 26 (Dec. 28, 1944), pp. 830-831.
8. A. R. Jumikis, *Mechanics of Soils: Fundamentals for Advanced Study*. Princeton. N.J.: Van Nostrand, 1964, p. 33.
9. Building Laws of the City of New York, Vol. 1 (1959).
10. Bethlehem Steel Corporation, *Steel H-Piles Handbook* 2196 Bethlehem, Pa.: (n.d.), pp. 27-32.
11. American Institute of Steel Construction, *Manual of Steel Construction*, 6th ed. New York: The Institute, 1963, pp. 1-16.

QUESTIONS

18-1. What is the primary function of piles? (One of load transfer).
18-2. How are piles classed based on their bearing capacity?
18-3. What is the purpose of pile driving?
18-4. Why is it necessary to drive timber piles below the lowest elevation of the groundwater table?
18-5. Why is it necessary to handle ready-made reinforced-concrete piles "academically"?
18-6. How are ready-made reinforced-concrete piles handled for storage, lifting and positioning in the frame of the pile-driving rig?
18-7. What are the objections to cast-in-place piles?

PROBLEMS

18-1. *Given*: (Refer to accompanying figure). Design load $P = 120$ kips and a regular H-pile section 10 WF 49, with a cross-sectional area $A = 14.4$ in.^{2} and web thickness of 3/8 in.,[1] for driving the pile to solid rock. The allowable bearing capacity σ_{all} of the given rock is $\sigma_{all} = 4{,}500$ psi. *Required*: increase the H-pile base area by adding to it sufficient area of steel plates and angles (lags) to bring up pressure between gross area of cross section of steel on rock to a range of from 3,000 to 6,000 psi. The lags are to be welded to the H-pile prior to driving.

Solution

Compressive stress σ in the given H-pile:

$$\sigma = P/A = 120{,}000/14.4 = 8{,}333 \text{ psi} > 4{,}500 \text{ psi}$$

hence lagging is needed.

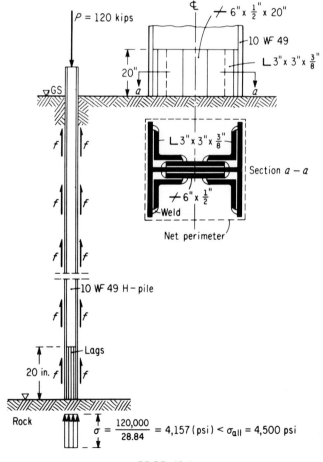

PROB. 18-1

Necessary built-up point area (assuming $\sigma_{all} = 4{,}500$ psi):
$$A_{nec} = P/\sigma_{all} = 120{,}000/4{,}500 = 26.67 \text{ (in}^2\text{)}$$
Use two plates and four angles with a cross-sectional area of:

10 WF 49............................	14.40 in.²
2 plates 6 × 1/2 in.....................	6.00 in.²
4L 3 × 3 × 3/8 in. at 2.11 in.².............	8.44 in.²
Total...........................	28.84 in.² > 26.60 in.²

∴ Lags increase considerably the bearing capacity of steel piles used in granular as well as in cohesive soils.

18-2. Provide for lagging of the given H-pile for the conditions shown in the figure. The minimum length a of the lagging depends upon the necessary length of the welds. However, in practice, the minimum length a_{min} of the lagging is taken as $a_{min} = 36$ in., the lags to be welded prior to driving.

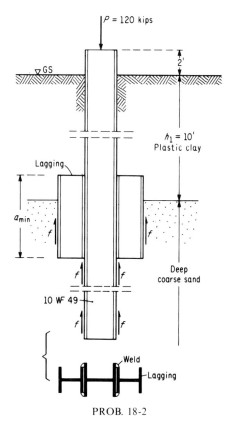

PROB. 18-2

chapter **19**

Bearing Capacity of Piles

19-1. METHODS OF DETERMINING BEARING CAPACITY OF PILES

One of the most important factors in the design and construction of a pile foundation is its bearing capacity.

The *Manual of Engineering Practice No. 27* published by the American Society of Civil Engineers[1] defines the term *bearing capacity* as follows: "Bearing capacity may be defined as that load which can be sustained by a pile foundation without producing objectionable settlement or material movement—initial or progressive—resulting in damage to the structure or interfering with its use."

The bearing capacity of a pile depends upon
1. Type and properties of the soil
2. Surface and/or groundwater regimen
3. Geometry of the pile (solid, hollow, rectangular, straight or tapered)
4. Pile material (timber, concrete, steel)
5. Size of pile (cross section, length)
6. Property of mantle surface of pile (rough or smooth)
7. Driving depth of pile
8. Method of embedding the pile into the soil (driving, jacking, jetting, vibrating, casting in place)
9. Position of pile (vertical or inclined)
10. Spacing of piles in a pile group.

The bearing capacity of pile foundations is estimated based on either the bearing capacity of a single pile, or that of a group of piles.

There are three main methods of determining the bearing capacity of a pile: (1) by static loading, (2) dynamic pile-driving tests in situ, and (3) analytically.

Using the data obtained from the pile-driving tests, such as the driving resistance to penetration of the pile into the soil, and other pertinent data,

Bearing Capacity of Piles

the pile bearing capacity is then calculated by means of the dynamic pile-driving formulas.

The ultimate pile bearing capacity may also be determined by static loading tests of piles in the field. Or else the pile bearing capacity may be determined analytically by means of static pile bearing-capacity formulas.

For preliminary estimates of pile bearing capacity sometimes empirical rules are made use of.

19-2. SOME NOTES ABOUT DYNAMIC PILE-DRIVING FORMULAS

The dynamic pile-driving formulas are derived analytically either by means of the work-energy relationship, or by the impact-momentum theory. To use the dynamic pile-driving formulas, one first must evaluate the resistance to driving the pile, and then relate this information to the allowable bearing capacity of the pile.

The bearing capacity of the pile is thus calculated from the energy necessary to drive the pile. Also this, in its turn, means that the bearing capacity of the pile depends not only upon the strength of the pile material itself, but also upon the strength of the pile-supporting soil.

19-3. PILE DRIVING

The concept of pile driving may be visualized from Fig. 19-1. Upon exerting a succession of blows (strokes) from a falling weight of the pile-driving hammer the pile penetrates into the ground. Theoretically, the kinetic energy E of the falling hammer is

$$E = Wh \qquad (19\text{-}1)$$

where W = weight of the falling hammer
 h = height of fall of hammer
 Wh = mechanical work

However, upon striking the pile, there occur considerable losses (friction, temporary elastic deformation of the pile-driving cap, the soil, and other losses) which reduce the available energy for performing a useful work. Hence the net available kinetic energy of the hammer for performing work may be written as

$$Wh - \Sigma(\text{energy losses}) \qquad (19\text{-}1a)$$

Along this thread of thought it is assumed that the resistance to penetration of the pile into the soil may be considered to constitute a single force of an average resistance R, and that with each stroke of the hammer the pile

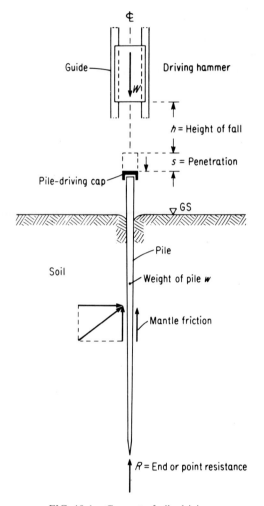

FIG. 19-1. Concept of pile driving.

penetrates a distance s against this resisting force R. Thus the mechanical work performed is Rs.

The factor s, usually expressed in inches per blow, is termed the *net penetration per blow* or simply the *set*. Sometimes s is recorded in number of blows per one inch.

The safe bearing capacity P of the pile is calculated as

$$P = \frac{R}{\eta} \tag{19-2}$$

where η is a certain factor of safety. Usually η is taken from 6 to 8.

Bearing Capacity of Piles

19-4. SUMMARY OF SOME DYNAMIC PILE-DRIVING FORMULAS

For a description of the various dynamic pile-driving formulas the reader is referred to Refs. 1, 2, 3, and 4. Here, for the purpose of design, only summaries of some of the most commonly known pile-driving formulas are given.

Pure, Classical, Complete Dynamic Pile-Driving Formula

This formula, as put forward by Redtenbacher[5] in 1859, is

$$\underbrace{\eta_e W h}_{1} = \underbrace{Rs}_{2} + \underbrace{\left[\eta_e W h \frac{w(1-e^2)}{W+w}\right]}_{3} + \underbrace{\left[\frac{R^2 L'}{2A'E'} + \frac{R^2 L}{2AE} + CR\right]}_{4\quad 5\quad 6} \qquad (19\text{-}3)$$

where 1 = total applied energy
          2 = useful work: energy used to move the pile a distance s
          3 = loss in impact
          4 = loss in cap due to elastic compression
          5 = loss in pile due to elastic compression
          6 = loss in soil due to elastic compression plus other losses
          η_e = efficiency of striking hammer (< 1.0)
          W = weight of hammer
          h = height of free fall of hammer
          R = dynamic resistance of soil (ultimate bearing capacity of pile in soil)
          s = amount of penetration of pile into soil per one strike of hammer on pile
          e = coefficient of restitution
          w = weight of pile
          L' = axial length of cushion block
          A' = cross-sectional area of cushion block
          E' = Young's modulus of elasticity of the cushion block
          L = driving length (depth) of pile
          A = cross-sectional area of pile
          E = Young's modulus of elasticity of pile
          C = a coefficient

Equation 19-3 must be solved for R, the ultimate pile bearing capacity.

The safe bearing capacity P of the pile is then calculated by dividing R by a factor of safety η:

$$P = R/\eta \qquad (19\text{-}4)$$

As far as is known, all other dynamic pile-driving formulas are obtained from Redtenbacher's classical formula by introducing various simplifications and specializations into it. For example, by assuming a completely inelastic impact ($e=0$), Eq. 19-3 becomes

$$R = \frac{AE}{L}\left[-s + \sqrt{s^2 + \frac{L}{A}\frac{2W^2h}{E(W+w)}}\right] \qquad (19\text{-}5)$$

Eytelwein's Formula (c. 1820)

is given for an inelastic impact ($e=0$) and ultimate bearing capacity R:

$$R = \frac{Wh}{s\left(1 + \frac{w}{W}\right)} \qquad (19\text{-}6)$$

The nominal safety factor in this formula is $\eta = 6.0$.

Weisbach's Formula (c. 1850)

Neglecting the impact loss, the ultimate bearing capacity of the pile is calculated as

$$R = -\frac{sAE}{L} + \sqrt{\frac{2WhAE}{L} + \left(\frac{sAE}{L}\right)^2} \qquad (19\text{-}7)$$

Engineering News Formulas

The so-called "*Engineering News* formulas" were published by the editor, A. M. Wellington, of *Engineering News* (New York) in 1888. These formulas allegedly give the safe bearing capacity of a pile for a factor of safety of $\eta = 6$. These formulas are

(a) For drop hammers:

$$P = \frac{Wh}{\eta(s+1.0)} \qquad (19\text{-}8)$$

(b) For single-acting steam hammers:

$$P = \frac{Wh}{\eta(s+0.1)} \qquad (19\text{-}9)$$

(c) For double-acting steam hammers:

$$P = \frac{(\vec{W}+ap)h}{\eta(s+0.1)} \qquad (19\text{-}10)$$

where all symbols, in consistent units, are the same as before, and
- s = average penetration of pile, per stroke, for last five strokes of a drop hammer; or 20 strokes of a steam hammer
- a = effective area of piston
- p = mean effective steam pressure

Stern's Formula (1908)[10]

$$R = \frac{AE}{L}\left\{-s + \sqrt{s^2 + \frac{2LWh(W + we^2) + s(W + w)^2}{AE(W + w)}}\right\} \quad (19\text{-}11)$$

McKay's Formula (U.S. Navy, Bureau of Yards and Docks)[2]

Safe bearing capacity P of pile:

$$P = \frac{2Wh}{s\left[1 + (0.3)\dfrac{w}{W}\right]} \quad (19\text{-}12)$$

where
- s = set, in.
- w = weight of pile, lb
- W = weight of hammer ram, lb

This formula has a nominal safety factor of $\eta = 6.0$.

Rankine's Formula[6]

$$R = \frac{2AE_L s}{36L}\left[\sqrt{1 + \frac{12E_n(12L)}{s^2 E_L A}} - 1\right] \quad (19\text{-}13)$$

where
- $E_L = 30 \times 10^6$ psi = modulus of elasticity of steel
- L = total length of pile, ft
- E_n = manufacturer's maximum rated energy, ft-lb
- A = net steel cross-sectional area of pile, in.^{2}
- s = set (values normally used are the average values obtained during the last 6 in. of driving, or for the actual final inch or less if sudden refusal conditions are encountered)

Compared with the general pile-driving formula for a perfectly elastic impact when $e = 1.0$, Rankine's formula is merely a special case of the general formula, which includes the Redtenbacher formula as another limit expression.

Hiley's Formula[6,7]

$$R = \frac{4E_n}{s + \dfrac{1}{2}(C_1 + C_2 + C_3)} \cdot \frac{W + e^2 w}{W + w} \quad (19\text{-}14)$$

where all symbols are as before.

The nominal factor of safety of this formula is $\eta = 3.0$. The factor $(1/2)(C_1 + C_2 + C_3)$ is analogous to the factor 0.1 in the *Engineering News* formula. The value C_1 equals the peak temporary (elastic) compression experienced in the pile load head and cap. The test values of C_1 are compiled by Chellis.[7] Depending upon the kind of hammer used, the value of $C_1 = 0.1$ was set for all hammers used in the Michigan pile-test program.[6] For some hammers, the value of $C_1 = 0.15$ was used.

The factor $C_2 + C_3$ is based on field measurements, and represents the combined temporary compression of pile and supporting ground. The Hiley formula is used extensively in Great Britain.

Pacific Coast Uniform Building Code Formula[6]

$$R = \frac{3E_n \frac{W + Kw}{W + w}}{s + \frac{48RL}{AE_L}} \qquad (19\text{-}15)$$

where the factor K is analogous to e^2 in the Hiley formula and is specified equal to 0.25 for ordinary practice with steel piles.

The term

$$\frac{48RL}{AE_L} \qquad (19\text{-}16)$$

is the longitudinal elastic shortening of the pile when subjected to the indicated ultimate driving stress ($4R = R_u$, where R_u is the computed ultimate pile load capacity, in kips, derived from the dynamic formula). The quadratic equation 19-15 may also be solved by the method of trial and adjustment. The nominal safety factor in Eq. 19-15 is $\eta = 4.0$.

Modified *Engineering News* Formula

The modified *Engineering News* formula,[6]

$$R = \frac{2Wh}{s + 0.1} \cdot \frac{W + e^2 w}{W + w} \qquad (19\text{-}17)$$

is a variation of the *Engineering News* formula, and contains a nominal factor of safety of $\eta = 6.0$. This formula is currently being used by the Michigan State Highway Department, supplanting the formerly specified *Engineering News* formula.

Gates Formula

Gates formula[6,8] is a strictly empirical relationship between hammer energy, final set, and measured design test load, with a safety factor of $\eta = 3.0$:

$$R = (2000)(1/7)\sqrt{E_n}\left(\log \frac{s}{10}\right) \qquad (19\text{-}18)$$

where E_n is the manufacturer's maximum rated energy, in ft-lb.

The Gates relationship, Eq. 19-18, does not have rational limits and does not apply to "refusal" conditions when $s \to 0$.

Rabe's Formula

Rabe's formula,[6,9] too, is an empirical one, is more complex than the others, and has been revised frequently. The 1960 version of Rabe's formula is

$$R = \frac{ME_n}{s + C} \cdot \frac{w}{w + \frac{W}{2}} \cdot B \qquad (19\text{-}19)$$

where M = a term incorporating the safety factor $\eta = 2$, the hammer efficiency factor, and a factor of 12 to cause the value of E_n to be expressed in inch-pounds.

$$C = \frac{C_1 + C_2 + C_3}{2} \quad \text{(by Hiley)} \qquad (19\text{-}20)$$

where $C_1 + C_2 + C_3$ = the compression loss in the cap and cushion, in the pile, and in the soil, respectively

B = a factor which relates dynamic to static pile-bearing capacity.

Brix Formula

Safe static load P [4,11]:

$$P = \frac{W^2 wh}{\eta(W + w)^2 s} \qquad (19\text{-}21)$$

This formula is to be used for $\eta > 3$, for sandy soils and drop hammers only.

Reference 2 by R. D. Chellis contains many domestic as well as foreign pile bearing capacity formulas.

Cummings[3] reviewed the various dynamic pile-driving formulas and found that some of them contain questionable assumptions, and that dynamic theories are mixed with static ones. However, the dynamic pile-driving formulas serve the engineer as a guide in estimating reasonably safe and uniform results over the entire pile-driving operation job on a construction site with reasonably uniform soil conditions.

For the evaluation of static bearing capacity of piles, the Building Laws of the City of New York recommends static loading tests of piles in situ.

It is well to remember that the bearing capacity of a pile group does not correspond always to that of a single pile. It is usually less than the sum of the bearing capacities of the individual pile bearing capacities in the pile group. In this respect, one merely recalls the combined effect of the overlap of pressure bulbs of adjacent piles on settlement of a compressible soil, viz., pile group. Pile spacing, therefore, is a factor affecting merger of adjacent pressure bulbs. However, overlapping pressure bulbs are not the only factors: they affect settlement, but not perimeter punching shear of a pile group.

19-5. MICHIGAN STUDIES ON PILES

One of the purposes of the so-called Michigan research on piles in the 1960's was "To determine the correlation between bearing capacity of the load-tested piles and estimated pile bearing capacity as obtained by (eleven) selected pile-driving formulas."[6]

In its final report (1965), *A Performance Investigation of Pile Driving Hammers and Piles,*[6] the following eleven dynamic pile driving formulas were studied:

1. *Engineering News* (EN)
2. Hiley
3. Pacific Coast Uniform Building Code (PCUBC)
4. Redtenbacher
5. Eytelwein
6. Navy-McKay
7. Rankine
8. Canadian National Building Code (CNBC)
9. Modified *Engineering News* (Modified EN)
10. Gates
11. Rabe

Before proceeding with the evaluation and comparison of the formulas in terms of conditions at the test sites, the report states a general note of caution, namely:

> It is the consensus of informed engineers that no one dynamic formula, relating dynamic to static resistance, affords a reliable means of estimating the

Bearing Capacity of Piles

longtime bearing capacity of piles in general. It has been reported that true safety factors may range from less than 1 to 17 when using *Engineering News* formula, and even for the more elaborate formulas of the Hiley type, true safety factors may range by a factor of 5. Thus, unwarranted conclusions should not be drawn concerning the general accuracy or uniformity of results obtained from the several formulas, on the basis of the particular date discussed here.

The evaluation of these dynamic pile-driving formulas revealed the following:[6] Bearing capacities for the several hammers, pile types, and test sites were compared to measured and interpolated pile bearing capacities. For the specific conditions considered, some of the more important and pertinent findings were:

1. In several instances the *Engineering News*, Navy-McKay, and Rankine formula design capacity had a true safety factor of less than unity with respect to test load capacity.

2. In several instances (Hiley, PCUBC, Redtenbacher, CNBC) formula design capacities had a true safety factor of 9 or more with respect to test load capacity.

3. In general, the Modified *Engineering News* and Gates formulas gave design capacity values with true safety factors substantially all falling in the range of 1.5 to 6 with respect to associated test load capacities.

The final, general conclusion from Michigan's pile research is that

... the estimates of supporting capacity by dynamic pile formulas, considering the entire group of eleven formulas, varied through a wide range and could not be used to predict pile capacity with any degree of certainty. ... The reader should examine the detailed data and discussion in Chapter 12 [of Ref. 6] and in the preceding summary to judge for himself the basis for these general conclusions.

The conclusion reads further:

While dynamic pile formulas leave much to be desired as a basis for predicting load capacity, it is strongly recommended that they be retained, as one method of rapid determination of capacity and controlling it under job conditions. Even though field loading tests are much more reliable as a measure of supporting capacity and soil test data much more satisfactory for predicting capacity, dynamic pile formulas will still be needed where load tests and adequate soil investigations are not available. Efforts should be directed to improvement of these formulas and selection of those most applicable to field conditions.

It is concluded that soil test data, consisting of standard penetration measurements taken during sampling and laboratory shear tests with a yield value from ring shear or equivalent yield value from unconfined compression, did provide a reliable and accurate basis of estimating static load capacity in this investigation. This conclusion is supported by a number of other investigations in which similar methods have been used, and there is much evidence to indicate that this is generally true. The Michigan State Highway Department has already moved in this direction on the basis of results from the current study, by adopting a modified *Engineering News* formula for specifications and job control.

Thus ends the 338-page Michigan final report, *A Performance Investigation of Pile Driving Hammers and Piles*, 1965.

19-6. THEORETICAL CALCULATIONS OF PILE STATIC BEARING CAPACITY

The design engineer is frequently in need of a simple method of calculation to determine in advance approximately the bearing capacity of a pile if the soil profile and the soil physical and strength properties are known, in order to determine the dimensions of the piles to use, as well as to compute the necessary number of piles for a foundation under design. This information is also needed for preparing the cost estimate of the foundation.

The static bearing capacity calculations are performed based on static equilibrium of forces partaking in the soil-pile system.

19-7. FORCES ON PILE

The bearing capacity P of a slender pile in soil is assumed (Fig. 19-2a) to be made up generally of two parts—pile mantle friction R_f and pile-tip resistance, R_t:

$$P = P_f + P_t = R_f + R_t \qquad (19\text{-}22)$$

The bearing capacity P of a bulbed-end pile is the soil reaction R on the bulb:

$$P = R = \sigma_{\text{all}} A = \frac{\pi}{4} \sigma_{\text{all}} D^2 \qquad (19\text{-}23)$$

where σ_{all} = allowable soil bearing capacity
  D = diameter of the horizontal projection of the bulb (perpendicular to R, see Fig. 19-2b)

The mantle friction R_f may be expressed as

$$R_f = E \tan \phi_1 \qquad (19\text{-}24)$$

where E = soil pressure on the mantle surface of pile
  ϕ_1 = angle of mantle friction
  $\mu_1 = \tan \phi_1$ = coefficient of mantle friction

As to the kind of lateral earth pressure E, it is chosen depending upon whether the lateral earth pressure on the pile mantle surface in these calculations is used as active earth pressure $E_a = (1/2)\gamma L_c^2 U K_a$, or distributed

Bearing Capacity of Piles

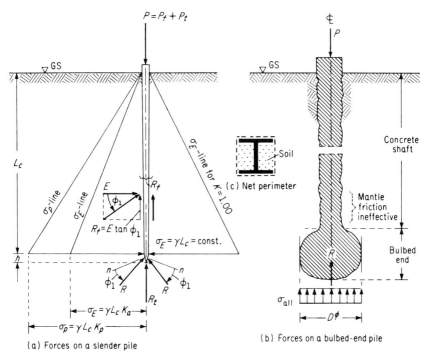

FIG. 19-2. Forces on piles.

by a simple hydrostatic rule, as $E = (1/2)\gamma L_c^2 U$, or as passive earth pressure (resistance) $E_p = (1/2)\gamma L_c^2 K_p$. Note that when $K_a = K_p = 1$, then $E_a = E = E_p = (1/2)\gamma L_c^2 U$.

Here γ = unit weight of soil
$\gamma = (1-n)G\gamma_w + \gamma_w nS$ = unit weight of soil above groundwater table   (19-25)
$\gamma_{sub} = (1-n)(G-1)\gamma_w$ is submerged unit weight of soil   (19-26)
γ_w = unit weight of water
n = porosity of soil
G = specific gravity of soil
S = degree of saturation of soil
L_c = thickness of soil layer in contact with pile
$K_a = \tan^2\left(\dfrac{\pi}{4} - \dfrac{\phi}{2}\right)$ = active earth-pressure coefficient
ϕ = angle of internal friction of soil
$K_p = \tan^2\left(\dfrac{\pi}{4} + \dfrac{\phi}{2}\right)$ = passive earth-pressure (resistance) coefficient

For K_a and K_p coefficients, see tables, Ref. 12.

U = length of average perimeter of pile cross section

$R = \sqrt{V^2 + E^2}$ = resultant soil reaction on tip of pile

$E = \sigma_E \pi rs \cos \alpha$ = total force of soil lateral resistance to penetration of pile (= horizontal component of R)

$\sigma_E = \gamma L_c$ = lateral stress on conical surface of pile at depth L_c below ground surface

πrs = area of conical surface of tip of pile

x = one-half of the central angle at tip of the pile

The mantle friction that prevails on any part of the pile is equal to (1) either the friction of the soil against the pile material, or (2) the shear strength of the soil immediately surrounding the pile, whichever is smaller.

As to the net perimeter of steel H-piles to use, the

> U.S.S. Steel H-piles driven into stiff clays usually trap soil between the flanges and web and compact the soil so that it becones a hard core carried down with the pile. This core aids in compressing the surrounding soil and building up resistance to further displacement—either by superimposed loads or driving. In laminated and stratified soil, a point build-up or lagging will assist in attaining desired resistance in relatively thin, stiff clay layers.[13]

The net perimeter of a steel H-pile to use is indicated by a dashed line in Fig. 19-2c. For example, an H-pile section designated as 8 WF 31 has a net perimeter U equal to $U = 4 \times 8$ (in.) = 32 in.

Steel H-piles into soft clays develop mantle friction resistance to penetration under loads nearly equivalent to the embedded surface area of the pile multiplied by the shearing strength of the soil.

Thus the magnitude of the mantle friction may be calculated by means of the earth-pressure theory as

$$R_f = E \tan \phi_1 A_m = (1/2) \gamma L_c^2 U \tan \phi_1 = U L_c \tau \qquad (19\text{-}27)$$

where

$$\tau = (1/2) \gamma L_c \tan \phi_1$$

or

$$R_f = (1/2) \gamma L_c^2 U \tan \phi_1 K_a \qquad (19\text{-}27a)$$

or

$$R_f = (1/2) \gamma L_c^2 U \tan \phi_1 K_p \qquad (19\text{-}27b)$$

depending upon the geometry of the pile and soil conditions.

Here $A_m = U L_c$ = mantle surface of embedded pile

$\tan \phi_1$ = coefficient of mantle friction

Bearing Capacity of Piles

The least mantle friction R_f is obtained, of course, with K_a, because $K_a < 1 < K_p$. Hence, by using K_a one would be, "on the safe side." Thus the bearing capacity P_f of a pile because of its mantle friction R_f alone is, by Eq. 19-27,

$$P_f = R_f = (1/2)\gamma L_c^2 U \tan \phi_1 \qquad (19\text{-}27)$$

Example

If $\gamma = 100$ lb/ft², diameter of pile $d = 12$ in., $L_c = 20$ ft, and $\phi_1 = 30°$, then $U = \pi d = (3.14)(12/12) = 3.14$ (ft); $L_c^2 = 20^2 = 400$ (ft²); $\tan \phi_1 = \tan 30° = 0.577$, and by Eq. 19-27 the mantle friction of this pile is

$$R_f = P_f = (1/2)(100)(400)(3.14)(0.577) = 36{,}200 \text{ (lb)} = \underline{18.1 \text{ (ton)}}$$

Mantle Friction for a Layered Soil System

The mantle friction R_f in a layered soil system (Fig. 19-3) is calculated theoretically as the weighted average τ_{ave} of the resistances of individual soil layers:

$$\tau_{\text{ave}} = \frac{\sum_1^n (\tau h)}{\sum_1^n h} = \frac{\tau_1 h_1 + \tau_2 h_2 + \tau_3 h_3 + \cdots + \tau_n h_n}{h_1 + h_2 + h_3 + \cdots + h_n} \qquad (19\text{-}28)$$

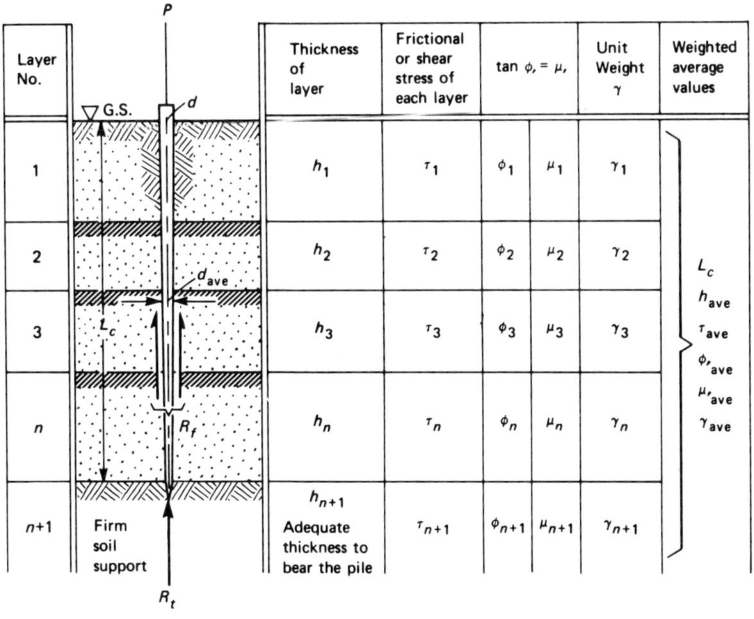

Layer No.		Thickness of layer	Frictional or shear stress of each layer	$\tan \phi_{,} = \mu_{,}$	Unit Weight γ	Weighted average values
1		h_1	τ_1	ϕ_1	μ_1	γ_1
2		h_2	τ_2	ϕ_2	μ_2	γ_2
3		h_3	τ_3	ϕ_3	μ_3	γ_3
n		h_n	τ_n	ϕ_n	μ_n	γ_n
$n+1$	Firm soil support	h_{n+1} Adequate thickness to bear the pile	τ_{n+1}	ϕ_{n+1}	μ_{n+1}	γ_{n+1}

Right-side bracket labels: L_c, h_{ave}, τ_{ave}, $\phi_{,\text{ave}}$, $\mu_{,\text{ave}}$, γ_{ave}

FIG. 19-3. Pile in a layered soil system.

Then

$$R_f = UL_c\tau_{ave} \tag{19-29}$$

where $U = \pi d_{ave}$ = perimeter of cross section of pile in feet (or in meters)
L_c = computation length of pile, in feet (or in meters)
τ_{ave} = average weighted specific frictional (shear) stress on pile, in lb/ft² (or in ton/m²);
$h_1, h_2, h_3, \cdots, h_n$ = thickness of individual layers of soil system, in feet (or in meters);
$\tau_1, \tau_2, \tau_3, \cdots, \tau_n$ = frictional force stress for each individual layer of soil, in lb/ft² (or ton/m²)

Handy's Soil Bore-Hole Direct-Shear Test Device

An expedient method for obtaining the individual shear strength values $\tau_1, \tau_2, \tau_3, \cdots, \tau_n$ of the individual layers of the soil profile is by the use of a portable soil bore-hole direct-shear test device.[14] This patented shear-test device was invented and developed by Prof. R. Handy of the Iowa State University of Science and Technology at Ames, Iowa. The principle of functioning of Handy's shear-test device is as follows:

A hole is bored into the soil, and two curved contact plates are expanded against the soil inside the bored hole to apply pressure n normal to its sides (see Fig. 19-4). A shearing stress τ is then applied to the soil by pulling up or pushing down the expanded curved device axially along the bore hole. Thus the test result is essentially a direct shear test on soil at the sides of the bore hole.

An important feature of Handy's soil bore-hole direct-shear test device is that tests may be precisely located in weak soil layers, because the tests are made after the bore hole is bored and the soil profile logged. As in most soil-boring operations, disturbed samples may be taken for correlation to other holes, or for moisture content or classifications tests.

During the course of development and testing of this new shear-test device, for purposes of comparison, conventional shear-test results of various kinds of soils showed good agreement among bore-hole, direct-shear and drained triaxial shear-test data. The soil bore-hole shear tests were possible to measure even in very soft soils with a friction angle as low as 15.2°.

This new device facilitates a complete determination of the maximum shear stress vs. normal stress failure envelope in a period from about 15 min to an hour, depending upon the consolidation time required for a particular soil.

Indeed, the device is a relatively simple and expedient instrument for measuring internal friction and cohesion of soil in situ.

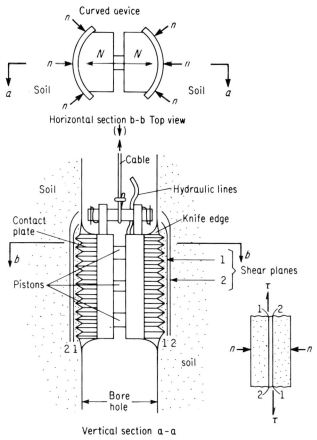

FIG. 19-4. Schematic of Handy's bore-hole direct-shear device (Ref. 14).

The average weighted unit weight of soil γ_{ave} is calculated as

$$\gamma_{ave} = \frac{\sum_{1}^{n}(\gamma h)}{\sum_{1}^{n} h} = \frac{\gamma_1 h_1 + \gamma_2 h_2 + \gamma_3 h_3 + \cdots + \gamma_n h_n}{h_1 + h_2 + h_3 + \cdots + h_n} \tag{19-30}$$

and the average weighted angle of internal friction of the soil ϕ_{ave} is calculated as

$$\phi_{ave} = \frac{\sum_{1}^{n}(\phi h)}{\sum_{1}^{n} h} = \frac{\phi_1 h_1 + \phi_2 h_2 + \phi_3 h_3 + \cdots + \phi_n h_n}{h_1 + h_2 + h_3 + \cdots + h_n} \tag{19-31}$$

If safe, allowable τ-values are used, then the allowable load P on the pile is calculated approximately as

$$P \leqslant R_t + R_f = \underbrace{(\sigma_{\text{all}} + \gamma_{\text{ave}} L_c) A}_{\substack{\text{tip} \\ \text{resistance} \\ R_t}} + \underbrace{U L_c \tau_{\text{ave}}}_{\substack{\text{mantle} \\ \text{friction} \\ R_f}} \qquad (19\text{-}32)$$

where σ_{all} = allowable soil bearing capacity at depth of tip of pile below ground surface
γ_{ave} = weighted average unit weight of soil
L_c = computation length of pile
A = cross-sectional area of pile

Other symbols are as before.

Equation 19-32 is merely a simplification. A major factor in oversimplification is unequal strain required to activate full tip and frictional resistance. For example, one may set a pile on a thin layer of bedrock, but before skin friction would develop, the pile might punch through. Hence, Eq. 19-32 does not pretend to great precision. However, the equation is simple to use in practice.

The computation length L_c of the pile is calculated by approximation from Eq. 19-32 as

$$L_c = \frac{P - \sigma_{\text{ave}} A}{U \tau_{\text{ave}} + \gamma_{\text{ave}} A} \qquad (19\text{-}33)$$

Some average values of mantle friction stresses τ are compiled in Table 19-1.[15] To convert these friction stresses into the British system of units of measurement multiply values in Table 19-1 by 2048.2 to obtain τ in lb/ft². Also, 1 kg/cm² $\approx$ 1 ton/ft². For other pertinent conversion factors see the Appendix.

19-8. BEARING CAPACITY OF A PILE TIP IN SOIL

One must distinguish between conical and pyramidal tips of piles. For sheet piling, the wedge-type tip also comes into consideration.

The ultimate resistance of the tip of a driven pile (viz., the bearing capacity of the tip of a pile) contributed by its tip support in certain types of soil may be determined analytically by

1. Krey's method of soil lateral resistance,[16] viz., classical earth pressure theory
2. The Fröhlich-Maag method of the critical pressure on soil[17,18]

TABLE 19-1

Average Mantle Friction Stresses τ in kg/cm^2

Kind of Soil	Driving Depth, meters	Timber Piles	Reinforced-Concrete Piles	Steel Piles
1. Noncohesive soils				
Fine sand	3.0-4.0			0.40-0.50
	4.0			0.50-0.80
Medium sand	3.0-4.0	0.40-0.60	0.40-0.50	0.30-0.40
	4.0	0.60-1.00	0.50-0.70	0.40-0.70
Coarse sand and gravel	3.0-4.0			0.30-0.40
	4.0			0.40-0.70
2. Cohesive soils				
Soft clay			0.15-0.25	
Stiff clay			0.30-0.45	
Sandy clay			up to 0.50	
3. Organic soils				
Peat, mud, swamp		0.07-0.10		

Note: See Ref. 15. These values are good for compression as well as for tension piles. These tabulated values are not good under conditions of vibrations of soils. Hence they are not good either for vibrating pile foundations.

3. The general method of the allowable soil bearing capacity
4. Other methods such as described in this book

Upon penetration of the pile and exhaustion of the soil frictional resistance the soil particles displace mutually. It may be assumed that the particles are in a state of liquid fluidity (plastic-flow condition), an assumption which has been frequently postulated in theoretical soil mechanics, for example, by Krey,[16] Fröhlich,[17-19] and others. However, in contradistinction to liquids, the displacing soil particles exert a resistance against their motion, a resistance which depends upon the pressure on the soil particles. This idea also underlies the theory of active and passive earth pressure.

Based on the idea of translocating soil particles and subsequent mobilization of resistance to displacement of the latter, Krey[16] reasoned that upon overcoming the soil's lateral resistance E the penetration of the pile takes place by occupying the empty space left behind by the displaced soil. Because of the assumption that the soil particles are a mobile fluid, and for the sake of simplifying the problem, it was assumed that the stress σ_E of lateral resistance of the soil is constant for the entire height of the cone of the tip, i.e.,

$$\sigma_E = \gamma L_c \tag{19-34}$$

where γ = unit weight of soil

L_c = computation length of embedded pile in soil (viz., driven length of pile, or total thickness of overburden layer of soil from ground surface to base of cone of pile tip)

The *average shear stress* τ_c (from pile load P) acting tangentially along the conical mantle surface area $A_m = \pi r s$ of the tip in contact with the soil for a partial penetration z of the tip (Fig. 19-5) can be calculated as

$$\tau_c = \frac{T}{A_m} = \frac{P \cos \alpha}{\pi \rho S} = \frac{P}{\pi} \frac{\cos \alpha}{z \tan \alpha \dfrac{z}{\cos \alpha}} = \frac{P}{\pi z^2} \frac{\cos^2 \alpha}{\tan \alpha} \tag{19-35}$$

where $T = P \cos \alpha$ = tangential force, and all other symbols are as shown in Fig. 19-5.

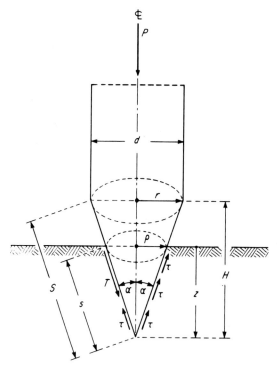

FIG. 19-5. Conical tip of pile.

For full embedment S of the conical tip the shear stress τ_c is

$$\tau_c = \frac{P}{\pi} \frac{\cos^2 \alpha}{S^2 \tan \alpha} \tag{19-36}$$

Bearing Capacity of Piles

19-9. BEARING CAPACITY OF A PILE TIP CONSIDERING SOIL LATERAL RESISTANCE

1. Conical Tip of Pile

Consider a fully embedded full cone of the tip of a pile (Fig. 19-6). The lateral displacement of soil particles, viz., penetration of pile, is resisted by the lateral soil resistance, the stress σ_E of which is assumed to be constant along the height h of the cone: $\sigma_E = \gamma L_c = $ const. The magnitude of the

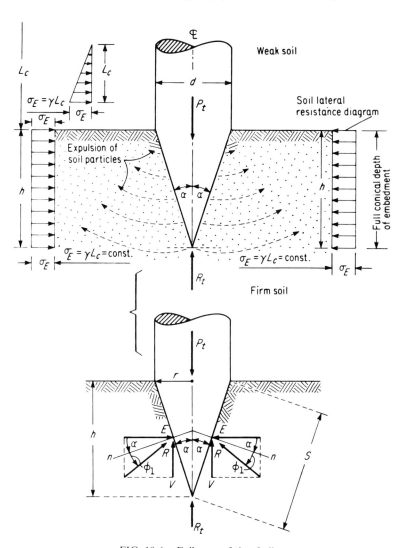

FIG. 19-6. Full cone of tip of pile.

force E of the soil lateral resistance on the conical surface of the cone on each side of the pile may be expressed approximately as

$$E = \sigma_E \pi r \frac{S}{2} \cos \alpha \qquad (19\text{-}37)$$

where $r = d/2 = $ radius of pile
$\pi r S = $ mantle surface of full cone
$S = r/\sin \alpha = $ length of slope of the cone to vertical axis of cone (pile)

In Fig. 19-6, V is the vertical component of soil reaction R (on the conical surface). Reaction R is offset from the normal n-n to the conical surface by the angle of friction ϕ_1 between the soil and the pile material.

The equilibrium of the pile load P_t (load + self-weight of pile) and the soil reactive forces are shown by a force diagram in Fig. 19-6. From the force diagram, the magnitude of V is written directly as a function of E:

$$V = E \tan (\alpha + \phi_1) = \sigma_E \pi r \frac{S}{2} \cos \alpha \tan (\alpha + \phi_1) \qquad (19\text{-}38)$$

For equilibrium,

$$P_t = R_t = 2V = \sigma_E \pi r S \cos \alpha \tan (\alpha + \phi_1)$$

$$= \pi \gamma L_c \frac{r^2}{\tan \alpha} \tan (\alpha + \phi_1) \qquad (19\text{-}39)$$

or, with $r = d/2$,

$$P_t = R_t = \frac{\pi}{4} \gamma L_c d^2 \frac{\tan (\alpha + \phi_1)}{\tan \alpha} \qquad \text{Q.E.D.} \qquad (19\text{-}40)$$

The bearing capacity of the pile, of course, is limited by the strength of the pile material, as well as by the strength of the soil.

Example

If $\gamma = 100$ lb/ft³, $d = 12$ in., $L_c = 20$ ft, $h = 10.5$ in., and $\phi_1 = 30°$, then

$$\alpha = \arctan (6/10.5) = \arctan (0.577) = 30°$$
$$\tan (\alpha + \phi_1) = \tan (30° + 30°) = \tan 60° = 1.732$$
$$\tan \alpha = \tan 30° = 0.577$$

and by this theory the ultimate bearing capacity of the tip of the round pile for the given conditions is, by Eq. 19-40,

$$P_t = R_t = (0.785)(100)(20)(12/12)^2 \cdot \frac{1.732}{0.577} = 4{,}710 \text{ (lb)} \approx \underline{2.35 \text{ tons}}$$

2. Pyramidal Tip of a Square Pile

Reasoning as for the conical tip, and observing that the square, pyramidal tip has four inclined side surfaces, the forces acting on the tip are (Fig. 19-7):

Total lateral resistance around tip:

$$E = 4\sigma_E \frac{d}{2} S \cos \alpha = 2\sigma_E hd \tag{19-41}$$

where S = height of the inclined side triangle.

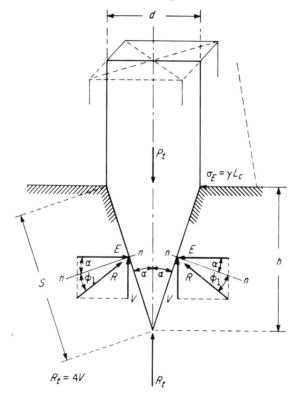

FIG. 19-7. Pyramidal tip of a square pile.

Total vertical resistance around tip:

$$V = E \tan (\alpha + \phi_1) = 2\sigma_E hd \tan (\alpha + \phi_1)$$

$$= 2\gamma h L_c d \tan (\alpha + \phi_1) = 2\gamma L_c \frac{d^2}{2 \tan \alpha} \tan (\alpha + \phi_1)$$

$$= \gamma L_c d^2 \frac{\tan (\alpha + \phi_1)}{\tan \alpha} \tag{19-42}$$

Bearing capacity of pyramidal tip:

$$P_t = R_t = V = \gamma L_c d^2 \cdot \frac{\tan(\alpha + \phi_1)}{\tan \alpha} \tag{19-42}$$

Example

If $\gamma = 100$ lb/ft^3, $d = 12$ in., $h = 10.5$ in., $L_c = 20$ ft, $\phi_1 = 30°$, then $\alpha = 30°$, $\tan \alpha = 0.577$, $\tan(\alpha + \phi_1) = \tan 60° = 1.732$

and by this theory the ultimate bearing capacity of the pyramidal tip of the square pile P_t for the given conditions is, by Eq. 19-42,

$$P_t = R_t = (100)(20)(12/12)^2 \cdot \frac{1.732}{0.577} = 6{,}000 \text{ (lb)} = \underline{3.0 \text{ tons}}$$

Of course the bearing capacity of the pile is limited by the strength of the pile material.

Example

If $\gamma = 100$ lb/ft^3, $d = 12$ in., $L_c = 20$ ft, $\tan \phi_1 = 0.577$, $\tan(\alpha + \phi_1) = 1.732$, then the total circular pile bearing capacity P is, by Eqs. 19-27 and 19-40.

$$P = P_f + P_t = R_f + R_t = 18.1 + 2.35 = \underline{20.45 \text{ (tons)}}$$

19-10. MINIMUM TIP RESISTANCE

The minimum resistance to penetration R_t by the tip of a circular pile is obtained by differentiation of R_t (Eq. 19-40) with respect to angle α, i.e., setting $dR_t/d\alpha = 0$:

$$\frac{dR_t}{d\alpha} = \frac{d\left\{\left(\frac{\pi}{4}\gamma L_c d^2\right)\left[\frac{\tan(\alpha+\phi_1)}{\tan\alpha}\right]\right\}}{d\alpha}$$

$$= \left(\frac{\pi}{4}\gamma L_c d^2\right)(\tan 2\alpha - \cot \phi_1) \tag{19-43}$$

Setting this derivative equal to zero, we obtain

$$\tan 2\alpha = \cot \phi_1 = \tan\left(\frac{\pi}{2} - \phi_1\right)$$

and

$$\alpha = \frac{\pi}{4} - \frac{\phi_1}{2} \tag{19-44}$$

Bearing Capacity of Piles

Because $\dfrac{d^2P}{d\alpha^2} = \left(\dfrac{\pi}{4}\gamma L_c d^2\right)\left(2\cos^2\alpha\right) > 0$, the R_t function has a minimum.

Substitution of $\alpha = \dfrac{\pi}{4} - \dfrac{\phi_1}{2}$ into the R_t equation renders

$$R_{t_{min}} = \left(\dfrac{\pi}{4}\gamma L_c d^2\right) \dfrac{\tan\left(\dfrac{\pi}{4}+\dfrac{\phi_1}{2}\right)}{\tan\left(\dfrac{\pi}{4}-\dfrac{\phi_1}{2}\right)}$$

$$= \left(\dfrac{\pi}{4}\gamma L_c d^2\right) \tan^2\left(\dfrac{\pi}{4}+\dfrac{\phi_1}{2}\right) \qquad (19\text{-}45)$$

If R_t is calculated considering K_a or K_p, then Eq. 19-45 must be multiplied by K_a or K_p, whichever is the case.

Similarly, the minimum tip resistance $R_{t_{min}}$ for a pyramidal tip of pile is

$$R_{t_{min}} = (\gamma L_c d^2) \tan^2\left(\dfrac{\pi}{4}+\dfrac{\phi_1}{2}\right) \qquad (19\text{-}46)$$

where d = length of side of base of the pyramid tip, or length of side of the square cross section of the pile

Example

If $\gamma = 100$ lb/ft³, $d = 12$ in. dia.; $L_c = 20$ ft, and $\phi_1 = 30°$, then

$$\tan^2(45° + 15°) = \tan^2 60° = (1.732)^2 = 2.999 \approx 3.0$$

Circular pile:

$$R_{t_{min}} = (0.785)(100)(20)(12/12)^2(3.0) = 4710 \text{ (lb)} \approx \underline{2.355 \text{ ton}}$$

Square pile:

$$R_{t_{min}} = (2.355)/0.785 = \underline{3.0 \text{ (tons)}}$$

19-11. BEARING CAPACITY OF A CONICAL PILE TIP CONSIDERING CRITICAL PRESSURE ON SOIL

According to Fröhlich,[17-20] the critical pressure (= proportional limit) on soil is that one which characterizes the instant in the loading of the soil by a foundation (without wall friction) when the lateral expulsion of the soil begins and plastic settlements commence.

According to Fröhlich-Maag,[18] the allowable pressure on soil σ_{all} at depth t below the ground surface when the motion of the soil particles at the edge of the base of the footing is just impending is, for a frictional-cohesive soil,

$$\sigma_{\text{all}} = \gamma t \left[\frac{(\cot \phi) + \left(\phi + \frac{\pi}{2}\right)}{(\cot \phi) + \left(\phi - \frac{\pi}{2}\right)} \right] + p_i \left[\frac{(\cot \phi) + \left(\phi + \frac{\pi}{2}\right)}{(\cot \phi) + \left(\phi - \frac{\pi}{2}\right)} - 1 \right] \quad (19\text{-}47)$$

$$= \gamma t A + p_i (A - 1) \quad (19\text{-}48)$$

where γ = unit weight of soil
 ϕ = angle of internal friction of soil
 $p_i = c \cot \phi$ = initial stress in soil due to cohesion
 c = cohesion of soil

The factors A and $(A - 1)$ were prepared by the author in Fig. 19-8.

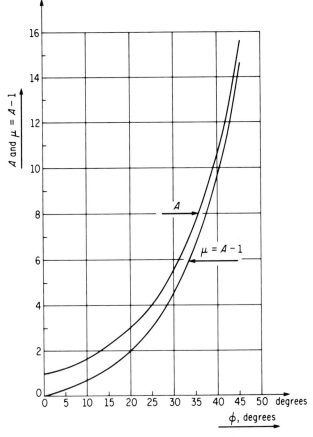

FIG. 19-8. Coefficients A and $\mu = A\text{-}1$ for allowable (critical) pressure in Fröhlich-Maag equations.

Bearing Capacity of Piles

The principle of calculating the bearing capacity of a pile by its tip resistance R_t is based on the equilibrium of the pile load (load + self-weight of the pile) with the upward-acting soil reaction V:

$$P_t = R_t = \sigma_{\text{all}} A = \frac{\pi}{4} \sigma_{\text{all}} d^2 \quad (19\text{-}49)$$

where d = pile diameter at base of cone of the tip.

19-12. STATIC PILE BEARING CAPACITY FORMULAS

The most commonly known static pile bearing-capacity formulas are those by Bénabenq (1913),[1] Dörr (1922),[21] Vierendeel (1927),[1] and Krey (1936)[16] (see Refs. 1, 2, 4). These formulas are based on earth pressure theory, and they consider mantle friction.

Bénabenq's Formula

This formula gives the ultimate bearing capacity as

$$R_u = \frac{md^2 \gamma L}{8 \sin \alpha} \tan^4\left(45° + \frac{\phi}{2}\right) + \frac{md\gamma L^2}{2}(\tan \phi) \tan^2\left(45° + \frac{\phi}{2}\right) \quad (19\text{-}50)$$

where m = ratio of perimeter of pile to its radius
d = diameter of pile
γ = unit weight of soil
L = embedded length of pile
α = one-half the central angle of the pile point
ϕ = angle of internal friction of the soil

Dörr's Formula

Dörr's formula gives the safe static bearing capacity P of a single pile by means of the following quadratic equation:

$$AL_c^2 - BL_c - P = 0 \quad (19\text{-}51)$$

as

$$\overbrace{A = R_f} \quad \overbrace{B = R_t}$$

$$P = R_f + R_t = (1/2)\gamma U(1 + \tan^2 \phi)\mu L_c^2 + a\gamma \tan^2(45° + \phi/2)L_c \quad (19\text{-}52)$$

or

$$L_c = \frac{-B + \sqrt{B^2 + 4AP}}{2A} \quad (19\text{-}53)$$

where R_f = total mantle friction of pile
R_t = point resistance of pile
a = cross-sectional area of pile
γ = unit weight of soil
ϕ = angle of internal friction of soil
L_c = computation length of pile
U = perimeter of pile
μ = coefficient of friction between soil and pile

Vierendeel's Formula

$$P = (1/2)\mu_1 \gamma \pi d L_c^2 \tan^2 (45° + \phi/2) \tag{19-54}$$

In this equation, the symbols are the same as before. For cast-in-place piles and driven timber and concrete piles with rough mantle surfaces, Vierendeel recommends using $\mu_1 = 0.33$. For all other types of piles in wet and plastic soils, $\mu_1 = 0.25$. The formula implies static friction piles.

Krey's Formula

The static bearing capacity P of a pile is here the sum of the tip resistance R_t and that of the mantle friction R_f:

$$P = \underbrace{\gamma a L_c \tan^2 (45° + \phi_1/2) K_p}_{R_t} + \underbrace{(1/2)\gamma \mu_1 U L_c^2 K_p}_{R_f} \tag{19-55}$$

where γ = unit weight of soil
a = cross-sectional area of pile
L_c = computation length of pile
ϕ_1 = coefficient of mantle friction between soil and pile material
ϕ = angle of internal friction of soil, in K_p:

$$K_p = \tan^2 (45 + \phi/2)$$

K_p = coefficient of earth resistance (= passive earth pressure)
$\mu_1 = \tan \phi_1$ = coefficient of friction between soil and pile material
U = perimeter of pile

19-13. EMPIRICAL RULES

One of the empirical rules of thumb for determining the safe, allowable load-bearing capacity of wood piles sometimes used by European foundation engineers for preliminary design purposes is that known as the "Hamburg Rule." According to this rule, the safe, allowable bearing capacity P_{all}

Bearing Capacity of Piles

of a wood pile is as many metric tons as there are centimeters of the average diameter d of the pile:

$$P_{\text{all}} \equiv d_{(\text{cm})} \quad \text{(metric tons)} \quad (19\text{-}56)$$

British engineers use the following simple, empirical rule for assessing approximately the safe bearing capacity of piles:

$$P_{\text{all}} \equiv \frac{A_{(\text{in.}^2)}}{3} \quad \text{(British tons)} \quad (19\text{-}57)$$

For example, a pile the cross-sectional area of which is $A = 12$ in. $\times$ 12 in. $= 144$ square inches would have a safe allowable bearing capacity of $P_{\text{all}} = 144/3 = 48$ (tons).

19-14 SAFE BEARING VALUES OF PILES

For information and preliminary design purposes, the estimated safe bearing capacities of piles singly or combined in small groups for piles 50 to 60 ft long are compiled for various types of soils in Table 19-2.[22] The values are in short tons per pile.

The allowable loads on piles are established by load tests, or by experience, or by manufacturers' recommendations, or they are set by engineering societies and governed by building ordinances. For example, the building laws of the City of New York (1948) limit wood-pile loads to a maximum working load of 20 tons on a pile having a 6-in. (15 cm) tip and 25 tons on a pile with an 8-in. tip. Other building codes set allowable pile loads (16 to 30 tons) according to the kind of wood and other considerations. The optimum load range of timber piles is 15 to 25 tons/pile, and the maximum load for usual conditions is 30 ton/pile.

Free-standing piles, as well as those driven through "soupy" layers of soil, must also be checked against buckling.

19-15. TENSION PILES

Tension piles are subjected to tensile forces or pull. Tension piles are distinguished from compression piles according to the direction of action of the applied force.

Tension piles resist tensile forces mainly by their mantle friction and their self-weight.

Suitable tension piles are those that
1. Have a large self-weight
2. Have a large mantle surface (large diameter and/or long length)
3. Have a rough mantle surface

TABLE 19-2
Estimated Safe Bearing Values of Piles Used Singly or in Small Groups
(Loads are in tons per pile. All piles are 50 to 60 ft. long, laterally supported)

Type of Action	Soil	Wood	Precast Concrete		Cast-in-Place Concrete, with Shells				Steel Pipe and Concrete		H piles, 14 in.
			12-in. square	18-in. octagonal	Tapered	Cylindrical		Button bottom			
						12-in.	18-in.		12-in.	18-in.	
End-bearing	Firm, fine sand	20–25	20–30	25–35	25	25	25–30	25–30	25–30	25–30	
	Coarse sand	20–30	25–35	30–40	25–30	25–30	30–35	30–35	30–35	30–35	
	Sand and gravel	20–30	30–40	40–50	30–40	25–35	35–40	35–40	30–35	35–40	40–60
	Hardpan or caliche	—		45–50		35	40–50	40–50	35–40	40–50	50–75
	Shale or disintegrated rock					35	40–50	40–50	35–40	40–50	70–100
	Bedrock					35	40–60	40–60	35–40	40–60	
Friction	Compacted silt	15	12		15	15–20	25–30				
	Soft clay	15–20	15–18		15–20	20–25	30–35				
	Medium clay	20–25	20–25	20–25	20–25	25–30	35–40				
	Stiff clay	20–25	25–30	30–35	25–30	25–30					
	Fine sand (confined)	25–35	25–30	30–35	25–30	25–30					

Bearing Capacity of Piles

4. Are driven butt down
5. Have enlarged ends or bulbs in the soil
6. Have a screw
7. Are used where there is a frictional soil of good bearing capacity available to develop the necessary friction

Conical piles driven with butts up are not suitable as tension piles. Concrete tension piles must be designed with reinforcement for tension.

Figure 19-9 shows a single, vertical, straight tension pile. The maximum or ultimate tensile force P_{max}, or pull-out capacity a straight tensile pile can take.is calculated as

$$P_{max} = R_f + W_p + \text{(eventual surcharge)} \quad (19\text{-}58)$$

where $R_f = \tau U L_{eff}$ = friction force for effective length L_{eff} of pile
τ = average, uniformly distributed assumed tangential friction stress along mantle surface of pile

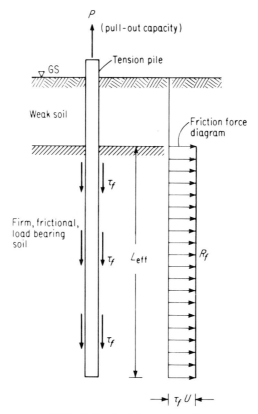

FIG. 19-9. Straight tension pile.

U = length of evolute of pile perimeter
L_{eff} = effective length of pile (see Fig. 19-9)
W_p = self-weight of pile

For τ-values see Table 19-1.

The adhesion of cohesive soils to the piles is not reckoned with here because it is very uncertain and difficult to determine.

For a single, vertical, cast-in-place, enlarged-end concrete pile (see Fig. 19-10) the maximum or ultimate tensile force P the pile can take is the self-weight W_p of the pile and the soil cone the tension pile would pull out of the soil:

$$P_{\max} = W_s + W_p \tag{19-59}$$

where W_s is the weight of the torn-out soil cone.

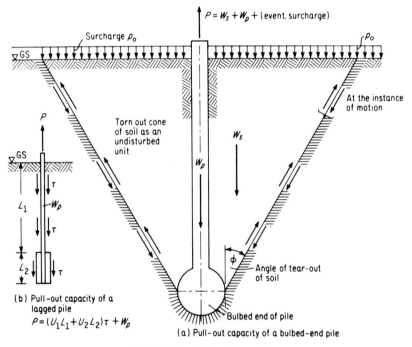

FIG. 19-10. Pull-out capacity.

The safe allowable tensile force P_{all} (working tensile load) on a tensile pile is usually established by dividing the ultimate tensile force $P_{\max}$ by a factor of safety of $\eta = 2$:

$$P_{\text{all}} = \frac{P_{\max}}{\eta} \tag{19-60}$$

Bearing Capacity of Piles

Problem

Refer to Fig. 19-11. Calculate the pull-out capacity P_{max} of a bulbed-end tension pile and the allowable tensile force P_{all} for a factor of safety of $\eta = 2$.

Solution

Self-weight of pile, W_p:

$$W_p = W_{shaft} + W_{sphere} = \left[\frac{\pi}{4}d_{sh}^2(L_1 + L_2) + \frac{4}{3}\pi r_3^3\right]\gamma_c$$

$$= [(0.785)(1.5^2)(36.0 + 14.0) + (1.33)(3.14)(2.0^3)](150)$$

$$= \underline{18,250 \text{ (lb)} = 18.25 \text{ k}}$$

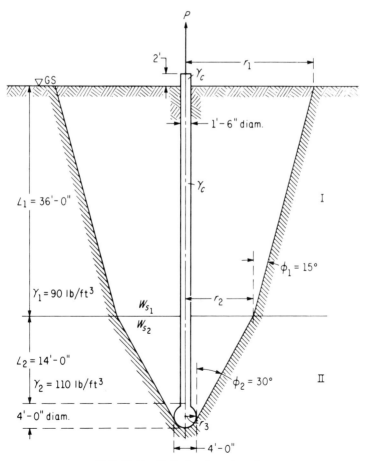

FIG. 19-11. Bulbed-end tension pile.

Auxiliary calculations:

$$r_2 = r_3 + (L_2 + r_3) \tan \phi_2 = 2.0 + (14.0 + 2.0) \tan 30°$$
$$= 2.0 + (16.0)(0.577) = \underline{11.23 \text{ (ft)}}$$

$$r_1 = r_2 + L_1 \tan \phi_1 = 11.23 + (36.0) \tan 15°$$
$$= 11.23 + (36.0)(0.2679) = \underline{20.87 \text{ (ft)}}$$

Weight of truncated soil cone W_{s_1}:

$$W_{s_1} = \left[\frac{\pi}{3}L_1(r_1^2 + r_1 r_2 + r_2^2) - \frac{\pi}{4}d_{sh}^2 L_1\right]\gamma_1$$

$$= \{(1.046)(36.0)[20.87^2 + (11.23)(20.87) + 11.23^2]$$
$$- (0.785)(1.5^2)(36.0)\}(90.0) = 2,700,632.7 \text{ (lb)} = \underline{2,700.63 \text{ k}}$$

Weight of truncated soil cone W_{s_2}:

$$W_{s_2} = \left[\frac{\pi}{3}(L_2 + r_3)(r_2^2 + r_2 r_3 + r_3^2) - \frac{\pi}{4}d_{sh}^2 L_2 - \frac{4}{3}\frac{\pi}{2}r_3^3\right]\gamma_2$$

$$= \{(1.046)(14.0 + 2.0)[11.23^2 + (11.23)(2.0) + 2.0^2]$$
$$- (0.785)(1.5^2)(14.0) - (2.092)(2.0^3)\}(110.0)$$

$$= 276,313.40 \text{ (lb)} = \underline{276.31 \text{ k}}$$

Total weight of truncated soil cones W_s:

$$W_s = W_{s_1} + W_{s_2} = 2,700.63 \text{ k} + 276.31 \text{ k} = \underline{2,976.94 \text{ k}}$$

Tensile force or pull-out capacity P_{max}:

$$P_{max} = W_s + W_p = 2,976.94 \text{ k} + 18.25 \text{ k} = \underline{2,995.19 \text{ k}}$$

Allowable tensile force on pile, P_{all}:

$$P_{all} = P_{max}/\eta = 2,995.19/2 = 1,497.595 \text{ k} \approx \underline{1,498 \text{ k}}$$

19-16. EMBEDMENT OF PILES BY VIBRATION

Piles and sheet piles may also be embedded into soil by vibration. In the report[23] on the exchange visit of an American delegation of civil engineers to the USSR in 1960 are briefly described methods of penetration of piles and "caissons" (shells) of reinforced concrete into soil by vibration, as sketched in Figs. 19-12, 19-13, and 19-14.

Vibration at resonance is reported to increase the efficiency of pile embedment. Also, vibration is less noisy than impact driving, creates and

Bearing Capacity of Piles

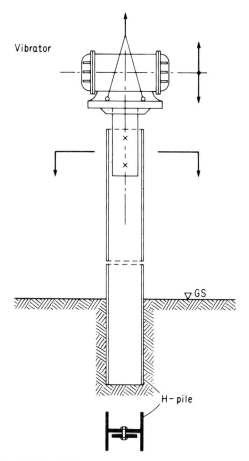

Fig. 19-12. Single-mass vibrator attached to an H-pile.

transmits no shock in the soil. The latter fact may be critical where pile work must be performed in a congested area.

Figure 19-13 shows a two-mass vibrator attached to a metal pile.

The vibration method for driving of open caissons (Fig. 19-14) and sheet piles into soil used in the USSR is also discussed in that report.

At this point, however, it is pertinent to note that in 1932 Professor A. August Hertwig, in collaboration with H. Lorenz, successfully drove the first timber pile into the soil by means of a vibrator at the site of the Technische Hochschule Berlin-Charlottenburg. In cooperation with Hertwig the firm Losenhausenwerk in Dusseldorf registered the German Reich's patent, DRP No. 611392, in which the principles of vibration driving were outlined.[24] The vibration method for penetrating piles by vibration

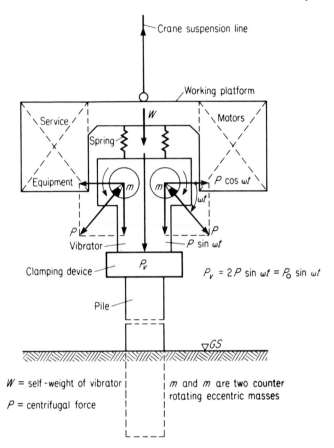

FIG. 19-13. Two-mass vibrator attached to a steel pile.

was not utilized much in Germany nor elsewhere until very recently. However, in the USSR this method was successfully developed and used in practice since 1949, according to the American exchange report.[23]

Today the vibration problem of piles has received a new look, and vibration equipment is now manufactured and sold not only in Germany, but also in France, England, and the United States. Vibration of piles is an efficient, expedient method of embedment in foundation engineering. The ultimate load R_u which a vibrated pile will support may be calculated by the following equation as derived by Davisson[25] for the Bodine Resonant Driver (BRD) is

$$R_u = \frac{(550)N + W \cdot r_p}{r_p + f \cdot s_L} \qquad (19\text{-}61)$$

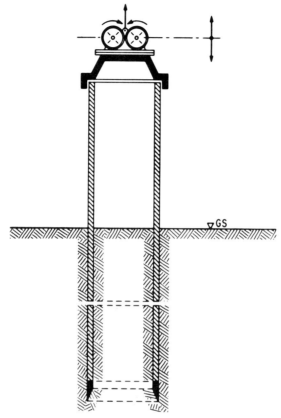

FIG. 19-14. Penetration of caisson into soil by vibration.

N = number of horsepower delivered to pile, in HP
$W = 22,000$ lb = weight of BRD – 1000 type of vibrator
r_p = final rate, of pile penetration, in ft/sec
f = frequency, cycles/sec
s_L = loss factor, ft/cycle, to be determined empirically. For closed-end pipe and H-piles in medium dense sand or sand and gravel $s_L = 0.025$ ft/cycle.

The formula can be modified to give the allowable pile load by applying a suitable factor of safety ranging from 2 to 3 based upon judgment, experience and the degree of accuracy of the loss factor value.

19-17: ELECTRO-OSMOTIC PILE DRIVING

The electrokinetic phenomenon of electro-osmosis may be used as a means to facilitate pile-driving operations in dense sandy-clayey soils.

Electro-osmosis is the electrokinetic phenomenon of moisture migration through a porous medium such as soil in an electric field under the influence of an electric potential difference of direct electric current (d.c.) from anode ⊕ to cathode ⊖.[4,26-28]

In electro-osmosis the mechanism for the translocation of soil moisture is possibly due to the existence of the electric diffuse double layer in a moist, colloidal soil system. The application of an electric potential difference to the soil system results in the displacement of the electrically charged mobile moisture films (surrounding the soil particles) relative to the immobile ones.

Because the mobile part of the moisture film is free to move, whereas the soil particles are not, a flow of soil moisture takes place from the higher potential [from ⊕ electrode (= anode) to ⊖ electrode (= cathode)]. At the cathode the electro-osmotically transferred moisture content in soil increases, thus temporarily softening up the soil. This would facilitate easy penetration of a pile driven into the soil.

To reduce the time "at rest" of the driven pile for recovering its bearing capacity after the pile has been "electro-osmotically" driven, the electric field is disrupted, the polarity of the pile is reversed (Fig. 19-15), the electric field in reverse is restored, the excess moisture from the driven pile is moved

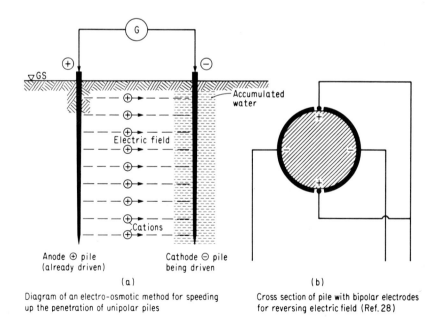

(a) Diagram of an electro-osmotic method for speeding up the penetration of unipolar piles

(b) Cross section of pile with bipolar electrodes for reversing electric field (Ref. 28)

FIG. 19-15. Electro-osmotic pile driving.

away electro-osmotically, and thus the shear strength of the soil, viz., the bearing capacity of the soil (pile), is quickly recovered.

According to Nikolayev,[28] the method of electro-osmotic pile driving is applicable to timber, reinforced-concrete, and steel piles.

Some of the advantages of electro-osmotic pile driving are
1. Accelerated pile-driving time (more than twice as compared with ordinary dynamic pile driving), resulting in savings of effort
2. Increased efficiency in pile-driving work
3. Improved quality in executed results of pile-driving projects
4. Upon driving, the cathode-piles become less damaged than those subjected to ordinary (hard) driving because of the increased moisture content in the soil around the driven (cathode) pile
5. Electric motors and generators are simple to operate
6. Method is economical[28]

REFERENCES

1. American Society of Civil Engineers, "Pile Foundations and Pile Structures," *Manual of Engineering Practice*, No. 27 (1946), p. 2.
2. R. D. Chellis, *Pile Foundations*. New York: McGraw-Hill, 1951, pp. 525-538.
3. A. E. Cummings, "Dynamic Pile Driving Formulas," in Boston Society of Civil Engineers, *Contributions to Soil Mechanics* 1925-1940. Boston, Mass.: The Society, 1940, pp. 392-413.
4. A. R. Jumikis, *Soil Mechanics*. Princeton, N.J.: Van Nostrand, 1962, pp. 213-225.
5. J. F. Redtenbacher, *Prinzipien der Mechanik und des Maschinenbaues*, cited in Ref. 1.
6. Michigan State Highway Commission, "A Performance Investigation of Pile Driving Hammers and Piles," Office of Testing and Research, Research Project 61 F-60. Final Report of a Study in Cooperation with the Bureau of Public Roads, U.S. Department of Commerce, Michigan Road Builders Association, Wayne State University, and Representative Hammer Manufacturers. Lansing: Michigan State Highway Commission, March 1965, p. 272.
7. R. D. Chellis, *Pile Foundations*, 2nd ed. New York: McGraw-Hill, 1961, p. 505.
8. M. Gates, "Empirical Formula for Predicting Pile Bearing Capacity," *Civil Engineering*, Vol. 27. No. 3 (March 1957), pp. 183-184.
9. W. H. Rabe, "Dynamic Pile-Bearing Formula with Static Supplement," *Engineering News Records*, Vol. 137, No. 26 (Dec. 26, 1946), pp. 868-871. Rabe's own imprint of 1960 shows revisions of his formula.
10. O. Stern, *Das Problem der Pfahlbelastung*, Berlin: Ernst, 1908, p. 161.
11. K. Zimmermann, "Die Rammwirkung im Erdreich: Versuche auf neuer Grundlage," *Forschungsarbeiten auf dem Gebiete des Eisenbetons*, No. 25, Berlin: Ernst, 1915.

12. A. R. Jumikis, *Active and Passive Earth Pressure Coefficient Tables*. Engineering Research Publication No. 43, Bureau of Engineering Research, College of Engineering, Rutgers University, New Brunswick, N.J., 1962.
13. United States Steel Corporation, *U.S.S. Steel H-Piles*. Pittsburg, Pa., ADUCO 25002-60 (n. d.), p. 14.
14. R. L. Handy and N. S. Fox, "A Soil Bore-Hole Direct-Shear Test Device," Special Report, Contribution No. 66-8a, Soil Research Laboratory, Engineering Research Institute, Iowa State University, Ames, Iowa, 1966.
15. *Grundbau Taschenbuch*, Berlin: Ernst, Vol. 1 (1955), p. 465.
16. H. D. Krey, *Erddruck, Erdwiderstand und Tragfähigkeit des Baugrundes*. Berlin: Ernst, 1936, p. 150.
17. O. K. Fröhlich, *Druckverteilung im Baugrunde*. Vienna: Springer, 1934, p. 142.
18. E. Maag, "Grenzbelastung des Baugrundes," *Strasse und Verkehr*, Vol. 24 (1938), pp. 349-357.
19. A. R. Jumikis, *Theoretical Soil Mechanics*, Princeton, N.J.: Van Nostrand, 1969.
20. A. R. Jumikis, *Mechanics of Soils: Fundamentals for Advanced Study*. Princeton, N.J.: Van Nostrand, 1964, pp. 128-140.
21. H. Dörr, *Die Tragfähigkeit der Pfähle*. Berlin: Ernst, 1922.
22. C. W. Dunham, *Foundations of Structures*, 2nd ed. New York: McGraw-Hill, 1962, p. 363.
23. A Report on Vibration in the USSR of an American Delegation, Sept. 14 to Oct. 5, 1959, published as the Highway Research Board Special Report No. 60, "Soil and Foundation Engineering in the Union of Soviet Socialist Republics." National Academy of Science-National Research Council Publication 806, Washington, D.C. (1960), pp. 21, 72-79, 112-115.
24. H. Lorenz. *Grundbau Dynamik*. Berlin: Springer, 1960, p. 292.
25. M. T. Davisson, "BRD Vibratory Driving Formula," *Foundation Facts*, Vol. 6, No. 1, 1970, pp. 9-11. Published by the Raymond Concrete Pile Division of Raymond International Inc., New York.
26. A. R. Jumikis, *Thermal Soil Mechanics*. New Brunswick, N.J.: Rutgers U. P., 1966, pp. 11-13, 85-88.
27. A. R. Jumikis, *Introduction to Soil Mechanics*. Princeton, N.J.: Van Nostrand, 1967, pp. 124-131.
28. B. A. Nikolayev, *Pile Driving by Electroosmosis*. New York: Consultants Bureau, 1962, pp. 1, 60.

QUESTIONS AND PROBLEMS

19-1. What is the function of bearing piles?
19-2. Why should timber piles be driven below the lowest position of the groundwater table?
19-3. Prepare a description of pile-driving hammers and their action.

Bearing Capacity of Piles 637

19-4. What are the advantages and disadvantages of piles made of various pile manufacturing materials?

19-5. Discuss briefly the determination of bearing capacity of end-bearing and friction piles.

19-6. Given: weight of drop hammer = 2,000 lb. Height of fall = 8 ft. Average penetration of pile under the last 25 strokes = 0.05 in. Calculate by means of the *Engineering News* formula the bearing capacity of the pile.

19-7. Given: weight of single-acting steam hammer = 1,500 lb. Height of fall = 4 ft. The average penetration of the pile under the last few strokes is 0.03 in. Calculate the bearing capacity of a pile by means of the *Engineering News* formula.

19-8. Determine the necessary weight W and the corresponding cross-sectional area A of the annular ring for a successful sinking of the round, open caisson as

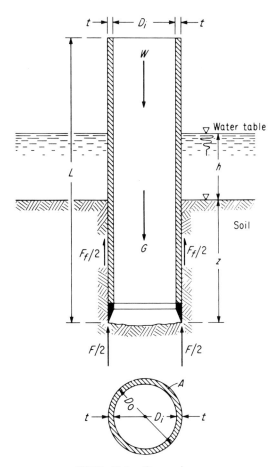

PROB. 19-8. Open caisson.

shown in the accompanying figure. Also, calculate the safe bearing capacity P of the caisson. Notations:

G = weight of caisson
$t \approx (0.1) D_o$ = wall thickness
$F = \sigma_{\text{allow}} A$ = reaction
$F_f = \mu_1 U z$ = total wall friction
$\mu_1 = \tan \phi_1$ = coefficient of wall friction
U = perimeter of caisson
z = penetration depth of caisson
D_i, D_o = inside and outside diameter of caisson, respectively

Hint for calculation:
$$W \approx \gamma_c A z, \text{ where } \gamma_c = \text{unit weight of concrete}$$

ΣV:
$$W + G - F - F_f = 0$$

Use buoyant weight where necessary. Solve the resulting quadratic equation.

chapter **20**

Pile Foundations

20-1. USE OF PILE FOUNDATIONS

The most commonly used kind of deep foundation is the pile foundation. Pile foundations belong to the oldest kinds of foundations. With recourse to the history of culture and the history of engineering, timber piles were used to support the structures (though primitive ones) of the lake dwellers, and to transmit these loads to firm soil, and the ancient bridge builders succeeded in building bridge piers on piles of considerable length and depth, thus avoiding difficult cofferdam work, pumping problems, erosion, scour, and other problems associated with laying of foundations in river beds. Caesar's *Commentaries* mention bridges founded on piles (across the Rhine) which were built by Roman legions prior to the Christian era.

Today the need for laying foundations on inferior building sites with weak, compressible soils dictates the use of piles and long tubular piles for transmitting structural loads to deep, firm geological formations, whether soil or rock. Frequently piles are also used to transmit lateral loads to soil.

Pile foundations are generally used

1. Where a shallow foundation is out of the question because too large a settlement is expected
2. Where soil erosion around and underneath an ordinary foundation can be anticipated
3. Where inclined forces of considerable magnitude act on the foundation
4. For densification of soil in order to increase its bearing capacity.

It is interesting to note that in Venice, the Netherlands, and in many other countries wooden piles were and still are used for foundations of structures.

Piles are especially used for laying of foundations of architectural, civil, and industrial engineering structures, bridge piers and abutments, earth retaining walls, machine foundations, locks, docks, sluices, piers, landings, moles, and lighthouses, for bracing and strutting of excavation walls, and other hydraulic and waterfront structures. Generally, it is more economical to use piles than piers.

All in all, it is sound to say that piles should be used as foundation elements only when it is uneconomical and infeasible to construct ordinary kinds of foundations directly supported on the soil.

20-2. ARRANGEMENT OF PILES

In a pile foundation, piles are arranged in a group tied together by a pile cap at the heads of the piles.

Piles should be arranged as symmetrically as possible relative to the line of action of the vertical load component. Also, the piles should be properly spaced according to the building code in question. Some of the most common patterns in arrangements of piles under column and wall footings for various numbers of piles are shown in Fig. 20-1.

Pile-spacing patterns may be in rows, in checkerboard pattern, or staggered to secure the minimum spacing allowed.

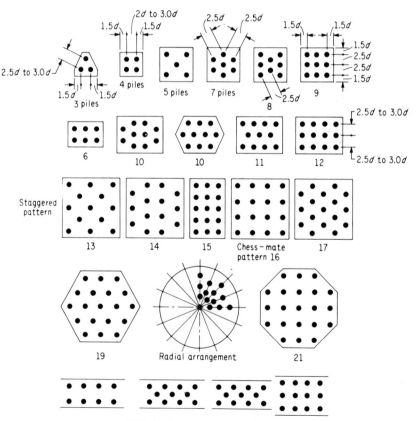

FIG. 20-1. Arrangement of piles.

Pile Foundations

The number of piles under column and wall footings and other foundations depends upon the following factors:
1. Magnitude of the load transferred to the piles
2. Safe allowable bearing capacity of the pile
3. Strength of the soil
4. Minimum allowable spacing of piles (by code and/or driving conditions of piles)
5. Maximum spacing of rows of piles considering settlement problems of pile foundations

Spacing of Piles

By spacing of piles is understood the distance between vertical axes of adjacent piles (Fig. 20-1). The spacing of piles, in its turn, depends upon several factors such as

1. Consideration of allowance in driving for small variations in position of driven piles; they may get out of plumb and out of theoretical alignment in rows,
2. Size of the pile-driving hammer. The hammer should not interfere with adjacent piles already driven; therefore the spacing of adjacent piles should be such as to allow the hammer to operate freely,
3. In the case of cast-in-place concrete piles, their spacing should be such that the adjacent empty shells or concrete filling not yet set and cured are protected against damage while working at the adjacent piles. This condition thus requires a larger spacing than that needed for timber or steel piles,
4. The need to allow for a reasonable spacing of piles, consistent with economy, in order to attain adequate stability against overturning forces, viz., moments,
5. Density, viz., strength of the soil.

In general, the spacing of timber piles is usually to $2\frac{1}{2}$ ft, and 3 ft for concrete piles, except where driven in staggered rows; then the spacing of rows may be reduced 2 or 3 in. for wood and concrete piles, respectively.

The spacing L of end-bearing piles center-to-center driven to rock, or to a firm layer of gravel and sand, and through soft clay is usually assessed in practice as $L_{min} = 2.5d$ (d = pile diameter of the butt or side of a square pile), and that for friction piles driven through soft clay into dense sand, or driven into loose sand is taken from $L = 3.0d$ to $L = 3.5d$. The minimum spacing of piles from the edge of a pile cap is usually taken as $1.5d$.

Because clay is remolded upon driving of a pile and because remolding of the clay depends largely on the spacing of the piles, the minimum distance of spacing apart, center-to-center, should be 3 times the side (or diameter) of the pile.

Upon being driven into the soil the pile displaces a volume of soil. Thus the volume of voids of the adjacent soil decreases by an amount equal to the volume of the embedded part of the pile, bringing about a densification of soil. If we accept that densification reduces the volume of soil by about $n = 20$ percent, then the radius R of the zone of soil densification around the pile may be calculated from the equality of the following volumes:

$$\frac{\pi}{4}d^2 z = \frac{\pi}{4}nD^2 z \tag{20-1}$$

where d = average diameter of pile
z = depth of embedded pile
D = diameter of extent of densification of soil by a pile
n = percent (or, rather, in decimal fractions) of possible decrease in soil volume from densification

With $n = 0.20$, $d^2 = (0.20)D^2$, $D^2 = (5)d^2$, and $D_{min} = (2.236)d$.

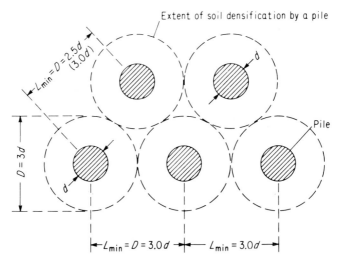

FIG. 20-2. Extent of densification of soil by piles.

Because additional densification may also be brought about by impact and vibration at the site during pile-driving operations, one would assume that the minimum spacing between piles should be about $2.5d$ to $3.0d$ for average soil density conditions (Fig. 20-2).

Press[1] showed by experiments that piles spaced at a minimum distance of $6.5d$ to $8.0d$ center-to-center behaved like single piles.

Pile Foundations

Number of Piles

The number of piles n for a centrically loaded column footing (uniformly spaced piles) is calculated as

$$n = \frac{P + W}{P_{\text{all}}} \tag{20-2}$$

where P = load transmitted from the column
W = weight of footing
P_{all} = safe allowable bearing capacity of pile

The number n of piles for a wall footing is calculated either for one unit of length of run of wall, or for the length of a repetitional transverse pattern section or panel of pile pattern (Fig. 20-3).

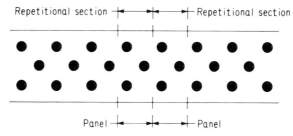

FIG. 20-3. Wall footing on piles (top view).

Relative to eccentrically loaded footings on piles, having decided upon the number r of rows of piles, say four, one would divide graphically the area of trapezoidal stress distribution diagram into r = four equal parts as shown in Fig. 20-4. Then each pile would be loaded equally. The centroids of each of the four ($=r$) partial trapezoidal areas determine the position of the axes of the r = four rows of piles. Then check pile spacing $L_{\min}$ in the rows; it should be that $L_{\min} \geqslant 3d$, or $L_{\min} > 2.5d$, whichever is the case. If this spacing condition is not satisfied, the number of rows must be reduced, or the width B of the pile cap (footing) must be increased.

Sequence of Driving Piles

The concept "pile driving" pertains to forcing or penetrating a pile into the soil in a definite position without prior excavation.

The principle of the order of pile driving is toward the line of the least resistance: away from an existing building or away from already driven piles; toward a body of water (lake, river) to avoid piles from being forced outward by later driving away from the water.

Piles for circular foundations are usually driven with the inner circle first to avoid heaving of the soil at the central part of the circle (which must

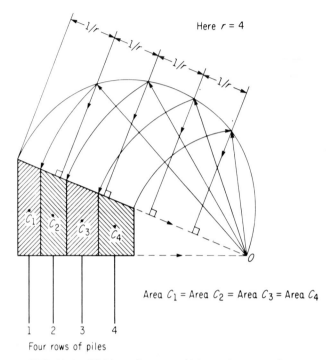

FIG. 20-4. Division of a trapezoidal area into r equal parts.

be hand-excavated between piles) and possible raising of piles already driven.

20-3. EFFICIENCY OF PILES IN A PILE GROUP

In the case of grouped end-bearing piles resting on bedrock, or on a stratum of dense sand and gravel overlying the bedrock, it is safe to assume that a group of n piles will safely carry n times the bearing value of a single pile. In the case of friction piles in a cohesive soil, with close spacing of piles intersection of pressure isobars of the individual piles builds up soil stresses which require reduction of the bearing value of individual piles in a group, by applying to the estimated safe bearing value of a single pile an efficiency factor η, the magnitude of which depends upon a number of variables. The Converse-Labarre formula for this efficiency factor η is contained in the Uniform Building Code of the Pacific Coast Building Officials Conference and in the specifications of the American Association of State Highway Officials.[2,3] The formula is

$$\eta = 1 - \phi \frac{(n-1)m + (m-1)n}{90mn} \qquad (20\text{-}3)$$

Pile Foundations

where m = number of rows of piles
 n = number of piles in a row
 ϕ = numerical value, in degrees, of angle whose tangent is d/s
 s = spacing, center to center, of piles, in feet or inches
 d = pile diameter, measured in the same units as pile spacing

Another formula giving considerably lower values for the efficiency factor is the Seiler-Keeny formula:[3,4]

$$\eta = \left\{1 - \left[\frac{11s}{7(s^2 - 1)}\right]\left[\frac{m+n-2}{m+n-1}\right]\right\} + \frac{0.3}{m+n} \quad (20\text{-}4)$$

where the terms are the same as in the Converse-Labarre formula but in which the term d does not appear.

For a discussion about η see Refs. 5 and 6.

20-4. SOME CONSIDERATIONS IN PILE-GROUP DESIGN

After the magnitude, point of application and direction of action of the loads and their resultant on the pile foundation have been determined, then the proper method of calculation of forces on piles is to be chosen. To do this one must have an idea about and give consideration to the following:

1. How to determine the number of piles in a panel of a pile system
2. How to distribute the piles in the system
3. What effect the vertical (V) and horizontal (H) forces and the resultant moments M have on the calculation of the forces acting on each of the piles
4. What methods of calculations there are available and how they differ among themselves
5. How actually to perform such calculations

If the central, symmetrical vertical (normal) structural load, including the weight of the pile cap or footing, is N; if the cross-sectional area of the pile is a, and if the allowable normal compressive stress of the pile material is σ, then the number n of piles is calculated as

$$n = \frac{N}{\sigma a} \quad (20\text{-}5)$$

where all quantities should be in consistent units.

With regard to what was said before, with centrical or symmetrical loading piles are distributed uniformly and symmetrically relative to load N, observing the prescribed or desired axial spacing of vertical piles, whichever is the case. The piles may be spaced in parallel rows or in rows staggered in parallel. When the resultant load N is applied at the centroid of a pile group, each pile receives an identical load, so the pile load P will be

$$P = N/n \qquad (20\text{-}6)$$

where $n =$ number of piles in the pile group. Each row receives $n_r = n/r$ piles, where r is the number of rows in the group of piles, and $P_r = n_r P =$ total load on n_r piles in each row.

The pile load P should be equal to or less than the safe allowable bearing capacity P_{all} of the single pile, i.e., $P \leqslant P_{\text{all}}$. Thus the number of piles n is recalculated to have

$$n \geqslant \frac{N}{\sigma_{\text{all}}} \qquad (20\text{-}7)$$

In practice, the self-weight of the pile is usually not reckoned with in determining n, viz., N. However, with large and heavy concrete and reinforced-concrete piles their weight is considered in calculating the number of necessary piles. The calculated number n of piles should be increased for even distribution in the group.

In distributing pile loads in a pile group for a pile foundation, the general idea is that the point of application of the resultant N of the superimposed loads should coincide with the resultant of the pile reactions (or center of gravity of the pile loads). Otherwise an eccentricity is introduced, resulting in a bending moment applied to the pile-group system. This, depending upon the magnitude of the moment, may put some piles in undesirable tension. Uneven loading may also result in uneven, differential settlement.

20-5. TENSION PILES AND BATTER PILES

When gas-filled, gas holders are subjected to an uplift force and must be held down by anchoring them to a massive, heavy foundation or to tension piles. In localities subject to floods, empty storage tanks may float because of the upward-acting hydrostatic pressure against the bottom of such tanks. This situation is coped with by using anchor or tension piles, or by adding massive dead weight in the form of thick, heavy concrete mats to which the tanks are anchored.

Abutments and earth-retaining structures on piles should be provided with compression as well as with tension piles. This gives the structure greater safety than without tension piles because the direction of earth

Pile Foundations

pressure may vary at times, and the earth-retaining structure should also be stable without the backfill.

Batter piles are driven at inclination in order to transmit vertical as well as horizontal loads from the structure to the soil or rock. The inclination of slopes of batter piles varies, depending upon the inclination of the resultant force on the pile; for example, $\tan \alpha = v : h = 6 : 1$, $12 : 5$, or whatever the slope is (Figs. 20-5 and 20-6).

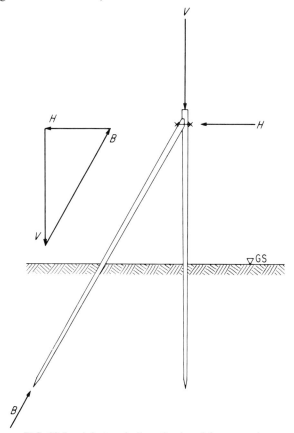

FIG. 20-5. A battered pile acting in axial compression.

Batter piles may have a practical maximum slope of $3 : 1$, although by special pile-driving rigs it is possible to drive batter piles at slopes $1 : 1$.

Batter-pile uplift (Fig. 20-6) should be watched for.

A batter pile acting in axial compression is shown in Fig. 20-5.

Batter piles are used in engineering structures such as railroad trestles to take care of lateral forces like wind forces W, centrifugal forces C on railroad curves, and seismic forces F_s, for example (Fig. 20-7).

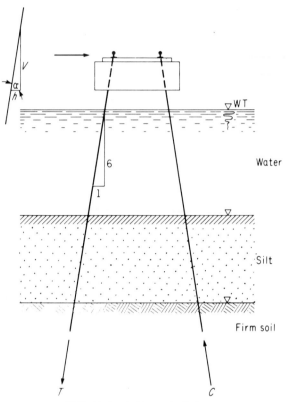

FIG. 20-6. A battered pile bent.

As a rule, foundation piles should be subjected to axial loads only, and never, if possible, to bending loads. However, in practice there are cases encountered when bending forces on piles are unavoidable. Bending forces are encountered in the design of high-pile clusters such as quays and wharves built on piles and other waterfront structures where one must cope, among other things, with lateral earth pressures. Bending of inclined piles may set in after many years' consolidating and settling of cohesive soils. Reinforced-concrete piles are especially sensitive to bending if they are not adequately reinforced. Steel piles perform satisfactorily against bending.

Figure 20-8 shows a batter pile subjected to an inclined load R. Resolution of R into axial load N and a horizontal load H shows that the pile is subjected to axial compressive load N and to a horizontal bending load H.

Figure 20-9 illustrates how to avoid bending stresses in grillage piles by rigid connection of compressive and tension piles. The compression pile is compressed, the tension pile is pulled (out of the ground).

Pile Foundations

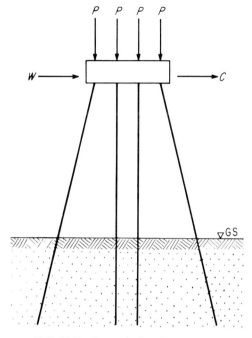

FIG. 20-7. Battered pile railroad trestle.

The magnitudes of the forces C in the compression pile and T in the tension pile are calculated respectively as

$$C = \frac{H \cos \beta + V \sin \beta}{\sin (\alpha + \beta)} \tag{20-8}$$

$$T = \frac{H \cos \alpha - V \sin \alpha}{\sin (\alpha + \beta)} \tag{20-9}$$

These equations are obtained from the force triangles as set forth (see Figs. 20-9a and 20-9b). From the rectangular-force triangle, obtain:

$$\frac{V}{R} = \sin \gamma \tag{20-10}$$

$$\frac{H}{R} = \cos \gamma \tag{20-11}$$

From acute-force triangle obtain by sine law, Fig. 20-9b, and substitute $\sin \gamma$ and $\cos \gamma$ from Eqs. 20-10 and 20-11:

$$\frac{R}{\sin (\alpha + \beta)} = \frac{T}{\sin (90 - \gamma - \alpha)} \tag{20-12}$$

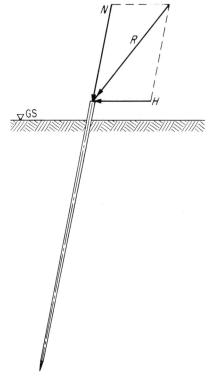

FIG. 20-8. Battered pile subjected to an inclined load R.

$$\frac{R}{\sin(\alpha+\beta)} = \frac{C}{\sin(90°+\gamma-\beta)} \qquad (20\text{-}13)$$

$$C = \frac{H\cos\beta + V\sin\beta}{\sin(\alpha+\beta)} \qquad (20\text{-}8)$$

$$T = \frac{H\cos\alpha - V\sin\alpha}{\sin(\alpha+\beta)} \qquad \text{Q.E.D.} \qquad (20\text{-}9)$$

20-6. STATIC PILE-GROUP SYSTEMS

Before commencing pile load calculations one should also become familiar with the two terms of pile systems—*statically determinate* and *statically indeterminate*

Strictly speaking, statically determinate pile foundations are those whose panels consist of two arbitrary piles. One of these piles must be assumed as fixed at one end, whereas the other three supports are hinged (Fig. 20-10),

Pile Foundations

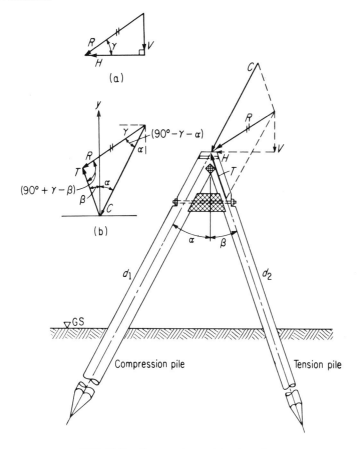

FIG. 20-9. Compression and tension piles.

or they contain three nonparallel, hinge-supported piles. The piles should not intersect at one point in the pile footing or grillage (Fig. 20-11).

In a statically determinate pile system, the pile loads are independent of stiffness of piles or of that of the superstructure.

From the practical point of view in performing pile load calculations, a statically determinate pile system is one whose forces can be determined by a simple static calculation, or by a simple resolution of forces.

A statically indeterminate pile system is one whose forces on piles depend not only upon the superimposed structural load but also upon the dimensions of the piles and the stiffness of the superstructure.

20-7. STATICALLY DETERMINED PILE SYSTEMS

Some statically determined pile systems are shown in Figs. 20-10 to 20-12.

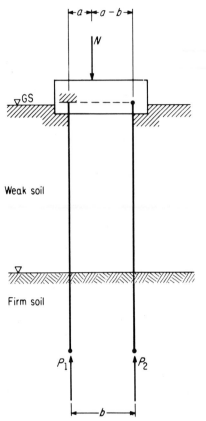

FIG. 20-10. Forces in piles or rows of piles.

(a) Forces in piles or rows of piles (Fig. 20-10):
By simple proportion:

$$P_1 b = N(b - a) \tag{20-14}$$

$$\therefore \quad P_1 = N\left(\frac{b-a}{b}\right) \tag{20-14a}$$

$$P_2 b = Na \tag{20-15}$$

$$P_2 = N\frac{a}{b} \tag{20-15a}$$

Here N is taken either per unit of length per run of foundation, or for a repetitional panel. Or N can be the total load on a rectangular or square footing.

Pile Foundations

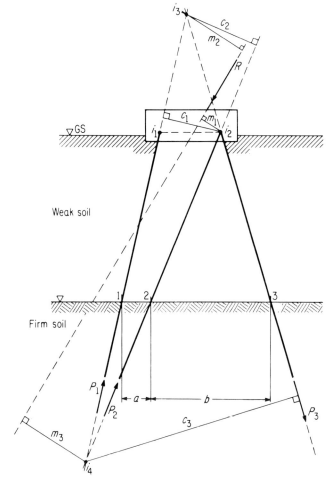

FIG. 20-11. Forces in a three-pile system.

(b) Calculation of pile loads by moments (Fig. 20-11):
Relative to intersecting point i_2 of pile axes 2 and 3, it should be that

$$\Sigma M_{i_2}: \quad P_1 c_1 - R m_1 = 0; \quad P_1 = R \frac{m_1}{c_1} \quad \text{(C)} \quad (20\text{-}16)$$

$$\Sigma M_{i_3}: \quad -P_2 c_2 + R m_2 = 0; \quad P_2 = R \frac{m_2}{c_2} \quad \text{(C)} \quad (20\text{-}17)$$

$$\Sigma M_{i_4}: \quad P_3 c_3 - R m_3 = 0; \quad P_3 = R \frac{m_3}{c_3} \quad \text{(T)} \quad (20\text{-}18)$$

If R is taken per linear foot (linear meter) of wall, then the pile loads are per linear foot (linear meter) of wall, and the spacing of piles may be determined on the basis of the allowable axial load per pile.

(c) Calculation of pile loads by force projections and moments (Fig. 20-12):

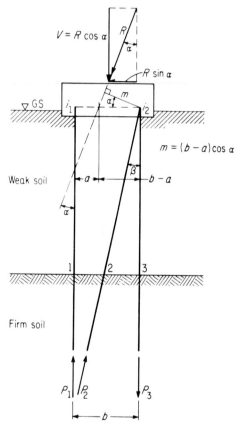

FIG. 20-12. A three-pile system.

Consider that two out of three piles are parallel. The sum of the projections of horizontal and vertical forces on the horizontal and vertical axes of the system and the sum of the moments render:

$$\Sigma H: \quad P_2 \sin \beta - R \sin \alpha = 0; \qquad P_2 = R \frac{\sin \alpha}{\sin \beta} \tag{20-19}$$

$$\Sigma V: \quad R \cos \alpha - P_1 + P_3 - P_2 \cos \beta = 0 \tag{20-20}$$

ΣM_{i_2}: $P_1 b - Rm = 0$; $m = (b-a)\cos\alpha$;

$$P_1 = R\frac{m}{b} = R\frac{(b-a)\cos\alpha}{b} \tag{20-21}$$

$$P_3 = R\left(\frac{\sin\alpha}{\tan\beta} + \frac{m}{b}\cos\beta\right) \tag{20-22}$$

(d) Calculation of forces by simple proportion and resolution into components (Fig. 20-13, a pile foundation on two bents of four piles):

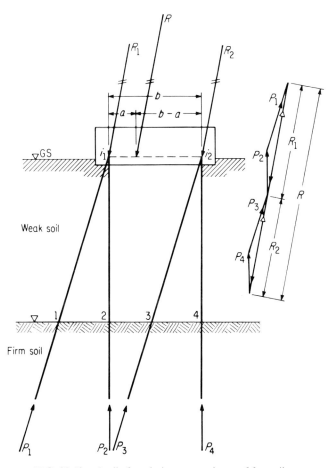

FIG. 20-13. A pile foundation on two bents of four piles.

Resultant force on each bent:

$$\bar{R}_1 = \bar{P}_1 + \bar{P}_2; \qquad \bar{R}_2 = \bar{P}_3 + \bar{P}_4$$

$$R_1 = R\frac{b-a}{b} \tag{20-23}$$

$$R_2 = R\frac{a}{b} \tag{20-24}$$

Resolve graphically resultant force on each bent into components: along the directions of the longitudinal pile axes (see force triangles in Fig. 20-13), and scale off the pile loads P_1, P_2, P_3, and P_4. Or calculate component forces trigonometrically by means of the sine law.

(e) Determination of pile loads by Culmann's graphical method (Fig. 20-14):

Proceed as set forth:

(i) Intersect load R with pile 1 to obtain point of intersection i_2.

(ii) Connect point i_2 with point O by means of the so-called Culmann's line C (a force at O).

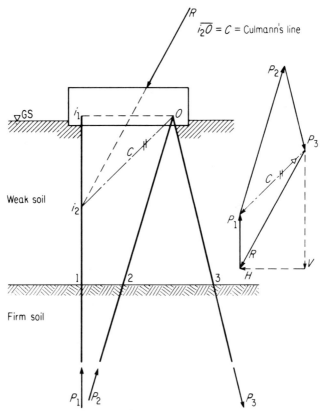

FIG. 20-14. Culmann's graphical method.

Pile Foundations

    (iii) Construct the force polygon and scale off the pile load P_1.
    (iv) Resolve C into components P_2 and P_3 and scale off the pile loads P_2 and P_3.

Point O at which the resultant of the external forces act is called the *center or zero point* of the system. Now that three forces, C, P_2, and P_3, pass through a common point of intersection O (system's center), these three forces are in static equilibrium.

This method renders good practical results.

20-8. STATICALLY INDETERMINATE PILE SYSTEMS

The pile loads in statically indeterminate pile systems may be calculated principally by two methods:
1. By approximative calculations
2. By several methods of analyses based on the theory of elasticity

Among the approximative methods of calculations are to be mentioned
    (a) Culmann's method
    (b) Method of stress trapezoid
    (c) Method of moment of inertia of pile system

20-9. CULMANN'S METHOD

The superimposed load on pile systems in Figs. 20-15 and 20-16 can best be distributed on piles graphically by Culmann's method. If the pile system contains not more than three different pile directions, then the piles, having the same inclination, may be combined into imaginary resultant piles as indicated in Figs. 20-15 and 20-16. Thus piles 2 and 3 in Fig. 20-15 are mentally replaced by one pile ($P_{2,3}$) spaced in the middle between piles P_2 and P_3. This pertains also to Fig. 20-16. The loads on the resultant piles can now be determined by Culmann's method as outlined for statically determinate pile systems to obtain from the force polygon (or analytically) the resultant pile loads ($P_{1,2}$), ($P_{3,4}$), and ($P_{5,6}$) (Fig. 20-16). The real loads on each pile of the 6-pile system are obtained by resolving the load on the imaginary resultant pile into two components: they then have half the magnitude of each of the corresponding imaginary resultant piles, i.e.:

$$P_1 = P_2 = \frac{(P_{1,2})}{2} \quad (20\text{-}25)$$

$$P_3 = P_4 = \frac{(P_{3,4})}{2} \quad (20\text{-}26)$$

$$P_5 = P_6 = \frac{(P_{5,6})}{2} \quad (20\text{-}27)$$

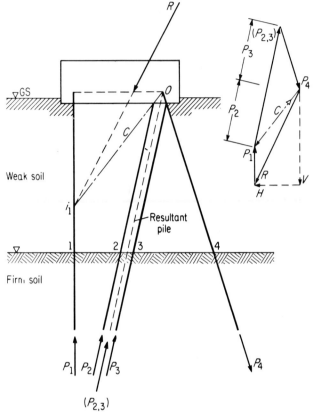

FIG. 20-15. Determination of forces graphically.

Comments

Among the approximative methods for determing pile loads in a statically indeterminate system, Culmann's method is of great advantage when large horizontal forces act on the pile footing. The approximative methods are also satisfactory for determining pile load distributions for small structures. They are often used in preliminary design of large constructions.

Because the different piles in a group are not all spaced the same distance from the resultant, they probably do not transmit equal loads to the soil.

More precise results in pile-foundation analyses may be expected from using methods which consider elastic deformations of piles, footings and settlement of soils.

Problem. Given a pile-foundation system as shown in Fig. 20-17. Determine by Culmann's method the loads on each pile.

Problem. Given a pile-foundation system as shown in Fig. 20-18. Determine loads on each pile.

Pile Foundations

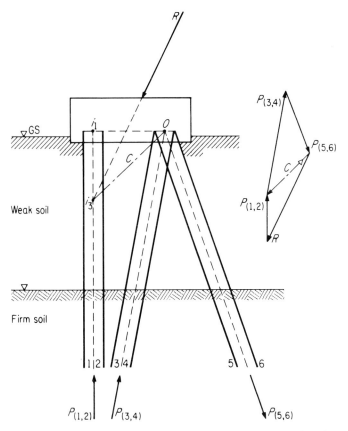

FIG. 20-16. A pile system with not more than three different pile directions.

20-10. TRAPEZOIDAL METHOD FOR DISTRIBUTING LOADS ON PILES

Vertical Piles

As already discussed, in designing the spacing of piles an important consideration is the uniform load transfer by all piles on the soil in order to avoid intolerable differential settlement. It was also mentioned that in order to attain these objectives, the resultant N of the vertical component of the superimposed loads should coincide with the resultant of the pile reactions (pile resistance, or pile loads). A very convenient method of spacing piles, accordingly, is the so-called trapezoidal method. By this method the superimposed load is distributed equally on each pile, and pile spacings are obtained graphically.

In this method, proceed as set forth.

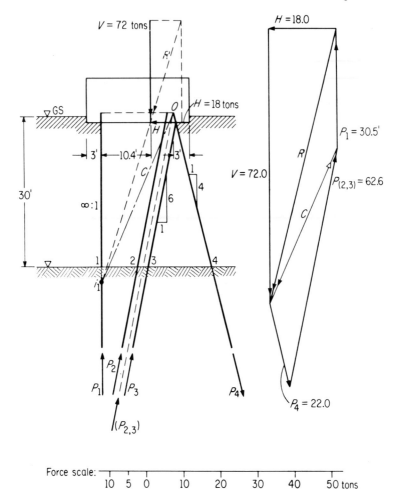

FIG. 20-17. Application of Culmann's method to calculation of forces in piles.

Suppose the piled footing is subjected to an inclined resultant load $\bar{R} = \bar{V} + \bar{H}$ at an eccentricity e (Fig. 20-19). For the width B of the footing, calculate and plot graphically the edge pressures σ_{max} and σ_{min}, and thus the trapezoidal pressure distribution diagram for the vertical component V. The edge pressures are

$$\sigma_{\substack{max \\ min}} = \frac{V}{BL}\left(1 \pm \frac{6e}{B}\right) \tag{20-28}$$

where L = length of footing perpendicular to drawing plane.

Pile Foundations

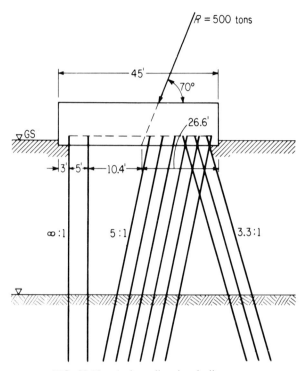

FIG. 20-18. A three-directional pile system.

The trapezoidal pressure area (which represents the total vertical load V) will now be subdivided, preferably graphically, into n equal parts—as many parts as the number of rows in this pile system. Each of the equal parts of the physical trapezoidal pressure diagram means the load P on the pile, viz., pile row, in the repetitive panel of the footing, i.e., $P = V/n$.

The position of the centroids of each part of the trapezoidal pressure diagram establishes the position of the rows of piles, hence also the spacing of the equally loaded piles, viz., pile rows.

Before continuing the discussion about spacing of piles or pile rows by the trapezoidal method, it would be expedient for the student to acquire an effective routine of the subdivision of a trapezoidal area into equal parts. Therefore, for the benefit of the student, a corresponding step-by-step worked-out example as to the mechanics of the subdivision is given here.

The routine may be described as follows.
1. Construct the trapezoidal pressure-distribution diagram ABB_1A_1 (Fig. 20-20).
2. Extend lines AB and A_1B_1 to obtain the point of intersection U, thus obtaining a triangle with its vertex at U.

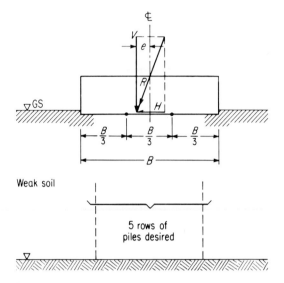

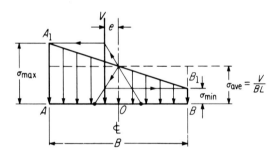

FIG. 20-19. A piled footing subjected to an inclined resultant load.

3. Draw a semicircle whose diameter is AU.
4. With a radius $UB = Ut$ and U as the center, transfer point B on the circle to obtain point t.
5. Project point t on line BU to obtain point T.
6. Divide a segment of line AT into the requisite number of parts, i.e., into $n = 5$ equal parts: $(AT)/n = AG = GH = HK = KL = LT$.
7. Project points G, H, K, and L vertically down (perpendicularly to AU), intersecting the circle at points g, h, k, and l respectively.
8. Transfer points g, h, k, and l (draw arcs with radii gU, hU, kU, and lU, respectively) on the base AB of the trapezoid to obtain the desired points of division g_1, h_1, k_1, and l_1.
9. Through points g_1, h_1, k_1, and l_1, draw lines g_1g_2, h_1h_2, k_1k_2, and l_1l_2

Pile Foundations

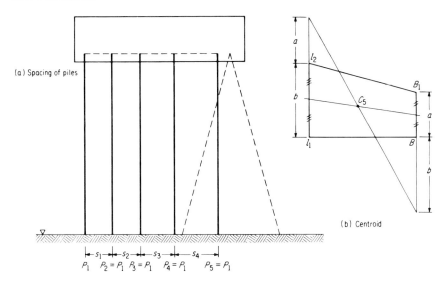

(a) Spacing of piles

(b) Centroid

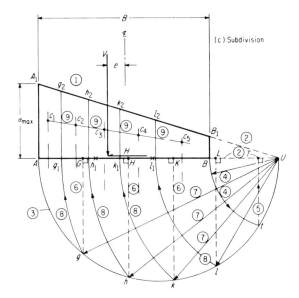

(c) Subdivision

FIG. 20-20. Subdivision of a trapezoidal area in five equal parts.

parallel to AA_1 and BB_1. These lines divide the trapezoidal area, viz., the superimposed load N into $n = 5$ equal parts.

The position of longitudinal pile axes is located at the centroids of each of the small trapezoids. These centroids C can be conveniently found graphically, as shown in Fig. 20-20b. The pile spacings are now $s_1 < s_2$

$< s_3 < s_4$. The load P on each pile (viz., row piles) is now one fifth of the total superimposed load N: $P_1 = P_2 = P_3 = P_4 = P_5 = N/n = N/5$.

These spacings and the assumed number of rows of piles must now be checked by trial and adjustment to satisfy code spacing requirements for pile embedment operations and/or settlements.

In this example, and in practice, the horizontal load H, if not large, is assigned equally to all piles to be resisted by horizontal shear resistance or bending, depending upon the lateral restraint of the piles. If the horizontal force is large, it must be taken up and transferred to the soil by inclined piles or by inclined pile bents, as shown in the following discussion (Fig. 20-21).

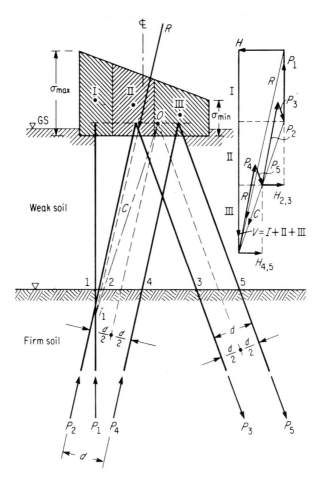

FIG. 20-21. Trapezoidal loading on a bent-type pile system.

Vertical and Inclined Piles

To continue, suppose that a given pile-foundation system is loaded with an inclined, eccentric load R as shown in Fig. 20-21. Using Culmann's method, the magnitude of the load in each pile can be most readily determined from the force polygon as shown in Fig. 20-21.

Again, one prepares a trapezoidal pressure-distribution diagram and subdivides it into three equal parts I, II, and III. The vertical pile and each of the two pile bents are located in the vertical lines through the centroid of the small trapezoids. Thus, the trapezoidal pressure-distribution diagram permits one to determine the distribution of the vertical loads: $V = I + II + III$. Then find the position of Culmann's vector $\overline{C} = \overline{R} - \overline{P}_1$ (see pile system and force polygon). From the force polygon, the load $P_1 = I$ on pile ① is now determined.

The horizontal load component H must be distributed to the inclined piles. This is accomplished by means of the force polygon (Fig. 20-22) by distributing the horizontal load proportionally to the vertical loads of the single piles (should there be such) and of the pile bents. Thus from the force polygon it can be seen that the vertical pile does not receive any horizontal load. The pile bent 4, 5 (on the right) receives a horizontal load of $H_{4,5}$, and the one in the middle receives a horizontal load of $H_{2,3}$, so that $\overrightarrow{H}_{2.3} + \overrightarrow{H}_{4.5} = \overleftarrow{H}$.

20-11. ECCENTRIC LOADINGS ON VERTICAL PILES FROM RESULTANT VERTICAL LOAD

When it is impossible to apply the external, superimposed resultant vertical load V at the center of gravity C of the pile group, eccentricity e results with respect to the center of gravity of the pile group. The vertical load V may be eccentric with respect to two coordinate axes through C or with respect to one coordinate axis through C. In all cases, the x and y principal axes must be directed through the centroid C of the pile group.

The problem is to calculate the load on each pile and on the most heavily loaded one.

Eccentricity about Two Axes

The load P on the pile in a group is calculated by the method of superposition. In this method, the individual pile loads are computed by determining separately the effects of eccentricity, and then the results may be summed up algebraically.

Compute the moments of inertia I_x and I_y of the pile group about the x and y axes respectively. Note that in general the moment of inertia I_s of

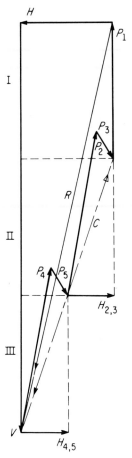

FIG. 20-22. Force polygon.

the pile system is equal to the sum of moments of inertia I of each individual pile relative to a common axis, i.e.,

$$I_s = \sum_1^n I_i \qquad (20\text{-}29)$$

Now let y_i be the distance of each pile from a common axis. Then the moment of inertia of each pile is written as

$$I_i = I_{io} + a_i y_i^2 \qquad (20\text{-}30)$$

where I_{io} = central moment of inertia of the cross-sectional area a of a pile
$a_i y_i^2$ = transfer of the central moment of inertia I_{io} of the pile cross-sectional area with respect to another axis distant y_i from pile

Pile Foundations

Because the term I_{io} in Eq. 20-30 is small as compared with $a_i y_i^2$, it may be neglected. Then the general moment of inertia I_s of the pile system (Eq. 20-29) is rewritten as

$$I_s = \sum_1^n (a_i y_i^2) \qquad (20\text{-}31)$$

Referring now to two axes, the system's moments of inertia are

$$I_x = a_1 y_1^2 + a_2 y_2^2 + a_3 y_3^2 + \cdots + a_n y_n^2 \qquad (20\text{-}32)$$

$$I_y = a_1 x_1^2 + a_2 x_2^2 + a_3 x_3^2 + \cdots + a_n x_n^2 \qquad (20\text{-}33)$$

where I_x = moment of inertia of the pile system with respect to its x-axis
I_y = same with respect to its y-axis

In our problem we are interested in the load on a pile and not in the stress. And if it can be assumed that all piles in the group are of the same diameter, viz., the same cross-sectional area, then all a_i's are equal, i.e.,

$$a_1 = a_2 = a_3 = \cdots = a_n = a \qquad (20\text{-}34)$$

and it may be assumed that a = unity (= 1). In such a case

$$I_x = y_1^2 + y_2^2 + y_3^2 + \cdots + y_n^2 = \sum_1^n y_i^2 \qquad (20\text{-}35)$$

$$I_y = x_1^2 + x_2^2 + x_3^2 + \cdots + x_n^2 = \sum_1^n x_i^2 \qquad (20\text{-}36)$$

The magnitude of the load P on any pile in the group is now calculated by the method of superposition as

$$P = \frac{V}{n} \pm \frac{V y_c y_n}{I_x} \pm \frac{V x_c x_n}{I_y} \leq P_{\text{all}} \qquad (20\text{-}37)$$

where V = resultant vertical load on the pile group
n = number of piles in the pile group
y_c = distance from the point of application of the resultant vertical load to the x-axis
y_n = distance of the selected pile from the x-axis
x_c = distance from the point of application of the load V to the y-axis
x_n = distance of the selected pile from the y-axis
P_{all} = safe allowable load on pile

The most heavily loaded pile in the group is the outermost pile on the same sides of the x- and y-axes as the point of application of the resultant loads, and the most lightly loaded pile the one diagonally opposite it.

A negative P means the uplift of the pile, and should be avoided by redesign.

If the values of pile loads vary among themselves greatly, the spacing of piles should be changed to a closer one, and by trial and adjustment arranged until all pile loads are reasonably equal.

Eccentricity about One Axis

If there is an eccentricity about one axis, say, the y-axis, then the maximum and minimum loads on a pile are calculated as

$$P_{\substack{max \\ min}} = \frac{V}{n} \pm \frac{V x_c x_i}{I_i} \leqslant P_{all} \tag{20-38}$$

Vertical Load and Moment

If, besides a resultant vertical load V, the pile group is also subjected to a resultant moment M, then, by analogy to the stress equation 10-8, the maximum and minimum loads on the most heavily and most lightly loaded piles, respectively, are calculated by the following equation:

$$P_{\substack{max \\ min}} = \frac{V}{n} \pm \frac{M x_n}{I_y} \leqslant P_{all} \tag{20-39}$$

In this case, equalization of pile loads is impossible.

Position of Center of Gravity of a Pile Group

Assume a pile system as shown in Fig. 20-23. The cross-sectional areas of the piles are $a_1, a_2, a_3, \ldots, a_k, a_n$. The distance of centers of the piles 1, 2, 3, $\ldots$, k, n from the x-axis (perpendicular to the drawing plane) at the left edge of the pile cap are $s_1, s_2, s_3, \ldots, s_k, s_n$.

The sum of the static moments M_x of pile cross-sectional areas (a) with respect to the x-axis is

$$M_x = a_1 s_1 + a_2 s_2 + a_3 s_3 + \cdots + a_k s_k + a_n s_n = \sum_1^n (as) \tag{20-40}$$

The static moment of the resultant cross-sectional areas of all piles the moment arm of which is s_c is $(na) s_c$. From the equilibrium of these two moments the position s_c of the center of gravity C of the pile system is

$$s_c = \frac{\sum_1^n (as)}{na} \tag{20-41}$$

If all piles have the same cross-sectional area

$$a = a_1 = a_2 = a_3 = \cdots = a_k = a_n$$

Pile Foundations

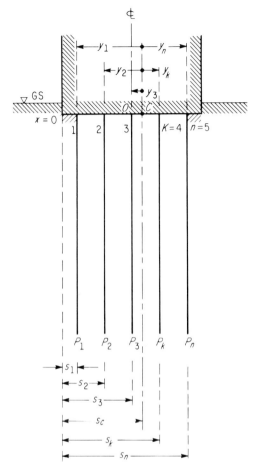

FIG. 20-23. Position of center of gravity of a pile group

then

$$S_c = \frac{\sum\limits_{1}^{n} s}{n} \tag{20-42}$$

Summary

When the eccentricity e is so large that piles cannot be loaded equally within the limits of the pile cap, then one may be forced to load the piles nonuniformly and heavily and to anticipate a nonuniform settlement of the pile foundation.

In designing an eccentrically loaded pile foundation,
1. Strive for a uniform distribution of pile loads.
2. The loaded piles should not exceed their bearing capacity.
3. Strive for such a distribution of loads and distribution of piles that the pile cap is minimum in size.
4. For correct and uniform distribution of pile loads, arrange for the point of application of the resultant load on a pile foundation to coincide with the point of application of the resultant of the pile reactions. This point of coincidence is called the *center of gravity* of the system of the pile group.

20-12. BEARING CAPACITY AND STABILITY OF PILE FOUNDATIONS

One of the requirements of pile foundations is that the latter should be so designed that all of the vertical structural loads should be transmitted to the soil by the piles only (the base of the pile footings not to take part in load transmission to the soil). Another requirement is that of arranging the piles so that each pile is as evenly loaded axially as possible. Also, it is well to space piles as widely as the design permits. Preferably the piles should be spaced so that the spacing at their tips should be equal to or larger than $3d$. Lateral forces should be coped with by means of batter piles or any other appropriate construction element (for example, by ties in a hinged frame structure where it applies).

In the design of a pile foundation each pile is to be checked to be sure that it does not exceed the allowable pile load.

When the structural load is transmitted to the soil by a number of piles, then they form a pile group.

Studies in soil mechanics and foundation engineering on bearing capacity of a pile group have shown that the bearing capacity of a pile group is less than the sum of the bearing capacities of the individual piles in the group. The bearing capacity of a pile group depends upon the spacing of the piles. This may be understood from the overlapping stress isobars from adjacent piles. The mantle friction of a loaded pile brings about a stressed zone around the pile. Hence the soil is more stressed at the lower end of the pile than at its upper end. Thus, in the case of a narrow spacing of piles, the stress isobars (viz., the pressure-influence zones) from mantle friction overlap, producing a great pressure concentration in the overlap zones. Therefore, under a pile group with narrow spacing of piles, the soil at the tips is subjected to a greater load than under a single pile carrying the same load as each of the piles in a pile group. This in its turn results in larger settlement of the

Pile Foundations

pile group than that of a single pile if all piles are loaded equally. In other words, for equal settlement the bearing capacity of a pile group is less than that of a single pile. Also, it should be observed that friction piles have a more unfavorable effect on the magnitude of settlement than end-bearing piles.

In order to utilize the bearing capacity of the single pile and to reduce the expected settlement, the requirement for a standardized spacing comes to the fore, the intention being to provide a well-utilized bearing capacity of the single piles by means of wide spacing of piles consistent with economical size of the superstructure and bracing. Close spacing of piles in granular soils may increase the bearing capacity of the soil because of its densification.

In practical foundation engineering the ultimate bearing capacity of a pile group is performed for an idealized foundation resting on the plane h-h imagined as passed through the tips of the piles (Fig. 20-24). Of course, on the average, the sum of the pile loads should not subject the soil at the level h-h of the pile tips to more than a corresponding unpiled foundation on the same plane or elevation.

All in all, the ultimate bearing capacity of a pile group is obviously the bearing capacity of the stratum at the tips of the piles. The soil loaded by end-bearing or friction piles must be checked for its bearing capacity against shear failure or groundbreak, as well as for its consolidation and maximum and differential settlement (where consolidation of cohesive soil applies).

20-13. THEORETICAL MINIMUM SPACING OF FRICTION PILES

It is assumed that the length L of the embedded pile is subdivided into a number of equal parts, the mantle surface of each part transferring by friction at the angle of mantle friction ϕ_1 a certain part of the load P on the pile and distributing the load over a circular area perpendicular to the tip of the pile (Fig. 20-25). The friction is assumed to be distributed uniformly along the pile. The diameter of this area, or the diameter of the pressure influence, is $d + 2z \tan \alpha$. Summation of all of the stress distribution diagrams I, II, III, IV ... results in a total curvilinear pressure-distribution diagram the maximum pressure ordinate of which is σ_{max}, which may be converted to an equivalent paraboloidal stressed body with the same maximum stress ordinate σ_{max}:

$$P = (1/2)\pi \rho^2 \sigma_{max} \qquad (20\text{-}43)$$

where ρ = radius of paraboloid
σ_{max} = height of paraboloid

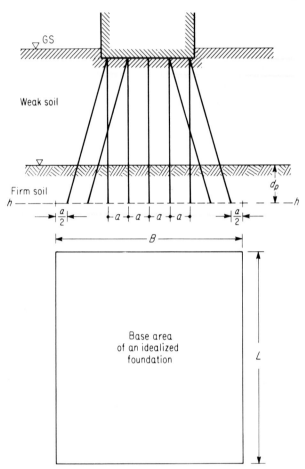

FIG. 20-24. Computation width B for an idealized foundation resting on the plane h-h through tips of the piles.

Analytically, each pile element transfers by friction to a circular area a fraction of the pile load P, namely

$$dP = \frac{P}{L_c} dz \qquad (20\text{-}44)$$

The stress on the circular area at the tip of the pile is

$$\sigma_{max} = \frac{4P}{\pi d L_c \tan \phi_1} \qquad (20\text{-}45)$$

where d = average diameter (or weighted diameter) of the pile the length of which is L_c.

Pile Foundations

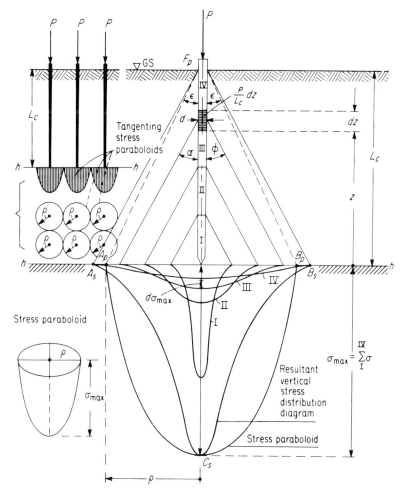

FIG. 20-25. Load transfer to soil by friction (after A. Bierbaumer, Ref. 7).

The friction pile load transfer to soil by friction underneath the tip of the pile and the radius ρ of influence of the transferred pressure to the soil may be conceived for $R_t = 0$ according to Bierbaumer[7] as illustrated in Fig. 20-25.

The stressed volume of the paraboloid, expressed in terms of σ_{max}, is approximately

$$(1/2)\pi\rho^2 \sigma_{max} = (1/2)\pi\rho^2 \frac{4P}{\pi d L_c \tan \phi_1} = P \qquad (20\text{-}46)$$

and

$$2\rho^2 = dL_c \tan \phi_1$$

$$\rho = \sqrt{\frac{dL_c \tan \phi_1}{2}} \qquad (20\text{-}47)$$

and

$$\tan \varepsilon = \frac{\rho - \dfrac{d}{2}}{L_c} \approx \frac{\rho}{L_c} = \sqrt{\frac{d \tan \phi_1}{2L_c}} \qquad (20\text{-}48)$$

Or, to act as single piles, the spacing of the piles at elevation of tips should be

$$s_{max} = 2\rho = 2\sqrt{\frac{dL_c \tan \phi_1}{2}} = \sqrt{2dL_c \tan \phi_1} \qquad (20\text{-}49)$$

Thus the size of the idealized foundation resting on a plane through the tips of the piles is (Fig. 20-26):

$$A = BL = (ns_{max} + 2L_c \tan \varepsilon)(ms_{max} + 2L_c \tan \varepsilon) \qquad (20\text{-}50)$$

where
A = area of idealized foundation
B and L = lengths of sides of idealized rectangular foundation
n = number of spacing along width B of idealized foundation
m = number of spacings along length L of idealized foundation

The compressive stress σ from the structure and foundation W (weight of piles and soil between piles excluded) on this area transferred to the soil by friction is thus

$$\sigma = W/A \leqslant \sigma_{all} \qquad (20\text{-}51)$$

The physical equation of the stress paraboloid from a differential element of the pile is

$$(1/2)\pi r^2 d\sigma_{max} = (1/2)\pi \left(\frac{d}{2} + z \tan \phi_1\right)^2 d\sigma_{max} = \frac{P}{L_c} dz \qquad (20\text{-}52)$$

From this the maximum stress $d\sigma_{max}$ is

$$d\sigma_{max} = \frac{2P}{\pi L_c} \frac{dz}{\left(\dfrac{d}{2} + z \tan \phi_1\right)^2} \qquad (20\text{-}53)$$

The total pressure of the circular plane at the elevation of the tip brought about by the pile load P is attained by summation of the stress paraboloids induced by the various pile elements of length dz. Thus the resultant

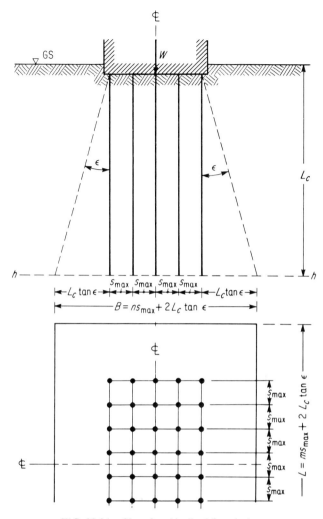

FIG. 20-26. Size of an idealized foundation.

stress diagram (imagine a volume of stress) is encompassed by the curved stress surface $A_s C_s B_s$. The maximum stress ordinate of this stressed body $A_s C_s B_s$ is calculated as

$$\sigma_{max} = \frac{2P}{\pi L_c} \int_0^{L_c} \frac{dz}{\left(\frac{d}{2} + z \tan \phi_1\right)^2} = \frac{4P}{\pi d} \left(\frac{1}{\frac{d}{2} + L_c \tan \phi_1}\right)$$

$$\approx \frac{4P}{\pi d L_c \tan \phi_1} \qquad (20\text{-}54)$$

The stressed body $A_s C_s B_s$ can now be converted into an equivalent paraboloid $A_p C_s B_p$ of the same physical height σ_{max}:

$$\underbrace{(1/2)\pi\rho^2 \sigma_{max}}_{\text{Paraboloid}} = (1/2)\pi\rho^2 \underbrace{\frac{4P}{\pi d L_c \tan \phi_1}}_{\sigma_{max}} = \underbrace{P}_{\substack{\text{load} \\ \text{on} \\ \text{pile}}} \quad (20\text{-}55)$$

or

$$\rho = \sqrt{\frac{dL_c \tan \phi_1}{2}} \quad (20\text{-}47)$$

and the maximum spacing between the centers of two adjacent paraboloids in the plane of the tips of the piles is $s_{max} = 2\rho$.

In principle, the pile load P is then transferred by friction and distributed over a circular area A_t through the tip of the pile:

$$A_t = (1/2)\pi d L_c \tan \phi_1 \quad (20\text{-}56)$$

The average pressure intensity σ_{ave} over area A_t under one pile is

$$\sigma_{ave} = \frac{P}{A_t} = \frac{2P}{\pi d L_c \tan \phi_1} \quad (20\text{-}57)$$

The contact pressure σ_o on an area $A = BL_c$ at the tip of the piles is

$$\sigma_o = N/A \quad (20\text{-}58)$$

where $N = \Sigma P =$ total structural normal loads on piles.

Pile foundations must be so designed that only piles transfer the superimposed load to the soil, because the piles are of greater rigidity than the soil. Therefore the soil intervening between the piles cannot be expected to carry any load directly. Also, in time the soil between the piles underneath the pile cap may settle, thus being no longer in contact with the base of the footing. Consequently the soil cannot be assigned to carry any load directly. The structural load is thus being transferred to the soil directly by the piles.

20-14. SETTLEMENT OF A PILE FOUNDATION

In a pile group or cluster of piles of a pile foundation the heads of all piles are joined together by a grillage, or a slab, or a cap to form a unit. By this means it is hoped that all uniformly loaded piles will settle uniformly, provided the pile-supporting soil is uniform.

Like any other foundation, pile foundations must be checked for their probable settlements for comparison with tolerable ones. Settlement

Pile Foundations

analysis of a pile group, viz., pile foundation, is still a relatively novel problem. Even today no firm and universally accepted methods of procedure for calculating consolidation and settlement of pile foundations have been put forward. Therefore, settlement of pile foundations can be estimated approximately only by the well-known consolidation theory of Terzaghi-Fröhlich[8,9] under the assumption that structural load is transferred to the soil uniformly at the elevation h-h of the tips of the piles resting on compressible soil (Fig. 20-26).

The settlement of end-bearing piles under design loads will about equal the settlements of test piles under similar loads.

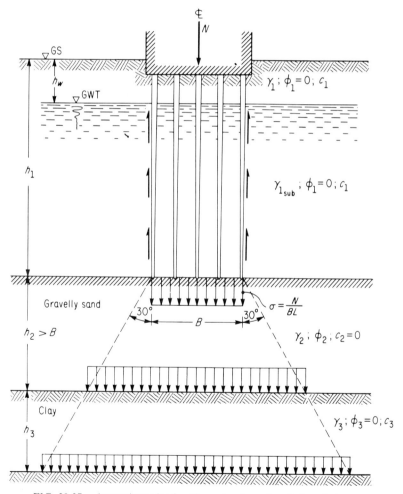

FIG. 20-27. Approximate load-settlement system of a pile foundation.

Settlement analyses of floating foundations on friction piles is not so simple, and there is in fact no exact method of analysis of such settlement. In practice, therefore, some engineers assume a uniform distribution of pile loads on an imaginary, assumed horizontal plane passed through the tips of the piles and perform settlement analysis, based on consolidation test results, of the expected, approximative settlement that for an ordinary foundation (unpiled) would occur if the foundation were a mat of the same depth and dimensions at the assumed plane. In such a case the method of settlement analysis of pile-supported foundations may be the same as that used for unpiled foundations (Fig. 20-27).

Figure 20-28 illustrates an approximate load-settlement system of a pile group of a floating-pile foundation. Note that in this system the seat of settlement pertains to a compressible layer of soil $H = (1.5)B$ units thick below the elevation of the tips of the piles, and encompased by the $(0.2)\sigma_o$-isobar.

Based on consolidation test results of the soil in question the settlement s at 100 percent consolidation is then calculated for a layer of compressible soil H units thick as

$$s = \frac{e_i - e_f}{1 + e_i} H \qquad (20\text{-}59)$$

where e_i = initial void ratio of the soil tested
e_f = final void ratio of the soil tested (see Ref. 9, pp. 353-450, 539-543).

20-15. SUMMARY ON SETTLEMENT OF A PILE GROUP

1. The settlement analysis of a pile foundation may be performed along the same lines as that performed for soils under loaded bearing areas (unpiled foundations) resting directly on soil.
2. The settlement of a pile group depends mainly upon the compressibility of the soil below the tips of the piles.
3. It is assumed that the soil between the piles in the pile group does not deform.
4. The stress distribution in soil below the tips of the piles is assumed to take place either according to Boussinesq, or according to the usual straight line, whichever method is used.
5. The depth of the stressed zone $[H \approx (1.5)B]$ below the plane of the tips is assumed to be the same as that used in settlement analysis of foundations without piles.
6. The settlement of a pile group is larger than that of a single pile (each pile in the group and the single pile loaded with the same load).

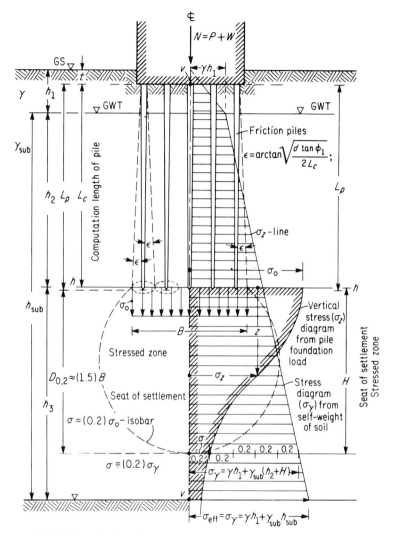

FIG. 20-28. Load-settlement system of a pile group of a floating pile foundation.

7. With equal loads on piles, settlement of a pile group increases with an increase in the number of piles in the group.
8. With an increase of spacing between piles in a pile group its settlement decreases, approaching the settlement of a single pile. With a spacing of about $6d$ the settlement of a pile group is approximately equal to the settlement of a single pile.

Besides settlement, the soil-foundation system must also be checked for its strength in soil bearing capacity. If the shear strength of the soil should become exhausted, a groundbreak in the soil would set in and might cause the piled structure to collapse. The stability analyses of soil-pile foundation systems are performed by the same methods as used for unpiled foundations.[9-11]

20-16. ELASTIC DEFORMATIONS OF PILES

The longitudinal elastic deformation ΔL (shortening) of an axially loaded end-bearing pile is calculated by Hooke's law as

$$\Delta L = \frac{PL}{AE} = \sigma \frac{L}{E} = \frac{1}{AE} PL_{\text{eff}} \tag{20-60}$$

where P = load on pile
$L = L_{\text{eff}}$ = here actual and effective length of pile
A = cross-sectional area of pile
E = Young's modulus of elasticity of pile material
$\sigma = P/A$ = compressive stress in pile

The elastic deformation of a friction pile loaded with an axial load P is calculated by the following considerations [12] (Fig. 20-29).

In a pier construction, a pile may be considered as being connected to the pier by means of a frictionless hinge H_u transmitting its load to the soil through a frictionless hinge H located at some point on the longitudinal axis of the pile. It is assumed that the frictional (mantle) resistance τ_z per linear foot of pile increases linearly from zero at the top of the pile to a maximum value $\tau_L = \tau_{\max}$ at the tip of the pile (Fig. 20-29):

$$\tau_z = \frac{z}{L} \tau_L \tag{20-61}$$

in which z = depth coordinate from base of pier to point where frictional resistance is τ_z
L = actual length of pile

Further, at equilibrium, the total frictional resistance $\left(\frac{0+\tau_L}{2} L\right)$ must be equal to the load P applied to the pile at its top:

$$\frac{\tau_L}{2} L = P \tag{20-62}$$

Then, by analogy to Hooke's law, and using Eqs. 20-61 and 20-62, the elastic deformation ΔL is calculated as

Pile Foundations

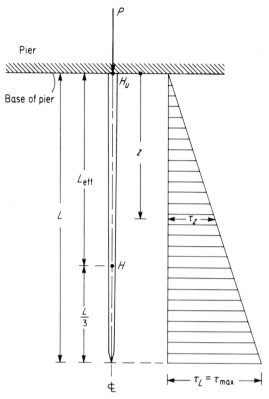

FIG. 20-29. Linear variation of frictional resistance along a friction pile.

$$\Delta L = \frac{1}{AE} \int_0^L \left(P - \frac{\tau_z}{2} z \right) dz = \frac{1}{AE} \int_0^L \left(P - \frac{z^2}{L^2} \right) dz$$
$$= \frac{1}{AE} P \left(L - \frac{L}{3} \right) \qquad (20\text{-}63)$$

or

$$\Delta L = (2/3) \frac{1}{AE} PL = \frac{1}{AE} P \left(\frac{2}{3} L \right) = \frac{1}{AE} P L_{\text{eff}} \qquad (20\text{-}64)$$

Therefore the elastic deformation (shortening) of a friction pile is thus the same as that of a pile of two-thirds length, driven to firm ground, and its effective length L_{eff} is two-thirds of the actual length L.

For a uniform, constant variation of frictional resistance along a friction pile, the shortening is $\Delta L = \frac{1}{AE} P \left(\frac{1}{2} L \right)$ and that for a parabolic distribution of frictional resistance is $\Delta L = \frac{1}{AE} P \left(\frac{3}{4} L \right)$.

20-17. HORIZONTAL DISPLACEMENT OF PILE BENT TOPS

In the foregoing discussions it was not considered that the top of a pile bent whose piles are axially loaded may undergo a horizontal displacement. Such a displacement Δh may be computed as follows (Figs. 20-30 and 20-31).

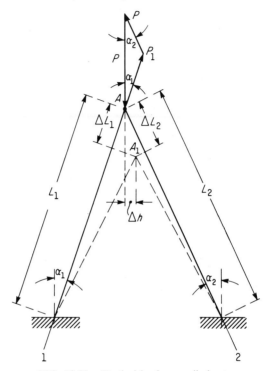

FIG. 20-30. Vertical load on a pile bent.

Let a vertical load P act on top of the pile bent (at point A). The axial loads P_1 and P_2 on piles 1 and 2, respectively, are

$$P_1 = P \frac{\sin \alpha_2}{\sin (\alpha_1 + \alpha_2)} \tag{20-65}$$

$$P_2 = P \frac{\sin \alpha_1}{\sin (\alpha_1 + \alpha_2)} \tag{20-66}$$

The corresponding linear deformations of the piles from loads P_1 and P_2 are, by Hooke's law,

$$\Delta L_1 = \frac{P_1 L_1}{A_1 E} = \frac{P}{E} \frac{\sin \alpha_2}{\sin (\alpha_1 + \alpha_2)} \frac{L_1}{A_1} \tag{20-67}$$

Pile Foundations

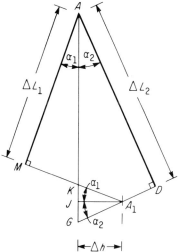

FIG. 20-31. Horizontal displacement of pile bent tops.

$$\Delta L_2 = \frac{P_2 L_2}{A_2 E} = \frac{P}{E} \frac{\sin \alpha_1}{\sin (\alpha_1 + \alpha_2)} \frac{L_2}{A_2} \qquad (20\text{-}68)$$

Because of these deformations the tops of the piles displace from their initial position at point A to a new position, say point A_1.

The horizontal distance dH between points A and A_1 is shown in Fig. 20-31 as $JA_1 = \Delta h$. It is calculated according to Jacoby[13] as follows:

$$KG = AG - AK = \frac{\Delta L_2}{\cos \alpha_2} - \frac{\Delta L_1}{\cos \alpha_1} \qquad (20\text{-}69)$$

$$KA_1 = (KG)\frac{\cos \alpha_2}{\sin (\alpha_1 + \alpha_2)} = \left(\frac{\Delta L_2}{\cos \alpha_2} - \frac{\Delta L_1}{\cos \alpha_1}\right)\frac{\cos \alpha_2}{\sin (\alpha_1 + \alpha_2)} \qquad (20\text{-}70)$$

$$JA_1 = \Delta h = (KA_1) \cos \alpha_1 = \left(\frac{\Delta L_2}{\cos \alpha_2} - \frac{\Delta L_1}{\cos \alpha_1}\right)\frac{\cos \alpha_1 \cos \alpha_2}{\sin (\alpha_1 + \alpha_2)} \qquad (20\text{-}71)$$

Substituting ΔL_2 and ΔL_1 expressions from Eqs. 20-67 and 20-68, respectively, into Eq. 20-71, obtain horizontal displacement Δh as

$$\Delta h = \frac{\cos \alpha_1 \cos \alpha_2}{\sin^2 (\alpha_1 + \alpha_2)} \frac{P}{E} \left(\frac{\sin \alpha_1}{A_2 \cos \alpha_2}L_2 - \frac{\sin \alpha_2}{A_1 \cos \alpha_1}L_1\right) \qquad (20\text{-}72)$$

This computation is of academic value only. From the practical viewpoint, however, such computation for horizontal displacement of the tops of a pile bent not only complicates the general pile calculations, but adds to it an unnecessary precision. Such precision is not useful because of imprecise assumptions underlying the pile theory and those by which pile loads are calculated, and because the final position of the driven piles almost never coincides with their positions on the blueprint.

20-18. PRECISE CALCULATIONS OF PILE GROUPS

Precise calculations, for most of these methods, are based on the theory of elasticity, and pile loads in the pile group are determined by means of axial, elastic deformations of the piles (Hooke's law). This method was put forward by Gullander (Göteborg) in 1902.[17] His theory was extended by Jacoby (Riga) in 1909,[18] Ostenfeld (Copenhagen) in 1921,[19] Nökkentved (Copenhagen) in 1924,[20] and Wünsch (Stuttgart) in 1927,[21] dealing with piles fixed at their heads and/or tips.

Nökkentved's work is clear and easy to follow, hence it has been acclaimed in practice since its publication. The assumptions underlying Nökkentved's theory are:
1. Axial deformations of a pile increase linearly with load, and so do its deflection (bending) and the rotational twist of the pile head.
2. Elastic deformations of the superstructure (a massive pile footing, for example) are considered negligible, hence are omitted from consideration.

Other authors contributing methods to precise pile foundation calculations are: Westergaard (United States) in 1917,[22] Krey (Germany) in 1932,[23] Labutins (Riga, Latvia) in 1933,[24] Agatz (Germany) in 1936,[25] Vetter (United States) in 1939,[26] and others.

20-19. OFFSHORE TOWERS

Offshore towers are now being constructed to replace outdated lightships near principal seaports along the Atlantic coast. These structures are four-legged steel towers fixed to the ocean floor by piles driven through a prefabricated trussed template set in position on the ocean bottom and standing in 40 to 80 ft of water. Once this template is firmly anchored, the superstructure of shop-fabricated sections is erected on it.

Possible wave heights of 40 to 50 ft, and smaller waves that set up periodic motion against more than one leg simultaneously to cause fatigue, are design considerations of such towers.[14] Hurricane winds along the Atlantic coast may exceed a velocity of 125 mph, equivalent to a pressure of 50 lb/ft^2 against a flat surface. A schematic representation of wind and wave actions and reactions is shown according to Ref. 14 in Fig. 20-32.

Pile Foundations

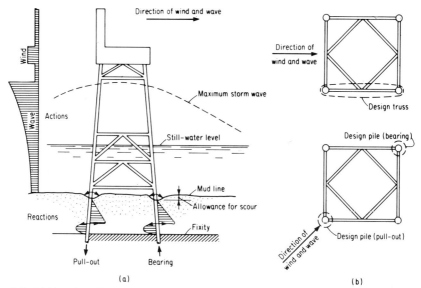

FIG. 20-32. An offshore tower (Ref. 14). (a) External actions from storm waves and reactions of the structure. (b) Assumed maximum load conditions for jacket and pile design.

The specifications for the construction of the Chesapeake Light Station, an offshore tower, required that the jacket legs penetrate a minimum depth of 12 ft into the ocean bottom. Then a pipe pile 33 in. in diameter was driven through each of the four jacket sleeves, which had a diameter of 39 in. The piles were driven open-end to a tip elevation of 228 ft below low water. The soil penetrated was largely sand, sand-clay and clay.[15]

Texas Tower Foundations

These towers, resembling oil-well platforms constructed in the Gulf of Mexico, are equilateral triangular radar platforms 180 ft on a side to accommodate early aircraft warning systems. An idea of the three-leg tower-supporting foundations is shown in Fig. 20-33. The permanent leg caissons of the tower are steel cylinders 10 ft in diameter and approximately 185 ft long. The lower end of the cylindrical caisson is flared out to form a cutting edge 15 ft in diameter. An outside steel shell encases the lower 63 ft of each caisson.[16]

20-20. FOUNDATIONS OF TALL STRUCTURES

A foundation of a chimney, tower, mast, or any other tall structure is a structural element by which the superimposed vertical load ΣV and wind moment $M = Wy$ are safely transmitted to the soil (Fig. 20-34).

686   **Deep Foundations**

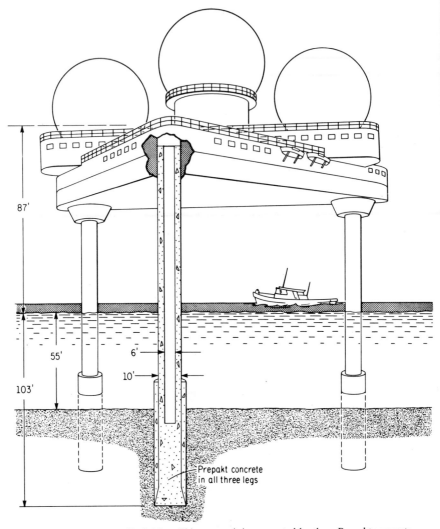

FIG. 20-33. Texas tower (Ref. 16). This type unit is supported by three Prepakt concrete reinforced legs anchored into sand.

The chimney foundation is thus subjected (1) to its self-weight, ΣG, and (2) wind load W (to be considered as acting in any direction) and the moment it forms.

The magnitude of the wind force depends upon the velocity of the wind, upon the shape, plan form, and position of the structure hit by the wind.

Tall structures, in general, must be designed with ample degree of stability to resist the maximum wind force on record for the particular construction site under consideration.

Pile Foundations

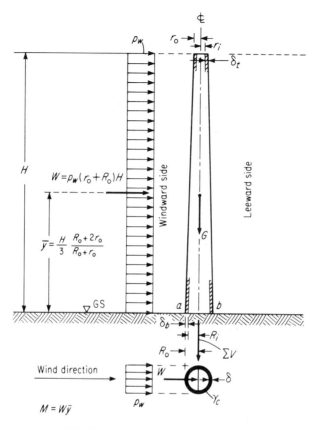

FIG. 20-34. Wind pressure on a chimney.

The wind pressure p_w in pounds per square foot on a vertical plane is

$$p_w = (0.003)v^2 \quad (\text{lb/ft}^2) \tag{20-73}$$

where v = wind velocity in miles per hour.[27]

The effect of wind on a structure is determined by the combined action of direct pressure and suction or wind drag parallel to the wind on the leeward side of the structure, the effect of the combined action of which is included in Eq. 20-73.

Designers normally assume not less than 75 mph, and with a 100 mph wind in storm-swept areas, these figures are considered conservative.

The wind pressure is assumed to act horizontally and is calculated on the entire area of the vertical projection of the structure. The wind moment must be calculated with respect to the base contact plane with soil (or piles) of the footing. It is recommended that wind-tunnel tests should be performed for open towers, frames, and structures which are large or of unusual design. Uplift of structures by wind must also be taken into account where applicable.

With reference to Fig. 20-34, the weight G of the tapered part of the chimney ($= $ normal load N) above section a-b may be calculated as

$$G = N = \gamma_c \pi \frac{H}{3}[r_o^2 + r_o R_o + R_o^2 - (r_i^2 + r_i R_i + R_i^2)] \qquad (20\text{-}74)$$

The footing of the chimney foundation must be so designed that no uplift on the windward side occurs. Also, the foundation must be laid deep enough to avoid the influence of precipitation water, frost and drought. Without an accurate knowledge of the soil conditions no trustworthy foundation design is possible.

The stability of a chimney foundation and the check of soil contact pressure should be calculated for the chimney's (1) maximum self-weight, (2) maximum self-weight and maximum wind moment, and (3) minimum self-weight (no lining while under construction) and maximum wind moment.

The factor of safety η against overturning of the chimney about plane a-b is

$$\eta = \frac{GR_o}{M} = \frac{NR_o}{Ne} = \frac{R_o}{e} \qquad (20\text{-}75)$$

For the plane of the base of the footing the factor of safety is usually required to be $\eta \geqslant 2.0$ for wind from any direction. Of course, the local building code must be consulted as to the required factor of safety.

In Eq. 20-75 e is the eccentricity of the resultant normal load N.

Prudent design requires that in all of the above three calculation requirements the pressure on soil should not exceed the allowable bearing capacity of the soil, and there should always be a compression of soil under the entire contact area of the base of the footing subjected to eccentric loading. This applies equally to all footing slabs.

Tall chimneys should have an adequate foundation to avoid intolerable settlement and leaning or collapse under overturning moment caused by wind and its vibratory racking (a compaction effect on the soil on the leeward side of the tall structure such as a chimney, for example). Therefore they are to be built on firm and frost-free soil or rock. In static and strength calculations and design the weight of the overburden soil over the footing should not be taken into consideration.

The economic form of a footing of a tall, circular structure (chimney) is a slab circular, or hexagonal, or octagonal in plan offering equal resistance to overturning in all directions.

Pile Foundations

For preliminary design, the diameter D of a plain concrete footing is empirically calculated as

$$D = \frac{H}{10} + d_i \quad \text{(ft)} \tag{20-76}$$

where H = height of chimney
d_i = inside diameter at top of chimney

The uniform thickness δ of the plain-concrete footing is taken empirically in preliminary design as

$$\delta = \frac{H}{25} \quad \text{(ft)} \tag{20-77}$$

The footing of the chimney is normally laid on a concrete cleanliness layer 4-6 in. thick.

The shear stresses in the foundation should be very carefully determined, particularly at the periphery of the chimney shaft where it joins the footing.

In earthquake-prone regions the tall structures like chimneys and towers should also be designed to resist seismic forces (earthquake shocks).

If the chimney on its circular footing is to be founded on piles, then the pile arrangement is determined as follows. One determines the necessary number of piles to use, assumes a certain number of concentric circles in the circular footing, and distributes the piles along these circles by taking into account the minimum spacing of piles (see Secs. 20-2 to 20-4, and 20-11 on eccentric loading of a vertical pile group). Now remember that for wind moment from any direction, the soil pressure-distribution diagram, figuratively speaking, "rotates about the vertical axis" of the chimney. Hence the pile spacing must take care of the heavy loaded piles on the leeward side in any direction. Now, assuming an equal spacing of piles in each ring, the necessary pile arrangement results. The number of piles to use depends, of course, upon the magnitudes of the normal load N and the magnitude of the moment, as well as upon the allowable bearing capacity of the pile. The load on each pile is then calculated by the method of moment of inertia of the pile foundation (Sec. 12-11). The routine of calculating loads on piles in a circular pile system is described in the following example.

Example

Given a small, circular pile system of $n = 12$ timber piles arranged as shown in Fig. 20-35. The pile system is loaded with a total vertical load of $\Sigma V = 120$ tons $= 240$ kips. The wind load on the superstructure is given as 15 kips, bringing about a static moment at the elevation of the pile heads of $\Sigma M = 562.75$ kip-ft (Fig. 20-36).

Assume that for the given condition (fine sand and small group of piles) the safe end-bearing value of timber piles to use is 25 tons = 50 kips (see Table 19-2). Assume further that the allowable load in tension is given as 10 kips per pile. The allowable shear force per pile in this example is given as 2 kips. Determine the magnitude and nature of pile loads.

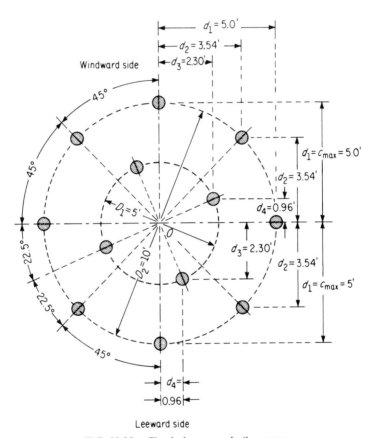

FIG. 20-35. Circularly arranged pile system.

Solution. With reference to Eq. 20-39 (eccentricity about one axis), we determine the section modulus Z of the pile groups as follows.

Minimum Moment of Inertia I_{min} of Pile Group:

$$I_{min} = 2(d_1^2) + 4(d_2^2) + 2(d_3^2) + 2(d_4^2)$$
$$= 2(5.0^2) + 4(3.54^2) + 2(2.30^2) + 2(0.96^2) = \underline{112.54 \ (ft^4)}$$

Section Modulus Z of Pile Group:

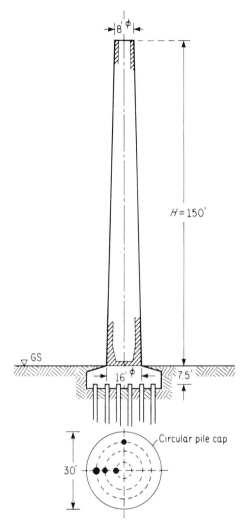

FIG. 20-36. Smoke stack on piles.

The maximum spacing c_{max} of the outermost pile from center O of the pile system is

$$c_{max} = \frac{D_2}{2} = \frac{10.0}{2} = 5.0 \text{ (ft)}$$

and

$$Z = \frac{I_{min}}{c_{max}} = \frac{112.54}{5.0} = \underline{22.51 \text{ (ft}^3)}$$

Magnitude of Load P on Any Pile in the Group:

$$P = \frac{\Sigma V}{n} \pm \frac{\Sigma M}{Z} = \frac{240}{12} \pm \frac{562.75}{22.51} = \underline{20.0 \pm 25.0} \text{ (kips)}$$

where $n = 12$ is the number of piles in the pile group.

Maximum Compression Load P_{max} in the Outer Piles on Leeward Side:

$$P_{max} = 20.00 + 25.00 = \underline{45.00 < 50 \text{ (kips per pile)}} \qquad \text{(C)}$$

Minimum Load P_{min} per Pile on the Windward Side Diagonally Opposite

$$P_{min} = 20.00 - 25.00 = \underline{-5.00} \text{ (kips)} < 10.0 \text{ kips} \quad \text{(T)}$$

Horizontal Shear Force F_H at Pile Heads per Pile:

$$F_H = H/n = 15/12 = \underline{1.25} \text{ (kips)} < 2.0 \text{ kips}$$

REFERENCES

1. H. Press, "Druckverteilung am Einzelpfahl und Einfluss benachbarter Pfähle," *Die Bautechnik*, No. 4 (1941), p. 45.
2. H. W. Bolin, "The Efficiency Formula of the Uniform Building Code," *Building Standards Monthly*, Vol. 10, No. 1 (January 1941).
3. R. W. Abbett (ed.), *American Civil Engineering Practice*. New York: Wiley, 1956, Vol. 1, Sec. 8, pp. 8-28.
4. J. F. Seiler and W. D. Keeney, "The Efficiency of Piles in Groups," *Wood Preserving News*, Vol. 22, No. 11 (November 1944), pp. 109-118.
5. F. M. Masters, "Timber Friction Pile Foundations." *Trans. ASCE*, Vol. 108, Paper 2174 (1943), pp. 115-173.
6. R. D. Chellis, *Pile Foundations*. New York: McGraw-Hill, 1951, pp. 135-143.
7. A. Bierbaumer, "Vorschläge für die Beurteilung von Flach- und Pfahlgründungen," *Oesterreichischer Ingenieur und Architekten Verein*, Vienna, 1929.
8. K. Terzaghi and O. K. Fröhlich, *Theorie der Setzung von Tonschichten*. Leipzig and Vienna: Franz Deuticke, 1936.
9. A. R. Jumikis, *Soil Mechanics*. Princeton, N.J.: Van Nostrand, Inc., 1962, pp. 353-450.
10. A. R. Jumikis, *Mechanics of Soils: Fundamentals for Advanced Study*. Princeton, N.J.: Van Nostrand, 1964, pp. 143-215.
11. A. R. Jumikis, *Stability Analyses of Soil-Foundation Systems: A Design Manual*. Engineering Research Publication No. 44, New Brunswick, N.J.: Rutgers University, College of Engineering, Bureau of Engineering Research, 1965.
12. C. P. Vetter, "Design of Pile Foundations," *Trans. ASCE*, Paper No. 2031, Vol. 104 (1939), pp. 758-778.
13. E. Jacoby, *Design of Quays* (in Russian). St. Petersburg: Ministry of Ways and Communications, 1916, pp. 86-87.

Pile Foundations

14. J. V. Rufin, "Steel Offshore Towers Replace Lightships." *Civil Engineering*, Vol. 35, No. 11 (November 1965), pp. 72-75.
15. J. W. Fowler, "Construction of the Chesapeake Light Station," *Civil Engineering*, Vol. 35, No. 11 (November, 1965), pp. 76-77.
16. G. F. A. Fletcher, "Heavy Construction Goes to Sea," *Civil Engineering* (January 1956), pp. 59-64.
17. P. Gullander, "Theorie der Pfahlgründungen," *Die Bautechnik*, 1928, p. 818.
18. E. Jacoby, *Jahrbuch der Gesellschaft für Bauingenieurwesen*. Berlin: VDI, 1925, p. 42.
19. A. Ostenfeld, *Teknisk Tidskrift* (Copenhagen), No. 1 (1921).
20. C. Nökkentved, *Beregning of Paelevaerker*. Copenhagen, (1924).
21. H. Wünsch, *Statische Berechnung der Pfahlsysteme*. Stuttgart: Wittwer, (1927).
22. H. M. Westergaard, "The Resistance of a Group of Piles" (Method of Center of Rotation), *Journal of the Western Society of Engineers*, Vol. 22, No. 10 (December 1917), pp. 704-713.
23. H. D. Krey, *Erddruck, Erdwiderstand und Tragfähigkeit des Baugrundes*. Berlin: Ernst, 1932, p. 162.
24. A. Labutins, "Die graphische Berechnung von Pfahlrosten für Kaimauern," *Acta Universitatis Latviensis*, Vol. 1, No. 7 (1933), pp. 333-368.
25. A. Agatz and E. Schultze, *Der Kampf des Ingenieurs gegen Erde und Wasser im Grundbau*. Berlin: Springer, 1936.
26. C. P. Vetter, "Design of Pile Foundations," *Trans. ASCE*, Vol. 104 (1939), p. 758.
27. F. D. C. Henry, *The Design and Construction of Engineering Foundations*. New York: McGraw-Hill, 1956, p. 414.

QUESTIONS

20-1. What is the purpose of static investigation of pile groups or pile systems? (The determination of the most favorable position and inclination of piles in a pile group with reference to the pile loads for the most unfavorable, favorable, and average cases of loading. Also, to predict probable settlement of the pile group foundation).

20-2. What is an advantageous position of a pile?
(One in which all piles are loaded at least approximately equally when not too large variations in loads result from the various loading conditions.)

20-3. Refer to the accompanying figure.
(a) What kinds of piles does the sketched pile foundation represent? (End-bearing and friction piles.)

(b) Is this foundation laid correctly or defectively, and why? (Defectively.)
(c) What can generally be said about the settlement of this structure? (Large differential settlement.)
(d) What is the effect of an intolerable differential settlement on the mat (or footing) and the columns of the structure? (Cracking and tilting.)

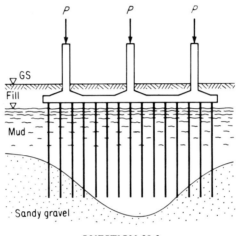

QUESTION 20-3

20-4. Analyze the nine sketches shown, and report whether the use of piles is justified by the conditions shown. Pile spacing, 3.5 diameters. Diameter of piles, 12 in.

20-5. What is understood by the term "moment of inertia of a footing"? How is this term used in foundation design, and why? Illustrate. (*Ans.*: The term comes into use whenever the footing is subjected to a moment. Relative to piles, the concept of the moment of inertia is applied in order to find maximum and minimum loads on piles.)

PROBLEMS

20-1. The distribution of contact pressure on soil under an eccentrically loaded wall foundation is represented by a soil pressure diagram as sketched. Prob. 20-1. Determine eccentricity e, the magnitude of the vertical load N and its position, and pile spacing for three rows of piles. All piles are to be equally loaded. Also, determine final width of footing with regard to pile spacing, if necessary.

Ans. $e = 0.64$ (ft). $N = 18.9$ (tons).

Determine pile diameter and check minimum spacing of piles at edges of footing, recheck width of footing, and redesign if necessary.

Pile Foundations

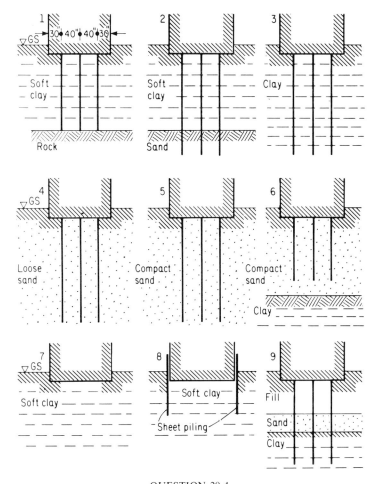

QUESTION 20-4

20-2. A long strip footing, 6 ft wide, is loaded eccentrically with a normal, concentrated load of $N = 6$ ton/ft of run of footing. The eccentricity of N is 12 in. to the right of the center line of the footing. The ultimate bearing capacity of the soil is 0.5 ton/ft.^{2}
  (a) Determine graphically and analytically edge pressures on soil.
  (b) Are edge pressures safe?
  (c) Use timber piles (end-support) to support the structure. Spread piles so that each pile is loaded with the same load. Select diameter, spacing, and number of rows of piles.
     Determine a repetitional panel of footing. All answers to this problem should refer to the repetitional panel of the wall footing.

20-3. A repetitional section of a bulkhead anchorage is subjected to a horizontal anchor pull of 10 tons as illustrated. A pair of batter piles sloping in opposite

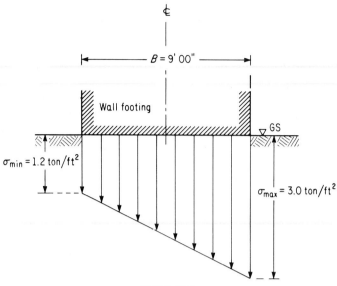

PROB. 20-1

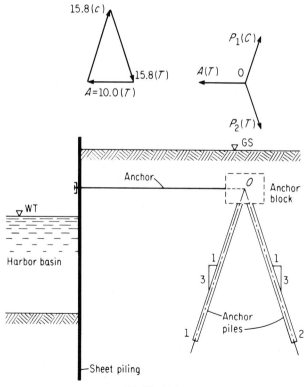

PROB. 20-3

Pile Foundations

directions is to be used to resist the pull. The axes of the piles and the anchor meet at the same point O. Both piles are 50 ft long.
(a) Draw to scale a free-body diagram of the anchorage.
(b) Determine graphically the nature and magnitude of axial forces in each of the piles.
(c) Determine pile material to be used and the cross-sectional size of the piles.
(d) Report whether or not pile 2 should be driven vertically, and why.

Ans. For (a) see Prob. 20-3. (b) $P_1 = 15.8$ tons (C), $P_2 = 15.8$ tons (T). (d) Discuss bending of pile; discuss the horizontal component of pile 2.

20-4. Distribute pile loads shown in the accompanying figure.

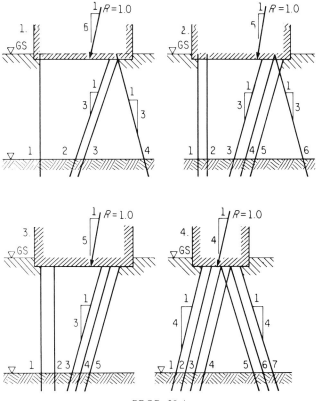

PROB. 20-4

20-5. Given a circular, truncated cone industrial smokestack as shown in Fig. 20-36. The outside top diameter of the smokestack is $d_o = 8.0$ ft in diameter. The outside diameter D_o at the ground surface is 16 ft. The weight of the smokestack is given as $G = 520$ tons, and rests on a reinforced circular concrete footing supported by timber piles. The maximum safe load per pile is given as 25 tons

per pile. The smokestack is subjected to a horizontal wind load at a pressure of $p_w = 25.0$ lb/ft² on the projected vertical cross-sectional area of the smokestack. Design the pile foundation.
1. Compute the forces acting at the elevation of the pile heads.
2. Calculate the number of piles necessary, and arrange them in a symmetrical arrangement in concentric circles in the circular footing.
3. Determine pile loads on the leeward and windward sides by the method of moment of inertia of the pile foundation system.
4. If necessary, redesign the pile foundation in order to avoid uplift of piles.
5. Check the horizontal shear at pile heads.

part V

SPECIAL TOPICS

chapter **21**

Scour

21-1. BRIDGE FOUNDATIONS

Bridge piers and abutments are generally built as monoliths of concrete or reinforced concrete to support the superstructure and transmit the loads to the soil or rock. The foundations of bridge piers and abutments may be footings laid on rock, spread footings at safe depth against scour (where it applies); they may be supported on acceptably firm soil, or they may be founded on piles, or on open or pneumatic caissons.

Some types of massive bridge foundations are shown in Figs. 21-1 and 21-2.

21-2. LOADS ON BRIDGE PIERS AND ABUTMENTS

The various loads of importance acting on bridge piers and abutments are dead loads, live loads, pressure from flowing water, wind, ice, lateral earth pressure, seismic forces, buoyant forces, friction, and reactions.

Pressure P_w on bridge piers exerted by flowing water may be calculated by Greiner's formulas:[1]

$$P_w = (1.5)v^2 \quad \text{for flat surfaces} \qquad (21\text{-}1)$$

$$P_w = (0.75)v^2 \quad \text{for rounded surfaces} \qquad (21\text{-}2)$$

where P_w = pressure exerted by flowing water on each square foot of vertical projection of upstream surface of pier, lb/ft²
 v = velocity of flow, ft/sec

Wind loads on high piers are estimated at 30 lb/ft² of exposed vertical surface. Wind loads on low piers may usually be disregarded.

Footings or piers and abutments exposed to the pressure of ice formed on water in direct contact with the concrete must be designed to sustain a force of at least 50 kips per linear foot, that being the crushing strength of block ice 1 ft thick.[2] If the exposed face slopes, the force may be reduced by half. The value of 25 kips per linear foot is usually used in the design of dams with sloping faces.

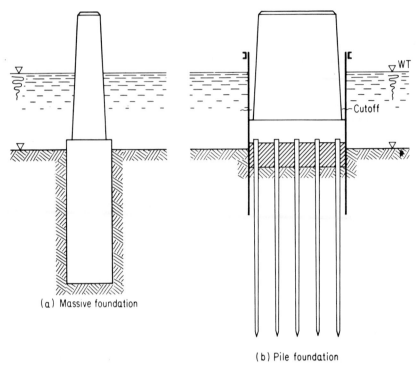

(a) Massive foundation

(b) Pile foundation

FIG. 21-1. Bridge piers.

Several methods of laying deep foundations, among them bridge foundations, have already been discussed in Parts I through IV. Hence they are not repeated here. However, a few words will be said here about scour under bridge foundations.

21-3. SCOURING OF RIVERBEDS

After choosing the site of a bridge one should resort to bridge hydraulics in order to ascertain whether there is a sufficient waterway (flow channel) to carry away floodwater and thus insure the safety of the bridge foundation before proceeding with the laying itself. Bridge piers and flow area (viz., cross-sectional area of the flow channel) obstruct river flow. Constriction of the cross section of the flow channel of a river by bridge piers and the swelling of water in the stream during high velocity flow and flood periods are factors bringing about undermining of bridge piers by scour if the foundations are not laid to a proper depth.

Scouring is especially intense with increase in stream velocity where the flow channel is narrow or where it is confined between high river banks.

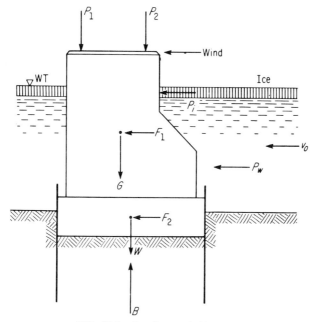

FIG. 21-2. Loads on a bridge pier.

Reduced flow area and a high flow stage both increase flow velocity, imparting to the water energy to erode, lift, wash out and transport away suspended soil particles from the immediate vicinity of bridge piers (Figs. 21-3 and 21-4), thus producing deep scours. Washing out or undermining of piers is a common cause of bridge failures. Under these conditions, if the riverbed is fine sand or alluvial silt, scouring may attain colossal depths. For example, an increase in depth of 40 ft was reported alongside the sand-island cofferdams in which caisson foundations of the Mississippi River Bridge at Baton Rouge, Louisiana, were built.[3]

Pile foundations and fairly deep footings in gravel beds are by no means immune to failure by scour.

The extent and intensity of scour depends upon the type of soil in the

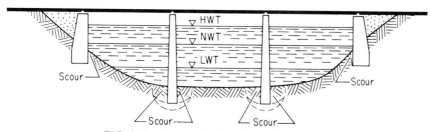

FIG. 21-3. Scour at bridge piers and abutments.

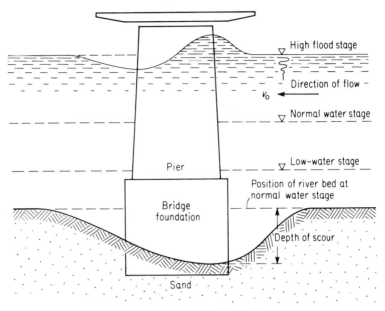

FIG. 21-4. Scoured river bed.

river bed, flow velocity, backwater caused by piers, shape of pier, constriction of river bed, angle of attack of flowing water striking the pier, and other factors.[4] Figure 21-5 illustrates the water regimen of a swollen river at bridge piers.

FIG. 21-5. Water regimen of a swollen river at bridge piers. (Photo by author.)

21-4. PROTECTION OF FOUNDATIONS AGAINST SCOUR

The necessary depth of embedment of a bridge pier foundation to cope with scour is an engineering problem of paramount importance. Unfortunately this depth is extremely difficult to ascertain. Avoidance of scour may be attained by carrying the pier below the line of scour, or by carrying the pier foundation down to rock (if attainable and economically feasible), or by using sheet piling around the base of the footing (Fig. 21-2).

The foundation may also be protected by means of rock-filled riprap around the pier (Fig. 21-6). Because of the relatively large size and the weight of the rock fragments in the riprap, it protects the foundation against washing away of the sand and thus prevents undermining and exposure of the foundations.

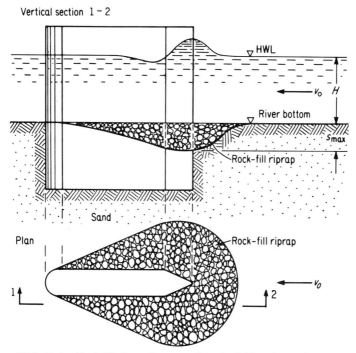

FIG. 21-6. Rock-fill riprap for protection of a bridge pier against scour.

If large rocks are difficult to procure, scour in sand may be checked by placing under the foundation fascine mats at proper depth of laying foundations, and by weighting the mats down with a belt of pitching stones around the foundation. In any case, for reasons of safety against scour, the bases of foundations of bridge piers must be laid deep enough—say several meters—below the expected depth of scour. The policy of the Atchison, Topeka and Santa Fe Railway of keeping bridge footings at least 25 ft below the stream bed is worth recording as the safety record of the company's bridges is good.[5]

21-5. FLOW VELOCITIES FOR SCOUR

Up to now there is no real general formula available to fit all scouring conditions.

The *American Civil Engineers' Handbook*[6] contains a formula for estimating roughly the probable scour:

$$v_s = md^{0.64} \qquad (21\text{-}3)$$

where v_s = critical mean velocity, in ft/sec, for a cross section of the flow channel under consideration

m = 0.82 for fine, light, sandy silt
     0.90 for coarse, light, sandy silt
     0.99 for sandy loam
     1.07 for a rather coarse silt, such as debris of hard soils

d = depth of water

With this information and a table (Table 21-1) of limiting velocities, some idea may be formed of the probable scour when a tentative velocity is ascertained.

TABLE 21-1
Scouring Bottom Velocities

Soil Material	ft/sec	m/sec
Soft soil	0.25	0.076
Clay and silty clay of medium density	0.50-2.95	0.15-0.90
Silt	0.40	0.12
Fine sand	1.00	0.30
Sand	1.00	0.30
Coarse sand	2.00-2.65	0.60-0.80
Gravel	2.00	0.60
Sea pebbles (1.06 in. = 2.71 cm dia.)	2.20	0.67
Slate (9.06 in.^{3} = 15.85 cm^3)	2.75-3.00	0.84-0.92
Broken stone	4.00	1.22

The bottom velocity v_b of flow will be about three fourths of the main velocity, or

$$v_b = (0.75)md^{0.64} \qquad (21\text{-}4)$$

Some values of the bottom velocities in meters per second for various soil materials for scouring of riverbeds are given in Table 21-1.

Scour

Based on experimental results, Mavis and Laushey[7] proposed a formula for "competent" velocity (material in a stream about to move with the current):

$$v_b = (0.50)d^{4/9}(s-1)^{1/2} \quad (21\text{-}5)$$

where v_b = competent velocity, ft/sec
d = diameter of soil particles, mm
s = specific gravity of soil material

In the words of these authors, this formula may "only suggest an answer."

Laursen and Toch (1956) in their report (see Ref. 4) present charts for estimating the depth of scour caused by bridge piers in natural streams with a continuous supply of sediment from upstream.

According to Ref. 7, the safe velocity v in ft/sec over different riverbed materials is compiled in Table 21-2.

TABLE 21-2
Safe Velocity over Riverbed Materials

Material	Velocity, ft/sec
Soft alluvial deposits	0.42
Clay	0.67
Sand and silt	1.00
Gravelly soil	2.00
Strong gravelly shingle	3.00
Shingle	4.00
Shingle and rock	5.00

Shingle is, to speak loosely and commonly, any beach gravel which is coarser than ordinary gravel, especially if consisting of flat or flattish pebbles and cobbles.

21-6. DEPTH OF SCOUR

According to Boldakov and Andreyev,[8] based on laboratory experiments, the maximum depth of scour s_{max} in meters is calculated as

$$s_{max} = C_1 C_2 C_3 v_0^2 - 30d \quad (21\text{-}6)$$

where C_1 = nose form coefficient of bridge pier (Fig. 21-7a)
C_2 = coefficient depending upon design width B_1 of pier and velocity of flow v_0, to be determined from graph in Fig. 21-8
C_3 = a coefficient depending upon depth H of approaching flow (outside the scour funnel) and design width B_1, to be determined from graph in Fig. 21-9

v_o = average stream-flow velocity over depth of water H before the pier, m/sec

d = diameter of soil particles, in consistent units of measurement

1. The coefficients C_1 as a function of angle of attack α and the ratio f/H are for the pier form in Fig. 21-7a (see Table 21-3):

TABLE 21-3
C_1

$\alpha°$ \ f/H	0	0.2	0.4	0.6	0.8	1.0
0	0.87	1.01	1.17	1.23		
10	0.89	1.03	1.18	1.23	1.27	
20	0.92	1.05	1.19	1.25		
30	1.08	1.15	1.23	1.27		
40	1.15	1.22				

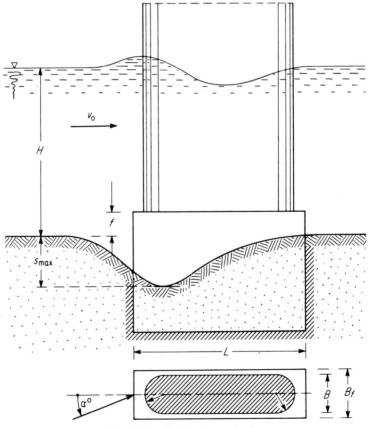

FIG. 21-7a. Bridge pier on footing.

Scour

Design width for normal flow approach ($\alpha = 0$):
$$B_1 = B + (B_f - B)\frac{f}{H} \qquad (21\text{-}7)$$

Design width B_1 for acute flow approach when $(f/H) \leq 0.3$:
$$B_1 = (L - B_o)\sin \alpha + B_o \qquad (21\text{-}8)$$

where
$$B_o = B + (B_f - B)\frac{f}{H} \qquad (21\text{-}9)$$

When $(f/H) > 0.3$,
$$B_1 = L \sin \alpha + B_o \cos \alpha \qquad (21\text{-}10)$$

where
$$B_o = B + (B_f - B)\frac{f}{H} \qquad (21\text{-}9)$$

2. Pier form with semicircular noses at both ends (Fig. 21-7b):

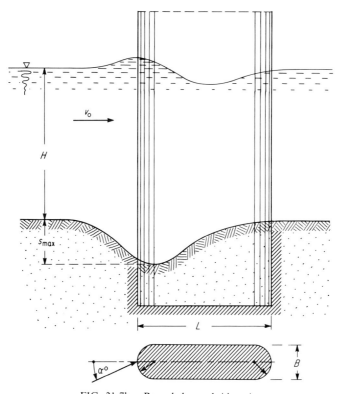

FIG. 21-7b. Rounded-nose bridge pier.

The C_1 coefficients for this pier form are compiled as follows (Table 21-4):

TABLE 21-4
C_1

$\alpha°$	0°	10°	20°	30°	40°
C_1	0.87	0.89	0.92	1.05	1.15

For normal approach, $B_1 = B$.
For acute approach $\alpha°$,

$$B_1 = (L - B) \sin \alpha + B \tag{21-11}$$

C_1 for angles $\geqslant 20°$ can be taken from this table when $(H/B) \geqslant 2$.

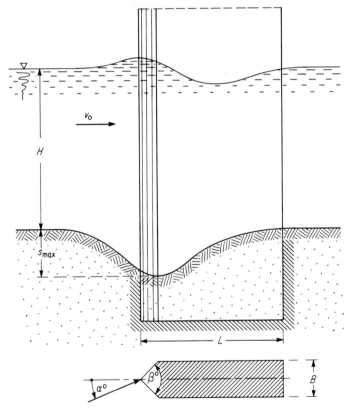

FIG. 21-7c. Sharp-nosed bridge pier.

3. Pier form with sharp nose at one end (Fig. 21-7c and Table 21-5): For this pier form the C_1 coefficient depends upon the nose angle β of the pier:

TABLE 21-5
C_1

$\beta°$	60°	90°	120°
C_1	0.76	1.00	1.25

For normal flow approach, the design width $B_1 = B$.
For acute angle of attack, the design width B_1 is

$$B_1 \approx (L-B)\sin\alpha + B \quad (21\text{-}12)$$

4. Form coefficient C_1 for a circular pier (Fig. 21-7d):

$$C_1 = 1.02$$

and

$$B_1 = D$$

5. Form coefficient C_1 for rounded nose at both ends, the pier resting on a cluster of high piles (see Fig. 21-7e and Table 21-6):

Scour must be checked at every pile, considering the pile as a circular pier.
For normal angle of attack, the design width is $B_1 = B$.
For acute angle of attack,

$$B_1 = (L-D)\sin\alpha + B \quad (21\text{-}13)$$

The safe depth t of laying of a foundation should be

$$t \geqslant \eta t_{\text{crit}} \quad (21\text{-}14)$$

where $\eta = 1.5 =$ factor of safety
$t_{\text{crit}} =$ critical depth at which expulsion of soil from underneath base of pier is just impending

Based on his research, Laursen[9] gave recommendations for design relationships for predicting depth of scour d_s at a bridge pier as shown in Fig. 21-10. This relationship pertains to a scour depth for a pier at a zero angle α of attack. Here b is the breadth of the pier projected at right angles

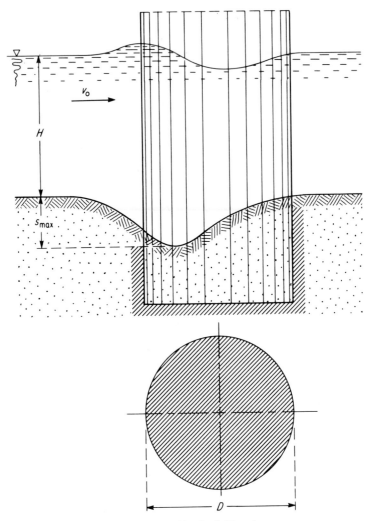

FIG. 21-7d. Circular bridge pier.

to the direction of attacking flow, and y_o is the upstream depth of water.

The relationship was obtained from a combination of an approximate analysis and laboratory experiments, and depends on knowledge of the flow conditions at the bridge site. The relationship pertains to the case in which sediment is supplied to the scour funnel.

If the pier is placed at an angle α to the flow (angle of attack), the scour increases substantially. The effect $K_{\alpha L}$ of angle of attack on depth of scour as determined by Laursen is shown in Fig. 21-11.

Scour

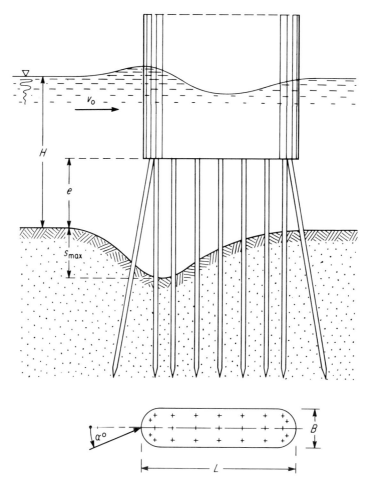

FIG. 21-7e. Rounded-nose bridge pier on piles.

TABLE 21-6

C_1

$\alpha°$ \ L/B	0	2	4	8	12
0	0.87	0.77	0.69	0.61	0.55
10	0.89	0.79	0.70	0.62	0.56
20	0.92	0.80	0.71	0.63	0.57
30	1.05	0.88	0.76	0.65	0.58
40	1.14	0.94	0.81	0.68	0.60

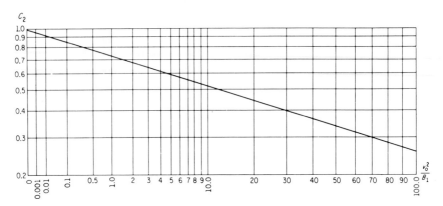

FIG. 21-8. Graph of coefficients C_2 for calculating local scour at bridge piers (Ref. 8).

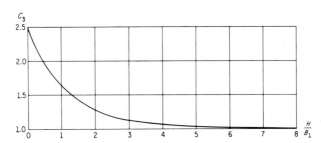

FIG. 21-9. Graph of coefficients C_3 for calculating local scour at bridge piers (Ref. 8).

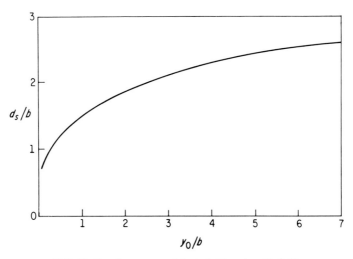

FIG. 21-10. Scour around basic bridge pier (Ref. 9).

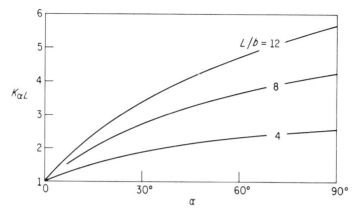

FIG. 21-11. Effect of angle of attack (Ref. 9).

The angle α of attack is the angle between the direction of the longitudinal axis of the pier and the direction of the flow in the river at the pier. According to Laursen, the design curve $K_{\alpha L}$ in Fig. 21-11 was drawn with conservatism and with due regard for the reliability of the various data.

Laursen found that the most important aspect of the geometry of the pier was the angle of attack between the pier and the flow, coupled with the length-width ratio (L/b) of the pier. A family of curves is shown in Fig. 21-11 as a multiplying factor $K_{\alpha L}$ to be applied to the depth of scour d_s. Thus the scour depth for a pier at an angle α of attack is obtained by multiplying the scour depth d_s at a pier at a zero angle of attack by the angle-length factor $K_{\alpha L}$.

The effect of pier nose shape on scour depth is shown in Table 21-7, as the shape coefficient K_s for nose forms of the piers. The factor K_s is applied to the basic scour depth d_s by multiplying d_s with K_s.

TABLE 21-7 *

Shape coefficients K_s for nose forms.
(To be used *only* for piers aligned with flow.)

Nose Form	Length-width Ratio	K_s
Rectangular		1.00
Semicircular		0.90
Elliptic	2:1	0.80
	3:1	0.75
Lenticular	2:1	0.80
	3:1	0.70

*After Ref. 9

All of the above data apply to an isolated pier in which the flow pattern is entirely dependent on the pier shape and any adjacent piers have no influence on the pier under consideration.

If the pier is set at an angle to the flow, Laursen recommends that no allowance be made for shape. If it is certain that the pier is aligned with the flow and will remain so during the service life of the bridge, a shape factor K_s can be applied as shown in Table 21-7.

REFERENCES

1. W. M. Angas and C. S. Proctor, "Foundations," in R. W. Abbett (ed.), *American Civil Engineering Practice*. New York: Wiley, 1956, Vol. 1, Sec. 8, pp. 8-50.
2. J. Feld, "Footings, Piers, and Abutments," in R. W. Abbett (ed.), *American Civil Engineering Practice*. New York: Wiley, 1957, Vol. 3, Sec. 26, pp. 26-52.
3. E. L. Erickson, "Some Measurements of Velocities and Scour at a Mississippi River Pier," Proceedings of the 26th Annual Meeting of the Highway Research Board, Dec. 5-8, 1946, pp. 124-128.
4. E. M. Laursen and A. Toch, "Scour Around Bridge Piers and Abutments," Iowa Highway Research Board, Bulletin No. 4, May 1956, pp. 42-43. Prepared by the Iowa Institute of Hydraulic Research in cooperation with the Iowa State Highway Commission and the Bureau of Public Roads.
5. R. W. Stewart, "Safe Foundation Depths of Bridges to Protect from Scour," *Civil Engineering*, Vol. 9, No. 6 (1939), pp. 336-337.
6. T. Merriman and T. H. Wiggin (eds.), *American Civil Engineers' Handbook*. 5th ed. New York: Wiley, 1947, Sec. 13, p. 1350.
7. A. P. Thurston (ed.), *Molesworth's Handbook of Engineering Formulae and Data*, 34th ed. London: Spon, 1951, p. 886.
8. E. V. Boldakov and O. V. Andreyev, *Crossings over Water Courses* (in Russian) Moscow: Scientific-Technical Publisher of Literature on Autotransportation, 1956, pp. 164-165.
9. E. M. Laursen, "Scour at Bridge Crossings," *Trans. ASCE*, Vol. 127, Part I (1962), p. 166.

Other Pertinent References

10. K. Terzaghi, "Failure of Bridge Piers Due to Scour," Proceedings of the First International Conference on Soil Mechanics and Foundation Engineering, Cambridge, Mass., June, 1936, Vol. 2, p. 26.
11. C. J. Posey, "Why Bridges Fail in Floods," *Civil Engineering*, Vol. 19 (February 1949), p. 94.
12. E. W. Lane and W. M. Borland, "River Bed Scour During Floods," *Trans. ASCE*, Vol. 119 (1954), p. 1072.
13. H. W. Shen, V. R. Schneider, and S. Karaki, "Local Scour Around Bridge Piers," *J. Hydraul. Div., Proc. ASCE*, Vol. 95, No. HY6 (November 1969).

PROBLEMS

21-1. *Given*: A bridge pier rounded at both ends on a rectangular footing. The pertinent plan dimensions of the pier and the footing are: $L = 30$ ft; $B_f = 19$ ft, and $B = 15$ ft. The pier is founded in sand. The angle of attack is $\alpha = 0°$. The depth of water before the scour is $H = 30$ ft, and the freeboard of the footing is $f = 1.0$ ft. The average flow velocity is given as $v_o = 2.0$ ft/sec. For purposes of stability, it is given that the embedment depth of the pier must be 12 ft into the soil. *Required*: calculate the maximum depth of scour.

21-2. Given the same problem as the preceding one, but the angle of attack is given as $\alpha = 20°$. What is the maximum depth of scour?

21-3. For the same pier as in Prob. 21-1, and for various α and f/H, calculate maximum scours and compare results, make your observations, and report.

21-4. Given a circular pier on sand whose diameter $D = B_1$; $H = 60$ ft, and $v_o = 2.0$ ft/sec. What is the maximum scour?

21-5. Given a bridge pier with a semicircular nose, and the corresponding data as set forth. $L = 30$ ft; $b = 15$ ft; $y_o = 30$ ft; $\alpha = 20°$. Determine the maximum scour at the bridge pier according to Laursen.

21-6. The same conditions as in Prob. 21-4, but $\alpha = 0°$.

21-7. Compare the results obtained in Probs. 21-5 and 21-6 with those obtainable by Boldakov's and Andreyev's method.

chapter **22**

Lateral Pressure in Deep Foundations

22-1. INTRODUCTORY NOTES

In practice, a foundation is considered, arbitrarily, to be deep if its depth d exceeds about twice the width B of the base of the footing, i.e., if $d > 2B$.

Whereas with shallow foundations the magnitudes of lateral earth pressure are insignificant, with deep foundations soil lateral pressures (resistances) should not be ignored, for they add to stability of the foundation against rotation and thus to reducing its settlement.

A foundation, subjected to vertical and horizontal loads, generally can translate and rotate. Upon embedding a deep foundation into soil, besides contact pressure at the base of the footing ($=$ vertical soil reaction), lateral soil reactions induced by the rotating foundation also act on the vertical (or inclined) sides of the foundations.

Although upon rotation wall friction develops in the side surface and base contact areas of the deeply embedded foundation, these frictional forces are in practice considered to be small, as is the case in sheet-piling calculations. Hence they are ignored—this is then on the safe side in respect to the stability of the deep foundation.

Because the soil lateral compression properties are not too well understood as yet, and because the degree of fixity of a deep foundation in the soil is almost impossible to determine, one tries to solve this problem approximately by making various simplifying assumptions.

22-2. ASSUMPTIONS

As in calculating and designing shallow foundations, in designing deep foundations, all forces are combined into one resultant centrical, vertical (normal) force N and one resultant horizontal force H, as well as into a resultant driving moment (Fig. 22-1).

Lateral Pressure in Deep Foundations

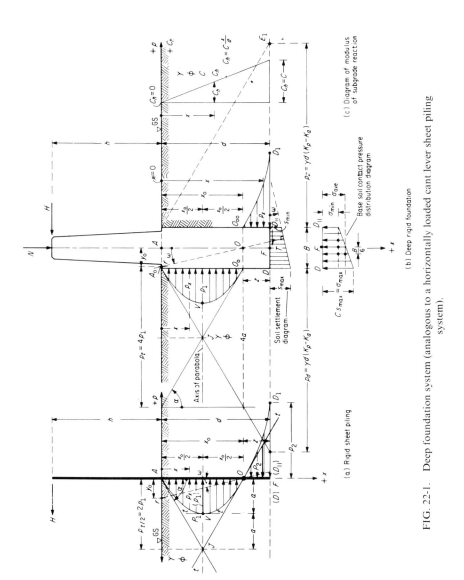

FIG. 22-1. Deep foundation system (analogous to a horizontally loaded cant lever sheet piling system).

The vertical, centrical force N induces a uniformly distributed soil contact pressure σ_o, whereas the horizontal force H tends to produce a rotation of the embedded foundation about the so-called pivot point O. If the rigid foundation rests in and on soil, this pivot point is a mechanical necessity in stability calculations against rotation. If the rigid foundation rests on bedrock, then the pivot point is situated at the bottom (base) of the foundation (Ref. 1).

The resultant horizontal load H, upon producing rotation, induces soil lateral resistances P_1 and P_2. These resistances are assumed to have a parabolic distribution with depth. The parabolic soil pressure distribution is here adopted from rigid sheet-piling theory.[1]

In this analysis of deep foundations it is further assumed that the foundation and its shaft above ground surface are rigid, i.e., they do not deflect when subjected to lateral loads and when they are rotating. Also it is assumed that the foundation-supporting soil deforms elastically, i.e., the soil is of Winkler's type, where settlement s of the soil is proportional to stress σ:

$$s = \sigma/C \quad \text{(cm)} \quad (22\text{-}1)$$

where C is the coefficient of subgrade reaction of the soil in kg/cm^3. Besides, it is assumed that the coefficient of subgrade reaction increases linearly with depth below ground surface. For example, the coefficient of subgrade reaction C_h for horizontal loads at elevation (depth) of base of the footing is equal to the coefficient of subgrade reaction for vertical loads. The coefficient of horizontal subgrade reaction C_h varies with depth: at the ground surface it is assumed that $C_h = 0$, and at the base of the footing $C_h = C$. Thus

$$C_h = C \, \frac{x}{d} \quad (22\text{-}2)$$

where x = depth below ground surface
d = depth of laying the foundation, measured from ground surface

As has already been stated, the lateral load H forms a moment with respect to the base contact area A of the foundation and thus redistributes the uniformly distributed soil contact pressure σ_o into a trapezoidal pressure distribution, resulting in maximum and minimum edge contact pressures σ_{max} and σ_{min}. Because of the soil lateral resistance to rotation of the foundation, it may be conceived that the maximum and minimum soil contact edge pressures, σ_{max} and σ_{min}, respectively, may be less than those from an eccentrically loaded foundation with no fixity (or restraint by the soil or rock) at all.

Lateral Pressure in Deep Foundations

Referring to Fig. 22-1, the angle of rotation ω is expressed as

$$\omega = \arctan\left(\frac{s_{max} - s_{min}}{B}\right) \quad (22\text{-}3)$$

where s_{max} and s_{min} are settlements at edge points of the base of the foundation, and B = width of foundation.

Substitution of $s = \sigma/C$ from Eq. 22-1 into Eq. 22-3 gives

$$\omega = \arctan\left(\frac{\sigma_{max} - \sigma_{min}}{CB}\right) \quad (22\text{-}4)$$

According to the method of reasoning above, the soil lateral pressure p_x at an arbitrary point distance x below the ground surface along the vertical wall, bringing about a soil lateral displacement of $s_h = \sigma/C_h$, may be written as

$$p_x = C_h s_h = C_h(x_o - x)\tan\omega \quad (22\text{-}5)$$

or, with Eqs. 22-2 and 22-4,

$$p_x = \frac{\sigma_{max} - \sigma_{min}}{Bd}(x_o - x)x \quad (22\text{-}6)$$

This p_x-function is a parabola starting at point A_o, where $x = 0$, and $p_x = 0$. This p_x-function is the general equation expressing soil lateral pressure distribution p_x along the vertical wall $A_o O_o$ and $O_{oo} D_1$ of a laterally (eccentrically) loaded deep foundation.

22-3. PARABOLIC SOIL LATERAL STRESS DISTRIBUTION

The parabolic soil lateral stress distribution equation p_x has a maximum value at $x = x_o/2$:

$$\frac{dp_x}{dx} = \frac{\sigma_{max} - \sigma_{min}}{Bd}(x_o - 2x) = 0 \quad (22\text{-}7)$$

Therefore

$$x = \frac{x_o}{2}$$

At this x,

$$p_{x\,max} = p_1 = \frac{\sigma_{max} - \sigma_{min}}{Bd} \cdot \frac{x_o^2}{4} \quad (22\text{-}8)$$

When $x = x_o$, $p_x = 0$ (at point O_o). When $x = d$,

$$p_x = p_2 = \frac{\sigma_{max} - \sigma_{min}}{B}(x_o - d) \qquad (22\text{-}9)$$

or

$$p_2 = -\frac{\sigma_{max} - \sigma_{min}}{B}(d - x_o) \qquad (22\text{-}9\text{a})$$

According to stability conditions, p_x should be less than the difference between passive and active earth pressures at that point of p_x:

$$p_x \leqslant \gamma x (K_p - K_a) = \lambda x \qquad (22\text{-}10)$$

where γ = unit weight of soil

$K_p = \dfrac{\pi}{4} + \dfrac{\phi}{2}$ = coefficient of passive earth pressure (resistance)

$K_a = \dfrac{\pi}{4} - \dfrac{\phi}{2}$ = coefficient of active earth pressure

ϕ = angle of internal pressure of soil
$\lambda = \gamma(K_p - K_a)$ = a coefficient
x = a running depth coordinate

At limit equilibrium, Eq. 22-10 is written as

$$p_x = \lambda x \qquad (22\text{-}11)$$

This equation is a straight line through point A whose coefficient of slope is $\lambda = \tan \alpha$. Thus, this straight line is a tangent to the parabola at point A.

By the following property of a parabola (Fig. 22-2), namely, that when the abscissa $x = x_o/2$, then the corresponding tangents intersect at a distance p_1 from the vertex V of the parabola (or at the distance of double vertex ordinate), we can write

$$2p_x = 2p_1 = \lambda \frac{x_o}{2} \qquad (22\text{-}12)$$

22-4. EQUILIBRIUM EQUATIONS

With the foregoing assumptions and auxiliary calculations fixed, these are the four unknowns necessary for performing static calculations of deep foundations: p_1, σ_{max}, σ_{min}, and x_o. These quantities are calculated by means of the well-known static equilibrium conditions, namely: $\Sigma F_x = 0$, $\Sigma F_y = 0$, and $\Sigma M = 0$.

With reference to Fig. 22-1, these general equilibrium equations are

$$\Sigma F_x = 0: \qquad N - \frac{\sigma_{max} + \sigma_{min}}{2} BL = 0 \qquad (22\text{-}13)$$

Lateral Pressure in Deep Foundations

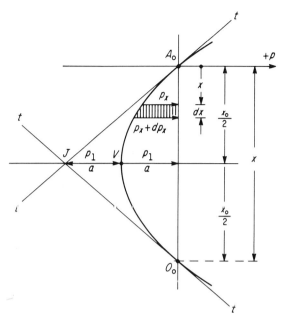

FIG. 22-2. Assumed soil lateral pressure distribution parabola.

$$\Sigma F_y = 0: \quad H + T - \int_0^d p_x\, dx = 0 \tag{22-14}$$

$$\Sigma M_A = 0: \quad Hh - Td + L\int_0^d p_x x\, dx - L\frac{\sigma_{max} - \sigma_{min}}{2} B\frac{B}{6} = 0 \tag{22-15}$$

Also,

$$\tan \omega = \frac{\sigma_{max} - \sigma_{min}}{BC} = \frac{p_2 B}{BC(d - x_o)} = \frac{p_2}{C(d - x_o)} \tag{22-16}$$

In these equations, L is the length of run of foundation perpendicular to the drawing plane.

22-5. MAXIMUM AND MINIMUM EDGE PRESSURES

The simultaneous solution of Eqs. 22-13 and 22-8, where in the latter

$$2p_1 = 2\,\frac{\sigma_{max} - \sigma_{min}}{Bd}\,\frac{x_o^2}{4} = \frac{\lambda}{2} x_o \tag{22-17}$$

resulting in

$$\sigma_{max} - \sigma_{min} = \frac{\lambda}{x_o} Bd \qquad (22\text{-}18)$$

renders the values for σ_{max} and σ_{min} as set forth

$$\left. \begin{array}{c} \sigma_{max} + \sigma_{min} - \dfrac{2N}{BL} = 0 \\[6pt] \sigma_{max} - \sigma_{min} - \dfrac{\lambda Bd}{x_o} = 0 \end{array} \right\} \qquad \begin{array}{c}(22\text{-}13)\\[6pt](22\text{-}18)\end{array}$$

$$\sigma_{max} : \sigma_{min} : 1 = \begin{vmatrix} 1 & 1 & -\dfrac{2N}{BL} \\[6pt] 1 & -1 & -\dfrac{\lambda Bd}{x_o} \\[6pt] + & - & + \end{vmatrix}$$

$$= \begin{vmatrix} 1 - \dfrac{2N}{BL} \\[6pt] -1 - \dfrac{\lambda Bd}{x_o} \end{vmatrix} : (-) \begin{vmatrix} 1 - \dfrac{2N}{BL} \\[6pt] 1 - \dfrac{\lambda Bd}{x_o} \end{vmatrix} : \begin{vmatrix} 1 & 1 \\[6pt] 1 & -1 \end{vmatrix} \qquad (22\text{-}19)$$

$$\sigma_{max} = \frac{N}{BL} + \frac{\lambda}{2} \frac{Bd}{x_o} \qquad (22\text{-}20)$$

$$\sigma_{min} = \frac{N}{BL} - \frac{\lambda}{2} \frac{Bd}{x_o} \qquad (22\text{-}21)$$

22-6. TANGENTIAL FRICTIONAL FORCE T AND HORIZONTAL FORCE H

Integration of Eq. 22-14 results in $H = f(\lambda, L, d, x_o, T)$

$$T = \lambda \frac{L}{x_o} \left(x_o \frac{d^2}{2} - \frac{d^3}{3} \right) - H \qquad (22\text{-}22)$$

This T should be $< \mu_1 N$, where μ_1 is the coefficient of friction between soil and base material of the foundation.

When $T = 0$, then

$$H = \lambda \frac{L}{x_o} \left(x_o \frac{d^2}{2} - \frac{d^3}{3} \right) = \frac{\lambda L d^2}{6} \left(\frac{3x_o - 2d}{x_o} \right) \qquad (22\text{-}23)$$

Lateral Pressure in Deep Foundations

22-7. POSITION OF PIVOT POINT O

Substitution of this H into the moment Eq. 22-15 when $T=0$, and its subsequent integration. renders the position x_o of the pivot O below ground surface:

$$Hh = \lambda L \left(\frac{d^2}{2} - \frac{d^3}{3x_o}\right) h = (1/2)\lambda L h d^2 - (1/3)\lambda L h \frac{d^3}{x_o} \qquad (22\text{-}24)$$

$$L \int_0^d p_x x \, dx = L \int_0^d \left[\frac{\sigma_{max} - \sigma_{min}}{Bd}\right](x_o - x)xx \, dx$$

$$= \lambda L \left(\frac{d^3}{3} - \frac{d^4}{4x_o}\right) = (1/3)\lambda L d^3 - (1/4)\lambda L \frac{d^4}{x_o} \qquad (22\text{-}25)$$

$$LB^2 \frac{\sigma_{max} - \sigma_{min}}{12} = (1/12)\lambda L d B^3 \frac{1}{x_o} \qquad (22\text{-}26)$$

Now substitute Eqs. 22-24, 22-25, and 22-26 into the moment Eq. 22-15 to obtain position x_o of the pivot point O as

$$x_o = \frac{d^2(3d + 4h) + B^3}{2d(2d + 3h)} \qquad (22\text{-}27)$$

When instead of a foundation a rigid sheet piling is given, then $B=0$, and the position of the pivot point O calculates as

$$x_o = \frac{d(3d + 4h)}{2(2d + 3h)} \qquad (22\text{-}28)$$

(See Ref. 1, Eq. 10-45. p. 352.)

22-8. SOIL LATERAL-PRESSURE ORDINATES

With x_o and σ_{max} and σ_{min} now known, the lateral parabolic pressure ordinates p_x, p_1, and p_2 can be calculated:

$$p_x = \lambda \frac{x_o - x}{x_o} x = \frac{6H}{Ld^2} \frac{x_o - x}{3x_o - 2d} x \qquad (22\text{-}29)$$

$$p_1 = (1/4)\lambda x_o = (1/4)\frac{6H}{Ld^2} \frac{x_o^2}{3x_o - 2d} \qquad (22\text{-}30)$$

$$p_2 = \frac{6H}{Ld^2} \frac{x_o - d}{3x_o - 2d} d \qquad (22\text{-}31)$$

Contact pressures:

$$\sigma_{max} = \frac{6H}{Ld^2(3x_o - 2d)}\left(\frac{2x_o N + \lambda dLB^2}{2\lambda BL}\right) \quad (22\text{-}32)$$

$$\sigma_{min} = \frac{6H}{Ld^2(3x_o - 2d)}\left(\frac{2x_o N - \lambda dLB^2}{2\lambda BL}\right) \quad (22\text{-}33)$$

22-9. CRITICAL LOAD H_c

Load H_c (critical load) at which a noncohesive soil will start to flow laterally, is derived in the following manner:

$$p_x = \frac{6H}{Ld^2(3x_o - 2d)}(x_o x - x^2) \quad (22\text{-}29)$$

$$\frac{dp_x}{dx} = \frac{6H}{Ld^2(3x_o - 2d)}(x_o - 2x) \quad (22\text{-}34)$$

This minimum load H_c occurs when the tangent (first derivative) to the parabola at point $x=0$ is equal to $\gamma x(K_p - K_a) = \lambda x$, Eq. 22-10:

$$\frac{6H}{Ld^2(3x_o - 2d)}x_o = \gamma(K_p - K_a) \quad (22\text{-}35)$$

From here,

$$H_c = \frac{\gamma Ld^2(3x_o - 2d)(K_p - K_a)}{6x_o}$$

$$= \frac{1}{2}\lambda Ld^2 - \frac{4\lambda Ld^5 + 6\lambda Lhd^4}{3[d^2(3d + 4h) + B^3]} \quad (22\text{-}36)$$

22-10. EMBEDMENT DEPTH

The corresponding depth d_c at which the soil flow begins is to be calculated by the following fifth-degree equation:

$$d_c^5 - \frac{18H}{\lambda L}d_c^3 + \frac{3\lambda LB^3 - 24Hh}{\lambda L}d_c^2 - \frac{6HB^3}{\lambda L} = 0 \quad (22\text{-}37)$$

When instead of a foundation a rigid sheet piling is given, then $B=0$, and the driving depth d of the sheet piling is calculated from the following cubic equation:

$$d^3 - \frac{18H}{\lambda L}d - \frac{24Hh}{\lambda L} = 0 \quad (22\text{-}38)$$

Lateral Pressure in Deep Foundations

which checks out with the rigid cantilever sheet piling theory (see Ref. 1, p. 357).

22-11. SOIL BEARING CAPACITY FOR DEEP FOUNDATIONS

In the case of a deep foundation laid in soil, the bearing capacity of the soil may be increased for calculation purposes because the lateral surcharge γz acts to densify the soil at depth z and also acts against lateral expulsion of soil from underneath the base of the foundation.

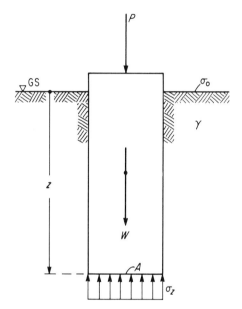

FIG. 22-3. Concerning soil-bearing capacity for deep foundations.

The soil bearing capacity σ_z at depth z below the ground surface may then be calculated (Fig. 22-3) as

$$\sigma_z = \sigma_o + \gamma z + \frac{\mu E_a U}{A} \quad (\text{ton/ft}^2 \approx \text{kg/cm}^2) \qquad (22\text{-}39)$$

where σ_o = soil bearing capacity at ground surface
γ = unit weight of soil (or weighted unit weight of soil if a stratified soil profile is present)
E_a = active earth pressure (or weighted active earth pressure)
μ = 1.0 to 3.0 t/m^2 = 0.10 to 0.30 kg/cm^2 = coefficient, or, rather,

unit frictional force between soil and foundation material (or weighted unit friction)
U = horizontal perimeter of foundation (average, or weighted)
A = base or contact area of base of foundation with soil

For all heavily loaded engineering structures founded in a weak soil the latter's bearing capacity must be determined by tests.

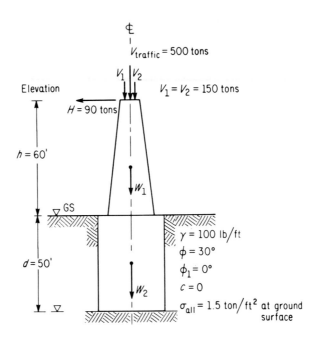

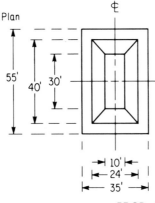

PROB. 22-1

REFERENCE

1. A. R. Jumikis, *Mechanics of Soils: Fundamentals for Advanced Study*. Princeton, N.J.: Van Nostrand, 1964, pp. 349-358.

PROBLEM

22-1. Given a rigid bridge pier as shown. Calculate base contact pressures σ_{max} and σ_{min}, and plot the soil lateral pressure-distribution diagram. If necessary, calculate the necessary depth d of foundation and redesign the foundation. The allowable soil pressure at the ground surface is given as $\sigma_{all} = 1.5$ ton/ft^2. Make all other reasonable engineering assumptions.

chapter **23**

Crib-Wall Cofferdam

23-1. ROCK-FILLED TIMBER CRIB

A crib-wall cofferdam consists of timber cribs built around a construction site to be enclosed against water of shallow depth (usually up to about 20 ft). The crib is built as a rectangular framework of creosoted, interlocking timbers. The timber (whole logs, or sawed material) are laid horizontally in alternate courses and piled up to form a sort of silo or bin, or cell (Figs. 23-1 and 23-2). These cells are then filled with clean rock fragments and noncohesive soil. Thus a crib wall is a gravity wall. The water side of the crib wall is provided with vertical sheeting. Because of their weight, crib-wall cofferdams are used in swiftly flowing waters.

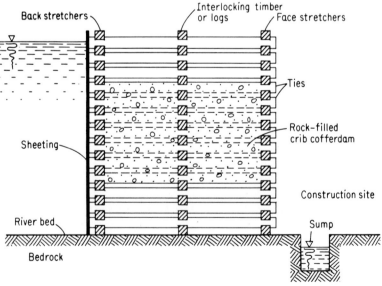

FIG. 23-1. Cross section of a gravity crib-type cofferdam in water on bare bedrock.

Crib-Wall Cofferdam

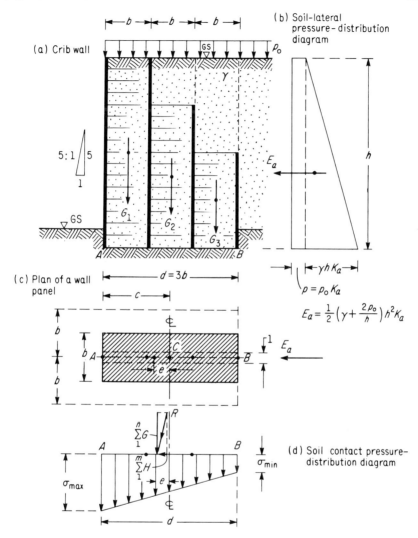

FIG. 23-2. A gravity crib-type earth retaining wall on land.

Crib-wall cofferdams are usually used up to about a 20-ft depth of water. However, crib-wall cofferdams to hold back water greater than 20-ft depths have been used, namely: (a) the width of the Conovingo Dam (Pennsylvania) crib-wall cofferdam was 30 ft, and its height 35 ft; (b) the width of the Bonneville Dam (Washington) crib-wall cofferdam was 60 ft, and its height 63 ft.

In general, the size of the crib-wall cofferdam must be such that when filled it should prevent sliding and overturning.

In hydraulic structures and foundation engineering timber cribs are usually used in temporary structures only.

The timber log crib-type cofferdam is used for foundation work in rivers or other kinds of water basins with hard, bare-rock bottom, in swift currents, and also where overtopping is a problem.

Also, timber crib cofferdams may be economical where timber is inexpensive. Otherwise one should resort to other, more economical types of cofferdams.

The cribs for use in water are usually built on land, then floated to the site, weighted with rocks, and sunk in position.

For the construction of a crib-wall type *retaining wall* on land, precast reinforced concrete and steel stretchers and ties are also used besides timber members. The cells of the crib-type retaining wall should be filled with clean rock and/or clean sand and gravel soil before backfilling the wall to prevent its lateral displacement out of its alignment.

Advantages of a Crib Wall

As an earth retaining structure, a crib wall is a relatively inexpensive method of construction as compared with massive gravity walls, or reinforced concrete walls, or anchored sheetpiling wall, and it is easy to construct requiring but a minimum of skilled labor. Also, a crib wall is free-draining, hence no danger to frost action, and therefore it need not be laid below the frost-penetration depth because of its flexibility. For the same reasons just stated a crib wall is relatively immune against differential settlement if placed on compressible soil.

However, it is advisable that high crib walls be built on proper foundations, which should be laid below the frost penetration depth.

23-2. STABILITY OF A CRIB WALL

The stability of a crib-type cofferdam can be checked against sliding and overturning with a satisfactory degree of accuracy. Its permeability, however, is of indeterminate nature and depends upon the density of its seams, watertightness of joints between the base of the crib-wall cofferdam and the crib-supporting soil, the backfill material, and the changing nature of the soil with depth.

For purposes of stability analysis, a crib-type retaining wall may be assumed to perform as a gravity wall.

The crib is usually filled with clean crushed stone, or rock fragments < 12 in. in size, and/or with a free-draining soil such as clean, coarse sand and gravel, to insure free drainage of the crib and thus its stability.

Crib-Wall Cofferdam

The stability calculations for a crib wall are performed in the same way as those presented for a box-type sheet-piling cofferdam supported on rock or soil (sheet piling not fixed into rock or soil).

To increase stability, sometimes the front face of the crib wall has a batter construction.

The stability of the crib wall against sliding is characterized by a factor of safety η as

$$\eta = \frac{\mu_1 \left(\sum_{1}^{n} V \right)}{\sum_{1}^{m} H} \tag{23-1}$$

where μ_1 = coefficient of friction between soil (rock) and fill material in crib
$\sum_{1}^{n} V$ = algebraic sum of all (n) vertical forces in crib system, including surcharge, if present
$\sum_{1}^{m} H$ = algebraic sum of all horizontal forces on crib
m = number of horizontal forces acting on crib

The weight of the crib is equal to that of the material within the cells of the crib, including the weight of the crib material.

23-3. FORCES ACTING ON CRIB ELEMENTS

When the crib cells are filled prior to backfilling the crib wall, then the crib as a free body performs like a silo: the rock/and or soil, filling the cells, exert a lateral pressure on the stretchers of the crib. The arching action of the inside fill material on the stretchers is then the basis for calculating the forces on the crib elements, viz., stretchers and ties.

The lateral pressure brings about a bending moment in the stretchers, but the reactions of the stretchers are transmitted to the ties; thus the latter are subjected to tensile forces.

The derivation of soil pressure in the crib is based on the theory of pressure on a tunnel roof, as shown in Ref. 1. The derivation is here based on the appropriate sketch, as shown in Fig. 23-3. The vertical pressure σ_z, according to this derivation, is

$$\sigma_z = \left(\frac{\gamma A - cU}{\mu K U} \right) \left(1 - e^{-\frac{\mu K U}{A} z} \right) + p_o \left(e^{-\frac{\mu K U}{A} z} \right) \tag{23-2}$$

where γ = unit weight of fill material in crib cell
A = horizontal cross-sectional area of cell of crib
c = cohesion of soil

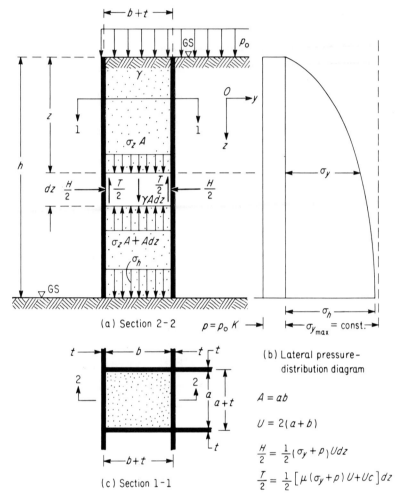

FIG. 23-3. Cell of crib.

U = perimeter of cell
$\mu = \tan \phi$ = coefficient of internal friction of soil (if applicable in the problem; if not, use μ_1 = coefficient of friction between fill material in cell and crib-wall material)
$K = \dfrac{\sigma_y}{\sigma_z} = \dfrac{\cos^2 \phi}{2 - \cos^2 \phi}$ = Krynine's soil lateral pressure coefficient for soil in a confined space like cofferdam or cell of crib[2,3]
e = base of natural logarithm system
p_o = vertical pressure from surcharge, if any
z = depth coordinate

Crib-Wall Cofferdam

The lateral compressive stress component σ_y against the inside wall is

$$\sigma_y = K\sigma_z \qquad (23\text{-}3)$$

Equation 23-2 can be specialized for $c = 0$ and for $p_o = 0$ as is the case for a pure frictional soil, and with no surcharge, respectively.

REFERENCES

1. A. R. Jumikis, *Mechanics of Soils: Fundamentals for Advanced Study*. Princeton, N.J.: Van Nostrand, 1964, pp. 385-386.
2. D. P. Krynine, discussion of Terzaghi's paper, "Stability and Stiffness of Cellular Cofferdams," *Trans. ASCE*, Vol. 110 (1945), p. 1083.
3. A. R. Jumikis, *Theoretical Soil Mechanics*, Princeton, N.J.: Van Nostrand, 1969, pp. 33-35.

PROBLEMS

23-1. Given a soil-filled, crib-type, earth-retaining wall as shown. The average unit weight of the compound crib-soil system is $\gamma_{ave} = 100$ lb/ft³. The unit weight of the backfill soil is $\gamma = 100$ lb/ft³. The angle of internal friction of soil is $\phi = 25°$. The angle of friction between backfill soil (outside the crib) and the crib wall is $\phi_1 = 0$. The coefficient $\mu_{b/s}$ of friction between the base of the filled-in crib and the crib-supporting soil is given as $\mu_{b/s} = 0.364$. The allowable bearing capacity of soil is $\sigma_{all} = 1.5$ ton/ft². The required factors of safety are:

(a) against rotation: $\eta_{rot_{all}} = 2.0$
(b) against sliding: $\eta_{sl_{all}} = 2.0$

Required: determine the stability of the crib wall against overturning (rotation) and sliding, and plot soil contact pressure-distribution diagram. Report whether pressures on soil are allowable. (*Hint*. To account for the weights of the concrete elements of the crib and the soil within the crib, as well as for the protrusions of the headers, it is practical to calculate the full weight W of the crib wall by using the outside plan dimensions $B \times B$, full height h, and an average unit weight of the soil-filled crib, here given as γ_{ave}.)

Ans.

$E_a = 1{,}950$ lb/linear foot of run of wall
$W = 10{,}800$ lb/ft
$x = 3.78$ ft; $e = 0.72$ ft
$\sigma_{max} = 0.90$ tons/ft²;
$\sigma_{min} = 0.32$ ton/ft²;
$\eta_{rot} = 6.2$; $\eta_{sl} = 2.0$.

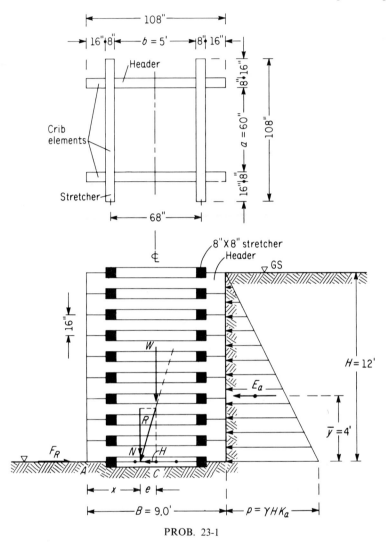

PROB. 23-1

23-2. Given a crib element whose length is $B = 9$ ft. The size of the crib element is 8 in. × 8 in. × 9 ft. Refer to Prob. 23-1, and Prob. 23-2. Calculate horizontal and vertical loads on stretchers and headers. Check the strength of the stretchers and headers.
(*Hint.* The axial load on header is the reaction from stretcher (S_h). The vertical load on header is the same as for stretcher (S_v). $S_v = S_h \tan \phi_1$.)

Crib-Wall Cofferdam

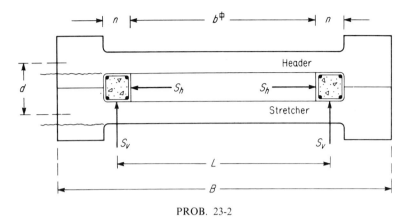

PROB. 23-2

23-3. Given a two-cell crib wall the base width of which is $2B = 18$ ft. The height of the outside cell is $h_o = 25$ ft, but the height of the cell on the backfill side is $h_b = 17.0$ ft. (See the accompanying figure.) All other data and requirements are the same as in Prob. 23-1. Verify whether or not the two-cell crib wall is stable.

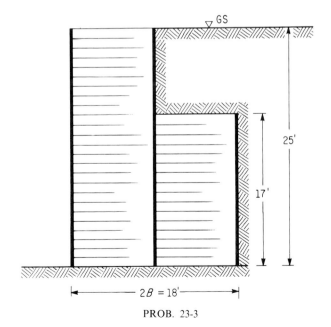

PROB. 23-3

chapter **24**

Seismic Effects on Foundations

24-1. EARTHQUAKE ACCELERATION

In evaluating the seismic effect on a structure it has become customary to indicate the earthquake shock or acceleration ratio λ of the horizontal seismic force F_h to the vertical component V of the resultant load of the structure:

$$\lambda = \frac{F_h}{V} = \frac{\alpha_{max}}{g} \qquad (24\text{-}1)$$

where α_{max} = maximum horizontal acceleration of earthquake
$g = 981$ cm/sec^2 = acceleration of gravity

In order to obtain the earthquake acceleration from seismographic records, the earthquake waves are regarded as sinusoidal oscillations, whose equation is

$$z = A \sin\left(2\frac{\pi}{T}t\right) \qquad (24\text{-}2)$$

where z = vertical displacement of an oscillating point P at time t (Fig. 24-1)
A = maximum amplitude at time $t = \pi/2$
T = period of the oscillation (or duration of vibration)

Differentiation of the z-equation twice with respect to time t renders acceleration α:

$$\alpha = \frac{d^2z}{dt^2} = -\frac{4\pi^2}{T^2} A \sin\left(\frac{2\pi}{T}t\right) \qquad (24\text{-}3)$$

Therefore α is at maximum when $\sin\left(\frac{2\pi}{T}t\right) = 1$

$$\therefore \quad \alpha_{max} = -\frac{4\pi^2}{T^2} A \qquad (24\text{-}4)$$

Seismic Effects on Foundations

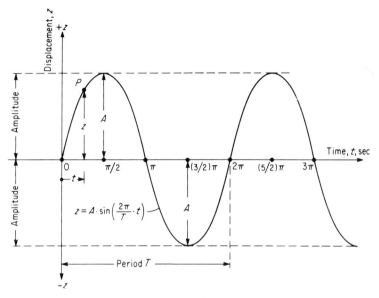

FIG. 24-1. Sinusoidal oscillation.

The A- and T-values are to be scaled off from the earthquake graphs. If, for example, period $T = 1.334$ sec, and maximum amplitude $A = 4.43$ cm (Tokyo, 1923), then $\alpha_{max} = \dfrac{(4)(10)}{(1.334)^2}(4.43) = 177.20/1.779 \approx 100$ (cm), and the earthquake shock ratio λ is

$$\lambda = \frac{\alpha_{max}}{g} = \frac{100}{981} \approx 0.10$$

24-2. EFFECT OF EARTHQUAKES ON STRUCTURES, FOUNDATIONS, AND EARTHWORKS

The foundation is the most important single factor in the design of an earthquake-resistant structure. This is because the earthquake shock is transmitted to any structure through the soil and/or rock upon which the structure is supported and thus through its foundations.

Depending upon the properties of the foundation and its design, the shock may be dampened or it may be magnified by the foundation before the shock is delivered to the superstructure.

When horizontal earthquake vibration shocks are transmitted through the soil to a foundation of a structure, the vibration of its upper part, because of its inertia, may lag behind the lower part, which may therefore suffer deformation because of the shock-induced stresses in the walls, columns, and other structural elements. These stresses may even destroy the structure.

Foundations supporting structures in regions which are prone to earthquakes are designed to resist a lateral seismic force. The magnitude of this seismic force F_s is usually assessed between one tenth and one fortieth of the weight W of the structure. This lateral seismic force results in large moments on the lowest flight of columns and large shear forces in the foundation. If the earthquake-induced shear forces exceed the strength of the foundation, the latter is destroyed—the foundation is unusable.

Because of earthquake oscillations the shear strength of the soil underneath foundations and in slopes of cuts and fills has frequently been exhausted.

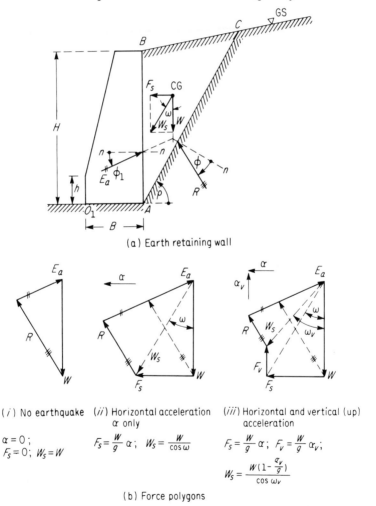

FIG. 24-2. Effect of a seismic force on the magnitude of active earth pressure.

Also too-steep earth slopes on earthworks may easily slide down from the effects of vibration on roads induced by all kinds of rolling stock.

24-3. EARTHQUAKE EFFECT ON EARTH RETAINING WALLS

Based on its studies, the Tennessee Valley Authority[1] reports that the weight of Coulomb's rupture wedge of soil and the seismic force combine into a resultant force. Hence the earthquake "increases" the active earth pressure on the wall (Fig. 24-2).

The length of the E_a-vector in the force polygon indicates clearly that the seismic force $F_s = (W/g)\alpha$ increases the magnitude of the active earth pressure E_a considerably. All symbols in Fig. 24-2 are the same as before, or as shown in Fig. 24-2.

24-4. EFFECT OF EARTHQUAKE ACTION ON BRIDGE PIERS

Lately bridges in the United States have, generally, performed well during earthquakes, as have tunnels (except for damage to their portals). However, some tunnels located in the fault zone have suffered badly in exceptionally strong earthquakes, such as those tunnels of the Southern Pacific Railroad in the July 12, 1952, earthquake.

Bridge piers, however, have been frequently destroyed by seismic forces, mainly those which could not cope with earthquake lateral shocks. Hence the necessity to select a site for bridge piers of the best-strength soil, to design them against horizontal shear, and to construct them as a massive structure of the best material. The nature of shearing a bridge pier by a seismic force is sketched in Fig. 24-3.

24-5. EFFECT OF VIBRATION OF SAND BEHIND BULKHEAD

In his large-scale earth-pressure tests with model flexible bulkheads, Tschebotarioff[2] found that "Exceptionally severe vibration of the sand backfill behind the bulkhead ... increased lateral pressures, so that the resulting bending moments in the bulkhead were increased by 60 percent and more, as compared with values obtained after normal backfilling. Subsequent vibration of the sand in front of the backfill decreased the bending moment somewhat."

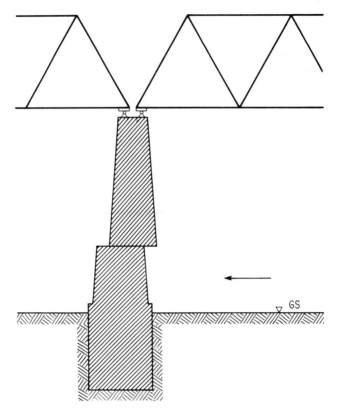

FIG. 24-3. Shearing a bridge pier by a seismic force.

24-6. MECHANICAL VIBRATIONS OF MACHINE FOUNDATIONS

A machine foundation differs from other structures mainly in that the former is subjected to dynamic loads which may vary quickly with time, bringing about vibration.[3] In stability analyses of foundations subjected to static loads it is enough to know the static strength characteristics of the soil. However, foundations loaded with dynamic loads present the designer with a much more complex problem than in the case of static loads. It should be realized that the elastic properties of the soil have a considerable influence on the design of dynamically loaded foundations. Hence careful dynamical soil tests are essential. Thus one is confronted with a subject called *soil dynamics*.[3]

To avoid resonance between a machine and its foundation, or that of an adjacent structure, the foundation must be designed, so to speak "out of tune" (rhythm) or out of resonance. In other words, the foundation should

Seismic Effects on Foundations

have a natural frequency differing by at least 20 to 30 percent of the speed of the machinery. Also, absorption of residual vibration should be attained by means of dampers incorporated in the foundation.

Another principle to observe in the design of machine foundations is that the amplitude of the foundation vibration should not exceed a certain allowable value. The above two principles complement each other mutually, and are based on dynamic characteristics of the soil.

Essentially, dynamic soil investigation consists of studying induced, controlled sinusoidal vibrations into the soil.

For basic elements of soil dynamics the reader is referred to the author's *Theoretical Soil Mechanics*.[3]

If a machine foundation is close to groundwater there exists the danger that vibrations may be propagated through the water to adjacent structures, even over long distances.

The walls of a vibrating structure undergo alternately compression and tension. Consequently symptoms of fatigue of the wall materials set in.

24-7. EFFECTIVE MEASURES IN ASEISMIC DESIGN

One of the most effective precautionary measures in designing and building earthquake-proof structures in earthquake-prone regions is construction on such firm ground that earthquake vibrations do not bring about any changes (densification of loose soil or overvibrating of dense soil, thus loosening, for example). Here, firm ground means unweathered rock, or geologically old, dense layers of soil. Especially one should avoid soil that can be thrown out of equilibrium by vibration (structures on slopes, steep river banks, or in landslide-prone terrain). Cases are known from the 1923 Tokyo earthquake, where a hotel built on a filled-up street collapsed completely, whereas structures founded on rock behind the fill survived the same quake with no cracks.

Thus it seems that the choice of an appropriate construction site and fireproof construction is of far greater importance than the application of an earthquake-proof method of construction (which does not render unconditional safety anyway). This, at least, is a point to observe in city planning.

However, man must frequently build on whatever site is available. Hence, in the interest of public safety, and to avoid damage to structures by resonance, special building codes for aseismic structural design are put into force. Such codes regulate the height of the buildings, seismic factors to use, kind of static systems of structures, fire protection, and other safety measures.

Under certain conditions some rocks and soils also contribute to the damage to structures if subjected to earthquake shocks.

The nature of the type of soil affecting the degree of danger of earthquake is summarized in Table 24-1, where Col. 3 shows the rating of the degree of danger (subsurface coefficient of danger), according to Sieberg.[4] Sieberg's coefficients are higher than the Mercalli scale degrees. By how many

TABLE 24-1

Earthquake Danger of Rocks and Soils (after A. Sieberg[4])

Rock and Soil	Danger	Sieberg's Subsurface Coefficient of Danger	Sieberg's Coefficients Higher than Mercalli Scale Degrees by
A. Soils of weathered solid rocks			
1. Weathered quartzite, slate, limestone, marble, dolomite	Dangerous because of the small quantities of decomposed residuum	1	0
2. Weathered sandstones, breccias, conglomerates, sand and gravel	Increasing danger with thickness of deposit and angularity of particles. Danger reduces with very large thickness	3-6	1-2
3. Weathered granites, quartz porphyries, diabase, gneisses; sandy and clayey soils; fine granite gravel	Same as under 2	3-6	1-2
4. Weathered basalt, shale, tuff clay soils	Same as under 2	2-10	1-3
B. Loose and alluvial soils			
1. Alluvial soils, transported sediments, sand, gravel, peat	Very, increasing with increase in water content	3-6	1-2
2. Clay soils, marl, loess, loam, boulder clay	Almost dangerless in dry condition and in massive, homogeneous state. Danger increases when soil is in a dry, crumbly, friable state, when saturated, and when in a plastic, kneadable, pasty condition	3-12	2-3
3. Marsh and swamp soils, silted-up lakes	Greatest degree of danger	8-16	3-4

Seismic Effects on Foundations

danger coefficients, in terms of Mercalli degrees, Sieberg's degrees are higher than Mercalli scale degrees is shown in Col. 4 of that table.

It thus seems that the sense of security man gets from the feel of solid earth's crust is ill-founded. It should be remembered that terra firma is nothing else but a changeable crust, about 120 km thin, and the uneasy innards of the globe which this crust covers never stop rumbling and stirring. In other words, earthquakes are possible in almost any region, but the range of probability varies enormously. When the probability exceeds a certain critical value, precautionary measures in design should be taken. Because precautionary measures entail additional expense, the criteria for determining whether they should be adopted are of paramount importance.

One also should be aware, therefore, that much more knowledge is needed for a better understanding of the behavior of man-made structures such as dams, hydraulic structures, hydropower plants, tunnels, bridges, transmission towers, and other structures subject to earthquake action.

If self-frequency of oscillations of tall structures such as tall buildings, smokestacks, and towers get in resonance with those of an earthquake, damage or even destruction of the structures may occur. Hence the necessity for determining the self-frequency of the structures.

The self-frequencies of smokestacks and towers are $f = 0.5$-1.0 cy/sec, i.e., similar to those of earthquakes. Hence the danger of resonance.

Each of the construction materials, too, has its own elastic properties which influence self-frequency. Hence it is desirable to construct a structure of seismically uniform materials. To reduce earthquake-induced large overturning moments, it is desirable that the structures be low and have solid, strong foundations and lightweight roofs. This means that the center of gravity of the structure (= point of application of the horizontal seismic force F_h) should be as low as possible. In the case of high towers, this may be accomplished by wide footings, or the foundation may be built as a closed box filled with a ballast. The structures should be effectively braced horizontally as well as vertically.

With pile foundations, sometimes batter piles are used for the protection of piled structures against earthquakes (Fig. 24-4). The purpose of the batter piles is to take up horizontal seismic forces from any direction and to transmit them to the soil or rock. The pile footing (cap) or grillage, too, must be laterally strong enough in order that it be not sheared off at the level of the pile tops at the level of the soil.

Slopes of cuts and fills, waterfront structures, bridge piers, water mains and earth-retaining structures should also be designed and constructed to cope with seismic forces and earthquake oscillations.

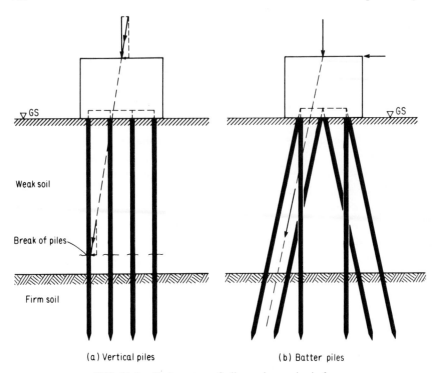

(a) Vertical piles   (b) Batter piles

FIG. 24-4. Performance of piles under a seismic force.

24-8. EARTHQUAKE CONSIDERATIONS IN THE DESIGN OF THE SAN FRANCISCO-OAKLAND BAY BRIDGE PIERS

Based on Kármán's concept of "apparent mass"[5] exerting a pressure from water and fluid bay silt on a bridge pier brought about by an earthquake, Raab and Wood[6] calculated the horizontal dimension of "apparent mass" b at distance y above the plane of slippage as

$$b = (0.7)\sqrt{h^2 - y^2} \qquad (24\text{-}5)$$

see Fig. 24-5.

The pressure p caused by earthquake, if the pier were surrounded to a depth h by water only, is given as

$$p = \alpha \gamma_w b = (0.7)\alpha \gamma_w \sqrt{h^2 - y^2} \qquad (24\text{-}6)$$

The additional pressure p' below the mud (silt) line is given as

$$p' = \alpha' \gamma_m b - \alpha \gamma_w b \qquad (24\text{-}7)$$

Seismic Effects on Foundations

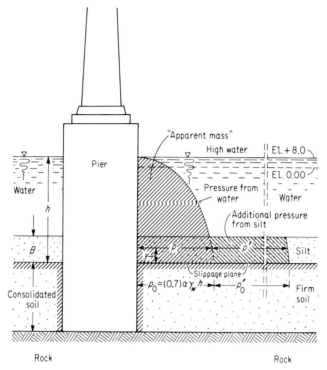

FIG. 24-5. Water and silt pressures acting on pier due to earthquake (Ref. 6).

where $\alpha' \gamma_m b = p + p'$ (24-8)

γ_w = unit weight of water
γ_m = unit weight of silt (mud)
a = acceleration of the pier relative to that of water (i.e., the assumed earthquake acceleration)
a' = acceleration of the pier relative to that of silt (or mud)
$\alpha = a/g$
$\alpha' = a'/g$
$g = 9.81$ m/sec² $= 32.2$ ft/sec² = acceleration of gravity

From Eqs. 24-6 and 24-7, obtain

$$\frac{p'}{p} = \frac{\alpha' \gamma_m b - \alpha \gamma_w b}{\alpha \gamma_w b} = \frac{\alpha' \gamma_m}{\alpha \gamma_w} - 1 = k \qquad (24\text{-}9)$$

and p is given by

$$p' = kp = (0.7)\alpha \gamma_w k \sqrt{h^2 - y^2} \qquad (24\text{-}10)$$

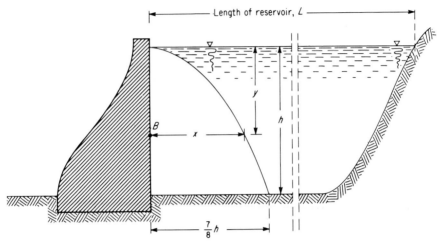

FIG. 24-6. Westergaard's pressure parabola.

At the plane of slippage, where $y = 0$,

$$p_o = (0.7)\alpha\gamma_w h \tag{24-6a}$$

$$p'_o = (0.7)k\alpha\gamma_w h \tag{24-10a}$$

The earthquake forces on the bridge were assumed as those resulting from a ground motion with a horizontal acceleration of 10 percent of gravity, a period of $T = 1.5$ sec, with a corresponding amplitude of $A = 2.2$ in. According to these authors,[6] this is a longer period than is possible.

24-9. EFFECT OF SHOCK OF WATER ON DAMS

The specific pressure P on dams and moles by water due to an earthquake is calculated by Kelen[7] as

$$P = (1/2)\gamma_w h^2 ag \tag{24-11}$$

where γ_w = unit weight of water
h = depth of water
a = coefficient for acceleration: $0.075\,g$ for Golden Gate bridge,
$\quad\quad\quad\quad\quad\quad\quad\quad\quad\quad\,\,\, 0.1\,g\,\,$ for Oakland bridge
g = acceleration of gravity

The hydrodynamic pressure of water imposed on hydraulic structures by an earthquake is calculated according to Westergaard[8] as a parabola (Fig. 24-6) defined by the equation

$$x = (7/8)\sqrt{hy} \tag{24-12}$$

Seismic Effects on Foundations

The inertial force of water contained by this parabola during earthquake oscillations is equal to the mass of water m multiplied by the acceleration of the earthquake. For any point B at a depth y,

Intensity of pressure p_B is

$$p_B = (54.6) h^{0.5} y^{0.5} \frac{\alpha}{g} \quad (\text{lb/ft}^2)/\text{ft}$$

Total force F_B is

$$F_B = (36.5) h^{0.5} y^{1.5} \frac{\alpha}{g} \quad (\text{lb/ft})$$

Moment M_B is

$$M_B = (14.6) h^{0.5} y^{2.5} \frac{\alpha}{g} \quad (\text{ft-lb})/\text{ft}$$

At the base $y = h$, hence

$$P = (54.6) h \frac{\alpha}{g}$$

$$F = (36.5) h^2 \frac{\alpha}{g}$$

$$M = (14.6) h^3 \frac{\alpha}{g}$$

REFERENCES

1. U.S. Tennessee Valley Authority, "The Kentucky Project," Technical Report No. 13, 1951.
2. G. P. Tschebotarioff, *Soil Mechanics, Foundations and Earth Structures*. New York: McGraw-Hill, Inc., 1951, p. 301.
3. A. R. Jumikis, *Theoretical Soil Mechanics*. Princeton, N.J.: Van Nostrand, 1969.
4. A. Sieberg, *Erdbebenforschung*. Jena: Fischer, 1933, p. 30.
5. T. von Kármán, discussion on H. M. Westergaard's paper, "Water Pressures on Dams during Earthquakes," *Trans. ASCE*, Vol. 98 (1933), pp. 434-436.
6. N. C. Raab and H. C. Wood, "Earthquake Stresses in the San Francisco-Oakland Bay Bridge," Paper No. 2123, *Proc. ASCE*, Vol. 67, No. 8 (1941), Part 2, pp. 1363-1384.
7. N. Kelen, "Erdbebenwirkung." in *Gewichtsstaumauern und massive Wehre*. Berlin: Springer, 1933, p. 17.
8. H. M. Westergaard, "Water Pressures on Dams During Earthquakes," *Proc. ASCE*, Vol. 57 (November 1931), pp. 1303-1318.

OTHER USEFUL REFERENCES

9. J. R. Freeman, *Earthquake Damage and Earthquake Insurance*, New York: McGraw-Hill, 1932.
10. A. C. Ruge, "Earthquake Resistance of Elevated Water-Tanks," *Trans. ASCE*, Vol. 103 (1938), pp. 889-938.
11. H. M. Westergaard, "Earthquake-Shock Transmission in Tall Buildings," *Engineering News-Record*, Nov. 30, 1933.
12. D. B. Gumensky, "Earthquakes and Earthquake-Resistant Design," in *American Civil Engineering Practice*, R. W. Abbett (ed.). New York: Wiley, 1957, Vol. 3, Sec. 34, pp. 34-01—34-34.
13. C. M. Harris and C. E. Crede (eds.), *Shock and Vibration Handbook*, 3 vols., New York: McGraw-Hill, 1961.
14. F. E. Richart, Jr., "Foundation Vibrations," *Proc. ASCE, J. Soil Mech. and Foundations Div.*, Vol. 86, No. SM4 (August 1960), pp. 1-34.
15. Proceedings of the Second World Conference on Earthquake Engineering, Seismic Council of Japan, Tokyo and Kyoto, 3 vols., 1960.

PROBLEMS

24-1. Given an earth-retaining wall as shown in Fig. 24-2.
$H = 5.0$ m; $h = 0.5$ m; $a = 1.0$ m; $B = 2.0$ m; $\gamma_{soil} = 1.8$ ton/m^3; $\gamma_c = 2.4$ ton/m^3; $\delta = 5°$; $\phi = 30°$; $\phi_1 = 20°$; $\alpha_h = (0.1)(g)$; $g = 9.81$ m/sec^2; $\alpha_v = (1/6)(g)$.

Required: Calculate the magnitude of active earth pressure E_a on the wall
(a) with no consideration of earthquakes
(b) considering a horizontal acceleration α_h of the seismic force on the backfill material
(c) considering horizontal and vertical accelerations, α_h and α_v, respectively

Compare the active earth pressures for the above three cases, and report the effect of earthquakes on the stability of the earth-retaining wall.

24-2. Recalling that the effect of a seismic force F_s on any horizontal section through a structure would be that of the force F_s applied at the center of gravity of all forces above the given section, check whether the horizontal section a-b of the massive concrete strip foundation, as yet without backfill, as shown, is safe against shearing off by a seismic force F_s. The horizontal acceleration of the seismic wave to reckon with is given as $\alpha = g/10$. The unit weight of concrete is 150 lb/ft^3, and the direct shear strength of the concrete is given as 200 psi. Also, check the concrete against crushing in section a-b. Would the entire foundation block slide because of the horizontal seismic force F_s? The coefficient of friction between concrete and soil is here given as $\mu = 0.58$.

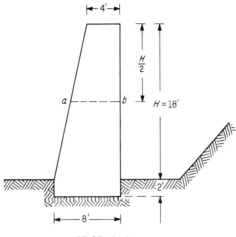

PROB. 24-2

24-3. By means of Westergaard's parabola, determine the hydrodynamic pressure distribution and total force of water imposed by an earthquake on a massive water-impounding dam. The point of application of the total horizontal seismic force is the centroid of the parabola. The depth of the water behind the dam is given as 30 ft; $\alpha = g/10$.

appendix **I**

Greek Alphabet

A	α	alpha
B	β	beta
Γ	γ	gamma
Δ	δ	delta
E	ε	epsilon
Z	ζ	zeta
H	η	eta
Θ	θ	theta
I	ι	iota
K	κ	kappa
Λ	λ	lambda
M	μ	mu
N	ν	nu
Ξ	ξ	xi
O	o	omicron
Π	π	pi
P	ρ	rho
Σ	σ, ς	sigma
T	τ	tau
Y	υ	upsilon
Φ	ϕ	phi
X	χ	chi
Ψ	ψ	psi
Ω	ω	omega

appendix **II**

Key to Signs and Notations

Symbol	Description
A	amplitude; area; anchor pull; base contact area of footing; force length; reaction
A_M	area for moment
ACI	American Concrete Institute
AASHO	American Association of State Highway Officials
ASCE	American Society of Civil Engineers
A_{crit}	critical area for shear
A_f	contact area for self-weight of footing
A_m	conical mantle surface area
A_{nec}	necessary area
A_s	cross-sectional area of soil; area of reinforcement steel for tensile forces
A_t	circular area perpendicular to the tip of a pile
A_v	area of shear in concrete design
Ave	average
A_1, A_2	footing areas
A'	cross-sectional area of pile cap or cushion block
a	acceleration; area of cross-section of a pile; cross-sectional area of a soil sample; displacement; distance; length; length of cantilever; moment arm; offset; a projection
$a\text{-}a, a\text{-}b$	sections
atm	atmosphere (pressure)
a_i	cross-sectional area of the ith pile
a_p	area of parabola
a_r	length
a_t	area of triangle

Appendix II

a_1, a_2, a_3, a_n	area
a'	acceleration
B	coefficient; buoyant force; width of cofferdam; width of footing; half-width of footing; reaction; length
B_f	computation width of bridge pier
B_{nec}	necessary width of footing
B_1, B_2	sides of a trapezoid
B_1, B_2	length of side of a square
b	dimension of "apparent mass"; distance; half-length of side of column; length; spacing of piles; thickness of wall; width of cofferdam, excavation, foundation; width of compression face of flexural member
$b\text{-}b$	a section
b_d	width
b_o	circular shear perimeter; length of critical section; length of periphery of critical section in shear
b_p	width of pedestal of bridge pier
C	centrifugal force; centroid; coefficient; coefficient of subgrade reaction; Culmann's line; reaction
$\overline{C}$	Culmann's vector
C.G.	center of gravity
$\mathcal{C}$	centerline
C_A	adhesion
C_B	center of buoyancy
C_W	center of displaced volume of water
C_h	coefficient of horizontal subgrade reaction
C_g	center of gravity
C_1, C_2, C_3, C_4	coefficients
c	cohesion; a constant; distance of farthest-spaced fiber from central axis; length; moment arm; safety factor
c'	coefficient
c_{adh}	adhesion of soil to foundation
cm	centimeter
const	constant
c_x, c_y	distance of extreme fiber
c_1, c_2, c_3	moment arms
D	dead load; diameter of circular well; diameter of extent of soil densification; diameter of footing;

	diameter of frozen column of soil; dielectric constant of soil-water system; distance; driving (embedment) depth; penetration depth of sheet piling; seepage pressure; diameter of bulbed-end pile.
D_f	depth of surcharge
D_i	inside diameter
DIN	Deutsche Industrie Normen (German Industrial Norms)
D_{min}	minimum diameter
D_o	outside diameter
D_s	length
d	depth of submergence; depth of foundation; diameter of pile; diameter of soil particle; differential; draft of caisson; driving depth; effective thickness of footing length of section of cofferdam; spacing of wells; thickness of footing
dA	differential of area
d_e	effective size of soil particle
dia.	diameter; distance of centroid
d_i	distance of centroid; inside diameter
d_{min}	minimum edge thickness of footing; minimum driving depth
d_{nec}	necessary thickness of footing
d_o	outside diameter
d_s	depth of scour
d_y	distance of centroid of pile from y-axis
d_1, d_2	depth
d_{10}	effective size of soil particle, or 10 percent diameter
d_{60}	60 percent diameter of soil particle
E	lateral earth pressure; Young's modulus of elasticity
E_a	active earth pressure
E_{inb}	inboard soil pressure
E_p	passive earth pressure (resistance)
E_s	silo pressure
E_1, E_2	Young's moduli of elasticity
E'	modulus of elasticity of pile cap
e	base of natural logarithmic system (2.7182...); eccentricity; Newtonian coefficient of restitution; void ratio
e_f	final void ratio

Appendix II

e_i	initial void ratio
e_{max}	maximum void ratio
e_{min}	minimum void ratio
e_p	eccentricity of column load
e_W	eccentricity of the weight of footing
F	force; frictional force; seismic force
F_D	driving force
F_R	resisting force
F_d	dynamic force
F_f	frictional force on cutting edge
F_h	horizontal seismic force
F_s	seismic force
F_v	vertical seismic force
f	self-frequency; tapping capacity of well; coefficient of friction
f-f	floating surface
f_1, f_2, f_3, f_4	coefficient of friction
f_c	allowable compressive stress in concrete
f_c'	compressive (ultimate) strength of concrete
f_i	coefficient of interlock friction
f_s	stress of steel
f_t	allowable tensile stress in concrete design; bending stress in concrete for fibers in tension
f_y	yield stress in steel
ft	foot
G	self-weight of cofferdam; specific gravity of soil particles; weight
GS	ground surface
GWT	groundwater table
G_{eff}	effective weight of cell ballast
G_{sh}	weight of sheet piling
G_z	weight of cofferdam
g	acceleration: $g = 981$ cm/sec^2 = 32.2 ft/sec^2; gram
H	distance; H-pile; height of retaining wall; horizontal force; hydrostatic head; length; thickness of compressible layer; thickness of water-bearing stratum; a vector
HWT	high water table

H_D	horizontal (tangential) driving force
H_L	total head (lift)
H_R	tangential frictional resistance
H_{crit}	critical depth
Hp	horsepower
H_s	safe resistance to sliding
H_t	total height of caisson
H_w	depth of water; hydrostatic pressure
$H_{2,3}, H_{4,5}$	horizontal loads
h	distance of heel point of wall to ground surface, or its extension; height of cofferdam; height of fill; height of free cantilever sheet piling; height of sheeting; total thickness of footing
h_f	depth
h_o	height of wetted filter
h_s	height
$h_1, h_2, h_3, \ldots, h_n$	depth of water; head of water; thickness of soil layers
hr	hour
I	electric current; moment of inertia
I_i	moment of inertia
I_{io}, I_o	central moment of inertia
I_s	moment of inertia
I_x, I_y	moments of inertia
I_1, I_2	moments of inertia
i	gradient; hydraulic gradient; pile number
i_{crit}	critical hydraulic gradient
i_{max}	maximum hydraulic gradient
i_{min}	minimum critical hydraulic gradient
i_1, i_2, i_3, i_4	points of intersection
in.	inches
J	point of intersection
j	a coefficient to d in concrete design
K	coefficient of lateral earth pressure; Krynine's coefficient of lateral earth pressure
K_a	coefficient of active earth pressure
K_p	coefficient of passive earth pressure (resistance)
K_1, K_2, K_3	coefficients

Appendix II

k	kip: 1 k = 1,000 lb
k	decimal fraction of thickness d; hammer coefficient; modulus of subgrade reaction; permeability coefficient
k_e	electro-osmotic coefficient of permeability
k_{ave}	average coefficient of permeability
k_h	horizontal coefficient of permeability
k_v	coefficient of vertical permeability
k_1, k_2	coefficients
k_{10}	coefficient of permeability at 10°C
k_{20}	coefficient of permeability at 20°C
kg	kilogram
kw	kilowatt
L	force; length; length of cell; length of filtration; length of foundation; length of pile; length of seepage path; length of wall; live load
L-L	a plane
L_c	computation length; depth; thickness of soil layer
L_{eff}	effective length of pile
L_{max}	maximum length
L_{min}	minimum spacing of end-bearing piles
L_p	embedment length of piles
L_1, L_2	length of piles
l	length; height
lb	pound
lg_{10}	decimal logarithm
ln	natural logarithm
log	logarithm to any base of (say base 50)
M	mass; mass of hammer; metacenter; moment; total working moment
M_A	moment about point A
M_B	total moment of change in buoyant force
M_C	resultant moment about C
$M_{\mathcal{C}}$	moment with respect to centerline
M_D	dead load moment; driving moment; overturning moment; static moment about point D
M_{E_a}	moment from active earth pressure
M_{E_p}	moment of forces acting on the tailside of cofferdam; moment from passive earth resistance

M_L	bending moment in longitudinal direction; live load moment
M_M	bending moments with respect to point M
M_R	resisting moment
M_S	bending moment ($=$ couple bS); bending moment in lateral (short) direction
M_W	restoring couple; static moment
M_{all}	allowable moment
M_c	metacenter
M_{fi}	total moment pertaining to interlock friction
$M_{i_1}, M_{i_2}, M_{i_3}, M_{i_4}$	moments with respect to points of intersection i_1, i_2, i_3, i_4, respectively
M_{m-m}	moment with respect to m-m axis
M_r	radial moment
M_{rt}	torsional moment
M_t	tangential moment
M_u	ultimate moment
M_{v-v}	moment about v-v
M_x	moment about critical section x
M_{x-x}	moment about x-x
M_y	static moment
M_{y-y}	static moment about y-y
M_z	moment about z-axis
M_1	static moment
$M_{1\max}, M_{2\max}$	maximum moments
$M_{\Delta B_O}$	static moment about point O
M_μ	moment of frictional resistance
m	meter; Poisson's number $= 1/\mu$
max	maximum
mi	mile
min.	minute(s)
min	minimum
mm	millimeter
m-m	critical section
m_1, m_2, m_3	moment arms
m_s-m_s	a section
N	capacity of electrical installation; number of blows
NWT	normal water table
N_c, N_q, N_γ	soil bearing capacity factors

Appendix II

N_c', N_q', N_γ'	soil bearing capacity factors
N_e	reaction
N_m	motor horsepower
N_p	pump horsepower
n	decimal fraction; an integer; a number; number of piles in a row; number of spacing of piles; porosity of soil; ratio of modulus of elasticity of steel to that of concrete
$n\text{-}n$	a normal
n_1, n_2	number of piles
O	center; origin of coordinates; pole
°	degree (of temperature or angle)
o	index (subscript) for various quantities
P	force; load
P_E	earth pressure ordinate; buoyant pressure from soil + hydrostatic pressure
PS	metric (continental) horsepower
P_c	column load
P_f	pile bearing capacity from mantle friction
P_i	ice pressure; load on any pile
$P_{i\max}$	maximum load on pile
$P_{\max}, P_{\min}$	loads on piles; pull-out capacity of tension pile
P_L	ultimate bearing capacity of pier
P_o	wave pressure
P_t	pile bearing capacity from tip of pile
P_u	ultimate load; total uplift pressure; ultimate bearing capacity of pile
P_{ui}	ultimate load on piles without weight of pile cap
$P_{u1}, P_{u2}, P_{u3}, P_{u4}$	ultimate loads
$P_{u\max}$	maximum ultimate load on pile
P_{ult}	ultimate load
P_w	pressure of water
$\bar{P}_1, \bar{P}_2, \bar{P}_3, \bar{P}_4$	vectors
$P_{1,2}, P_{3,4}, P_5,$	resultant pile loads
p	pressure on soil; mean effective steam pressure; pressure; ratio of area of tension reinforcement to effective area of concrete; steel ratio; surcharge intensity

p_a	compressed air pressure
p_b	pressure from ballast
p_f	pressure from weight of footing
p_h	horizontal pressure ordinate
p_i	initial stress in soil: $p_i = c \cot \phi$
p_o	surcharge intensity; pressure
p_s	pressure from surcharge
p_{tot}	total pressure
p_w	hydrostatic pressure; pressure from weight; uplift pressure; wind pressure
P_{w1}, P_{w2}	pressure diagram ordinates
p_x	parabolic function; pressure ordinate
p_1	pressure ordinate
psf	pounds per square foot
psi	pounds per square inch
Q	discharge; flow of water; force; shear force
Q.E.D.	*quod erat demonstrandum* (which was to be proved or shown)
Q_{eff}	effective discharge
Q_i	frictional force
Q_r	shear force in radial direction; shear resistance; vertical shear force
Q_t	shear force in tangential direction
Q_μ	total shear resistance in a vertical, neutral plane
q	rate of water pumped; ultimate bearing capacity of clay
q_{all}	allowable soil bearing capacity
q_{net}	net soil bearing capacity
q_u	ultimate, unconfined compressive strength of clay
q_{ult}, q'_{ult}	ultimate soil bearing capacity
R	bearing capacity of pile; ohmic resistance; radius of circular footing; radius of influence; radius of circular footing; vector; resultant
R_B	vertical projections of sheet piling
R_b	reaction
R_d	dynamic resistance of soil
R_f	mantle friction of pile, or caisson; total frictional resistance

Appendix II

R_i	inside radius
R_o	outside radius
R_s	vertical component of soil reaction
R_t	tip resistance of pile
$R_{t_{min}}$	minimum tip resistance
R_τ	shear resistance of clay soil
R_u	coefficient of internal resistance (ultimate) in concrete design
R_1, R_2	radii; resultant forces; soil reactions
$\bar{R}, \bar{R}_1, \bar{R}_2$	vectors
r	radius; length; moment arm
r_i	inside radius
r_{max}	maximum radius
r_o	radius; outside radius
r_{st}	radius of well for which stage y_{st} is computed
r_1, r_2	radii
S	degree of saturation of soil; embedment; force; spacing of freezer pipes; surcharge; suspension forces
S_i	shear force
Sta.	station
S_1, S_2	forces in struts
S_{3-4}	spacing of piles
S'_{u_1}	shear force in firm clay
s	distance; draw-down; penetration; settlement; spacing of piles; specific gravity of soil particles
s-s	critical shear plane
s_a	moment arm
s_h	horizontal displacement
s_{max}	maximum spacing of piles; maximum draw-down
s_1, s_2, s_3, s_4	distances; draw-downs; lengths
sec	second
T	period tangential force; temperature; tensile force; tension in tie rods; t-pile
T-T	a tie rod
T_f	freezing temperature
T_i	ice temperature
T_s	soil temperature; surface temperature
t	metric ton: 1 t = 1,000 kg; thickness of lateral sur-

	charge; thickness of seal; time; depth of sheet piling below bottom of excavation
$\tan \phi$	coefficient of internal friction of soil
$\tan \phi_1$	coefficient of friction between soil and foundation material
ton	short ton: 1 ton = 2,000 lb = 2 k
$t\text{-}t$	a tangent
U	coefficient of nonuniformity (or uniformity); perimeter; ultimate design load capacity
USS	U.S. Steel (Corporation)
V	shear force; vertical force; load; volume
$\overline{V}$	vector
V_L	shear force in longitudinal direction
V_x	pile reaction
V_s	stereometric volume
V_u	shear force
V_{1234}	volume
V_e	"stress volume"
v	critical vertical shear stress; shear stress; velocity; wind velocity
$v\text{-}v$	a section
v_b	bottom velocity
v_c	shear stress carried by concrete
v_{cu}	shear stress in concrete
v_{max}	maximum shear stress
v_o	fall velocity
v_s	critical mean velocity
v_u	ultimate shear strength
$v_{u_{all}}$	allowable shear strength in concrete; allowable shear stress for beam action
W	water pressure; weight; wind load
WT	water table
W_c	self-weight of caisson
W_b	weight of ballast
W_n	weight
W_o	self-weight of caisson and ballast
W_p	weight of pile
W_s	weight of backfill due to seismic force; weight of soil

Appendix II

W_w	weight of water
W_z	water pressure
W_1	apportioned weight of footing per pile
W_{1u}	ultimate apportioned weight of footing per pile
w	moisture content of soil; weight per pile
X	coordinate; distance
x	an unknown quantity; coordinate; thickness of seal
$x\text{-}x$	a section
x_b, x_g	coordinates
x_n	distance
x_o	coordinate; distance; depth of pivot point O below ground surface; moment arm
y	a coordinate
y_c	distance
y_i, y_n	distances
y_o	a coordinate
y_{st}	stage of water in a well
Z	section modulus
Z_p	pull-out force
Z_u	ultimate pull-out capacity
z	a coordinate; depth; depth of embedded pile; depth of foundation; moment arm
z_1, z_2	depths
α (alpha)	acceleration; angle; angle of batter of a retaining wall and piles; angle of rupture
α_{max}	maximum acceleration
α_h	horizontal acceleration
α_v	vertical acceleration
α_1, α_2	angles
α'	acceleration
β (beta)	angle; angle of batter of piles; angle of inclination: coefficient of load characteristic; a ratio
γ (gamma)	angle; unit weight of concrete; unit weight of soil
γ_b	buoyant weight of soil
γ_c	unit weight of concrete

Appendix II

γ_d, γ_{dry}	dry unit weight of soil
γ_m	unit weight of mud (silt)
γ_1	surcharged unit weight of soil
γ_s	unit weight of soil; unit weight of solids of soil by absolute volume
γ_{sat}	saturated unit weight of soil
γ_{sub}	submerged (buoyant weight) of soil
γ_w	unit weight of water
Δ (delta)	increment
ΔB	change in buoyant force
$\Delta L, \Delta L_1, \Delta L_2$	longitudinal elastic deformation
Δh	change in h
Δy	a tilt
δ (delta)	thickness of caisson wall; thickness of plain concrete footing; thickness of shell
ε (epsilon)	angle
ζ (zeta)	uplift intensity factor; zeta potential
η (eta)	coefficient of efficiency; dynamic viscosity of water; efficiency factor of a pile group; factor of safety
η_D	load factor for dead weight
η_L	load factor for live load
η_{all}	allowable factor of safety
η_e	efficiency of hammer
η_m	coefficient of mechanical efficiency of motor
η_p	coefficient of mechanical efficiency of pump
θ (theta)	angle; angular displacement
λ (lamdba)	acceleration ratio; coefficient; electroconductivity of water; seismic factor; ratio of short side to long side of caisson
μ (mu)	a coefficient; coefficient of internal friction of soil: $\mu = \tan \phi$; micron: 1 micron $= 1$ $\mu = 1 \times 10^{-4}$ cm $= 1 \times 10^{-3}$ mm; Poisson's ratio: $\mu = 1/m$
$\mu\mu$	millimicron: 1 $\mu\mu = 10^{-7}$ cm $= 10^{-6}$ mm

Appendix II

$\mu_{b/r}$	coefficient of friction between ballast and rock
μ_1	$\mu_1 = \tan \phi_1$; coefficient of friction between soil and wall material
ξ (xi)	distance; frost penetration depth; thickness of ice wall
π (pi)	geometric ratio of circumference of a circle to its diameter: $\pi = 3.14159...$
ρ (rho)	angle of rupture; radius; specific ohmic resistance
$\rho_{r/b}$	a ratio of radius to width
$\sum$ (sigma)	summation of
$\sum_{1}^{n}$	summation of ... from 1 to n
ΣF	mantle friction
ΣL	sum of resistances
σ (sigma)	stress
$\sigma_A, \sigma_B, \sigma_C$	corner stresses
σ_E	lateral stress on conical surface of soil
σ_W	pressure of soil from weight of footing
σ_{all}	allowable stress in sheet piling; allowable stress in soil
σ_{ave}	average pressure (stress)
σ_b	bending stress
$\sigma_{b_{all}}$	allowable bending stress
$\sigma_{b_{max}}$	maximum allowable bending stress
σ_{b_t}	allowable tensile stress in bending
σ_c	compressive stress in concrete
σ_{crit}	critical edge pressure
σ_d	stress in diaphragm
σ_e	edge pressure from moment
σ_h	average horizontal pressure on vertical mantle surface
σ_{max}	height of stress paraboloid; maximum stress
σ_{min}	minimum stress
σ_n	normal stress
σ_{net}	net upward soil contact pressure
σ_o	contact pressure; soil bearing capacity at the ground surface
$\sigma_{s_{max}}$	soil reaction, maximum stress from
$\sigma_{s_{min}}$	minimum stress from soil reaction
σ_{sp}	silo pressure

σ_t	flexural strength of concrete: tensile stress
$\sigma_{t_{all}}$	allowable tensile stress in interlock
$\sigma_{t_{max}}$	maximum tensile stress
σ_u	ultimate soil pressure
σ_x	stress at critical section x-x
σ_y	horizontal stress on soil
σ_z	soil bearing capacity at depth z; vertical pressure
σ_1	major principal stress; soil pressure
σ_2	soil pressure
σ_3	minor principal stress
$\sigma_{1_{all}}, \sigma_{2_{all}}, \sigma_{3_{all}}$	allowable soil pressures
$\sigma_\text{I}, \sigma_\text{III}$	principal stresses
τ (tau)	shear stress
τ_L	frictional mantle resistance on a pile
τ_{ave}	average shear stress
τ_c	shear stress on conical tip of pile
τ_{max}	maximum shear stress
τ_s	shear stress in concrete
τ_u	frictional stress
τ_{ult}	ultimate shear strength
τ_z	frictional mantle resistance
ϕ (phi)	angle of internal friction of soil; angle of tear of soil; a capacity reduction factor in concrete design; a coefficient in electro-osmosis; symbol for indicating a round cross-sectional area
ϕ'	reduced angle of friction
$\phi_{b/r}$	angle of friction between ballast and rock
ϕ_s	angle of internal friction of cofferdam-supporting soil
ϕ_steel	diameter of steel rods
ϕ_1	angle of friction between soil and foundation material
ψ (psi)	a coefficient
ψ_1	a coefficient
Ω (omega)	angle; angle of rupture
ω (omega)	angle; angle of rupture; amplitude
$\triangle$	a symbol for a geometric triangle

Appendix II　　　　　　　　　　　　　　　　　　　　　　　　　　　　**769**

▽, ▼　　　　　　symbols to indicate a particular surface elevation, for example ground surface, groundwater table, or any elevation on elevation plans of structures and soil profiles

⌀　　　　　　　symbol to indicate a square cross-sectional area
╤　　　　　　　plate-like material
∴　　　　　　　therefore

appendix **III**

Conversion Factors of Units of Measurement

Multiply	By	To Obtain
Length		
centimeters (cm)	0.01	meters
	10	millimeters
	0.03281	feet
	0.3937	inches
meters (m)	100	centimeters
	1,000	millimeters
	3.28	feet
	39.37	inches
	1.094	yards
micron (μ)	$0.001 = 10^{-3}$	millimeters
	$0.0001 = 10^{-4}$	centimeters
	$0.000001 = 10^{-6}$	meters
millimeters (mm)	0.1	centimeters
	0.001	meters
	1,000	microns
	0.00328	feet
	0.03937	inches
feet (ft)	12	inches
	1/3	yards
	30.481	centimeters
	0.3048	meters
inches (in.)	0.08333	feet
	2.54	centimeters
	0.0254	meters
	25.4	millimeters
yards (yd)	3	feet
	91.44	centimeters
	0.9144	meters

Appendix III

Area

square centimeters (cm²)	$0.0001 = 10^{-4}$	square meters (m²)
	100	square millimeters (mm²)
	0.001076387	square feet (ft²)
	0.155	square inches (in.²)
square meters (m²)	$10,000 = 10^4$	square centimeters (cm²)
	$1,000,000 = 10^6$	square millimeters (mm²)
	10.764	square feet (ft²)
	$(2.471)(10^{-4})$	acres
	$(3.861)(10^{-7})$	square miles (mi²)
	1.196	square yards (yd²)
square millimeters (mm²)	0.01	square centimeters (cm²)
	$(1.550)(10^{-3})$	square inches (in.²)
square feet (ft²)	144	square inches (in.²)
	0.111111	square yards (yd²)
	929.0304	cm²
	0.0929	m²
	$(2.296)(10^{-5})$	acres
	$(3.587)(10^{-8})$	mi²
	1/9	yd²
square inches (in.²)	$(6.944)(10^{-3})$	ft²
	$(6.452)(10^{-4})$	m²
	6.452	cm²
	645.2	mm²
square yards (yd²)	$(2.066)(10^{-4})$	acres
	9	ft²
	8,361.273	cm²
	0.8361	m²
	$(3.228)(10^{-7})$	mi²
acres	0.405	hectares (ha)
	4,046.849	m²
	43,560	ft²
	4,840	yd²
hectares (ha)	10,000	m²
	2.471	acres
	$(1.076)(10^5)$	ft²
square kilometers (km²)	247.104	acres
	100	ha
	10^6	m²
	$(10.76)(10^6)$	ft²
	0.3861	mi²
	$(1.196)(10^6)$	yd²

square mile (mi^2)	640	acres
	$(27.88)(10^6)$	ft^2
	259	ha
	2.590	km^2
	$(3.098)(10^6)$	yd^2

Volume

cubic centimeters (cm^3)	10^{-3}	liters (l)
	10^{-6}	m^3
	10^3	mm^3
	$(3.531)(10^{-5})$	ft^3
	$(6.102)(10^{-2})$	in.^{3}
	$(2.642)(10^{-4})$	U.S. gallons
cubic meters (m^3)	10^6	cm^3
	10^9	mm^3
	35.3148	ft^3
	61,024.044	in.^{3}
	1.308	yd^3
	264.173	U.S. gallons
	10^3	liters (l)
cubic feet (ft^3)	$(2.8316)(10^4)$	cm^3
	0.028316	m^3
	1,728	in.^{3}
	0.03704	yd^3
	7.48053	gallons
	28.316	liters (l)
cubic inches (in.^{3})	16.387	cm^3
	$(1.639)(10^{-5})$	m^3
	$(5.787)(10^{-4})$	ft^3
	$(2.143)(10^{-5})$	yd^3
	$(4.329)(10^{-3})$	gallons
	$(1.639)(10^{-2})$	liters (l)
cubic yards (yd^3)	27	ft^3
	46,656	in.^{3}
	0.7646	m^3
	202.0	gallons
	764.6	liters (l)

Capacity

U.S. gallons (gal)	3,785	cm^3
	3.785	liters (l)
	$(3.785)(10^{-3})$	m^3
	0.133680	ft^3
	231	in.^{3}
	$(4.951)(10^{-3})$	yd^3
	0.833	British Imperial gallons
British Imperial gallons	1.201	U.S. gallons
	4.545	liters (l)

Appendix III

liters (l)	10^3	cm^3
	10^{-3}	m^3
	0.035315	ft^3
	61.025	$in.^3$
	$(1.308)(10^{-3})$	yd^3
	0.22007	British Imperial gallons
	0.2642	U.S. gallons

Weight

grams (g)	10^3	milligrams (mg)
	10^{-3}	kilograms (kg)
	0.03527	ounces (oz)
	$(2.205)(10^{-3})$	pounds (lb)
kilograms (kg)	10^3	grams
	10^{-3}	metric tons (t)
	$(9.84206)(10^{-4})$	long tons
	2.205	pounds (lb)
	$(1.102)(10^{-3})$	short tons
metric tons	10^3	kilograms (kg)
	0.984206	long tons
	2,204.6223	pounds (lb)
	1.1023112	short tons
kips (k)	1,000	pounds
	0.5	short tons
	453.592	kilograms (kg)
	0.453592	metric tons (t)
long tons	1,016	kilograms (kg)
	1.016	metric tons (t)
	2,240	pounds
	1.120	short tons
ounces (oz)	28.349527	grams (g)
	0.0625	pounds (lb)
pounds (lb)	453.592	grams (g)
	0.453592	kilograms (kg)
	$(4.464286)(10^{-4})$	long tons
	$(4.535924)(10^{-4})$	metric tons (t)
	16	ounces
	0.0005	short tons
	0.001	kips
short tons	907.18486	kilograms (kg)
	0.90718486	metric tons (t)
	2	kips
	0.89287	long tons
	32,000	ounces (oz)
	2,000	pounds (lb)

1 lb of water fills 27.68 in.^{3} = 0.01602 ft^3 = 0.1198 U.S. gallons
1 kg of water fills 1 liter = 1,000 cm^3 = 10^{-3} metric tons
1 U.S. gallon of water at 62°F (16.65°C) weighs 8.3364 lb
1 ft^3 of water at 60°F (15.54°C) weighs 62.4 lb
1 cm^3 of distilled water at 4°C weighs 1 g
1 liter of distilled water at 4°C weighs 1 kg

Weight per Unit Length

short tons per yard	0.992111	tons per meter (t/m)
short tons per foot	2.976333	tons per meter (t/m)
short tons per inch	35.715996	tons per meter (t/m)
pounds per yard (lb/yd)	0.49605554	kilograms per meter (kg/m)
pounds per foot (lb/ft)	1.48816662	kilograms per meter (kg/m)
pounds per inch (lb/in.)	0.178579995	kilograms per centimeter (kg/cm)
kilograms per centimeter (kg/cm)	5.599731	pounds per inch (lb/in.)
kilograms per meter (kg/m)	0.67196777	pounds per foot (lb/ft)
	2.0159033	pounds per yard (lb/yd)
metric tons per meter (t/m)	1.007952	short tons per yard
	0.335984	short tons per foot
	0.0279987	short tons per inch

Stress

atmospheres (physical), (atm)	1.01325	bars (b)
	760	mm column of Hg at 0°C
	29.92	inch column of mercury at 0°C
	10.33228	meters column of water
	406.78	inch column of water
	33.89	foot column of water
	1,033.228	g/cm^2
	1.033228	kg/cm^2
	(1.033228)(10^4)	kg/m^2
	10.33228	t/m^2
	2,116.8	lb/ft^2
	14.70	lb/in.^{2}
	1.058	ton/ft^2
bars (b)	750.062	mm column of Hg
	1.019716	kg/cm^2
	14.503	lb/in.^{2}
	0.98692	physical atmospheres
feet of water	0.02949	atmospheres (atm)
	0.88265	inch column of Hg
	2.241931	cm column of Hg
	30.481	cm column of water
	0.030479	kg/cm^2
	0.30479	t/m^2
	62.427	lb/ft^2
	0.43352	lb/in.^{2}

Appendix III

grams per square centimeter (g/cm^2)	0.001	kg/cm^2
	0.01	t/m^2
	$(3.937)(10^{-1})$	inch column of water at 4°C
kilogram per square centimeter (kg/cm^2)	10	t/m^2
	0.967814	physical atmospheres (atm)
	0.980665	bars
	32.8093	feet of water
	2.0481614	kip/ft^2
	2,048.1614	lb/ft^2
	14.223293	$lb/in.^2$
	735.559	mm column of Hg
	1.024	ton/ft^2
kilograms per square meter (kg/m^2)	$(9.678)(10^{-5})$	atm
	$(7.356)(10^{-3})$	cm column of Hg at 0°C
	$(2.896)(10^{-3})$	inch column of Hg at 0°C
	$(1.0000488)(10^{-3})$	meter column of water
	$(3.281)(10^{-3})$	feet column of water
	$(3.9372)(10^{-2})$	inch volumn of water at +4 °C
	$(1.0)(10^{-4})$	kg/cm^2
	$(2.0481614)(10^{-4})$	kip/ft^2
	0.2048	lb/ft^2
	$(1.422)(10^{-3})$	$lb/in.^2$
	$(1.024)(10^{-4})$	ton/ft^2
metric tons per square meter (t/m^2)	0.1	kg/cm^2
	0.20481614	kip/ft^2
	204.81614	lb/ft^2
	1.4223	$lb/in.^2$
	0.102408	ton/ft^2
kips per square foot (kip/ft^2)	0.488244	kg/cm^2
	1,000	lb/ft^2
	6.94445	$lb/in.^2$
	0.5	ton/ft^2
	4.88247	t/m^2
pounds per square foot (lb/ft^2)	$(4.725)(10^{-4})$	physical atmospheres
	0.48829	cm column of water
	0.01602	foot column of water
	0.1922	inch column of water at +4 °C
	$(3.591)(10^{-2})$	cm column of Hg at 0°C
	$(1.414)(10^{-2})$	inch column of Hg at 0°C
	$(1.0)(10^{-3})$	kip/ft^2
	$(6.944)(10^{-3})$	$lb/in.^2$
	0.0005	ton/ft^2
	0.4882	g/cm^2
	0.0004882	kg/cm^2
	4.882	kg/m^2
	0.004882	t/m^2

pounds per square inch ($lb/in.^2$)	0.06804	physical atmospheres
	0.7031736	meter column of water
	2.307	foot column of water
	27.68	inch column of water
	5.171	cm column of Hg
	2.036	inch column of Hg
	0.070307	kg/cm^2
	703.07	kg/m^2
	0.70307	t/m^2
	0.144	kip/ft^2
	144	lb/ft^2
	0.072	ton/ft^2
tons (short) per square foot (ton/ft^2)	0.945	physical atmospheres
	9.7663	m column of water at 4°C
	384.5	inch column of water at 4°C
	71.83	cm column of Hg at 0°C
	28.28	inch column of Hg at 0°C
	976.5	g/cm^2
	0.97648 $\approx$ 1.00	kg/cm^2
	9,764.8	kg/m^2
	9.7648	t/m^2
	2,000	lb/ft^2
	2	kip/ft^2
	13.889	$lb/in.^2$

Unit Weight

grams per cubic centimeter (g/cm^3)	0.001	kg/cm^3
	1,000	kg/m^3
	1	t/m^3
	62.4	lb/ft^3
	0.03613	$lb/in.^3$
kilograms per cubic meter (kg/m^3)	0.001	g/cm^3
	0.001	t/m^3
	0.062427	lb/ft^3
	$(3.613)(10^{-\ })$	$lb/in.^3$
tons per cubic meter (t/m^3)	62.4	lb/ft^3
pounds per cubic foot (lb/ft^3)	0.01602	g/cm^3
	0.00001602	kg/cm^3
	16.018548	kg/m^3
	0.0160185	t/m^3
	$(5.787)(10^{-4})$	$lb/in.^3$
pounds per cubic inch ($lb/in.^3$)	27.68	g/cm^3
	$(2.768)(10^4)$	kg/m^3
	27.68	t/m^3
	1,728	lb/ft^3

Appendix III

Velocity

cm/sec	864	m/day
	36.0	m/hr
	0.6	m/min
	0.01	m/sec
	0.036	km/hr
	0.0006	km/min
	0.00001	km/sec
	2,835.36	ft/day
	118.140	ft/hr
	1.9686	ft/min
	0.03281	ft/sec
	1,034,906.4	ft/yr
	34,015.68	in./day
	1,417.32	in./hr
	23.622	in./min
	0.3937	in./sec
	$(3.728)(10^{-4})$	mi/min
	$(6.213)(10^{-6})$	mi/sec
cm/min	0.0166	cm/sec
	14.4	m/day
	0.6	m/hr
	0.01	m/min
	0.000166	m/sec
	0.03281	ft/min
	0.00054683	ft/sec
	0.3937	in./min
	0.00653542	in./sec
m/min	1.6667	cm/sec
	$(1.6667)(10^{-2})$	m/sec
	3.281	ft/min
	0.05468	ft/sec
	0.06	km/hr
	0.001	km/min
	$(3.728)(10^{-2})$	mi/hr
	$(6.214)(10^{-4})$	mi/min
m/sec	100	cm/sec
	60	m/min
	3.60	km/hr
	0.06	km/min
	196.851	ft/min
	3.281	ft/sec
	2.237	mi/hr
	0.03728	mi/min

km/hr	$(1.6667)(10^{-2})$	km/min
	27.7778	cm/sec
	16.67	m/min
	0.2778	m/sec
	54.68	ft/min
	0.9113	ft/sec
	$(1.036)(10^{-2})$	mi/min
	0.6214	mi/hr
ft/min	0.5080	cm/sec
	0.01829	km/hr
	$(3.048)(10^{-4})$	km/min
	0.3048	m/min
	$(0.5080)(10^{-3})$	m/sec
	0.01667	ft/sec
	525,600	ft/yr
	0.01136	mi/hr
	$(1.892)(10^{-4})$	mi/min
ft/sec	30.481	cm/sec
	1.097	km/hr
	$(1.829)(10^{-2})$	km/min
	18.29	m/min
	0.30481	m/sec
	60	ft/min
	0.01136	mi/min
	0.6818	mi/hr
ft/yr	$(9.665164)(10^{-7})$	cm/sec
	$(5.79882)(10^{-7})$	m/min
	$(1.9025)(10^{-6})$	ft/min
	$(2.16203)(10^{-8})$	mi/hr
in./sec	2.54	cm/sec
	1.524	m/min
	0.0254	m/sec
	4.9998	ft/min
	0.08333	ft/sec
	60	in./min
mi/hr	44.7041	cm/sec
	1.609	km/hr
	$(2.682)(10^{-2})$	km/min
	$(4.46)(10^{-4})$	km/sec
	0.44704	m/sec
	26.82	m/min
	5,280	ft/hr
	88	ft/min
	1.467	ft/sec
	$(1.667)(10^{-2})$	mi/min
	$(2.778)(10^{-4})$	mi/sec

Appendix III

Rates

cm^3/sec	0.0010	liter/sec
	$(1.0)(10^{-6})$	m^3/sec
	$(3.5314667)(10^{-5})$	ft^3/sec
	$(2.642)(10^{-4})$	U.S. gallon/sec
m^3/sec	$(1.0)(10^6)$	cm^3/sec
	1,000	liter/sec
	35.314667	ft^3/sec
	264.2	U.S. gallon/sec
liters per minute	1,000	cm^3/min
	16.444	cm^3/sec
	$(5.886)(10^{-4})$	ft^3/sec
	$(4.403)(10^{-3})$	gallon/sec
liters per second	1,000	cm^3/sec
	$(1.0)(10^{-3})$	m^3/sec
	0.03531	ft^3/sec
	0.2642	gallon/sec
cubic feet per minute (ft^3/min)	472.0	cm^3/sec
	0.472	liter/sec
	0.1247	gallon/sec
	62.43	lb of water/min
cubic feet per second (ft^3/sec)	0.02832	m^3/sec
	28.32	liter/sec
	448.831	gallon/min
	7.48052	gallon/sec
gallons per minute (gallons/min)	0.06308	liter/sec
	8.0208	ft^3/hr
	$(2.228)(10^{-3})$	ft^3/sec
	6.0086	short tons of water/24 hours
gallons per second (gallons/sec)	3,785	cm^3/sec
	3.785	liter/sec
	0.003785	m^3/sec
	0.1337	ft^3/sec

Mechanical Work; Energy

kilogram-meters (kg-m)	10^5	g-cm
	$(2.724)(10^{-6})$	kilowatt-hours
	$(2.724)(10^{-3})$	watt-hours
	9.807	watt-seconds
	7.233	ft-lb
kilowatt-hours (kw-hr)	$(3.671)(10^5)$	kg-m
	$(2.655)(10^6)$	ft-lb
	1,000	watt-hours

foot-pounds (ft-lb)	$(1.383)(10^4)$	g-cm
	0.1383	kg-m
	$(3.766)(10^{-7})$	kilowatt-hours
	$(3.766)(10^{-4})$	watt-hours

Power

kilowatts	1.341	horsepower [550 (ft-lb) sec system]
	737.6	(ft-lb)/sec
foot-pounds per sec	$(1.818)(10^{-3})$	horsepower
	$(1.356)(10^{-3})$	kilowatts
	$(1.843)(10^{-3})$	metric horsepower
	1.356	watts
horsepower (metric)	75	(kg-m)/sec
horsepower (U.S.)	550	ft-lb/sec
	1.014	metric horsepower
	745.7	watts
horsepower (metric)	542.5	(ft-lb)/sec
	0.9863	horsepower [550 (ft-lb) sec system]
	0.7355	kilowatts
	735.5	watts
(kg-m)/sec	0.013333	metric horsepower

appendix **IV**

Commonly Used Soil Tests and Their Applications

No.	Soil Properties or Tests	Symbol or Description	Test Results	Practical Applications of Test Results
1	Bearing capacity	σ (lb/in.^{2}, ton/ft^2, kg/cm^2, or t/m^2)	Load-carrying capacity	Evaluation of strength of soil for supporting structures
2	Capillarity	H (cm)	Capillary height in sand and silt	(a) Frost penetration depth in soils (b) Water loss from reservoirs by capillarity (c) Soil moisture transfer (d) Soil susceptibility to frost
3	Soil color tests	Soil color	Color coordinates	Identification of soils
4	Combustion (ignition) test	b (g or lb)	Weight of burned matter	(a) Content of organic matter (b) Settlement of structures on organic soils
5	Compaction	W_{d_3} (lb/ft^3 or kg/m^3)	Maximum dry density at optimum moisture content	Compaction of subgrade and fills
6	Consistency (Atterberg) limits of cohesive soils	L.L. (percent)	Liquid limit; moisture content in percent at 25 blows exerted by a test cup	(a) Soil classification (b) Clue to shear strength of soil (c) Stability of soil mass

Appendix IV

No.	Soil Properties or Tests	Symbol or Description	Test Results	Practical Applications of Test Results
		P.L. (percent)	Plastic limit; moisture content in percent upon cracking of soil threads $\frac{1}{8}$ in. in diameter	(a) Soil classification (b) Soil plastic properties
		P.I. (percent)	Plasticity index; moisture in percent	(a) Soil classification (b) Thickness of subbase courses (c) Impermeable core material for earth dams
		S.L. (percent)	Shrinkage limit; percent of moisture at attained constant volume	(a) Soil classification (b) Evaluation of possibility of saturation (c) Evaluation of usefulness of soil for earthworks
7	Consolidation test	s (cm)	Settlement analysis	(a) Compression of soil (b) Soil swelling pressures on floors, earth-retaining structures, and tunnels (c) Expansion (d) Permeability (e) Preconsolidation load (f) Settlement analysis of soils and structures
8	Direct shear test	τ (kg/cm² or lb/ft²); ϕ c	Shear strength of soil Angle of internal friction; cohesion	(a) Direct shear strength (b) Bearing capacity of soils (c) Depth of laying foundations (d) Earth pressure calculations (e) Stability analysis of earth slopes
9	Freezing tests of soils	Q ξ Δh		(a) Amount of water absorbed by soil (b) Soil performance under freezing and thawing (c) Artificial freezing and thawing of soils for foundation engineering purposes (d) Frost penetration depth (e) Frost heaves (f) Frost susceptibility of soils

Appendix IV

No.	Soil Properties or Tests	Symbol or Description	Test Results	Practical Applications of Test Results
10	Granulometry: mechanical analysis of soil (sieving)	Percentage of various soil fractions	Particle size accumulation curve for particles $\geqslant 0.074$ mm (+ No. 200 material)	(a) Compaction of soils (b) Effective particle size, d_{10} (c) Evaluation of filters (d) Frost susceptibility of soil (e) Identification and classification of soil (f) Permeability (g) Soil mixtures for earthwork purposes (h) Soil texture (i) Uniformity coefficient, $U = \dfrac{d_{60}}{d_{10}}$
11	Hydrometer analysis	Percentage of fractions	Particle size, < 0.1 mm	Same as No. 10
12	Large-scale tests			Large-scale tests in the laboratory and/or the field
13	Model experiments		Similarity of model and prototype	Small-scale laboratory tests on various subjects pertaining to soil and foundation engineering
14	Moisture content of soil	w (percent) $n; e$ D	Moisture content in percent by dry weight of solid particles of soil	(a) Classification of soil (b) Comparison of soils of similar nature (c) Porosity; void ratio (d) Relative density (e) Soil compaction (f) Soil-freezing tests (g) Volumetric and gravimetric relationships
15	Permeability	k (cm/sec or ft/day)	Coefficient of permeability at constant head or at falling head	(a) Frost penetration depth of soils (b) Soil consolidation (c) groundwater flow (d) Lowering of groundwater table (e) Grouting and injection (f) Seepage through earth dams
16	Physicochemical properties	σ_o ζ, E_s	Surface tension Electro-osmosis Thermo-osmosis Electrokinetic potentials	(a) Frost action in soils (b) Soil-moisture migration (c) Chemical stabilization of soil

No.	Soil Properties or Tests	Symbol or Description	Test Results	Practical Applications of Test Results
		ε	Dielectric constants	
			Ion exchange	
		D	Vapor diffusion	
			Wetting; swelling	
17	Porosity	n	Volume of voids of solid particles, water, and air	(a) Calculations involving physical properties (b) Evaluation of densification of sand (c) Permeability (d) Settlement calculations (e) Soil-freezing tests (f) Volumetric and gravimetric relationships
18	Relative density	D	Porosity, viz., density attained in various degrees of compaction; wet and dry	Density control of earthworks made of noncohesive soils, and degree of compaction of fills, or state of density of natural granular deposits; for noncohesive soils only
19	Specific gravity	G	Ratio of density of soil solids to the density of water at $+4°C$	(a) Buoyant weight of soil (b) Compaction of soil (c) Density calculations (d) Hydrometer analyses (e) Soil-freezing tests (f) Volumetric and gravimetric relationships
20	Specific surface	A (cm^2)	Surface area of soil particles in a unit weight (1 g) of soil or in a unit volume (1 cm^3) of soil	(a) Electrokinetic phenomena in soil (b) Frost susceptibility (c) Amount of moisture (d) Soil-moisture migration
21	Triaxial compression test	α ϕ c τ	Angle of rupture Angle of internal friction Cohesion Shear strength	(a) Bearing capacity of cohesive soils (b) Depth of laying foundations (c) Stability analyses of earth slopes (d) Earth pressure calculations (e) Shear strength (f) Modulus of elasticity

Appendix IV 785

No.	Soil Properties or Tests	Symbol or Description	Test Results	Practical Applications of Test Results
22	Unconfined compression test	α τ ϕ c	Angle of rupture Angle of internal friction Cohesion	(a) Same as No. 21 (b) Unconfined compressive strength of soil (c) Unconfined shear strength
23	Unit weight	γ	Weight per unit volume of soil	(a) Bearing capacity of soil (b) Classification of soils by density (c) Density of a soil deposit (d) Soil-freezing tests (e) Lateral earth pressures on walls and sheet piling (f) Settlement analyses (g) Stability analyses of earth masses and slopes
24	Void ratio	e	Ratio of volume of voids to volume of solid particles of soil	(a) Calculations involving physical properties (b) Settlement analyses (c) Permeability (d) Soil-freezing tests

appendix V

Bibliography for Further Reading

1. H. P. Aldrich, "Precompression for Support of Shallow Foundations." *J. Soil Mech. and Foundations Div., Proc. ASCE*, Vol. 91, No. SM2 (1965), pp. 5-20.
2. R. W. Abbett (ed.), *American Civil Engineering Practice*. New York: Wiley. 1956-1957.
3. Anon., "Tower Forms 10 Stories High are Sunk for Newport Bridge." *Engineering News-Record*, Mar. 2, 1967, pp. 28-32.
4. D. D. Barkan, *Dynamics of Bases and Foundations*, trans. from the Russian by L. Drashevska. New York: McGraw-Hill, 1962.
5. L. Bjerrum, A. Casagrande, R. B. Peck, and A. W. Skempton, *From Theory to Practice in Soil Mechanics: Selections from the Writings of Karl Terzaghi*. New York: Wiley, 1960.
6. E. W. Brand, "Comparative Analysis of Data from Pumping Tests in an Unconfined Aquifer." *Proceedings of the Institution of Civil Engineers (London)*, Vol. 38 (October 1967), pp. 267-284.
7. A. Casagrande, "Role of the 'Calculated Risk' in Earthwork and Foundation Engineering," *J. Soil Mech. and Foundations Div., Proc. ASCE*, Vol. 91, No. SM4 (1965), pp. 1-40.
8. R. D. Chellis, *Pile Foundations—Theory, Design, Practice*. New York: McGraw-Hill, 1951.
9. M. P. Cloyd, "Monopod." *Civil Engineering*, March 1968, pp. 55-57.
10. C. W. Dunham, *Foundation of Structures*. 2nd ed., New York: McGraw-Hill, 1962.
11. G. A. Leonards (ed.), *Foundation Engineering*, New York: McGraw-Hill, 1962.
12. H. Q. Golder, "State-of-the-art of Floating Foundations." *J. Soil Mech. and Foundations Div., Proc. ASCE*, Vol. 91, No. SM2 (1965), pp. 8-88.
13. Bent Hansen, *A Theory of Plasticity for Ideal Frictionless Materials*. Copenhagen: Teknisk Forlag, 1965.
14. A. R. Jumikis, "Soil Vertical Stress Influence Value Charts for Rectangles." *Proc. ASCE*, Vol. 97, No. SM2, February, 1971, Paper No. 7860.
15. A. R. Jumikis, *Stability Analyses of Soil-Foundation Systems*. Engineering Research Publication No. 44, College of Engineering, Rutgers University, New Brunswick, N.J., 1965. Contains complete logarithmically spiraled rupture surface tables.
16. A. R. Jumikis, *Stress Distribution Tables for Soil Under Concentrated Loads*. New Brunswick, N.J.: Bureau of Engineering Research, College of Engineering, Rutgers University, Engineering Research Publication No. 48, 1969.

17. A. R. Jumikis, *Thermal Soil Mechanics*. New Brunswick, N.J.: Rutgers U. P., 1966, pp. 119-179. Chapters 7-12 on freezing-in foundation excavations.
18. A. R. Jumikis, *Vertical Stress Tables for Uniformly Distributed Loads on Soil*. New Brunswick, N.J.: Bureau of Engineering Research, College of Engineering, Rutgers University Engineering Research Publication No. 52, 1971.
19. J. Lysmer and F. E. Richart, Jr., "Dynamic Response of Footings to Vertical Loadings," *J. Soil Mech. and Foundations Div., Proc. ASCE*, Vol. 92, No. SM1 (1966), pp. 65-91.
20. G. G. Meyerhof, "Shallow Foundations." *J. Soil Mech. and Foundations Div., Proc. ASCE*, Vol. 91, No. SM2 (1965), pp. 21-31.
21. H. A. Mohr, "The Gow Caisson," *J. Boston Soc. Civil Eng.*, Vol. 51, No. 1 (1964), pp. 75-94.
22. G. J. Murphy and D. N. Tanner, "The Bart Trans-Bay Tube." *Civil Engineering*, December 1966, pp. 51-55.
23. N. M. Newmark, "Effects of Earthquakes on Dams and Embankments." *Géotechnique* (London), Vol. 15, No. 2 (1965), pp. 139-160.
24. L. Obert and W. I. Duval, *Rock Mechanics and the Design of Structures in Rock*. New York: Wiley, 1967.
25. J. D. Parsons, "Piling Difficulties in the New York Area." *J. Soil Mech. and Foundations Div., Proc. ASCE*, Vol. 92, No. SM1 (1966), pp. 43-64.
26. R. B. Peck, "Records of Load Tests on Friction Piles," Highway Research Board's Special Report No. 67. Washington, D.C., 1961.
27. E. A. Prentis and L. White, *Underpinning*, 2nd ed. New York: Columbia University Press, 1950.
28. Proceedings of the Sixth International Conference on Soil Mechanics and Foundation Engineering, Sept. 8-15, 1965, Montreal, Canada. Vol. 1-3. Toronto: U. of Toronto Press, 1966.
29. Proceedings of the Symposium on Soil-Structure Interaction, University of Arizona, September 1964, Tucson, Arizona.
30. A. DeF. Quinn, *Design and Construction of Ports and Marine Structures*. New York: McGraw-Hill, 1961.
31. C. M. Harris and C. E. Crede (eds.), *Shock and Vibration Handbook*. New York: McGraw-Hill, 1961.
32. *Research Conference on Shear Strength of Cohesive Soils*. Held June, 1960, University of Colorado, Boulder Colorado, Soil Mechanics and Foundations Division of the *ASCE*: New York, 1960 (reprinted in 1966).
33. F. E. Richart, Jr., J. R. Hall, Jr., and R. D. Woods, *Vibrations of Soils and Foundations*. Englewood Cliffs, N.J.: Prentice-Hall, 1970.
34. G. B. Sowers and G. F. Sowers, "Failures of Bulkhead and Excavation Bracing." *Civil Engineering*, January 1967, pp. 72-77.
35. M. G. Spangler, *Soil Engineering*, 2d ed. Scranton, Pa.: International Textbook, 1960.
36. K. Terzaghi, "Dam Foundation on Sheeted Granite." *Géotechnique* (London), Vol. 12, No. 3 (1962), pp. 199-208.

37. K. Terzaghi and R. B. Peck, *Soil Mechanics in Engineering Practice*, 2nd ed. New York: Wiley, 1965, Chapter 9 on Foundations, pp. 472-588.
38. J. H. Thorney, *Foundation Design and Practice*. New York: Columbia U. P., 1951.
39. L. C. Urquhart (ed.), *Civil Engineering Handbook*. New York: McGraw-Hill, 1959.
40. A. B. Vesić, "Bearing Capacity of Deep Foundations in Sand," *Highway Research Record*, No. 39, pp. 112–153, 1963.
41. A. White, J. A. Cheney, and C. M. Duke, "Field Study of a Cellular Cofferdam." *J. Soil Mech. and Foundations Div., Proc. ASCE*, Vol. 87, No. SM4, Part I (August 1961), pp. 89-124.
42. R. V. Whitman and F. E. Richart, Jr., "Design Procedures for Dynamically Loaded Foundations." *J. Soil Mech. Foundations Div., Proc. ASCE*, Vol. 93, No. SM6 (November 1967), pp. 169-193.
43. R. L. Wiegel (Coordinating Editor), *Earthquake Engineering*. Englewood Cliffs, N.J.: Prentice-Hall, 1970.

Indexes

AUTHOR INDEX

Abbett, R. W., 174, 175, 692, 716, 786
Agatz, A., 684, 693
Aldrich, H. P., 786
Andrew, C. E., 500
Andreyev, O. V., 707, 716
Andrus, F. M., 99
Angas, W. M., 564, 716

Bacon, Sir Francis, 4
Baker, A. L. L., 100
Barkan, D. D., 786
Bell, A. L., 98
Benjamin, S. G. W., 18
Beyer, K., 461, 463, 469, 477
Bierbaumer, A., 673, 692
Biot, M. A., 455
Bjerrum, L., 786
Blum, H., 136, 137, 141, 142, 174, 176
Boldakov, E. V., 707, 716
Bolin, H. W., 692
Bolni, H. W., 99
Borland, W. M., 716
Brand, E. W., 786
Brennecke, L., 209
Briske, R., 53, 99, 175

Casagrande, A., 251, 252, 302, 786
Casagrande, L., 275, 302
Chellis, R. D., 51, 98, 595, 605, 637, 692, 786
Cheney, J. A., 788
Chibaro, A., 97, 100
Cloyd, M. P., 786
Corso, J. M., 97, 100
Coulomb, C. A., 38, 39, 42, 119, 120, 123, 132, 201, 503, 741

Crede, C. E., 750, 787
Crieger, W. P., 47, 98
Cummings, A. E., 605, 637

Davis, R. P., 564
Davisson, M. T., 632, 638
De Simone, R. C., 500
De Simone, S. V., 455
Dörr, H., 623, 638
Drashevska, L., 786
Duke, C. M., 788
Dunham, C. W., 564, 636, 786
Dunn, C. P., 18
Duval, W. I., 787

Elston, J. P., 303
Engelund, A., 544
Erickson, E. L., 716
Erickson, H. B., 288, 303

Fadum, R. E., 251, 252, 302
Feld, J., 716
Fletcher, F. A., 693
Ferguson, P. M., 100
Flügge, W., 463, 477
Föppl, A., 463, 477
Föppl, L., 477
Fowler, J. W., 693
Fox, N. S., 99, 637
Freeman, F. R., 750
Freudenthal, A. M., 96, 97, 100
Fröhlich, O. K., 90, 95, 100, 107, 125, 342, 344, 348, 614, 615, 621, 622, 637, 677, 692

Gales, R. R., 517

Gates, M., 605, 637
Girkmann, K., 463, 477
Glossop, R., 302
Golder, H. Q., 786
Granger, F., 18
Gullander, P., 684, 693
Gumensky, D. B., 750
Gwilt, J., 13, 18

Hager, K., 209, 222, 246
Hall, J. R., 787
Handy, R. L., 99, 289, 303, 586, 612, 613, 637
Hansen, Bent, 786
Hansen, J. Brinch, 57, 99, 175
Hanson, W. E., 99
Harper, R. F., 18
Harris, C. M., 750, 787
Harza, L. F., 98
Hazen, A., 252, 254, 302
Hayashi, H., 356, 380, 455
Heck, N. H., 99
Helmers, N. F., 518
Henry, F. D. C., 693
Hertwig, A. A., 631
Hetényi, M., 455
Hoffmann, R., 302
Hool, G. A., 564

Jacoby, E., 683, 684, 692, 693
Jacoby, H. S., 564
"J. E.", 18
Johnson, A. I., 100
Johnson, A. W., 303
Johnson, S. J., 303
Joosten, H. J., 285, 303
Jumikis, A. R., 18, 32, 98, 99, 100, 125, 174, 209, 302, 303, 348, 517, 595, 637, 638, 692, 729, 735, 749, 786, 787
Just, L. H., 500

Karaki, S., 716
Kármán, von, Th., 746, 749
Karol, R. H., 303
Keeny, W. D., 645, 692
Kelen, N., 748, 749
Keller, J., 278
Kinne, W. S., 564
Kirkbride, W. H., 517
Klenner, C., 120, 125, 128

Krey, H., 147, 509, 510, 517, 614, 615, 623, 637, 684, 693
Krieger, S., 463, 477
Krynine, D. P., 201, 221, 223, 227, 242, 734, 735
Kyrieleis, W., 302

Labutins, A., 684, 693
Lambe, T. W., 287, 303
Lane, E. W., 716
Laursen, E. M., 707, 711, 712, 715, 716
Leonards, G. A., 175, 786
Levy, I. G., 500
Li, Shu-t'ien, 595
Lohmeyer, E., 209
Look, A. D., 303
Lorenz, H., 631, 638
Lysmer, J., 787

Maag, E., 614, 621, 622, 638
Maclean, D. J., 302
Masters, F. M., 692
McCoy, B. O., 47, 98
Melan, E., 463, 477
Merriman, T., 716
Meyerhof, G. G., 787
Michell, J. H., 463, 477
Mohr, H. A., 787
Monfore, G. E., 98
Moore, Jr., J. T., 303
Moran, D. E., 497, 597
Müller-Breslau, H., 44
Murphy, G. J., 786
Muss, M., 302

Napier, A. S., 517
Newmark, N. M., 787
Nikolayev, B. A., 635, 638
Nökkentved, C., 684, 693

Obert, L., 787
Obrcian, V., 500
Ohde, J., 455
Ostenfeld, A., 684, 693

Paproth, E., 564
Parsons, J. D., 787
Peck, R. B., 71, 99, 100, 120, 125, 128, 251, 302, 786, 788
Perrault, C., 13, 14, 18

Indexes

Peterson, H. R., 595
Platner, S. R., 18
Posey, C. J., 716
Potter, J., 348
Prandtl, L., 57
Prentis, E. A., 787
Press, H., 642, 692
Proctor, C. S., 500, 564, 716
Purcell, C. H., 500
Pynnonen, R. O., 303

Quinn, A. DeF., 787

Raab, N. C., 98, 99, 746, 749
Rabe, W. H., 605, 637
Rawlinson, G., 18
Redtenbacher, J. F., 637
Reissner, H., 463, 477
Richart, Jr., F. E., 750, 787, 788
Riggs, L. W., 500
Rinne, J. E., 99
Rowe, P. W., 155, 156, 174, 175
Rufin, J. V., 693
Ruge, A. C., 750

Schaper, G., 544, 595
Schleicher, F., 455, 463, 477
Schneider, V. R., 716
Schoklitch, A., 265, 302, 559, 564
Schultze, E., 693
Schütz, 303
Seely, H. R., 500
Seelye, E. E., 564
Seiler, J. F., 645, 692
Seydel, K. H., 303
Shen, H. W., 716
Shulits, S., 302
Sichardt, W., 263-265, 302
Sieberg, A., 52, 53, 744, 745, 749
Simmonds, A. W., 302
Skempton, A. W., 59, 80, 99, 786
Sowers, G. B., 787
Sowers, G. F., 787
Spangler, M. G., 787
Steinfeld, K., 509, 517

Steinman, D. B., 12, 18
Stern, O., 603, 637
Stewart, R. W., 716
Streck, O., 539, 544
Svensson, N. L., 96, 100

Tanner, D. N., 787
Tench, R., 587, 595
Terzaghi, K., 16, 18, 55-57, 60, 64-66, 71, 76, 78, 99, 100, 120, 121, 125, 128, 174, 175, 209, 222, 240, 246, 251, 302, 455, 677, 692, 716, 787, 788
Thornburn, T. H., 99
Thorney, J. H., 787
Thurston, A. P., 716
Tiedemann, B., 18
Timoshenko, S., 463, 477
Toch, A., 707, 716
Trefethen, J. M., 53, 99
Tschebotarioff, G. P., 163, 164, 175, 176, 303, 741, 749

Urquhart, L. C., 788

Vesić, A., 455, 788
Vetter, C. P., 684, 692, 693
Vitruvius, 12-14, 18

Watson, S. R., 12, 18
Westergaard, H. M., 99, 684, 693, 748-750
White, A., 788
White, L., 787
White, R. E., 564
Whitman, R. V., 788
Wiegel, R. L., 788
Wiggin, T. H., 716
Williams, W. W., 289, 303
Winkler, E., 455
Winterkorn, H. F., 303
Woinowsky-Krieger, S., 463, 477
Wood, H. C., 98, 99, 746, 749
Woodruff, G. B., 500
Woods, R. D., 787
Wünsch, H., 684, 693

Zimmermann, H., 455, 637

SUBJECT INDEX

Abutments, loads on, 701
Acceleration
　earthquake, 51-53, 738
　gravity, 51
　horizontal, 51-53, 739
　maximum, 53
　San Francisco-Oakland Bridge, 17, 53
　seismic, 51-53, 180, 738, 739, 740, 741
　vertical, 51, 53, 740
Acceleration scale, Sieberg's, 52, 53
Accumulation curves, soil particle size, 248, 254
ACI Building Code, 328, 348, 354, 359-361, 374, 380, 382, 383, 385, 390, 395, 399, 407, 408, 421, 422, 432
Active earth pressure, 38, 39, 119, 120, 122, 128, 201
　coefficients of, 139, 206, 242
　diagrams of, 119-123, 126
Active and passive earth pressure coefficient tables, 201, 503
Acts of nature, 4
Additives to grout, 283
Adhesion of soil
　to foundation, 43, 60
　to piles, 232
Aeroembolism, 552
Allowable bearing capacity; see Bearing capacity
Allowable bearing values; see Bearing values, allowable
Allowable pressure; see Pressure, allowable
Allowable settlement, effect of vibration on, 74
Alphabet, Greek, 753
Aluminium Company of America, 15
Ammonia refrigeration plant, schematics of, 294
Anchor, 128, 132, 164-174
　length of, 165, 167, 174
　tensile force in, 148, 150, 164, 207, 209
Anchor blocks, 128, 165, 166
　forces on, 168-174
　placing of, 165
　position of, 165, 167
　spacing of, 166, 171
Anchor piles, 114, 166
Anchor pull, 168, 169
Anchor rod (tie-rod), 114, 128, 150
　resistance of, 169
Anchor slab, 165
Anchor tension, 148, 150, 164, 207, 209

Anchor wall, 165, 166
　forces acting on, 168-174
Anchorage, 164-174
　dead-man, 114, 132, 166
　exterior, sheet piling, 113, 114, 166
　factor of safety of, 170
　failures of, 165, 172, 173
　foundations, 28
　modes of failure of soil, 172, 173
　piled, 165, 166
　of piles, modes of, 403, 404
　placing of, 114, 168
　safety requirements of, 172-174
　of sheet piling, 114, 132, 134
　systems of, 132, 133, 165, 166
　tie-rod, 114, 172
　wale of, 174
　zones of, 114, 132, 165, 167, 168
Anchored bulkhead, 133, 163, 164
Anchored sheet piling, 132, 134, 146-155, 163
　examples on calculating of, 152-155
　methods of analyses, 147-155, 159-162
　stability of, 134
　　driving depth, 163, 164
　　fixed earth support, 146-164
　　free earth support, 146-155
　　Tschebotarioff's method, 163, 164
Anchoring of piles into pile caps, 403, 404
Ancient Greeks, 12
Ancient times, foundation engineering in, 10-12
Apparent mass, Kármán's, 746, 747
Applications of commonly used soil tests, 781-785
Aprons, seepage-retarding, 49, 50
Aquifer, 50
Arch, inverted, 315, 316
Artesian water pressure, 36, 50
　effect on stability of bottom of excavation, 50
ASCE, 598
Aseismic design, 53, 743-746
ASTM, 99
Attack, angle of, 708-713, 715

Babylon, 12
　tunnel in, 12
Backfilling methods, 133
Ballast
　for cellular cofferdam, 202, 203, 214
　central vertical shear plane in, 215, 220, 234, 238, 241, 243
　coefficient of internal friction of, 242

Indexes 793

Ballast *(continued)*
 in cofferdam, 196
 confined, 201
 drainage of, 242
 effective weight of, in cell, 219, 243, 244
 in floating caissons, 532, 540, 541
 material, 203, 213, 214, 218, 223
  failure in shear, 224
  frictional resistance, 225
  unit weight of, 235, 242
  weight of, 200
 in open caissons, 486
Bamboo mats, 12
Bamboo piles, 12
Banks of excavation, protection of, 108-111
Batter piles, 569, 646-650, 745, 746
Beams
 equivalent, concept of, 158, 159
 grade, 452-454
Bearing capacity, definition of, 598
Bearing capacity of clay, 56-60, 68-74, 76, 79, 81, 82, 233
 ultimate for cofferdams, 233
Bearing capacity of conical pile, 621-623
Bearing capacity factors, 56, 57, 61
 diagram (graph) of, 56
Bearing capacity of foundations
 for caissons, 73, 80, 81, 86, 88, 89
  allowable, 73, 81
  ultimate, 73, 80
 for circular footings, 58, 59
 for continuous (strip) footings, 29, 56-58
 for deep foundations, 59, 60, 73, 74, 80-82, 86, 88, 89; *see also* Deep foundations
  allowable, 60, 61, 73, 81
  determination of, 59, 71, 72, 74, 78, 79, 85, 86, 88, 598
  ultimate, 59, 60, 73, 80
 for mats (rafts), 71, 72, 76; *see also* Mat foundations
  ultimate, 71, 72, 76, 78, 85, 86, 88
 of piles, 598, 599, 625, 626, 670, 671
 for piers, 59-61, 73, 74; *see also* Pier foundations
  allowable, 60, 61, 73, 74, 81
  ultimate, 59-61, 73, 74, 80, 81, 85, 86
 for pile foundations, 82, 83, 598, 670, 671; *see also* Pile foundations
  pile groups, 87
 of piles, static formulas of, 623, 624
 for rectangular footings, 56-59
 for rigid mats, 432
 of screw piles, 588, 589
 for shallow foundations; *see also* Shallow foundations
  allowable, 59, 77
  ultimate, 55, 64-67, 84, 85

Bearing capacity of foundations *(continued)*
 for square footings, 58
 for strip footings, 29, 56-58
Bearing capacity, net, 59, 67
 Skempton's formula, 59, 80
Bearing capacity of pile tips, 614-623
Bearing capacity of rocks
 presumptive, 55
 safe, 54, 55
Bearing capacity of soil, 31, 54, 55, 59-61, 77, 81, 82
 allowable, 59-61, 65-68, 71, 77, 79-82, 84
 for caissons, 73, 80, 81, 86, 88, 89
 of clay, 56-60, 68-74, 76, 79, 81, 82, 233
 of cohesive soils, 79, 81, 82, 84
  determination of, 56-60, 67-75, 77, 80, 81, 85
 for deep foundations, 60, 61, 81, 727, 728
 for mats on clay, 78, 79, 81
 for piers in clay, 81
 for pile foundations in clay, 82
 for pile foundations in silt, 87
 for rough footing base of strips, 56
 of sand, 60, 67, 68, 70, 71, 73
 of shallow foundations, 55, 59, 65-68, 77, 84
 of silt, 65-67, 84-87
 for smooth footing base of strips, 57
 net, 59, 67
 noncohesive soils, 55
 safe, 54, 55, 65-67
 ultimate, 29, 55-61, 64-67, 73, 74, 80-85
 for cofferdams on clay, 233
 for cofferdams on sand, 232
Bearing capacity of soil-foundation systems
 for deep foundations, 59-61
  in clay, 80
  in sand, 73
  in silt, 86
  on rock, 88
 for mat foundations
  on clay, 78
  on sand, 71
  on silt, 85-86
  on rock, 88
 net, 59, 67
 of piers, 59-61, 73, 74, 80, 81, 85, 86
 of piles, 75, 82, 83, 87
  determination of, 75
 of pile foundations in clay, 82
  in sand, 75
  in silt, 87
  on rock, 89
 of shallow foundations on clay, 76
  on sand, 67
  on silt, 84, 87
  on rock, 87
Bearing capacity, ultimate, 55-60, 67-76

Bearing capacity, ultimate *(continued)*
  for caissons, 73, 80, 81, 86, 88, 89
  for circular footings, 58, 59
  of clay, 76, 78, 80, 82
  for cofferdams, 233
  for continuous strip, 29, 56-59, 64
    rough base, 56
    smooth base, 57
  for deep foundations, 59-61, 73, 74, 80, 81, 86
  equations for, 55-59, 64-67
  examples for calculating of, 64-67
  for individual footings, 56-59, 64-67, 76, 84, 85
  for mats (rafts), 71, 72, 76, 78, 85, 86, 88
  for piers, 59-61, 73, 74, 80, 81, 85, 86
  for pile foundations, 75, 82, 83, 87, 670, 671
  by Prandtl, 57
  for rectangular footings, 56-59
  of sand, 67, 69, 73
  for shallow foundations, 55-59, 64-67, 76, 78, 84, 85
  of silt, 84, 85
  of soil, 29, 55-61, 64-67, 69, 73, 74, 76, 78, 80-85
  specialization of equations, 57-59
  for square footings, 58, 64, 66
  for strip (continuous) footings, 29, 56-58, 64, 65
  Terzaghi equations, 55, 56
Bearing capacity values of soil, sources of, 54, 55
Bearing piles, 568
Bearing values, allowable, 54, 55
  proposed for mats on sand, 71, 72
  reduction of, 72
Belled-out caissons, 482, 484
Bell towers, 16
Bénabeng's pile bearing capacity formula, 623
Bending moment, maximum, in sheet piling, 136, 138, 139, 142-146, 162, 163
  diagrams for, 137, 138, 144, 152, 157, 158
  position of, 137-140, 142-144, 152, 154, 155, 157, 158, 162, 163
  reduction of, 155-157
Bending moment reduction factors, 155-157
Bending stress, 158
Bents, pile, 648, 682-684
Berms for cofferdams, 183, 189, 190, 203, 219, 220, 231, 232
Berms of earth slopes, 105, 183
Bethlehem Steel sheet piling, 129, 130
Bibliography for forther reading, 786-788
Billet plate, 328, 329
Block, anchor, 166

Blows N, number of, in SPR tests, 61-67, 69-71
Blum method for calculating of sheet piling, 136-139, 143-146
  examples of, 138, 139, 143-146
Bodine resonant pile driver, 632
Boldakov-Andreyev's depth of scour, 707, 711
Boiling of sand at cofferdam, 187
Bore-hole direct-shear test device, Handy's, 612, 613
Bottom scouring velocities of flow, 706, 707
Box-type cofferdams, 185-191, 195, 198, 201, 207-209
Braced sheet piling cofferdam, 187, 189
Braces
  for equal axial compression, 122, 123
  spacing of, 122, 123
  for trenches, 108, 123
Bracing for cofferdams, 187, 189
Bracing frame, horizontal, 111, 185, 186
Bracing for sheeting and sheet piling, 113
Braking forces, 36
Breakwater (mole), 12-14
Bridge design, models for, 16, 17
Bridge foundation caissons; *see* Caissons of bridge foundations
Bridge foundations
  Delaware Memorial Bridge, 493
  Little Belt Bridge, Denmark, 529-533
  Lidingö Bridge, Stockholm, 582-585
  Pont Neuf Bridge, Paris, 16
  Pont Royal Bridge, Paris, 17
  Rialto Bridge, Venice, 16
  sequence in construction of, 191, 192
  shearing by seismic force, 742, 746
Bridge piers, 701-705, 708-710, 712-716
  basic scour around, 714, 715, 717
  Chinese, 12
  form coefficients of, 707-716
  Greiner's pressure formula, 701
  loads on, 701
  nose forms of, 707-716
  sequence in construction of, 191, 192
Brooklyn Bridge
  foundations of, 17
  pneumatic caisson, 17
Brix pile driving formula, 605
Building Code, ACI; *see* ACI Building Code
Bulbs of pressure, 30, 31, 76, 572
Bulkheads, 129, 163, 164
  anchored, definition of, 133
"Bull's liver," 86

Indexes

Buoyancy of floating caisson, 522, 523, 527, 535, 541
Bureau of Yards and Docks, Design Manual, 156, 227, 232, 240
    calculations of stability of cofferdams, 227
    calculations of stability against failure in shear, 240
    pile driving formula, 603
    ultimate pull-out capacity of a pile, 232

Caesar's commentaries, 639

Caisson, definition of, 482

Caisson and pier foundations; see also Deep foundations
    allowable bearing capacity for
        in clay, 80
        on rock, 88, 89
        in sand, 73
        in silt, 86
    allowable load on, 81
    bearing capacity equations, Skempton's, 80
    bulbed-end piers, 81
    determination of bearing capacity
        of clay, 81
        of sand, 74
    embedment depth of piers, 80
    factor of safety, 73, 81
    groundbreak, 19, 50, 68, 71, 73, 75, 76, 81, 86, 203, 218, 219, 231, 249, 339, 481
    groundwater, 86
    hydrovibration, 74
    perimeter of pier, 59, 80
    position of groundwater table, 74
    rock flour, 86
    settlement of
        in clay, 81
        in sand, 74
        in silt, 86
    skin friction, 80, 81
    special considerations, 74, 81, 87
    ultimate bearing capacity
        of clay, 80
        of sand, 73
    ultimate loads on piers, 80
    use of, 73, 481, 482, 484, 521, 522, 546
    vibration of, into soil, 74

Caissoned foundations, 481, 529
    completion of, 529

Caissons, 17
    allowable bearing capacity of, 73, 80
    ballast of, 486
    belled-out, 482, 484
    of bridge foundations
        Brooklyn Bridge, N.Y., 17
        classification of, 482
        Delaware Memorial Bridge, 493
        dome-capped, 17, 497, 498

Caissons (continued)
        Little Belt Bridge, Denmark, 529-533
        Huey P. Long, New Orleans, La., 17
        Mackinac Straits Bridge, Michigan, 51, 481, 488, 498
        San Francisco-Oakland Bridge, California, 481, 488, 497, 498
        Tagus River Bridge, Lisbon, 481, 498, 499
        Verrazano-Narrows Bridge, N.Y., 481, 512, 514
    cutting edges of, 11, 17, 485, 490, 491, 493, 494, 508-511
    model of, 511
    definition of, 482
    depth of, 17, 488, 495
    design of, 533, 537-544
    determination of bearing capacity of, 59, 60
    dome-capped, 17, 497
    Moran's, 497
        Tagus River Bridge, 498
    Egyptian, 11
        cutting edges of, 11
    elements of, 485
    floating, 8, 521-544
        buoyancy of, 522, 523, 527, 535
        completion of, 529, 543, 544
        description of, 521, 522
        forces on, 523, 526, 528, 533-537
        Little Belt Bridge, Denmark, 529-533
        problem on, 537-544
        static calculations for, 533-537
    forms of, 483
    foundations, 481
    inclined, 484
    limestone cutting edge, 11
    open, 8, 11, 17, 481-500
        advantages of, 484
        ballast of, 486
        calculations of diameter, 487
        calculations of thickness of shell, 488
        circular, 488, 501, 509
        cutting edges, 11, 17, 485, 487, 490, 491, 493, 494, 509-511
        cylindrical, 501, 509
        definition of, 484
        depth limits, 481
        depth of sinking, 485
        description of, 481
        diameter of, 487
        disadvantages, 484
        Egyptian, 11
        elements of, 485
        excavation of, 486, 488-490
        forces on, 501-503
        forms of, 482-484, 486
        formulas for concrete seals, 491, 492
        knife-edges for, 17, 485, 490, 491, 509-511

Caissons *(continued)*
   lateral pressure on, 502, 503, 506, 507, 510
   limestone cutting edge, 11
   mantle friction, 503
   penetration of, 11, 504, 633
   penetration depth of, 488
   pressure on, 502, 503, 506, 507, 510
   problems on, 515-517
   rectangular, 507-509
   reduction of frictional resistance on, 11
   round, 501-507
   seal, concrete, 491-497
   shaft, 485
   sinking of, 17, 485, 488-490, 496, 514
   size of, 487
   special, 484, 514
   statics of, 501-520
   stepped, 484
   taper of, 487
   thickness of seal, 491-493
   thickness of shell (wall), 488, 495, 506-509
   types of, 482, 483
   ultimate bearing capacity of, 73, 80
   uses of, 481, 482, 484
  pneumatic, 8, 17, 27, 482, 545-567
   advantages of, 54, 563
   air lock, 546, 551
   bending moment in walls, 554, 556
   for Brooklyn Bridge, N.Y., 17
   calculations of, 553
   ceiling of, 549, 550, 556
   compressors, 552
   construction of, 547-549
   cutting edges, 547, 548, 550
   description of, 545
   disadvantages of, 563, 564
   disease, 552
   earth pressure on cutting edge, 558-563, 567
   elements of, 549
   endurance limit, 17, 547
   forces on, 553-556
   forces on cutting edge, 557-563, 567
   fore-chamber, 548, 551
   loads on walls, 555, 556
   locks, 546, 548, 549, 551, 552
   materials lock, 548
   max depth, 547
   men lock, 548
   parts of, 549
   principle of, 545, 546
   problems on, 564-567
   shaft, 548-552
   sickness, 547
   sinking of, 548
   sketch of, 548
   Triger's, 17, 546
   walls of, 549, 550

Caissons *(continued)*
   work in, 552
   working chamber, 545, 547, 549, 552, 554, 555
  special, 484, 514
  steps in sinking of, 514
Campanila of Pisa, 16
Cancani earthquake intensity scale, 52, 53
Cantilever sheet piling, free, 132-139, 181
  loaded with active earth pressure, 132, 133, 140-146
  loaded with a horizontal, concentrated load, 132-139
Capacity, tapping, of a wellpoint, 263
Capacity factor in footing design, 384
Caps, pier, 584, 585
Caps, pile, 401, 745
Cast iron piles, 568
Cast-in-place concrete piles, 568, 585, 586
  for Pentagon foundations, 586
Cedar piles, 11, 576
Cells
  box-type, 202
  circular, 178, 186, 211, 215, 230
   design of, 216
  cloverleaf, 211, 230, 241
  for crib walls, 730, 731, 733, 734
  diameter of, 216
  effective weight of, 232
  segment type of, 211, 212, 220, 230
  semicircular, 178, 186
Cell ballast, 202, 203
  sand, 218
  checking against shear on vertical planes, 222-226
  effective weight of, 219
Cellular circular cofferdam, analysis of, 218, 240
Cellular-type cofferdams; *see* Cofferdams, cellular
Cellular mats, 429
Cement grout, 191
Cement slurry, 195
Centric loading, 37
Centrically inclined loading, 37
Centrically loaded plain concrete footings; *see* Concrete footings
Centrically loaded reinforced concrete footings; *see* Concrete footings
Checking the pile cap, 408-412
Chemical grout, AM-9, 287
Chemical grouting, 279, 282-284
Chemical solidification of soil, 285-290
  Joosten's method, 285-290

Indexes

Chemically solidified cofferdam, 288, 290
  soil, 288-290
Cheops pyramid, 11
Chimneys, design of, 688-692
  wind pressure on, 687
Chinese bridge piers, 12
  cofferdams for, 12
Choosing the kind of foundation, 19-35
Chrom-lignin grouting process, 287
Circle, equivalent, 262, 263
Circle, friction, 510
Circle method, Krey's, 510, 511
Circular-type cellular cofferdams; see Cofferdams, circular
Circular footings, 58, 59, 459-471, 689-692
Circular piers, 59, 74, 712
Circular shear area, 408
Circular shear perimeter, 408
City of New York Building Laws, 589, 591, 625
Classical pile driving formulas, 601
Clay
  adhesion to piles, 232
  allowable compressive strength of, unconfined, 57
  allowable shear strength, 82
  angle of internal friction, 42, 43
  bearing capacity of, 56-60, 64-66, 68-74, 77
    safe, 54, 55
    ultimate, 55-59, 67, 76, 82, 84, 85
  cohesion of, 42, 56, 64-66, 233
  cofferdams on, 233
  core of, 178
  foundations in (on),
    deep foundations, 80
    individual footings of, 73-80
    mats on, 78, 80
    piers in, 80, 81
    piles in, 82, 83
    shallow foundations on, 67-71, 76-78
    summary of, 67, 76-78, 80-82
    ultimate bearing capacity for, 56, 59-61, 71-74, 76, 78, 80-83, 85-87
  shear strength (resistance), 42, 56, 76, 82
  ultimate bearing capacity, 76, 78, 80, 82, 233
  ultimate shear strength, 76, 82
  unconfined compressive strength, 57
  varved, 298
Clay blanket, 179
Clay grouting, 284, 285
Cleanliness course (layer), 322, 323, 326, 328, 356
Clearances, minimum, 402, 403
Clover-leaf cofferdams, 211, 240, 241

Cluster of piles, 583
Code, Building, ACI, see ACI Building Code
Coefficients of
  earth pressure, 38, 39, 242
    active, 38, 242
    Coulomb's, 120, 122
  efficiency of motor, 273
  efficiency of pile group, 644, 645
  efficiency of pump, 272
  Klenner's, 120
  Krynine's, 201, 221, 223, 227, 242
  passive, 38, 242
  Terzaghi and Peck, 120-122
  form, of bridge piers, 707-716
  friction
    between ballast and rock, 242
    between ballast and steel, 242
    at base of cofferdam, 197, 221
    on lubricated surfaces, 44
    of sand on solid surfaces, 43
  interlock friction, 227, 237
  internal friction of dry ballast soil, 242
  nonuniformity, 254
  nose, of bridge piers, 707-716
  permeability, 248, 250-255, 264, 273
    chart of, 248
    electro-osmotic, 275
    figure of, 252
    table of, 251, 264
  shape of nose forms, 707-716
  resistance, 384, 385, 387, 388
  soil, reduced, 56, 57
  wall friction, 43, 44
  uniformity, 254
Cofferdams, 13-16, 103, 177-209, 211-247
  application of, 177
  avoiding uplift in, 188
  ballast of, 196, 200-203, 213, 214, 218, 219, 223, 235, 242-244
  on bedrock, 177, 178, 183, 195-198
  berms for, 183, 189, 190, 219
  box-type, 185-191, 195, 198
  bracing of, 185-187, 189
  calculations of, 195-198, 201, 202, 207-209, 222, 227, 232-235
  cellular, 178, 211-247
    analysis of, 218, 240
    forces acting on, 216, 217
    on rock, 219-231
    stability against overturning, 197, 222, 244
    stability against sliding, 221, 231, 233, 244
  check of dimensions, 218, 219
  checking against shear on vertical plane, 222-225, 235, 240
  Chinese, 12
  circular, 211, 214, 216, 240, 241
    description of, 212-216
    strength of, 216

Cofferdams *(continued)*
 classification of, 177, 178, 211
 on clay, 219, 233
 clover-leaf type, 211, 240, 241
 computation length of, 200, 232
 computation width of, 223
 construction of, 129
 crib-type, 178, 198, 730-737
 definition of, 177
 deep, land, 185
 design of, 217, 218, 231
 dewatering of, 190, 191
 diameter of, 214, 216
 diaphragm of, 211, 212
 diaphragm stress of, 202
 double-row (wall) sheet piling, 12, 178, 185-191, 195
  braced, 187, 189
  drainage, 242
  strutted, 188
  uses of, 186, 187
 driving depth of sheet piling, 181, 231
 earth, 178
  stability requirements of, 180
 effective height of, 199, 200
 effective weight of, 219
 examples of calculations, 207-209, 240-245
  on clay, 233, 234-240
  on sand, 231
  on rock, 195-198, 219-231, 241
  shear forces in ballast, 224, 227
  static system of, 216, 217, 220
 for excavation in water, 184
 factor of safety, 197, 221
 fill for, 188, 212
 forces acting on, 216, 217
 freely draining material, 212
 height of, 197, 200, 213
 hydraulic gradient in, 203, 204
 inboard face, 196
 inclined frame, 183
 interlocks, 129, 130, 216, 230, 231
 land, 177, 184-186
 length of, 212-214
 materials (soil) for, 179, 213, 214, 218, 233
  use of, 185
 in open waters, 177
 outboard face, 196
 overturning of, 197, 219, 221, 222, 244
 on permeable soil, 177
 Pont Neuf, Paris, 16
 problems on, 210, 246, 247
 rectangular, 219
 river diversion, Roman, 14, 15
 rock-fill, 177-179
  definition of, 179
 on rock, 195-198, 219
 Roman, 13
  single-walled, 13
 on sand, 216-219, 231

Cofferdams *(continued)*
 ultimate bearing capacity of, 231
 sand boiling at, 187
 sand support, 231
 segmental type, 211, 212
 shear stresses in, 199, 200
 sheet piling
  braced, 187, 189
  steel, 178, 212, 215
  timber, 111, 112, 178, 182, 187
 sheet piling interlocks, 129, 130, 216, 231
 shell of, 215, 216
  weight of, 244
 single-wall sheet-piling, 13, 177, 180-183
 stability of, 180, 189, 195, 197, 198, 203, 216, 219, 226, 231, 232-234
 strength of, 195
 strutted, 181, 183, 188
 supported by earth fill, 182, 183, 215
 timber, 178, 187
 types of, 211
 use of, 177, 180, 186, 187, 211
 weight of shell, 244
 width of, 200, 212, 213, 216, 219, 223
 World Trade Center, New York, 129

Cofferdam system for driving depth, 295

Cohesion, 41, 42, 56, 131, 233
 of clay, 233
 in footing calculations, 64-66
 in sheet piling calculations, 42
 variations in, 42, 131

Cohesive soils, 58, 131, 179
 bearing capacity of, 57, 58, 233
 sensitive property of, 42, 131
 shear strength of, 42

Columns, steel, basis for (billet plate), 328, 329, 354

Column footings on piles, 404-412

Column footings, 324, 325, 361, 396
 centrically loaded, 382
 deformation of, 361
 dimensions of, 324
 eccentrically loaded, 396
 example on design, 324
 loads on, 8

Combined footings, 8, 312, 315, 419-425
 asymmetrically loaded, rectangular, 421, 422
 piled, 404
 symmetrically loaded, rectangular, 420
 trapezoidal, 315
 trapezoidal, at property line, 417-419

Combined foundations, 8, 312

Composite piles, 568, 593-595

Compression piles, 651-657

Concept of equivalent beam, 158, 159, 164

Indexes

Concrete
  prepact, 687
  tremie, 192
Concrete footings (foundations), 323, 350-381
  circular, 58, 59, 459-471
  design procedure, 350-353
  hollow, square, 363, 366
  mats, 382
  plain, 323, 350-381
  rectangular, 350-381, 382-458
  reinforced, 382-458
  square, 350, 351, 357-360, 362-366
  strip (wall), 352, 353
  stepped, 374-380
Concrete piles, 568, 578, 579
  cast-in-place, 578, 579, 585, 586
  classification of, 586
  handling of, 581
  hoisting of, 581, 582
  precast (prefabricated), 578-581
  reinforced, 578, 580, 581
  in situ, 578, 579, 585, 586
  sizes of, 579
  storage of, 581, 582
  use of, 578
Concrete seals, 192
  pouring of, 192
  thickness of, 192, 193, 491-493
Concrete wall footings, plain, 323
Concreting, underwater, 13, 191-195
  horizontal method, 193, 194
  vertical method, 193
Conical tip of piles, 616-618
Connected footings, 422
Consistency of soil, 62
Consolidation of soil, 27
Consolidation tests, 27, 31
Construction
  of bridge piers, 191, 192
    sequence in, 191-195
  of cofferdam, 129
  of floating caissons, 521-544
    launching of, 527-529
    open caissons, 481-520
    pneumatic caissons, 547-549
    a river diversion monolith, 15
    a Roman mole (breakwater), 12-14
Contact area, 6
Contact pressure, 6, 29, 30
  distribution diagram of, graphically, 6, 30, 332-336
  distribution on soil, 6, 30, 332-337
  maximum, minimum, 332-337
Continuous footings (foundations), 315, 426-442
  ultimate bearing capacity for, 55-58

Continuous footings *(continued)*
  on c soil, 58
  on $(\phi\text{-}c)$-soil, 57
  on ϕ soil, 58
Contribution of piles to shear force in concrete, 410, 411
Conventional method for calculating free-cantilever sheet piling, 134-139
Conventional method for design of mat foundations, 431-442
Conventions in sheet piling, 133
Converse-Labarre pile efficiency formula, 645
Conversion factors
  for motor and pump capacities, 267
  of units of measurement, 40, 267, 614, 770-780
Conversion of numbers of blows for fine sand below water table, correction formula for, 70
Correction (shape) factors, 57
Correction formula from N' to N, 70
Correlation of N with soil properties, 68
Corrosion of metal piles, 591, 592
Coulomb's earth pressure theory, 42, 119, 120, 201
  linear pressure distribution diagram, 119, 120
Creosote-treated wood piles, 574
Crib, 730, 734, 737
  cell of, 733, 734
  elements of, 733
  forces acting on, 733
Crib-type wall, timber, 730
  problems on, 735-737
  rock filled, 730
Crib-wall cofferdams, 178, 187, 198, 730-737
Criterion for frost penetration depth in soil, 321
Critical planes (sections) for diagonal tension, 352, 354, 360, 362, 365, 409, 410
  for moments, 353, 354, 357, 358, 362, 365
Critical sections, 353, 354, 358, 365, 382, 383, 384, 386, 389, 390, 393, 409
  forces at, 357, 358, 409, 410
  for moments, 353, 354, 357, 358, 362, 365
  for shear, 353, 354, 360, 362, 365, 409, 410
Critical settlement, 180
Cross section of piles, 569
Culmann's method for pile analyses, 656-658, 660, 661
Curves for bending moment reduction factors, 156
Curves, depression, 250

Cutoff, 48
  underseepage, 231
Cutting edges
  acute, 559, 560
  blunt, 558, 561, 562, 567
  caisson, 11, 17, 485, 490, 491, 493, 494, 509-511
  calculations of, 510, 550, 558
  Egyptian, 11
  of limestone, 11
  model of, 511
  of open caisson, 11, 485, 490, 491, 509-511
  of pneumatic caissons, 550, 558-563
    forces on, 557-563
    pressure on, 510, 555-557
  of reinforced concrete, 490
  soil pressure distribution around, 509-511
  of steel, 490, 493
  all welded, 490, 493, 494

Dams
  diversion, 14
  earth, 297
  homogeneous, 48
  massive, 47, 49
  water impounding, 47
Dampproofing, 445, 447, 448
Darcy's law of filtration, 249, 253
Dead man, 114, 132
  anchorage of, 114, 166
Deep excavations, 103
  classification of, 103
  definition of, 309
  dewatering of, 103
Deep foundations, 8, 59-61, 73-76, 80-83, 86-89, 481-500, 501-520, 718-729
  allowable bearing capacity, soil, for, 59, 60, 61, 73
    determination of, 74
  caissons for, 73, 481-500; see also Caissons and Pier foundations
  critical loads on, 726
  definition of, 481
  depth of, 59, 481, 726, 727
    description, 103
    embedment, 59, 60
    excavation for, 103
    floating caissons, 521-544
  lateral pressure on, 501-503, 506, 512, 518, 718-729
    diagrams of, 119-123, 227, 228, 231
    distribution of, 119-123, 227, 228, 231
    equations, 39, 722, 723
  mantle friction, 75
  open caissons, 8, 481-520; see also Caisson and Pier foundations
    allowable bearing capacity, 60, 61
    perimeter of, 59

Deep foundations (continued)
    skin resistance of, 59
    perimeter of pier, 59
  piers, 59, 60, 61, 73, 74
    ultimate bearing capacity of, 59, 60, 73, 83
  piles, 8; see also Pile foundations
  pneumatic caissons, 8, 17, 27, 545-567
  position of groundwater table, 74
  problems on, 515-522, 525-527, 538-544
  settlement of, 74
  skin resistance, 59, 60, 75
  ultimate bearing capacity, 59, 60
    on clay, 83
    on sand, 73
  uses of, 481, 482
Deep land cofferdams, 185
Deep-well pumps, 266
Deflection of sheet piling, 147, 158
  in cofferdams, 206, 207
Deformation
  of column footings, 361
  of piles, 680, 681
Delaware Memorial Bridge, 493
Densification piles, 642
Densification of soil
  by compaction, 277
  by hydrovibration, 277-279
  by piles, 642
  by vibration, 277, 281
Density
  of sand, 55
  of soil, 39, 40
Density, relative, 62, 68, 70, 71
  values of (N), 62, 68-71
Depression curves, 250
Depth
  of caisson, 17, 488, 495
  critical, of excavation banks, 105-109
  of deep foundations, 59, 60, 726, 727
  driving, of sheet piling for cofferdam, 204-206
  embedment of sheet piling, 114, 135, 137, 139-143, 145, 150, 154, 162
Depth of footings, shallow (of foundations), 17, 28, 31, 56, 317
  critical, 321, 322
  discussion on, 29
  example on, 29-31
Depth of frost penetration in soil, 29, 317, 319, 320, 321
Depth of influence of contact pressure, 6, 30
Depth limits of frost penetration in foundation engineering, 291, 292
Depth of piers, 59, 60
Depth of scour, 706, 711

Indexes 801

Depth of scour *(continued)*
  by Boldakov and Andreyev, 707, 711
  by Laursen, 707, 711, 712, 715, 716
Depth of sheet piling, 113, 136, 137, 181, 204-206
  minimum driving, 134, 135, 139, 140, 142, 145, 150, 154, 162
  maximum driving, 143, 145
Depth of strip footings, 30, 56
Design
  aseismic, 53, 743-746
  Design of cofferdams, 216-218, 231
    of chimneys, 688-690
    example of, 689-692
  of column footings, 324
  of earth cofferdams, 180
  of floating caissons, 533, 537-544
  of footings on clay, 76-80
    on sand, 67-71
  of foundations, 28, 384-477
    important points in, 4
    mistakes made in, 4
    procedures in, 28
  of Golden Gates Bridge, 51-53
    earthquake considerations in, 746-749
  of mats (rafts), 431-442
  of piers on clay, 81
    in sand, 73, 74
  of pile groups, 645, 646
  rigid mat foundations, 431-442
Design Manual of Bureau of Yards and Docks, 156, 227, 232, 240
Design moment, 155-157
Design problems of sheet piling, 132
Design procedure for designing
  example on, 29, 30
  foundations, 3, 4, 28-30, 73, 80, 81, 432-442
    mats on clay, 80
    mats on sand, 73
    rectangular, plain concrete footings, 351-353
    rectangular, reinforced concrete footings, 382-458
    square footings, 329-331
Design seepage line, 220
Dewatered zone, 256-260
Dewatering
  of cofferdams, 188, 190, 191
  of excavations, 103, 248-277
    chart for, 248
    example of, 267-273
    multiple stage, 257, 258
    single stage, 256, 257
    wellpoint arrangement, 260-262, 264, 268, 271
  wellpoints, 256-260
  motor capacity, 266, 267
  pump capacity, 266, 267

Dewatering *(continued)*
  radius of influence, 265
Dewatering, electrical, 273-277
Dewatering systems
  for different soils, 248
  a chart for, 248
  electro-osmotic, 276, 277
  multiple stage, 257, 258, 266
  pumping from multiple wells, 258-266
  single stage, 256, 257, 266
Diagonal tension, critical planes for, 361
Diaphragm stress in box-type cofferdam, 202
Diaphragms
  in cofferdams, 211, 223
  of sheet piling, 178, 186
Dielectric constant, 275
Differential settlement, 20, 31
Diffuse double layer, electrical, 273, 275
Dikes, earth, 177
DIN, 49, 757
Dipole water molecules, 27
Direct footing (foundation), 6, 26, 310
Direct-shear (bore-hole) testing device, Handy's, 612, 613
Discharge into excavation, 253-255
Displacement, horizontal, of pile bent tops, 682-684
Distribution of loads on piles, 656-664
Diversion dams, 14
Division of trapezoidal area into n equal parts, 643, 644
Division of triangular area into n equal parts, 123, 184
  into two equal parts, 149, 150
Dome-capped caisson (Moran's), 17, 497
  for Tago River Bridge, Lisbon, 498
Dome, Königsberg, 16
Double layer, electric, diffuse, 273, 275
Double-row (wall) cofferdams, 12, 178, 185, 187-191, 195
  advantages, 187
  disadvantages, 187
  use of, 186, 187
Draft of floating caissons, 528
Drainage
  of ballast, 242
  galleries for, 179
Drains, intercepting, 445, 446
Drains, sand, 300, 301
Drawdown, 250
  maximum, 259
  rapid (sudden), 133, 249
  in wells, 263

802   Indexes

Dredged line, 132, 140, 165
Dredging wells of open caissons, 493-495
Driving depth, 181
  of sheet piling
    for cofferdam, 231
    minimum, 134, 135, 137, 139, 141, 142, 145, 148, 150, 151, 162, 204-206
    practical, 136, 137, 139, 141, 142, 145, 162
    theoretical, 136, 137, 149, 141, 142, 145, 150-152
Dry unit weight of soil, 40
Dwellers, lake, 11, 573, 639
Dynamic force, 36
Dynamic pile dricing formulas, 75, 599-606, 632
Dynamic viscosity, 275
Dynamics of soil, 742

Earth cofferdams, 177, 178
  design of, 180
  forces on, 180
  materials to use for, 179
  seepage pressure in, 180
  stability requirements of, 180
Earth dikes, 177
Earth embankments, 177
Earth pressure
  active, 38, 39, 119, 120, 123, 128, 132, 171, 201
  Coulomb's, 38, 39, 119, 120, 123, 128, 132
  lateral, 36-39
    on anchor blocks (walls), 169-172
    on cutting edges, 509-511
    on earth retaining walls, 126, 127
    on open caissons, 509-511
    passive (resistance), 39, 134, 137, 138, 140-144, 148, 152, 160-162, 201
    on pneumatic caissons, 566
    on sheeting, 119-122
    on sheet piling, 135, 137, 138, 140, 142, 144, 148, 152, 158, 161, 164
Earth pressure coefficients, 38, 39, 206
  active, 139, 201, 206, 242
  Krynine's, 201, 221, 223, 227, 235, 242
  passive, 139, 206, 242
Earth pressure coefficient tables, 201, 503
Earth pressure diagrams, 120
  active, 119-123, 126
  comparison of, 120
  Coulomb's, 120, 122, 132
  Klenner's, 120
  lateral pressure, 132, 135, 137, 138, 140, 143, 144, 148, 152, 158, 161, 164, 169, 170
  Terzaghi-Peck's, 120-122

Earth pressure diagrams *(continued)*
  for dense and medium dense soil, 121
  for loose sand, 121
  for soft and medium soft clay, 122
Earth pressure ordinate, passive, reduced, 149, 150
Earth pressure tables, 201
Earth pressure theory, Coulomb's, 42, 119, 120
Earthquake loading, 385
Earthquakes, 51-54
  acceleration, 51-53, 738
  danger of, re. rocks and soils, 744, 745
  effects
    on bridge piers, 741, 742, 748
    on earth retaining walls, 739, 741
    on structures and foundations, 51-53, 730-751
  forces, 36, 51-53, 738, 740, 748
    calculation of, 51
  intensity scale, 52-54
  problems on, 749, 750
  seismic factors, 51-53
  water and silt pressure on pier, due to, 746-748
  Westergaard's pressure parabola, 748
Earth retaining structures, 126, 127
  bulkheads, 129, 133, 163, 164
  crib-type, 178, 198, 730-737
  massive, 126, 127
  sheet piling, 50, 109-111, 129, 165
  walls, 50, 109-111, 126, 127, 129, 163-165, 178, 198, 730-737
Earth retaining systems, 133
Earth slopes, 104, 105
  berms, 183
Earth's crust, thickness of, 481
Eccentric loading, 37
  of pile groups, 665, 668
  of piles, 665
Eccentric wall footing, 413-419
  at property line, 413-419
Eccentricity, equivalent, 395-397
Eccentricity
  one-way, 331-333
  two-way, 345-348
Edge pressure, critical, Fröhlich's, 90, 95, 344, 345
Edge pressure on soil under footings, 332-337, 723, 724
Edges, cutting, 11, 17, 485, 490, 491, 493, 494, 509-511
Effects, electrokinetic, 285
Effects, seismic
  on foundations, 738-751

Indexes 803

Effects, seismic *(continued)*
  on structures, 53, 739, 741
  on wall friction, 43
Efficiency coefficients of motors, 373
Efficiency coefficients of pumps, 272
Efficiency of piles in a pile group, 644, 645
Egyptian cutting edge, limestone, 11
Elastic deformation of piles, 680, 681
Electric diffuse double layer, 273, 275
Electric field, water migration in, 273-276, 634
Electric gradient, 275
Electrical dewatering, 273-277
  principle of, 274
Electro-chemical stabilization of soil, 277
Electroconductivity, 275
Electrokinetic effects, 285
Electrokinetic phenomena, 273, 634
Electro-osmosis, 273-277, 633-635
  definition of, 273, 634
Embedment of piles, 568
  by driving, 577, 590, 643, 644
  by vibration, 75, 568, 569, 630-633
Embedment of sheet piling, 571
Empirical rules for static pile bearing capacity, 624, 625
Emulsions, bituminous, 285
Endurance limit for working in pneumatic caissons, 17, 547
Engineering News pile driving formulas, 602-604
Engineering structures, failures of, 4
Equilibrium of a floating body, 523
Equivalent beam
  concept of, 158, 159, 164
  steps in analysis of, 159-162
  Tschebotarioff's, 163, 164, 175
Erosion, 231
  characteristics of, 217
  of soil particles, 204
Excavations, 8, 50, 103, 104, 178
  below groundwater table, 8
  bracing of, 108, 111, 113, 122, 123, 185, 186
  classification of, 103
  critical depth of, 107, 108
  dewatering of, 103, 248-297
  horizontal bracing frame, 111
  important elements in laying foundations, 103
  influx of water in, 249
    example on, 254
    problem of, 255
  maintenance of slopes, 104, 105

Excavations *(continued)*
  maximum unsupported depth of, 105-107
    calculations of, 105-107
  methods of dewatering, 248-277
  predrainage of, 249
  pressure relief by, 65
  protection of banks of, 108-111, 113, 114
  sheeting of, 50, 108
  sheet piling, 108, 109, 113, 114
  size of, 103
  slopes of, 103, 104
  stability of bottom of, 50
  stepped, 105-107
  struts, 108
  strutting, 108
    inclined, 110
  through open water, 103
  timbering of, 111, 112
Expulsion of soil from underneath a cofferdam, 203
Exterior sheet piling anchorage, 113, 114
Extreme frost penetration depth, maximum, 319, 320

Face of footing, 6
Faces of cofferdams, inboard and outboard, 196
Factor of safety, 59, 61, 65-68, 70, 71, 73, 77, 79, 84-86, 89-98, 136, 149, 150
  anchorage, 170, 338, 339, 341
  assumptions of, 91, 92
  calculations of, 92, 170, 197, 229, 232, 240
  for cofferdams, 197, 221, 222, 229, 231, 232, 240, 244, 245
  in earthworks, 227
  against erosion, 204
  in foundation engineering, 89-98, 227
  against groundbreak, 231, 339
  Navy's Yards and Docks, 228
  numerical values of, 94, 95
  against pull-out of sheet piling, 232
  seismic, 51-53
  against shear failure
    of ballast in the vertical plane, 224, 227, 240
    of interlocks, 226, 227
  against tension failure in interlocks, 224
  values of, 94, 95
  variables in, 90
Failure
  of anchorage, 165, 172, 173
  of ballast in shear, 223
  in bearing capacity of cofferdams, 233
  of engineering structures, 4
  of foundations, 17
  by slippage of interlocks, 235
  of soil ballast in shear, 224
    between ballast and sheet piling, 231

Failure (*continued*)
  in general shear, 56-58
  in local shear, 56-58
  of sheet piling—anchor-wall system, 172, 173
  in tension of interlocks, 224
Fascine, 493
  definition of, 493, 494
  use of, 493, 494
Fascine mattress, 493, 494, 705
Feasibility of designs for foundations, 28
Fill material for cofferdams, 188
Filter, loaded, inverted, 231
Filtration, Darcy's law of, 249, 253
Film of moisture (water), 274, 275
Fine sand and silt, 85
  liquefaction of, 71
  quick condition of, 74
Fixed earth support
  concept of equivalent beam, 158, 159
  sheet piling, 158-163
  stability analysis of, 158-159
  static conditions at, 158
Fixity, position of, 206
Flat mat, 428
Flexibility number, 156
Floating body, 523, 526
  buoyant force, 524, 526
  draft (submergence), 524
  equilibrium of, 523
  of irregular cross section, 526
  stability of, 523-527
  submergence (draft), 524
  tilting of, 524
Floating caissons (boxes), 482, 521-544
  advantage of, 521, 522
  ballast in, 532, 540, 541
  buoyancy, 522, 523, 527, 535, 541
  completion of, 543, 544
  description of, 522, 523
  design of, 533, 537-544
  draft of, 528
  example of calculation, 537-544
  flotation of, 528, 535, 538, 539
  forces on, 523, 526, 528, 533-537
  forces on, upon launching, 523, 526, 528
  hydrostatic pressure on, 541
  loads on, 543
  launching of, 527-530, 532
  for Little Belt Bridge, Denmark, 529-533
  metacenter of, 523-527, 540
  metacentric height, 524, 527, 540, 543
  in open-cut dock, 538-539
  problem on, 537-544
  rip-rap underneath of, 542
  sequence in turning of, 532, 533

Floating caissons (boxes) (*continued*)
  in service condition, 537
  sinking of, 541, 542
  static calculations of, 533-537
  stresses at base of, 544
  in sunken position, 536
  tilting of, 539
  turning of, 532, 533
  use of, 521, 522
Floating foundations, 315
Floatation, 528, 535, 538, 539
Flow line, uppermost, 241
Flow net theory, 251
Flow velocity equation, Hazen's, 252-255
Flow velocity for scour, 706, 707
  competent, 707
  reduction of, 189, 190
  seepage, 189, 190
Flow of water, electro-osmotic, 275-277
Fluctuating groundwater table, 93, 94, 319, 574
  zones of, 573, 574
Footings; *see also* Shallow foundations
  base of, 6
    rough, 56
    smooth, 57
  for bent columns, 339-341
  calculations of, 351, 353, 364-366
  centrically loaded, 350-381
    rectangular, reinforced concrete, 388-392
    square, reinforced concrete, 382-388
    strip, 392-394
  classification of, 8, 309
  on clay, 76-80
  circular, 58, 59, 459-477
  column, 313, 324, 325, 361
    deformation of, 361
  combined, 8, 312, 315, 419-422
    definition of, 8
    piled, 404
    trapezoidal at property line, 422
  critical section in, 353, 354, 383, 386, 388-390
  concrete
    cutoff of piles in, 327
    plain, 350
    reinforced, 382-458
  connected, 422
    at property line, 423-425
  continuous, 56-58, 315
  definition of, 5, 6, 8, 310
  depth of, 31, 56, 59, 317-321
  dimensions of, 324
  direct, 6, 26, 310
  eccentrically loaded, 331-334, 340, 343
    on piles, 400
    rectangular, reinforced concrete, 395-398

Indexes 805

Footings *(continued)*
    strip (wall), 399, 400
    effect of width of, on stress distribution, 29-31
    effective thickness of, 356, 357, 359, 360, 364, 382, 384
    enlarged, 422
    examples on, 351-353, 364-366
        centrically loaded, 385-388, 463-467
        eccentrically loaded, 339-343, 346-348, 404-412, 470-477
    face of, 6
    heavy wall, 323, 324
        stepped, 324
    individual (isolated), 8, 57-59, 87, 313, 413
        in clay, 76-80
        on rock, 87-88
        in sand, 67-71
        in silt, 84-87
    massive, 352, 374-380
    parameters of, 57
    pedestal type, 327, 328, 354
    on piles, 327, 400-412, 643, 690-692
    plain concrete, rectangular, 350-381
    plain concrete, wall, 323, 324, 352, 353, 367-373, 380
    problems on, 351-353, 364-366, 404-412
    at property lines, 412-419
    protection of reinforcement in, 326
    pyramidal, 312, 374-380
    rectangular
        on (ϕ-c)-soils, 57-59
    reinforced, rectangular
        centrically loaded, 382-394
        eccentrically loaded, 395-425
    rigid, 352, 353
    on rock, 87, 88
    on sand, 67-73
    shallow, 8, 26, 55-59, 67-73, 76-80, 84-87, 309, 313
        on (ϕ-c)-soils, 57-59
    on silt, 84-87
    simple, 312
    sloped, 312, 313
    square, 58, 64, 66, 67
        plain, centrically loaded, 350, 351, 357-360, 362-366
        reinforced, centrically loaded, 382-388
        reinforced, eccentrically loaded, 395-399, 404-412
    stepped, 6, 312, 324, 325, 374-380
        stresses in, 374-380
    strapped, 8, 422-425
    strip (wall), 6, 29-31, 56-58, 64-66, 312, 313, 352-355, 412-415
        with rough base, 56
        with smooth base, 57
    thickness of edges of, 326
    thickness, effective, of, 356, 357, 359, 360, 364, 382, 384

Footings *(continued)*
    trapezoidal, 315
    at property line, 417-419, 423
    ultimate strength design of, 362
    uplift of, according to Hayashi, 356
    use of, 21-25
    wall, 313, 323, 352, 353, 413-417
        centrically loaded, 6, 352, 353, 392-394
        eccentrically loaded, 399, 400, 413-417
        at property line, 413-417
    widened, 6
    width of, 6, 56
    working stress design, 360
Force majeure, 4
Force polygon, 173, 666, 740
Forces, 36
    acting on anchor block (wall), 168-170
    acting on cutting edges, 557-563
    acting on a foundation, 36, 37
    artesian water pressure, 36
    braking, 36
    buoyant, 524, 526
        on caissons, 501-503, 524, 526
    on cellular cofferdams, 216, 217
    circumferential, in interlocks, 235
    at critical sections, 357-359
    diagrams of, 158, 162
    dynamic, 36
    on earth cofferdams, 180
    earth pressure, 36
    earthquake, 36, 51, 738, 740, 741, 748
    on floating caisson, 523, 526, 528, 533-537
    friction, 36, 59, 170, 180, 197, 225
    hydrostatic, 36, 44, 442-444, 541
    ice pressure, 36, 51
    impact, 36
    mantle resistance of pier, 59
    on open caisson, 501-503
    on piles, 608-611, 655-658
    on pneumatic caissons, 553-556
    pore water pressure, 36
    pull-out, of sheet piling, 232
    seepage, 36, 50
    seismic, 36, 51, 738, 740
        effect of on retaining walls, 739, 741
        Westergaard's parabola, 748
    side, frictional, 171
    on sheet piling, 132, 133-138, 140, 142, 144-146, 148
    skin resistance, 59
    soil reaction, 36
    superior, 4
    swelling, 36
    tensile (in tie-rods, anchors), 148, 150, 164, 207, 209
    thermal, 36
    in a three-pile system, 653, 657
Form coefficients, nose, of bridge piers, 707, 715

Foundation design, 28, 385-388
  capacity factor in, 384
  important points in, 4
  mistakes made in, 4
Foundation engineering, 3, 4, 7, 10-18
  in ancient times, 10-12
  definition of, 3, 5, 7
  description of, 4
  discipline of, 3, 7
  factor of safety in, 89-98
  historical notes about, 10-18
  maxim in, 4
  in Roman times, 12-14
  why study, 3, 4
Foundation pits, 8, 103
Foundation requirements, 19
Foundation settlement, 20, 31, 69
Foundation terminology, 5, 7
Foundation types
  guide to tentative selection, 21-25
Foundation walls, 6
Foundations
  anchorage of, 28
  in ancient times, 10-12
  bearing capacity of
    for deep foundations, 59-61, 73-76, 80-83, 86, 87
    safe, 55
    of shallow foundations, 56-59, 64-69, 76, 84, 85
  box-type, 316
  bridge pier, 16, 17, 51, 481, 488, 493, 497-499, 529-533, 702, 713
  caissoned, 529
  caisson, 16, 17, 80, 86, 481-567
    in clay, 76, 80-83
    floating, 521-544
    open, 481-520
    pneumatic, 545-567
    in sand, 73-76
    in silt, 86-87
    on rock, 87-89
  calculating of critical depth of, 37
  choosing the kind of, 19-32
    examples on, 29, 30
    problems on, 32-35
    tables for, 21-25
  on clay, sand, silt and rock, 67-89
  circular, 58, 59, 459-471
    centrically loaded, 58, 461-467
    eccentrically loaded, 467-476
    uniformly loaded, 58, 461-467
  classification of, 7-9, 21-25, 309-313
  in clay, 76, 80-83
  combined, 8, 312
  continuous, 426
  dampproofing of, 444, 447

Foundations (continued)
  deep, 8, 59-61, 82, 83, 86, 87, 479-500, 718-729
    allowable bearing capacity for, 60, 61
    ultimate bearing capacity for, 59, 60, 85
  definition of, 5, 309
  depth of, 17, 28, 31, 37
  design of, 3, 4, 28-30, 73, 80, 81, 385, 388
    capacity factor in, 384
  design of mats, 432-442
  direct, 6, 26, 310
  embedment depth of, 59
  excavation for, 103
  factors affecting choice of, 27, 28
  factors of safety, 59, 61, 65-67, 70, 71, 73, 77, 84, 89-98
  floating, 315, 316
  floating caisson, 521-544
  footings of, 57-59
  forces on, 36, 37
  from the 15th to the 17th centuries, 16, 17
  from the 18th century to date, 17
  guide to tentative selection of, 21-25
  importance of, 4
  isolated (individual), 8, 57-59, 87
  laying of, 13, 15, 31
  Mackinac Straits Bridge, Michigan, 481
  mats, 31, 315, 316, 426-442
    design of, 431-442
    on clay, 76-80
    on sand, 71-73
    on silt, 84-87
    on rock, 88
  in medieval ages, 16
  mistakes in design, 4
  open caissons, 16, 17, 481-520
  ordinary, 8
  pedestal, 6, 327, 328
  pile, 8, 27, 31, 75, 82, 83, 87, 89; see also Pile foundations
    advantage of, 31
    bearing capacity of, 31, 87, 598
    in clay, 82, 83
    determination of, 75
    feasibility of design, 28
    in sand, 75, 76
    in silt, 87
    on rock, 89
    settlement of, 20, 31
  pier, 59, 60, 80, 81
    in clay, 80-83
    on rock, 88, 89
    in sand, 73-76
    in silt, 86-87
  pneumatic caisson, 17, 545-567
    for Brooklyn Bridge, N.Y., 17
  procedure for designing of, 28
  purpose of, 5, 8
  pyramidal, 374-380
  rectangular, 58, 59, 382-425

Indexes

Foundations *(continued)*
  requirements of, 19, 20, 25
  restricted by property line, 412-419
  of the Rialto Bridge in Venice, 16
  "right kind" to choose, 25, 26
  resultant load on, 37
  on rock, 67, 87-89
  in Roman times, 12-16
  San Francisco-Oakland Bridge, California, 481, 746, 747
  shallow, 8, 55-59, 76, 307-349; *see also* Shallow foundations
    ultimate bearing capacity of, 55-59, 67, 76
    in sheet piling enclosure, 13
    in silt, 67, 84-87
    caisson, 86, 87
    deep, 73-87
    individual, 84-87
    mat, 85, 86
    piers, 85, 86
    pile, 87
    shallow, 67-73, 76-80, 84-87
  sloped, 312
  soil-strip systems, 30, 57, 58
  special, 8, 9
  spread, 8, 310
    definition of, 8, 310
  square, 6, 58, 64-67, 329-331, 331-337, 350-388, 388-392
  stability requirements of, 19, 20, 25
  stepped, 6, 8, 312, 324, 325, 374-380, 484
  strapped, 8, 422-424
  strength of, 3, 4
  strip, 6, 29-31, 56-58, 312, 313, 352-355, 367-380, 392-394
  suitable kinds of, 21-25
  tables for choosing the kind of, 21-25
  Tagus River Bridge, Lisbon, 481
  for tall structures, 8, 685-682
  terminology of, 5
    illustrated, 6
  Texas towers, 685, 686
  underwater concreting (tremie), 13, 191-195
  for various site conditions, 21-25
  in water, 13
  waterproofing of, 444-451
  of the Verrazano-Narrows Bridge, N.Y., 481
  widened, 314

Foundations, deep, 481-500; *see also* Deep foundations, Caissons, Pier and Pile foundations
  in clay, 80-83
  on rock, 67, 87-89
  in silt, 86-87
  in sand, 67, 73-76

Foundations, pile; *see* Pile foundations

Foundations, shallow; *see* Shallow foundations

Frame
  bracing, 185, 186
  cofferdam, inclined, 184

Free cantilever sheet piling, 132-134, 181
  calculations of
    by Blum's method, 136-139
    by conventional method, 134
    method of minimum driving depth, 134, 135
    position of pivot point, 135
  definition of, 133
  equations for, 134-139
  examples on, 138, 139
  theoretical depth, 136, 136

Free earth support for sheet piling, 146-148

Free-standing piles, 569

Freezer pipes, spacing of, 294

Freezer point, 293

Freezing of soil
  artificial, 290-297
  depth limits of artificial freezing, 291
  refrigeration system, 294
  thickness of walls of frozen soil, 295

Frequency of tall structures, 745

Friction
  angle, internal, of soil, 38, 41-44
    coefficient of, 43, 44
  interlock, 223, 225, 227
    coefficient of, 235
  mantle, 572, 600, 609, 614, 615
  negative, 27, 572
  skin, 59, 81, 572, 600, 609, 614, 615
  wall, 41

Friction circle, 510

Friction coefficient
  of ballast material, 225
  at base of cofferdams, 197, 221
  on lubricated surfaces, 44
  between sand and rock, 221
  of sand on solid surfaces, 43

Friction piles, 27, 569, 681
  load transfer to soil, by, 673, 675
  min spacing of, 671-676

Frictional resistance, 11, 170
  between ballast material and sheet piling, 225
  of friction piles, 681
  side face, 171

Friction forces, 36, 59, 170, 180, 197, 225

Friction-pile foundations, 8, 679

Friction piles, 8, 27, 82, 569, 681

Frictional-cohesive soils
  ultimate bearing capacity of, 57-59

Fringe, capillary, 45
Fröhlich's critical edge pressure, 90, 95, 344, 345
Frost criteria, 321
Frost penetration depth, 29, 317, 319, 320
 chart of, 319
 extreme, 319, 320
 limits in foundation engineering, 292
Frost-prone soils, 321
Fundare, 5
Fundatio, 5
Fungi, 574
Funicular polygon, 158

Galleries, Persian, underground, 12
Gates' pile driving formula, 605
General shear, 56-58
Geometric forms of piles, 568, 569
Gradient, hydraulic, 203, 204
 decrease in, 219
 electric, 275
Greek alphabet, 753
Greeks, ancient, 12
Greiner formula, water pressure on piers, 701
Grillages, 310, 451-455, 745
 reinforced concrete, 310
 steel, 310, 311, 452, 453
 timber, 310, 311, 451, 452
Ground piles, 569, 570
Groundbreak, 19, 50, 68, 71, 73, 75, 76, 81, 86, 203, 218, 219, 231, 249, 339, 481
 at inboard toe of cofferdam, 231
Ground surface, 6
Groundwater, 31, 72, 85, 86
 conditions of, 50, 579
 in motion, 44, 46
 at rest, 44, 45
Groundwater table, 3, 6, 8, 27, 28, 32, 42, 67, 69, 70, 72, 74, 93, 256-260, 319, 574
 fluctuating (variation in), 93, 94, 319, 574
 lowering of, 188, 256
 position of, 27, 67, 69, 70, 72, 74, 579
Grout, 328
 cement, 283, 284
Grouting, 279, 283
 bitumen, 279, 284, 285
 cement, 279
 chemical, 279
 clay, 285
 underwater, 191, 195
GWT, 6, 32, 34, 45, 93, 113, 127

H-piles, 111, 590, 592, 593, 610
Hamburg rules for pile bearing capacity, 624, 625

Hammer, driving, 600
Hammurabi's laws, 12
Handling of concrete piles, 581, 582
Handy's soil bore-hole direct-shear test device, 612, 613
Hazen's coefficient of permeability, 252-255
 example of, 254
Hazen's flow velocity equations, 252-255
Headers, of crib walls, 730, 736, 737
Headwater, 46
Heavy-duty H-cores, 592, 593
Heavy-duty H-piles, 592
Heavy wall footings, 323
 example on, 323
 stepped, 324
Height of cofferdam, 196
 effective, 199, 200
High piles, cluster of, 583
Hiley's pile driving formula, 603, 604
Historical notes on foundation engineering, 10
 review of, 10, 17, 18
Hoisting of reinforced concrete piles, 581, 582
Hollow square footing, 363, 366
Homogeneous dam, 48
Horizontal acceleration, 51-53
Horizontal bracing frames, 111, 185, 186
Horizontal displacement of pile bent tops, 682-684
Horizontal loads (pressure), 36, 37, 60
Horsepower, 266, 267
Hütte, 302
Hydraulic gradient
 in cofferdam, 203, 204, 219
 decrease of, 219
 in wellpoint systems, 259
Hydrological conditions at site, 24
Hydrostatic pressure, 36, 44
 on a floating caisson, 541
Hydrostatic uplift pressure, 36
 mats resisting, 442-444
Hydrovibration, 74, 277-279
 definition of, 277
 equipment for, 277-279
 Keller's method of, 278
 process of, 277
 vibrator for, 277

Ice, 36
Ice pressure, 36, 51
 on cofferdams, 217

Indexes

Ice pressure (*continued*)
  on piers,
    of Mackinac Straits Bridge piers, 51
  tensile stress from, 230
Ice wall thickness, 294, 295
  formula for, 295
Island, sand, artificial, 17, 496, 512-514
Illustration of foundation terms, 6
Inboard face of cofferdams, 196
Inboard pressure, 228, 231
Inboard toe
  groundbreak at, 231
  piping at, 231
Inclination of piles, 569, 647, 648
Inclined frame cofferdam, 183
Inclined loading, 37
  example on, 37
Inclined piles, 583, 647-651, 653-661, 664
Inclined resultant load, 337, 338
Inclined strutting, 110
Individual footings, 8, 57-59, 67, 70, 87, 88; *see also* Shallow foundations, Shallow footings
  in clay, 73-80
  in (ϕ-c)-soils, 57-59
  on rock, 87, 88
  in sand, 67
  in silt, 84, 87
Induced interlock stress, 230
Influx of water in excavation, 249-250, 251, 254, 255
Influence radius, 265
Injections, 279, 285-287
Intercepting drains, 445, 446
Internal friction of soil
  angle of, 41, 105
  weighted average, 503
Installation of electro-osmotic dewatering system, 276, 277
Installation of sand drains, 301
Interior shoring and bracing, 113
Internal friction coefficient
  of ballast material (soil), 242
  between ballast and rock, 242
  between ballast (soil) and steel, 242
Intensity factors, uplift, 48, 49
Intensity scale, earthquake, 52, 53
Interlocks of sheet piling, 129, 130, 216, 231
  allowable tension in, 223
  friction in, 223, 225, 227, 235
  factor of safety, 226, 227
  failure by slipping, 235
  failure in tension, 224

Interlocks of sheet piling (*continued*)
  interlock tension, 229
  lubrication of, 242
  resistance of, 223
  rupture of, 216
  rupture in tension, 229
Interlock strength, 233, 242
Interlock stress
  circumferential tensile forces in, 235
  from ice pressure, 230
  induced, 231
  tensile, 230
  total, 231
Interlock tension, 231, 245
Intolerable differential settlement, 74
  effect of vibration on, 74
Inverted arch (vault), 315, 316
Inverted filter, loaded in cell, 231
Inverted T-beams, 315
Island, sand, artificial, 17, 496, 512-514
Isobars, 6, 21, 26, 29, 30, 31
  merger of, 30, 31
Isolated footings, 313, 413-417; *see also* Shallow foundations, Footings
  at property line, 413-417

Jetting of piles, 571
Joosten's chemical soil solidification (injection), 285-290

Keller's method of hydrovibration, 278-279
Key to signs and notations, 754-769
Klenner's earth pressure distribution diagrams, 120
Knife-edges for open caissons, 11, 17, 490, 491, 494
Königsberg Dome, 16
Krey's circle method, 510, 511
Krey's pressure system for sheet piling, 147
Krynine's lateral earth pressure coefficient K, 201, 221, 223, 227, 235, 734

Labarre-Converse pile group efficiency formulas, 644, 645
Lagging of metal piles, 591
Lake dwellers, prehistoric, 11, 573, 639
Land cofferdams, 177, 184-186
  braced, 186
Lateral earth pressure, 36-39, 119
  on caissons, 502, 503, 506, 507
  coefficients of, 38, 139, 201, 202, 242
  on cutting edges, 510
  on deep foundations, 501-503, 506, 512, 518, 718-729

Lateral earth pressure *(continued)*
  distribution diagrams
    Coulomb's, 119, 120, 123
    after Klenner, 120
    Navy's, 227, 228, 231
    after Terzaghi and Peck, 121, 122
  equations of, 39
  on retaining walls, 126, 127
  on sheeting, 119-124, 128
  on sheet piling, 132, 137-142
Lateral expulsion of soil from underneath a cofferdam, 203, 219
Lateral stress distribution diagrams, Navy's, 227, 228, 231
Lateral surcharge, 220
Launching of a floating caisson, 527-530, 532
  forces on, 523, 526, 528
Laursen's depth of scour, 707, 711, 712, 715, 716
Laws, Darcy, 249
  Hammurabi's, 12
Layered soil system, mantle friction of, 611, 612
Laying foundations, concept of, 5
Leaning Tower of Pisa, 16
Leibnitz, 10
Length of anchor, 165, 167
Length, computation, of cofferdam, 200, 241
  of piles, 612, 614, 616
Lidingö Bridge piers, Stockholm, 583-585
  reinforced concrete tubular piles, 582-585
Lift, 266
Lifting of piles, 581, 582
Light, plain concrete wall footings, 323
Limestone cutting edge, 11
Limits of depth in foundation engineering, 291, 292
Limits, endurance, for a man to work under pressure, 17, 547
Limnoria, 587
Liquefaction of sand, 71
  of silt, 85
Little Belt Bridge caissons, Denmark, 529-533
Load transfer to soil by friction piles, 673-676
Loaded inverted filter in cell, 231
Loading of piles, eccentric, 665
Loads, 36, 37
Loads on bridge piers, 701
Local scour, 714
Local shear, 56-58
Loess soils, 41, 42, 85, 86
  angle of internal friction of, 41

Loess soils *(continued)*
  cohesion of, 42
Locks for pneumatic caissons, 546, 548, 549, 551, 552
Long, Huey P., Bridge at New Orleans, 17
Losenhausenwerk, 631
Lowering of groundwater table, 256
  advantages of, 256
  in cofferdam, 188
  double stage, 257
  flow velocities, 256
  multiple stage, 257
  pumping from multiple wells, 258
  single stage, 256
  in trench, 256
Lubrication of interlocks, 242
Luke, 10

Machine foundations
  mechanical vibrations of, 742, 743
Mackinac Straits Bridge pier caisson, 51, 481, 488, 494, 496
  ice pressure on, 51
  dredging, 496
  sketch of, 495
Maintenance of excavation slopes, 104, 105
  improper, 105
Mantle friction, 503, 600, 609, 614
  negative, 27, 59, 75, 81, 83, 572
  weighted average, 503
Mantle resistance, 11, 59, 614, 615
Mantle surface, 11
Marine borers, 27, 574, 575
Mass, apparent, Kármán's, 746, 747
Materials
  for ballast, 203
    freely draining, for cofferdams, 212
  pile, 568, 578, 587, 595
  sheet piling, 129, 130
  soil, 91
Materials, river bed, scouring, 218, 702-704, 706, 707
Mat foundations, 8, 26, 29, 78-80, 315, 316, 426-442
  allowable bearing capacity for, 59, 71
    on clay, 79, 81, 82, 84
    on rock, 88
    on sand, 67, 68, 70, 71, 72
    on silt, 85, 86
    maximum, 79
    proposed values of, 71, 72
    safe, 79
  bamboo, 12, 31
  cellular, 429
  choice between strip and mat, 29
  definition of, 8

Indexes

Mat foundations *(continued)*
  design procedures for, 73, 80, 85, 88, 427
  determination of soil bearing capacity for, 71, 72, 80, 85, 88
    of c and ϕ, 85
  factor of safety, 71, 79, 86
  flat slabs, 428
  groundbreak, 71, 79, 86
  groundwater table, position of, under, 72
  mats, 382
    fascine, 493, 494, 705
    reinforced, 426-442
      resisting hydrostatic uplift pressure, 442-444
    ribbed type, 428
    rigid, 431-442
  on rock, 88
  on sand, 67, 68, 70-72
  settlement of, 72, 79
  on silt, 85, 86
  slab-type, 427, 428
  soil pressure distribution under, 430, 431
  special considerations, 73, 80, 86
  Terzaghi's equations, 78
  ultimate bearing capacity for, 71, 78, 79, 85
  unconfined compression test of soil for, 80
  of uniform thickness, 427
  uses of, 21, 22, 24
Mats, 382
Mattresses, fascine, 493, 494, 705
McKay's pile driving formula, 603
Mechanical vibrations of machine foundations, 742, 743
Medium, porous, 273
Medieval ages, foundations in the, 16
Mercalli-Cancani-Sieberg earthquake intensity scale, 52, 53, 744, 745
Merger of isobars, 30, 31, 572
Metacenter, 523-527, 540
Metacentric height (distance), 524, 527, 540, 543
Metal piles, 568, 587, 588
  lagging of, 591
  steel, 568, 589, 590
  steel H-, 590-592
  steel pipe, 590, 592
Metal base plate (billet), 328, 354
Michigan studies on piles, 606-608
Migration of soil moisture (water) in electrical field, 273-276
  through porous medium, 273
Mistakes in foundation design, 4
Model caisson cutting edge, 511
Models for bridge design, 16, 17
Modes for anchoring of piles into pile caps, 403, 404

Modes of failure of soil by anchorage, 172, 173
Modulus of subgrade reaction, 719-721
Moisture conditions, changes in, 42
Moisture (water) film, 274, 275, 634
Moisture (soil) migration in electrical field, 273-276, 634
Moisture proofing of basements, 448, 449
  of foundations, 448, 449
  waterstops, 446, 447
Mole construction in Roman times, 13, 14
  definition of mole, 14
  principle of construction, 12-14
Moments
  bending, maximum, 136, 138, 139, 142, 143, 155-157
  calculation of
    in sheet piling, 136, 137-139
    in footings, 356
  design, 157
  diagrams for sheet piling, 137, 138, 142, 143
  driving, 224, 226
  of inertia of pile groups, 666, 667
  overturning, 222, 224
  position of maximum bending moment in sheet piling, 137-140, 142-144, 152, 154, 155, 157, 158, 162, 163
  reduction of, 155-157
  resisting, 93, 225, 226
Monolith for river diversion, 14, 15
Moran's dome-capped caisson, 17, 497, 498, 499
Motor capacity, 266, 267
  coefficient of efficiency, 273
Multiple-stage dewatering, 257, 258

N-charts, 56, 68-69
$N = f(N')$, correction formula for, 70
N number of blows, 61-70
N-values, 61, 67
  bearing capacity, 56, 64-68, 70, 84
  blows, 61-72
  corrected, 57
  relative density of, 68-71
    for fine sand, 70
    for granular material, 62
Nature, acts of, 4
Natural opponents, soil and water, 3
Navy's factor of safety, 228
Navy's lateral stress distribution diagram, 227, 228
Navy's pile driving formula, 603
Negative mantle (skin) friction, of piles, 27, 59, 75, 81, 83, 572
Net penetration of piles, per blow, 600

Net perimeter
  of foundation, 254, 255, 408
  of pier, 59
  of pile group, 82
  around piles, 609, 610
Net soil bearing capacity, 59
Net soil pressure, 67
Non-cohesive soils, *see* Soils, non-cohesive
Nose forms, 715
  of bridge piers, 707, 715
  coefficients of shape of, 707-716
Notations, key to signs and, 754-769
New Orleans Bridge, Louisiana, 17
Number of blows, N, 61-66, 68, 69, 70
  conversion of, for below water table in fine sand, 70
  correlation with soil properties, 62
  in Standard Penetration Test (SPT), 62-67
Number of piles, calculation of, 643, 644

Oakland Bridge pier foundations, 17, 53, 481, 488
Obtuse stress wedge, volume of, 398
Offshore towers, 684, 685
  Texas Tower, 685, 686
Old Testament, 11
One-way eccentricity, 331-333
Open caissons: *see* Caissons, open and Deep foundations
Open excavations, 8, 50, 105, 106, 488-490
  dry, 8, 50, 103
  maximum unsupported depth of, 105
  between sheet piling, 50
  stepped, 105, 106
Open pumping, 249
Opponents
  natural, 3
  soil, 3
  water, 3
Ordinary foundations, 8; *see also* Foundations
Ordinary wells, 258, 267
Oscillation
  of groundwater table, 93, 94, 319
  sinusoidal, 739
Outboard face of cofferdam, 196, 220
Overlapping pressure bulbs (isobars), 30, 572
Overlapping rupture surfaces, 511
Overturning of cofferdam, 197, 219, 221, 222, 224
Overvibration of sand, 71, 743

Pacific Coast (Unified Building Code) pile driving formula, 604

Panel of piles, 643, 644
Parabola
  pressure, 556, 557, 719, 723, 748
  Westergaard's, 748
Parabolic stress, 671, 673, 721, 722, 723, 748
Paraboloidal stress, 671, 673, 676
Parameters
  of footings, 57
  tests for
    c, 41, 42
    ϕ, 41
Passive earth pressure (resistance), 39
Passive earth pressure coefficients, 139, 206, 242
Passive earth pressure ordinate, reduced, 149, 150
Patterns of pile spacing, 640
Pedestal, 6
  elongated, 422
Pedestal footings, 327, 328
  foundation, 328
Penetration, net, of pile per blow, 600
Penetration
  of open caissons, 488, 504, 512, 514, 633
    example on, 505, 515-517
  of piles
    by driving, 643, 644
    by electro-osmosis, 633-635
    by hydrovibration, 277-279
    by vibration, 75, 630-633
  of pneumatic caissons, 549
  of sheet piling, 113
    in clay, 113
    by driving
    by electro-osmosis, 633-635
    in sand, 113
    by vibration, 633
Penetration depth in soil
  of caissons, 488
  of frost, 29
  of piles, 570, 571, 598, 600, 601, 603
  of sheet piling, 113, 181-183
Penetration resistance test, standard (SPR test), 61-67; *see also* Standard penetration resistance tests
  advantages of, 62
  description of, 61-64
  disadvantages, 62
  examples on, 64-67
  soil pressure chart, 69, 70
Pentagon, piles for foundations, 586
Perimeter, circular, around piles, 408
Perimeter of caisson, calculations of, 254, 255
Perimeter net
  around piles, 609, 610
  of a pile group, 82

Indexes   **813**

Perimeter of pier, 59
Perimeter of shear, 411
Permeability, coefficient of, 248, 250-255, 264, 273
Persian underground galleries, 12
  tunneling of, 12
Pick-up-points of piles, 581, 582
Piers, 59, 60, 61, 73, 74, 80, 88, 89
  allowable bearing capacity of, 60, 61, 73, 81, 88, 89
  bridge, 493, 495, 701-705, 708-710, 712-716, 747
    basic, scour around, 714, 715
    Chinese, 12
    circular, 59, 74
    in clay, 80, 81
    Greiner's water pressure formula for, 701
    ice pressure on, 701
    loads on, 59, 701
    nose forms of, 707-716
  bulbed-end, 81
  caps for, 583, 585
  of Delaware Memorial Bridge, 493
  embedment depth of, 59, 60
  form coefficients of, 707-716
  Lidingö Bridge, Stockholm, 582-585
  Mackinac Straits Bridge, Mich., 495
  mantle resistance of, 11, 59, 60, 614, 615
  perimeter of, 59
  on piles, 582-585, 713
  on rock, 88, 89
  in sand, 73, 74
  in silt, 86, 87
  of San Francisco-Oakland Bridge, 17, 498
  skin resistance, 59, 60
  of Tagus River Bridge, Lisbon, 498, 499
  of Transbay Bridge, California, 17, 498
  ultimate bearing capacity of, 59-61, 73, 74, 80, 81, 85, 86
  use of, 481, 482
  of Verrazano-Narrows Bridge, N.Y., 495
  wind loads on, 701
Pile bent tops, horizontal displacement of, 682
Pile bents, 648, 682
Pile caps, 401, 745
  anchoring of piles into, 403, 404
  checking for shear, 408-412
  checking for punch, 408
  size of, 402
  spacing of piles in, 402
Pile, conical, bearing capacity of, 621-623
Pile cutoff in footing (cap), 327, 402
Pile driving, 75, 599, 643, 644
  concept of, 600
  electro-osmotic, 633-635
  shoes for, 577, 590

Pile driving *(continued)*
  by vibration, 75. 630-633
Pile-driving formulas, 75, 599-606
  Brix, 605
  Canadian National Building Code, 606
  classical, complete, 601
  dynamic, 599-606
  Engineering News, 602-604
    modified, 604, 606
  Eytelwein's, 602, 606
  Gates', 605, 606
  Hiley's, 603, 604, 606
  McKay's, Navy-McKay's, 603, 606
  Michigan studies of, 606-608
  Pacific Coast Uniform Building Code, 604, 606
  Rabe's, 605
  Rankine's, 603, 606
  Redtenbacher's, 601, 606
  Stern's, 603
  Weissbach's, 602, 603
Pile-driving shoes, 577, 590
Pile foundations, 8, 31, 75, 76, 82, 83, 87, 89, 639-698; *see also* Deep foundations
  advantage of, 31, 594
  allowable bearing capacity of, 82, 670, 671
    of piles, 75, 83, 87
  in clay, 82, 83
    allowable bearing capacity of, 82
    special considerations, 83
    ultimate bearing capacity, 82
  determination of
    bearing capacity of piles, 75, 81
    friction pile capacity, 83, 608
  embedded length of piles, 82, 583, 609, 614
  embedment of piles by vibration, 75, 568, 569, 630-633
  end-bearing piles, 8, 83, 568, 569
  factors affecting choice of, 27, 28
  feasibility of designs, 28
  friction piles, 8, 27, 82, 569
  negative mantle friction, 27, 83, 572
  performance of piles, 83
  perimeter of pile group, 82
  piles, end bearing, 8, 83
  piles, friction, 569, 679, 681
  piles in sensitive clay, 83
  pulling tests of piles, 83
  in sand, 75, 76
  settlement of, 75, 83, 87, 676-680
  shear strength of clay for, 82
  in silt, 87
  skin friction, 59, 75, 83, 87
    negative, on piles, 27, 59, 75, 81, 83, 572
  skin resistance, 59, 75, 83, 87
  spacing of piles, 83
  special considerations, 75, 83, 87
  stability of, 670, 671
  static loading tests, 75, 606

Pile foundations *(continued)*
  static pile bearing capacity
    empirical rules, 623, 624, 625
    of pile tips, 614-623
    of piles, 617-623
    of screw-piles, 588, 589
  ultimate bearing capacity, 82
    of piles, 87, 588, 589
  unconfined compression strength of clay, 82
  use of, 82
Pile group
  efficiency of piles in, 644, 645
  pressure bulbs in, 572
Pile group static systems, 650
  Culmann's method, 656-661, 664
  statically determined, 650
  statically indeterminated, 650
Pile groups, 583, 644-646
  analysis of, 668-669, 684
    graphically, 655-664, 666, 667
    summary of, 669, 670
    use of, 645, 646
  in clay, 676-680
  design of, 676-680
  efficiency of, 644, 645
  perimeter of, 82
Pile materials, 568, 573-595
Pile perimeter, 609, 610
Pile punch, 408
Pile requirements, 573
Pile screw, 587-589
Pile static bearing capacity formulas, 608-624
Pile tips, 576, 577, 614-623
  bearing capacity of
    conical, 616-618, 621-623
    pyramidal, 619
  minimum tip resistance, 620, 621
Pile trestles, 649
Pile vibration, 75
  embedment by, 630-633
Piled anchorage, 165, 166
Piled footings, 400-412, 690-692
  anchoring of piles into, 403
  caps, 401
  clearances, minimum, 402, 403
  combined, 404
  pile cutoff, 402
  problem on, 402-404, 690-692
  spacing of piles, 402, 640-642, 663
Piles, 12, 31, 568-638
  advantages and disadvantages, 594
  allowable load on, 614
  anchor, 166
  arrangement of, 166, 640, 641
  bamboo, 12

Piles *(continued)*
  batter, 569, 646-650, 745, 746
  bearing, 568, 569
  bearing capacity of, 598
    determination of, 598-599
    static, empirical formulas, 624, 625
  static formulas, 608-624
  bearing values of, 625
    safe, 625
    table for, 626
  bulbed-end, 609, 628, 629
  butt of, 569
  cast-in-place (situ), 568, 585, 586
  cast iron, 568, 587, 588
  cedar, 11, 573
  calculations of, 624, 625
  classification of, 568, 571
  cluster of, 583
  composite, 568, 593-595
  compression, 651-657
  computation length of, 612, 614, 616, 679
  concrete, cast in place, 585, 586
  concrete, cross section of, 569
  concrete footings on, 327
  concrete, handling of, 581-582
  concrete, precast (prefabricated), 578-582
  concrete, reinforced, 578-581
  concrete, use of, 578
  cylindrical, 568
  decayed tips of, 574
  definition of, 568, 573
  distribution of loads on, 656-664
  dimensions of, 575, 576, 580
  driving shoes for, 577, 590
  eccentric loading of, 665
  effective length of, 681
  elastic deformation of, 680, 681
  embedment of, 568, 630-633
    by driving, 643, 644
    by electro-osmosis, 633-635
    by vibration, 630-633
  end-bearing, 83, 568, 569
  end-bulbed, 609, 628, 629
  foot of, 569
  footings on, 400-412
  forces on, 608-611
  free-standing, 569
  friction, 8, 27, 82, 569. 681
    mantle (skin), 27, 59, 75, 81, 83, 503, 572, 600, 609, 611, 612, 614
    function of, 568
  geometric forms of, 568, 569
  ground, 569, 570
  H-, 111, 590, 592, 593, 610
  handling of, 581, 582
  head of, 569
  heavy-duty H-piles, 592, 593
  inclination of, 569, 647, 648
  inclined, 583, 647-661, 664
  jetting of, 571

Indexes 815

Piles *(continued)*
  lagging of steel piles, 591
  in a layered system of soil, 611, 612
  loading of, static, 606
  mantle friction, 27, 59, 75, 81, 83, 503, 572, 600, 609, 614
  material of, 568, 573, 587-595
  metal, 568, 587, 588
  method of embedment, 568, 630-633
  Michigan studies on, 606-608
  number and pattern of, 640, 643, 644
  pick-up points of, 581, 582
  pipe
    reinforced concrete, 582, 583
    steel, 589, 590
  precast, 578-580
    sizes of, 579
    use of, 579
  prefabricated, 578
    handling and hoisting of, 581, 582
  pressure bulbs from, 572
  ready made, 568, 578, 579
  reinforced concrete, 568, 578, 580
    sizes of, 579, 580
    use of, 579
  reinforced concrete pipe, 582-585
  safe bearing values of, 625, 626
  screw, 587-589
  sheet, 50, 111, 129, 568, 571
  soldier, 110
  spacing of, 402, 640-642, 661, 663
  species of trees for, 573
  splicing of, 575
  square, 619, 620
  steel, 568, 589, 590
    corrosion of, 591-593
    driving shoe, 590
    H-piles, 590-592
    load on, 589
    pipe, 589, 590
  storage
    of concrete, 581
    of timber, 578
  tapered, 568
  tension, 568, 625, 627-630, 646, 647, 651-657
    pull-out capacity, 627-630
  terminology of, 568-572
  timber, 568, 573
    basic requirements of, 573
    bearing capacity of, 626
    decayed tops of, 574
    dimensions of, 575
    economy of, 578
    impregnated, 573, 574
    length of, 575
    splicing of, 576, 577
    storage of, 578, 581
    tips of, 576, 577, 614-623
    uses of, 573

Piles *(continued)*
  tubular, 582-585, 590, 592, 593
  use of, 28
  wood
    cross section of, 575
    decay of, 574
    economy of, 578
    treated (impregnated), 573, 574
    types of, 570, 571
    untreated, 574
Pipe piles, reinforced concrete (tubular), 582-585
Pipes, steel, 591
Piping
  in soil, 218, 219, 231, 233
  at inboard toe, 231
Pisa, Leaning Tower of, 16
  reasons for leaning, 16
Pitching stones, 705
Pits, foundation, 8
  below groundwater table, 8
  in dry, 8
Pivot point, position of, 135, 137, 725
Plain concrete footings, 350-381
  centrically loaded, 350, 352-381
  design procedure, 350-352
  examples on, 352, 353, 364-366
  problems on, 351-353, 364-366
  rectangular, 350-381
  square, 350, 351, 357-366
  strip (wall), 352, 353, 355
Plain concrete wall footings, 323, 367-374
  example on, 323
  heavy, 323
  light, 323
  massive strip, 374-380
  pyramidal strip, 374-380
Planes, critical
  for diagonal tension, 352, 354, 360, 362, 365, 409, 410
  for moments, 353, 354, 357, 358, 362, 365
Plastic zones, 108
Plate, billet, metal, 328, 354
Pneumatic caissons; *see* Caissons, pneumatic
Polygon, force, 173, 666, 740
  funicular, 158
Poling boards, 108
Pont Neuf Bridge in Paris, 16
  foundations of, 16
Pore water pressure, 36
Porosity, 40
Pouring of concrete seal (tremie), 192
  requirements of, 192
Porous medium, 273
Potential, zeta, 275

Prandtl's ultimate bearing capacity of soil, 57
Precast concrete piles, 578-582, 594
Predrainage, 249
Prefabricated piles, 568, 573, 578-580, 582, 594
Prehistoric lake dwellers, 11, 573, 639
Prepact concrete, 687
Pressure
  active, earth, 39, 119-123, 126
  allowable on rock and soil, 55
  artesian, 36, 44, 50
  bulbs of, 30, 31, 76
  on caisson, 509-511
  contact, on soil, 6, 29, 30
  critical, on soil, 344, 345
  on cutting edges, 509-511, 558-563, 567
  distribution of, 31, 76
    by a rigid slab, 431, 432
    by a rigid structure, 31
    under mats, 430, 431
  earth, 36, 37, 39
  earth, passive (resistance), 39
  edge, Fröhlich's, 90, 344, 345
  horizontal, 60, 553, 555, 566
  hydrodynamic, 180
  hydrostatic, 36, 44, 180
  by ice, 36, 51, 217, 230
    on cofferdam, 217
    on Mackinac Straits Bridge piers, 51
    tensile stresses from, 230
  lateral, 36-39
    in deep foundations, 60, 718-729
    earth, 37-39
  net, 67
  parabola, Westergaard's, 748
  passive, earth, 39
  pore water, 36
  seepage, 36, 44, 50
    in earth cofferdam, 180
  uplift, 36, 44-50, 442-444
    diagrams of, 45-50, 442, 444
    reduction of, 50
  of water, 36, 44-51
    on bridge pier, 746-748
  Westergaard's parabola, 748
  wind, on structures, 36
Pressure bulbs, 30, 76, 572
Pressure distribution, 29-31
  Coulomb's diagram, 119, 120
  around cutting edge, 509-511
  under mats, 428, 430, 432, 434, 440, 442, 444
  diagrams of, 119-122, 332-337
Pressure ordinates on sheet piling, 135, 137, 138, 140, 142, 144, 148, 152, 158, 161, 164
Pressure parabola, Westergaard's, 748

Pressure relief, by excavation, 65
Pressure on pneumatic caissons, 553-556, 558-563, 566, 567
Pressure of silt and water on pier due to earthquake, 746-748
Pressure system by Krey, 147
Presumptive rock and soil bearing capacity, 54, 55
Principal stresses, 105, 106
Principle of electrical dewatering, 274
Principle of injections, 286
Principle of pneumatic caisson, 546, 547
Probability and statistics in factor of safety, 96
Problem on seepage, 231
Problems on
  box-type cofferdams, 210
  bracing and strutting of excavations and trenches, 125
  choosing the kind of foundation, 32-35
  circular-cellular type cofferdams, 246, 247
  circular footings, 463-367, 467-476, 476, 477
  crib-type walls, 735-737
  Culmann's graphical pile analysis, 658, 660, 661
  cutting edges, 566, 567
  deep foundations, 518-520, 537-544, 729
  dewatering of excavations, 304-306
  earthquakes, 749, 750
  floating caissons, 525-527, 538-544
  footings, 348, 349, 457
  influx discharge into excavation, 255
  isolated footings at property line, 414
  lateral earth pressure
    on sheeting, 128
    on wall, 125-127
  lateral soil pressure distribution diagrams, 125-128
  massive, pyramidal strip footing, 379-381
  massive, rigid strip footing, 352, 353, 379-381
  open caisson, 515-517, 518-522, 635-637
  open, single-wall cofferdam, 210
  pile foundations, 694-698
  piles, 596, 597, 635, 694-698
  piles and pile driving, 694-698
  plain concrete wall footings, 352, 353
  pneumatic caissons, 564-567
  rectangular, plain concrete footings, 351-353, 364-366
  reinforced concrete footings
    rectangular, 389-392
    square, 385-388
    square on piles, 404-412
    strip, 393, 394
  scour, 717

Indexes 817

Problems on *(continued)*
  shallow foundations, 456-458
  sheet piling, 128, 175-176
  sheet piling design, 132, 175-176
  sheeting, 128
  square, plain concrete footings, 329-331, 350-381
  statics of strutting, 125, 128
  tension piles, pull-out capacity, 629-630
Procedures, general
  for designing foundations, 28
  for footing design, 350-353
    problem on, 351-353
Profiles, soil, 32-35, 50
Properties of soil
  cohesive, 42
  correlation with N, 68
  methods of improvement, 279-282
Property line, 411-419
  foundations restricted by, 411-419
  trapezoidal footing at, 417-419, 422, 423
Proportional limit, Fröhlich's critical edge pressure, 90
Protection of excavation banks, 108-111
Protection against scour, 513, 705, 706
Protection of steel reinforcement in footings, 326
Public enemy No. 1 (water), 20
Puller, 36
Pull-out capacity, 232, 627-630
  angle of tear, 628
  of a pile, 232
  of tension piles, 627-630
Pull-out of sheet piling, 232, 233
  factor of safety, 232, 233
  stability against, 232, 233
Pump capacity, 266, 267
  coefficient of efficiency, 272
Pumping
  open, 249
  from multiple wells, 258
Pumps
  deep well, 266
  efficiency coefficient of, 272
  screw-type, 13
  submersible, 266
Punch of pile in pile cap, 408
Pyramid, Cheops, 11
Pyramidal footings, massive, stepped, strip footings, 374-380

Quick condition of sand, 74, 204, 231, 249

Rabe's pile driving formula, 605, 606
Radius of influence, 258, 259, 265

Raft (mat) foundations, 8, 315, 426-442
  definition of, 8, 427, 428
Rake, 110
Rankine's pile driving formula, 603
Rapid drawdown, 133, 249
Reaction, soil, 36
Rectangle, equivalent, 262
Redtenbacher's pile driving formula, 601, 602
References, 18, 32, 98-100, 125, 174, 175, 209, 246, 302, 303, 348, 380, 455, 477, 500, 517, 518, 544, 564, 595, 637, 638, 692, 693, 717, 729, 735, 749, 750
Refrigeration plant, ammonia, schematics of, 294
Reinforced concrete grillages, 310
Reinforced concrete footings, 382-458
  on piles, 400-412
    centrically loaded, 400, 402
    eccentrically loaded, 404-407
  rectangular, 382, 388-392, 395-399
    centrically loaded, 388-392
    eccentrically loaded, 395-399
  square
    centrically loaded, 382-388
    eccentrically loaded, 395-399
  strip (wall)
    centrically loaded, 392-394
    eccentrically loaded, 395-399
Reinforced concrete mats, 426-442
Reinforced concrete piles, 568
  hoisting of, 581, 582
  storage of, 581
Reinforced concrete slabs (mats, rafts), 426-442
Reinforced concrete pipe piles, 582-585
Relief of pressure by excavation, 65
Relative density, 62, 68, 71
  of sand, 62, 68, 70, 71
  values of N, 62, 68-71
Repetitional pile sections, pile panels, 643, 644
Requirements, stability
  of earth cofferdams, 180
  of foundations, 19, 20, 25, 338-345
  of piles, 573
Requirements for pouring of concrete seal (tremie), 192
Resistances
  to anchor pull, 169
  coefficients of, 384, 385, 387, 388
    passive earth, 39
  earth, 39
  of interlocks, 223
  line, 266
  mantle, 11, 59

Resistances *(continued)*
  to shearing of soil, 56, 57
  skin, 59, 60, 75
  soil, 172
  in standard penetration resistance tests, 61-67
    soil pressure chart, 69, 70
  of tips of piles, 620, 621
    minimum, 620, 621
Resonance of structures, 745
Responsibilities, legal, of civil engineers, 3
Resultant load on foundations, 37, 337, 338
Retaining walls, 126, 127
  bulkheads, 129, 163, 164
  crib-type, 178, 198
  earth, 126, 127
  massive, 126, 127
  sheet piling, 109-111, 129, 165
Retarding of seepage, 49, 50
Review, historical, of foundations, 10-18
Rialto Bridge foundations, 16
Ribbed mats, 428
Rigid mat foundations, 431, 432
  design of, 432-442
Rigid strip footings, 352, 353
Rigid structure, pressure distribution in soil, by, 31, 431-442
Rip-rap, 705
  under footing caisson, 542
  against scouring in riverbed, 705
  rock-fill, 179, 190, 705
River beds, scouring of, 218, 702-704, 706, 707
River diversion dam, Saguenay River, Canada, 14, 15
  monolith for, 14, 15
River, swollen, at bridge piers, 704
Rock-filled cofferdam, 177-179, 730-737
  definition of, 179, 730
Rock flour, 86
Rock and soil presumptive bearing capacity, 55
Rock and soil safe bearing capacity, 54, 55
  sources of, 54
Rocks
  bearing capacity of, 54, 55
  foundations on, 87-89
  mats on, 88
  piers on, 88, 89
  piles on, 89
  shallow, individual footings on, 87, 88
  summary of foundations on rocks, 87-89
  unit weight of, 39
Rods, anchor (tie), 148, 150
Roman cofferdam, 13

Roman engineer Vitruvius, 12-16
Roman method of laying foundations, 13
Roman mole, construction of, 13, 14
Roman times, 12-16
Rough base strip footing, 56
  ultimate bearing capacity of, 56
Rowe's moment reduction, 155
Rows of piles, 644, 652
Running sand, 27, 318
Rupture of interlocks
  of sheet piling, 216
  in tension, stability of, 229
Rupture surfaces in soil, 114-118
  angles of, 115-119
  equations of, 115-118
    active case, 115, 116
    passive case, 117, 118
  overlapping, 511

Safe bearing capacity of soil and rock, 54, 55, 64-67
Safety against shear in cofferdam, 200
  against motion, 93
  against rotation, 93
Safety against shear failure of soil, 20
  general shear, 56-58
  local shear, 56-58
Safety requirements of anchorage, 172-174
Safety factors, 59, 61, 65-68, 71, 79, 86, 136, 148, 149, 338, 339, 341, 600-606
  for anchorage, 170
  assumptions for, 91, 92
  calculations of, 92, 93
  description of, 89-98
  in foundation engineering, 89-98
  seismic, 51-53
  values of, 94, 95
Saguenay River, Canada, 14, 15
Samos, 12
Sand ballast
  in cofferdams, 196, 200, 219
  effective weight of, 219, 243, 244
  in floating caissons, 532
Sand drain installation, 301
Sand piles, 300, 301
San Francisco-Oakland
  Transbay Bridge pier, 17, 53, 746-749
  caissons, 481, 488
  dome-capped (Moran's) caissons, 497-499
Sand
  allowable bearing capacity, 60, 67, 68, 70, 71, 73
  artificial island, 17, 512-514
  berm, 231
  boiling at cofferdam, 187

Indexes 819

Sand *(continued)*
  caissons on, 73
  effect of vibration on, 71, 741
  footings on, 67-71
  foundations on, 67-71
  liquefaction of, 71
  mats on, 71-73
  pier foundations in, 73, 74
  pile foundations in, 75, 76
  quick condition of, 74, 204, 231
  relative density of, 62, 68
  running, 318
  settlement of, 70
  ultimate bearing capacity, 55-59, 64-67, 76-84
Sand ballast, 200, 203, 218, 223, 532, 540, 541
Sand drains, 300
  definition of, 301
Sand island, artificial, 17, 496, 512-512
Sand piles, 300
Sand wedge, 14, 15
Saturated soil, unit weight of, 40
Screw piles, 587-589
Screw-type pumps, 13
Scour, 29, 318, 701-717
  around basic bridge pier, 714
  at bridge, 703, 714
  depth of, 705, 706, 707, 711, 715
    according to Boldanov-Andreyev, 707-711
    according to Laursen, 707, 711, 712, 715, 716
  flow velocity, 706, 707
  local, 714
    graphs for calculation of, 714, 715
    problems on, 717
    protection against, 705, 706
  of river bed, 218, 318, 702-704, 706
Scouring material, in river bed, 706, 707
Scouring of river beds, 702, 704
Seal, caisson, 491-493
Seal, concrete
  grouting of, 191, 195
  thickness of, 192
  underwater, 191-195
Section, critical, 353, 354
  for moment, 356-360, 383
  for shear, 360, 383
Section moduli, 130, 136, 139
Seepage, 44
  curve, 220
  flow, reduction of velocity, 189, 190
  force, 36, 50
  intensity factors, 48, 49
  line, 220, 241

Seepage pressure, 36, 44, 50
  on earth cofferdams, 180
  problem, 231
  retarding aprons, 49, 50
  undercut, 231
Segment-type cellular cofferdam, 211, 212, 235
  description of, 211-214
  static system, 220
Seismic acceleration, 51-53, 738, 739
Seismic effects on foundations, 738-751
  on structures, 739
Seismic factor, 51-53
Seismic forces, 36, 51, 180, 738, 740-742
  acceleration of, 51-53, 180, 738
Self-frequency, 745
Sequence in construction of a bridge pier, 191-192
Sequence in driving piles, 643, 644
Sequence in underwater concreting
  horizontal method, 193, 194
  vertical method, 193
Set, 600
Settlement, 31
  critical, 180
  of deep foundations, 61, 74, 86
  differential, 20, 31, 72
    of footings, 31
    of foundations, 20, 31
  of individual footings, 70, 71, 78
  intolerable, 74
    of footings, 72
    of mats on sand, 31, 72, 79
  of pier foundations, 61
    in clay, 81
    on rock, 88, 89
    in sand, 73, 74
    in silt, 86, 87
  of pile foundations, 82, 83, 676-680
    in clay, 83
  of shallow foundations
    in clay, 69, 78
    on sand, 70, 71
  seat of, 29, 31
  of strip foundations (footings), 31
  tolerable, 20, 31
Shaft of open caisson, 485-485
Shaft of a pneumatic caisson, 548, 550, 551
Shallow excavations, 103
Shallow footings, 313; *see also* Shallow foundations
  on clay, 67-71, 76-78
  on rock, 88
  on sand, 67-71
  settlement of, 69-71, 78

Shallow footings (*continued*)
  on silt, 84-87
Shallow foundations, 8, 26, 55-59, 67-71, 76-80, 84-87, 307-349
  allowable bearing capacity for
    on clay, 76-80
    on rock, 88
    on sand, 67-71
    on silt, 84-87
  allowable net bearing capacity, 77
  allowable soil bearing capacity, 55, 59, 65-68, 72, 84
  classification of, 8, 309-313
  on clay, 58, 76-80
  definition of, 8, 309
  depth of, 56, 59, 317
  design load, 70, 71
  design procedures, 73
  determination of soil bearing capacity for, 68, 69, 85
  examples for calculating bearing capacity, 64-67
  factor of safety, 67, 77
  footings, 8, 26, 57-59, 64-73, 307-349
  groundbreak, 76-78
  groundwater table, position of, 27, 67, 69, 70, 74, 75
  individual footings, 8, 57-59, 67, 84-87
    on clay, 76-80
    on rock, 88
    on sand, 67-71
    on silt, 84-87
    in (ϕ-c)-soils, 57-59
  laying of, 5
  liquefaction
    of sand, 71
    of silt, 85
  mats, 8, 26, 29, 71-73, 78-80, 426-444
  N-charts, 56, 68, 69
  N for silts, 84
  N_c, N_q and N_γ coefficients, 76
  net soil pressure, 67
  ordinary, 8, 56-59
  overvibration of sand, 71, 743
  parameters, test, 76
  problems on, 456-458
  relative density of sand, 62, 68
  on rock, 87-89
  on sand, 58, 67-73
  settlement of, 69, 70-78
  shape factors for footings, 57
  shear in soil
    general, 56-58
    local, 56-58
  on silt, 84-87
  Skempton's net pressure on soil, 59
  soil pressure as $f(B, D, N)$, chart for, 70
  soil net pressure, for, 59, 67
  special considerations, 71, 78
  SPR-test, 84

Shallow foundations *(continued)*
  Terzaghi's equations, 76
  test parameters, 76
  ultimate bearing capacity for, 55-59, 64-67, 76-84
    calculations of, 64-67, 76, 77
    equations of, 55, 56
    shear strength, net, of clay, 76
    unconfined compressive strength of clay, 57, 65, 76, 77
  uniformly loaded, 57-59
  use of, 310
  vibration of sand, 71
Shape coefficients for bridge pier nose forms, 707-716
Shape factors for footings, 57
Shear area, circular, 408
Shear, beam, 411, 412
  critical sections for, 360
  general, 56-58
  local, 56-58
Shear failure of soil, 56-59
Shear forces
  in ballast, 224
  at critical sections, 352, 354, 360, 362, 365, 409, 410
Shear perimeter, 411
  circular, around a pile, 408
Shear in pile cap, 408-412
Shear from piles on, on footings, 410-412
Shear plane in ballast, central, vertical, 215, 220, 231, 234, 238, 241, 243
Shear by punch, 408
Shear strength of soil (resistance), 56, 76, 82, 611, 612
  allowable, 82
  of clay, 82
  of cohesive soil, 42
  ultimate, of clay, 76, 82
  weighted average, 611, 612, 616
Shear stress
  allowable, 361
  calculation of, 360-366
  in cofferdam, 200, 201
  weighted average, 611, 612, 616
Shear in vertical plane (cofferdam), 215, 330, 231, 234, 238, 241, 243
Shearing a pier by a seismic force, 742
Sheeting, 108-111, 119
  lateral earth pressure on, 119-124
  timber, 108-111
    horizontal, 110, 111
    vertical, 110
Sheeting walls, 109-111

Indexes

Sheet piling, 50, 111, 129, 568, 571
  analyses, 114, 131-134, 146, 147, 164-166, 168-174
    with berms, 215, 220
    Blum method, 136-139, 141-146, 152
    for cofferdam, 212, 215
    conventional method, 134-136, 140, 141
    equivalent beam method, 158-163
    Tschebotarioff's system, 163, 164
  anchored, 132, 134, 146-155, 158-164
    calculation of, 153-155
    definition of, 134
    example on, 152
    stability of, 134
  anchor tension, 148, 150
  anchorage, 114, 132, 134, 164-174
    zones of, 132
  bending moment diagrams, 137, 138, 144, 152, 157, 158
  bending moment, maximum, in, 136, 138, 139, 143-146
  Bethlehem Steel, 129, 130
  calculation of, 132-139
    Blum method, 136, 138, 139, 141-146
    conventional method, 134-139
  cantilever, free, 132, 134-146
    with active earth pressure, 140-146
  in clay, 113, 114
  in cohesive soil, 114
  computation length, 20, 206, 207
  conventions on, 133
  definition of, 129
  deflection of, 147, 158, 206
  design of, 132
  design moment for, 155-157
  diaphragm, 178
  dimensions and properties of, 130
  dredged line, 132, 140, 165
  driving depth of, minimum, 134-137, 139, 141, 142, 145, 150, 151, 162, 204-206, 231
  embedment of, 571
  embedment depth
    by electro-osmosis, 633-635
    by driving, 134, 135-140, 150-151, 162, 291
  equations, 135, 136, 140
    derivation of, 136, 140
  equivalent beam method, 159, 160
  examples, 138, 139, 143-146
  factor of safety, 136, 149, 150
  fixed end support, 146, 158-163
  forces on, 132-138, 140, 142, 144-146, 148
  free cantilever, 132, 133, 135, 137
    calculation of, 134-139
    definition of, 133
    equations for, 134-139
    factor of safety, 136
    minimum driving depth, 134, 135
  free earth support, 146-148

Sheet piling *(continued)*
  in impermeable stratum, 189
  interlocks of, 129, 130
    strength of, 233, 242
    tension in, 229, 245
  interlock strength, 233
  interlock stress, 231
    from ice pressure, 230
    total, 231
  interlock tension, 229, 231, 235, 245
  loaded with active earth pressure, 132, 134-146
  loaded with concentrated horizontal load, 132, 134-139
  loading diagrams, 132, 135, 137-141, 144, 148, 152, 158, 161, 164
  maximum bending moment in, 136, 138, 139, 143-146, 162, 207, 209
  method for calculating of, 132-155
  method of minimum penetration depth, 146
  minimum driving depth, 134, 135, 139, 143, 145, 148, 150, 151, 162
  moment reduction, 155
    curves (graphs) for, 156
  in noncohesive soils, 131, 132
  penetration depth of, 113, 134, 135, 181-183
  position of pivot point, 136, 137
  problems on, 132
  profiles of, 130
  pull-put capacity, ultimate, 232
  requirements of, 129
  section modulus, 130, 136, 139, 155
  sections of, 130
  simplifications in theory of, 133, 163, 164
  sizing of sheet piling, 136, 139, 141, 146
  stability calculations of, 133, 232
  stability against pull-out, 232, 233
  steel, 112, 113, 130, 131, 139
  strutted, 188
  systems of, for analyses, 132, 133, 135, 137-139
  tensile forces in tie-rods, 148, 150
  theoretical depth of, 136, 137, 139, 140, 145
  timber, 111, 112, 187
  Tschebotarioff's method for calculating of, 163, 164
  ultimate pull-out capacity, 232
  underseepage cutoff, 231
  use of, 129, 147, 571
  in waterlogged soil, 113
Sheet piling-anchor wall system failure of, 172, 173
Sheet piling cofferdams; *see* Cofferdams, sheet piling
Sheet piling interlocks, 129, 130
  rupture of, 216
  strength of, 233, 242

Sheet piling loading schemes, 132, 135, 137, 138, 140, 142, 144, 148, 152, 158, 161, 164
Sheet piling systems, 132, 133, 135, 137, 139, 163, 164
Sheet piling tie-rods, 114, 128, 150, 172
Sheet piling wall failure, 165
Shelf, 6
Shell of cofferdam, 215
Shell, thickness of, for open caisson, 483, 485, 488
Shingles, 707
Shoes, pile driving, 577, 590
Shoring, 108
Shoring and bracing, interior, 113
Sieberg's earthquake intensity scale, 52, 53, 744, 745
Sickness, caisson, 547
Signs and notations, key to, 754-769
Silt
   allowable bearing capacity of, 65-67, 84-87
      determination of, 72, 74, 75, 77, 80, 81
   "bull's liver," 86
   caissons in, 86
   individual footings in, 84
   foundations on, 84
   liquefaction of, 85
   loess soil, 41, 42, 85, 86
   mats on, 85, 87
   piers in, 86
   pile foundations in, 87
   pressure on bridge pier due to earthquake, 746-749
   rock flour, 86, 87
   shallow foundations on, 84, 85
   summary on, 84
   ultimate bearing capacity of, 84, 85
Simple footings, 312
Simplifications in sheet piling theory, 133
Single stage dewatering, 256, 257
Single-wall sheet-piling cofferdam, 13, 177, 180-183
   earth suported, 182, 183
   strutted, 181, 183
Sinking of caissons; see Caissons, sinking of
Sinusoidal oscillation, 739
Site conditions for foundations, tables, 21-25
Site hydrological conditions, 24
Sizing of sheet piling, 136, 139, 141, 146
Skempton's formulas, 59, 80.
Skin friction, 29, 59, 60, 75, 81, 83
   average values of, 81
   negative, of piles, 27, 59, 75, 81, 83, 572

Skin resistance, 59, 60, 75
Slab foundations, 315; see also Mat foundations
   cellular, 428
   circular, 58, 59, 459-476
   flat, 428
   ribbed, 428
   uniform, 427
Sliding on base, 218, 221, 341, 342, 343
   of cellular cofferdam, 221, 231, 233, 244
Sliding, stability calculations, 339-343
Sliding surface, 132
Sloped footings, 312
Slopes of berms and excavations, 104, 105
Slopes, designation of, 178
   maintenance of, 104, 105
   protection of, 104, 105
   stability of, 105
   unsupported depth, maximum of, 105, 107
Slurry, bentonite, 487
Slurry-trench method, 295, 297-300
Smoke stack on piles, 691
Smooth base, strip footing, 57
Slipping interlocks, 223
Slippage failure of interlocks, 235
Snow load, 36
Soil bore-hole direct-shear test device, Handy's, 612, 613
Soil coefficients, reduced, 56, 57
Soil densification by piles, 642
Soil dynamics, 742
Soil erosion, 204
Soil expulsion from underneath a cofferdam, 203
Soil failure, 56-58
Soil-foundation system, 19
   analyses of, 19
Soil material for use in cofferdams, 179
Soil mechanics, 743
Soil particles, effective size of, 254
Soil particle size accumulation curves, 248, 254
Soil pressure
   under mats, 430, 431
   net, 67
Soil profiles, 29, 32-35, 50
Soil properties, correlation with N, 68
Soil reaction, 36
Soil resistance, 172
Soil and rock safe bearing capacity, 53, 55
   sources of, 54
   unit weights of, 39

Indexes

Soil-strip-foundation system, 6, 29-31
Soil, Winkler's, 720
Solidified soil chemically, 288-290
  in cofferdam, 288
Soil densification by hydrovibration, 277-279
Soil solidification chemically, 285-290
  Joosten's method, 285-290
Solidification of soil chemically, 285-290
Soils
  adhesion to foundation, 43, 60, 232
  allowable bearing capacity, 29, 54, 55, 59-61, 67, 76, 77, 79-82, 84
  angle of internal friction of, 38, 41
  bearing capacity, 54, 55
    allowable, 29, 54, 55, 59-61, 67, 76-84
    net, 67
    presumptive, 54, 55
    safe, 54-55
    tests commonly used, 781-785
    ultimate, 55-59, 67, 76, 78, 82, 84, 85
  bearing capacity of deep foundations, 59, 60, 727, 728
  bearing capacity factors, 56-59
    graphs of, 56
  bearing capacity of shallow foundations, 55-59, 65-73, 76-80, 84-88
  clay, varved, 298
  clay; see Clay
  cohesion of, 41, 42, 56, 64-66, 131
  consistency of, 62
  consolidation of, 27, 31
  densification of
    by compaction, 277
    by hydrovibration, 74, 277-279
    by piles, 642
    by vibration, 277-279, 281
  failure in general shear, 56
    in local shear, 56
  frictional-cohesive, 57-59
    ultimate bearing capacity, 57-59
  frost-prone, 321
  loess, 41, 42, 85, 86
  piping in, 218
  profiles of, 32-35
  settlement of, 31
  shear
    general, 56
    local, 56
  soldier piles, 110
  solidification of
    chemically, 285-290
    by firing, 281
    by freezing, 290-297
    by grouting, 282-285
  stability of, 278-288
  stabilization of, 279-306
    chemically, 285-290

Soils *(continued)*
  electro-chemically, 277
  by freezing, 290-297
  by grouting, 282-285
  test of, commonly used, 781-784
  ultimate bearing capacity, 55-59, 67, 76, 82, 84-89
  unit weights of, 39
  waterlogged, 13, 113
Solomon's temple, 11, 573
Spacing
  of piles, 640-642, 663, 671-676
  of sheet piling in cofferdams, 188
  of wells, 263, 264
Special caissons, 484, 514
Special foundations, 8, 9
Spacing
  of anchor blocks, 166, 171, 172
  of braces for equal axial compression, 122, 123
  of freezer pipes, 294
  of piles, 640-642
  of wellpoints, 260, 261, 263, 264, 268, 270, 271
Splicing of piles, 575, 576
SPR-test, 61-64
Spread footings (foundations), 8, 310, 312
  definition of, 8, 310
Springs, sealing off, 21
Square footings, 58, 64-67, 350, 351, 357, 358-360, 362-366, 382, 383, 386
  bearing capacity of, 58, 64-67
  calculations of, 264-366
  centrically loaded, 329-331
  eccentrically loaded, calculations of, 331-337
  plain concrete, centrically loaded, 329-331, 350-381
  reinforced concrete
    centrically loaded, 382-388
    eccentrically loaded, 388-392
    thickness, effective, 382, 387
Square, hollow footings, 363, 366
Square piles, 619, 620
Stability
  of anchor walls, 172, 173
  of anchored sheet piling, 134, 172
  of bottom of excavation, 50
  of cellular cofferdams, 216, 219, 221, 231
    on clay, 234
  of cofferdams, 189, 195, 197, 198, 203, 216, 219, 225, 228, 229, 231-233, 234
  of earth cofferdams, 180
  of floating body, 523, 526
    caisson, 482, 521-544
  of pile foundations, 670, 671
  sheet piling, 133, 232, 233

824 **Indexes**

Stability analyses (calculations)
  of sheet-pile-anchor-wall system, 172, 173
  of sheet piling systems, 133
  of slopes, 105
  of soil-foundation system, 19
Stability requirements
  for cofferdams, 180
    on sand, 198
  for foundations, 19, 20, 25, 338-345
  for pile foundations, 670, 671
Stability against
  groundbreak, 339
  rotation, 339
  sliding, 338, 339
    example of, 339-343
Stability of sheet piling, 133
Stabilization of soil, 279-288
  bitumen, 279
  cement, 279
  chemical, 279, 285-287
  electrochemical, 277
  by electro-osmosis, 274
  by grouting, 282-285
    cement grouting, 279
    clay grouting, 279
  table for, 280-282
Stage in wells, 263
Standard penetration resistance test (SPR), 61-67
  advantages of, 62
  bearing capacity factors, 68
    chart, $f(\phi)$, 68
    diagram, 69
  correction formula from N' to N, 70
  description of, 61-64
  disadvantages of, 62
  examples on, 64-67
  N-values, 61
  number of blows, 61, 62
  soil pressure chart, 69, 70
State of consistency, 62
Static calculations for floating caisson, 533-537
Static conditions at fixed earth support, 158
Static conditions of open caissons, 511, 512
Static loading tests of piles, 606
Static pile bearing capacity formulas, 623, 624
  Bénabenq's, 623
  Dorr's, 623
  empirical rules, 624, 625
  Krey's, 624
  Vierendeel's, 624
Static systems of
  cellular cofferdams, 216, 217, 220
  circular footings, 459-461
  pile groups, 650-657

Statically determined pile systems, 650-657
Statically indetermined pile systems, 650-657
Statics of open caissons, 501-520
Statics of strutting, 122-124, 181
Statistics and probability re, factor of safety, 96
Steel columns, bases of, 328, 329
Steel grillages, 310, 311, 452, 453
Steel cutting edges, 490, 493, 494
Steel piles, 568, 589, 610
  corrosion of, 591-592
  H-piles, 590, 610
  heavy-duty H-piles, 592
  heavy-duty H-core, 593
  lagging of, 591
  pipe, 590
  splicing of, 589, 590
  tubular, 582-585
Steel pipe piles, 590, 592
  heavy-duty, concreted, 592, 593
Steel sheet piling, 112, 113, 130, 131, 139; see also Sheet piling
  cofferdams, 178; see also Cofferdams
  in waterlogged soils, 113
Stepped caissons, 484
Stepped excavation, 105, 106
  maximum unsupported depth of, 105, 106
Stepped footings, 6, 8, 312, 324, 325
  heavy wall, 324, 374-380
  massive, 379-380
Steps in sinking of caissons, 514
Steps for design of a rigid mat, 431-432
Stern's pile driving formula, 603
Stones, pitching, 705
Storage of piles, 578
Strap, definition of, 424
Strapped footings, 8, 422
  at a property line, 423, 424
Strength of cofferdam, 195, 216
Strength
  of clay, compressive, unconfined, 57
  of foundations, 3, 4
    dependence of, 37
  interlock, of sheet piling, 233
  shear, ultimate, 76, 363
  ultimate design method for foundations, 362
Strength of soil, unconfined, 57, 65
Stress
  bending, 158
  diaphragm, in box-type cofferdams, 202
  induced in interlocks, 230
Stress distribution in soil, effect of width on footing, 29-31

Stress distribution in soil *(continued)*
  isobars, 6, 26, 29-31
Stress, parabolic, 719-721-723
Stress paraboloid, 671, 673, 676
Stress volume under circular footings, 468
Stress wedge, volume of, 398
Stressed zone, 6, 26, 679
Stresses
  compressive, ultimate, on soil, 55, 56
  major principal, 105-107
  in massive, plain concrete pyramidal footings, 374-380
  in stepped footings, 374-380
  in strip footings, 374-380
  minor principal, 105-107
  shear, in cofferdams, 200, 201
  tensile, in interlocks from ice pressure, 230
  total, in interlocks, 231
  ultimate compressive, 55-57
    at base of floating caissons, 544
Stretchers, of crib walls, 730, 736, 737
Strip footings (foundations); *see* Footings, strip
Structures
  engineering failures of, 4
  medieval, 16
  rigid, pressure distribution by, 31, 431-442
Strutted double-wall box-type sheet piling cofferdam, 188
Strutted sheet piling, 188
Strutted single-wall cofferdam, 13, 181, 183
Strutting
  inclined, 110, 123, 124, 181, 183
  statics of, 122-124, 181
Subgrade reaction, modulus of, 719-721
Submerged unit weight of soil, 40
Submergence of a floating body, 524
Submersible pumps, 266
Substructure, 6
Suction height, 266
Sudden (rapid) drawdown, 249
Suitable kinds of foundations; *see* Foundations, suitable kinds of
Summary of foundations, 67
  on clay, 76-83
  on pile groups, 669, 670
  on rock, 87-89
  on sand, 67-76
  on silt, 84-87
Sumps, 178, 185, 249
Superstructure, 6
Supports for excavation walls, 108-114
Surcharge, 60, 70, 217
  lateral, 219

Surcharged unit weight of soil, 39
Surface of rupture, 114-118
  of sliding, 132
Swiss Association of Highway Specialists, 57

T-beams, inverted, 315
T-slab, 316
  inverted, 316
Tables,
  for active and passive earth pressure coefficients, 201
  for choosing the kind of foundation to use, 20-25
  for correlation of number of blows N with some soil properties, 62
  for mantle friction stresses, 615
  of presumptive bearing capacity of rocks and soils, 55
  of proposed allowable bearing values for mats on sand, 72
  for unit weight of rocks and soil, 39
  for coefficients of permeability, 251-252
  for stabilization of soil, 280-282
Tagus River Bridge dome-capped caisson, 481, 498, 499
Tailwater, 46
Tall structures, foundations of, 8, 685-692
Taper of open caissons, 487
Tapered piles, 568
Tapping capacity, wellpoint, 263
Telescoping steel pipes, 109
Temple, Solomon's, 11
Tensile forces (anchor tension) in sheet piling, 148, 150, 164, 207, 209
  in interlocks, 229, 245
Tension
  anchor, 148, 150, 164
  diagonal, 362
    critical planes for, 362
  in interlocks, 231
  stability of cell against rupture, 228
Tension piles, 625, 627-630, 646, 647
  bulbed-end, 568, 628, 629
Teredo, 574
Terminology
  of foundations, 5
    illustrated, 6
  of piles, 568-572
Termites, 575
Terra firma, 745
Terzaghi's equations (formulas)
  bearing capacity, 55, 56
  stability of cofferdam against failure in shear on center line, 240
  standard penetration resistance, 64-66

Terzaghi and Peck, lateral earth pressure diagrams for sheeting, 121, 122
Test parameters
  c, 41, 42
  ϕ, 41
Test, SPR, 61-64
Tests, commonly used, 781-785
Tests, consolidation, 31
Texas tower, foundations of, 685, 686
Thermal forces, 36
Thermal stabilization of soil, 281, 290-297
Three-pile system, forces on, 653-657
Tie-bars (rods), 150, 166, 172, 211, 223
Thixotropic fluid, 487
Tiers of wells
  multiple, 257
  single, 256
Tilting of a floating body, 524
Timber cofferdams, 178, 187, 198
Timber crib, 730-732
  problems on, 735-737
  rock-filled, 730
Timber grillages, 310, 311, 451, 452
Timber piles, 568, 573
  economy of, 578
  splicing of, 575
  storage of, 578
  termites in, 574
Timber sheeting
  horizontal, 110, 111
  vertical, 108, 110
Timber sheetpiling, 111, 112, 178, 187
Timber sheet piling cofferdams, 178, 187
Timbering
  of excavations, 108, 109
  of trenches, 108, 109
Tips of piles, 576, 577
  bearing capacity of, 614-623
  resistance of, 620, 621
  shapes of, 616, 617, 619, 621, 623
Tolerable settlement, 20, 31
Total compressive stress diagram on inboard wall, Navy's, 228
Total interlock stress, 231
Towers
  bell, 16
  Leaning, of Pisa, 16
  offshore, 684, 685
  Texas, 685, 686
Transbay Bridge pier, California, 17
Transient loads, 36
Transfer of loads to soil by friction piles, 673-676

Trapezoidal area, division into n equal parts, 643, 644, 663
Trapezoidal footings, 315
Trapezoidal footings at property line, 417-419
  combined, 422
  for unequal loads, 423
Tremie concrete, 192
Tremie method for underwater concreting, 192
Trenches, timbering of, 108, 109
Trestles, piles of, 649
Triangular area, dividing into two (n) equal parts, 123, 149, 150
Triger, French engineer, 17, 546
Tschebotarioff's anchored bulkhead system, 163, 164
  simplified equivalent beam method, 163, 164
Tubular piles, for Lidingö Bridge, Stockholm, 582-585
Tunnel in Babylon, 12
Tunneling of Persian underground galleries, 12
Turnbuckles, 109
Turning sequence of floating caisson, 532, 533
Two-way eccentricity, 345-348
  example on, 346-348

Ultimate bearing capacity; see Bearing capacity, ultimate
Ultimate compressive stress, 55, 56
Ultimate pull-out capacity
  of sheet piling, 232
  of tension piles, 627-630
Ultimate shear strength of clay, 76
Ultimate strength design of footings, 362
Ultimate stress, 55
  compressive, 55
Unconfined compressive strength of clay, 59, 65
Underground galleries, Persian, 12
Under-river tunnel in Babylon, 12
Underseepage cutoff, sheet piling, 231
Underwater seal, 191, 192, 193
  grouting of, 191, 195
  thickness of, 192, 193
Uniformly loaded shallow foundations, 57-59
Uniform slab (mat), 427
Unit weight of
  ballast, 235, 242
  concrete, 192
  formulas for soil, 40
  rocks, 39

Indexes

Unit weight of *(continued)*
  soils, 39, 40
    approximate, 39, 40
    average, 55
    buoyant, 40
    dry, 40
    formulas for calculations, 40
    inundated, 40
    moist, 40, 41
    saturated, 40
    solids, 41
    submerged, 40
    surcharged, 39
    table of, 39
    unsurcharged, 39
    weighted average, 503
  of solids, 41
Units of measurement, conversion factors, 40, 267, 614, 770-780
Unsupported depth of excavation, 105-107
  maximum, 105-107
Untreated wood piles, 573, 574
Uplift, 45
  avoiding in cofferdams, 188
  of footings, according to Hajashi, 356
Uplift intensity
  factors of, 48, 49
  under the base of a homogeneous dam, 48
  massive dam, 47
Uplift pressure, 44, 442-444
  calculations of, 45-48
  diagrams of, 45-49, 442, 444
  hydrostatic, 36, 44-50, 51
  mats, resisting, 36
  reduction of, 50
  underneath base of a massive dam, 47
Uppermost flow (seepage) line, 241

Variation in angle of internal friction of soil, 41
  cohesion, 41
Variation in groundwater table, 93, 94
Various site conditions for foundations, table for, 21-25
Varved clay, 298
Vaults (arches), inverted, 315, 316
Velocity of scour, 706, 707
Velocities
  bottom, scouring, 706, 707
  competent, 707
Velocities over river-bed materials, 707
Venice, Rialto Bridge, 16
Verrazano-Narrows Bridge, N.Y.
open caisson, 481
  artificial sand island, 512-514

Vibrations, 36, 105
  of caissons into soil, 74
  effect on wall function, 43
  for embedment of sheet piling, 571
  effect on settlement of soil, 74
  forces of, 36
  mechanical, 74
    of machine foundations, 742, 743
    of sand, 71, 741
  of piles into sand, 71, 74, 75, 630-633
  of sheet piling, 630
Vibrators, 631-633
Viscosity, dynamic, 275
Vitruvius, Roman engineer, 12-14
Void ratio, 40, 62
Volume of obtuse stress wedge, 398

Wale, anchorage, 110, 174
Wall
  anchor, 165, 166
  foundation, 6
  ice, thickness of, 294, 295
Wall footings; *see* Footings, wall
Wall friction, 41
  angle of ϕ, 43, 44
  coefficients of, 43, 44
  effects on, by vibration, 43
  weighted average, 503
Wall loads, 8
Walls
  anchorage of, 28
  foundation, 6, 28
  open caisson, thickness of, 488
  sheeting of, 108-111
Walls, retaining, 109-111, 126, 127, 129, 163, 164-178, 198
Water (moisture) film, 274, 275
Water impounding dam, 47
Water influx into excavation, 249
Water (moisture) migration in an electric field, 273-276
Water pressure
  artesian, 44, 50
  on bridge pier due to earthquake, 746-748
  hydrostatic, 36, 44
  pore, 36
Water table
  groundwater, 3, 6, 8, 27, 28, 32, 42, 67, 69, 70, 72, 74, 93, 256-260, 319
  lowering of, 188, 256
  position of, 27, 67, 69, 70, 72, 74
  variation in, 93, 94, 319
Waterlogged soil, 13, 113
Waterproofing of foundations, 444-451
  dampproofing, 445, 447, 448

Waterproofing of foundations *(continued)*
  of basements, 448
  drainage, 445
  insulation, 449
  membranes, 447, 449, 450
  waterstops, 446, 447
  waterproofing materials, 449, 450
  waterstops, 446, 447
Wedge, rupture, of soil, sand, 14, 15
Wedge, stress, obtuse volume of, 398
Weep holes, 218, 231, 242, 243
  of cell, 232
  of cell ballast, 219, 243, 244
  in cofferdam, 219
Weight, effective
  of cell ballast (soil), 219, 243, 244
  of cofferdam, 219
Weight, unit
  of rocks, 39
  of soils, 39, 40
  of structure, 36
  weighted average, 36
Weighted average of
  angle of internal friction of soil, 503
    of wall friction, 503
  shear strength, 611, 612, 616
  shear stress, 611, 612
  unit weight of soil, 503
Weight of cofferdam shell, 244
Weissbach's dynamic pile driving formula, 602, 603
Welded steel cutting edges, 490, 493, 494
Wellpoint systems
  advantages of, 256
  asymmetrical, 256
  multiple tiers, 257
  single stage, 256
  symmetrical, 256-260
Wellpoints, 256-260
  tapping capacity of, 263

Wells
  circular arrangement of, 259-266
  drawdown in, 263
  dredging (caisson), 193-195
  imperfect, 263
  multiple, pumping from, 258-266
  ordinary, perfect, 258, 267
  rectangular arrangement of, 261-263
  spacing of, 260, 261, 263, 264, 268, 270, 271
  stage in, 263
Westergaard's pressure parabola, 748
*Widened footings (foundations), 3, 4
Width of cofferdam, 200, 212, 213, 216, 219
  computational, 223
Width of footing, 6, 56
  effect of width on footing, 6, 29, 31
Wind pressure
  on chimneys, 687
  on structures, 687
Winkler's soil, 720
Winsol resin chemicals for grouting, 287
Wood piles, 573-578
  economy of, 578
Work in pneumatic caisson, 552, 553
Working chamber of pneumatic caisson, 548-550
  calculations of, 553-557
World Trade Center, N.Y.,
  cofferdam for, 129, 131
  construction of, 129, 131

Zeta potential, 275
Zones
  anchorage, 114, 132, 165, 167, 168
  dewatered, 256-260
  of fluctuating groundwater, 573, 574
  plastic, 108
  stressed, 6, 26, 679